High Fat Diet with Chronic Diseases

High Fat Diet with Chronic Diseases

Editors

Xiaoyu Wang
Frédéric Dutheil

MDPI Basel • Beijing • Wuhan • Barcelona • Belgrade • Novi Sad • Cluj • Manchester

Editors

Xiaoyu Wang
College of Food Science and Nutritional Engineering,
China Agricultural University
Beijing
China

Frédéric Dutheil
Occupational and Environmental Medicine,
Université Clermont Auvergne
Clermont-Ferrand
France

Editorial Office
MDPI
St. Alban-Anlage 66
4052 Basel, Switzerland

This is a reprint of articles from the Special Issue published online in the open access journal *Nutrients* (ISSN 2072-6643) (available at: https://www.mdpi.com/journal/nutrients/special_issues/highfat_Chronic).

For citation purposes, cite each article independently as indicated on the article page online and as indicated below:

Lastname, A.A.; Lastname, B.B. Article Title. *Journal Name* **Year**, *Volume Number*, Page Range.

ISBN 978-3-7258-1265-3 (Hbk)
ISBN 978-3-7258-1266-0 (PDF)
doi.org/10.3390/books978-3-7258-1266-0

Contents

About the Editors

Xiaoyu Wang

Xiaoyu Wang is an Associate Professor at the College of Food Science and Nutritional Engineering, China Agricultural University, specializing in nutrition and metabolism. She was selected for the 8th Young Elite Scientist Sponsorship Program by the China Association for Science and Technology (CAST). Additionally, she has led several research projects, including the National Natural Science Foundation of China and the National Key R&D Program of China. To date, she has co-authored more than 40 papers, with more than 20 as the first or corresponding author. She has received numerous honors, including the National Commercial Science and Technology Progress Grand Prize and the Scientific and Technological Special Grand Prize from the China Inspection and Testing Society. She is also an institute member of the Chinese National Committee of the International Dairy Federation and the Science and Nutrition Committee of Infant Formula Powder of the China Dairy Industry Association.

Frédéric Dutheil

Frédéric Dutheil is a Professor of Medicine and the Head of the Center for Professional Pathology within the Health Work Environment Department at Université Clermont Auvergne. His primary research focus is sports nutrition. Frédéric has actively participated in various projects, particularly in the DoWellB project, funded by the Agence Nationale de la Recherche, which is centered around the design of monitoring systems aimed at enhancing well-being.

nutrients

Editorial

New Insights into High-Fat Diet with Chronic Diseases

Xiaoyu Wang [1,2,*], Rui Song [1], Maëlys Clinchamps [3] and Frédéric Dutheil [3]

1 College of Food Science & Nutritional Engineering, China Agricultural University, Beijing 100083, China; songrui_123@cau.edu.cn
2 Key Laboratory of Precision Nutrition and Food Quality, Department of Nutrition and Health, China Agricultural University, Beijing 100193, China
3 CNRS, LaPSCo, Physiological and Psychosocial Stress, University Clermont Auvergne, F-63000 Clermont-Ferrand, France; maelysclinchamps@gmail.com (M.C.); fred_dutheil@yahoo.fr (F.D.)
* Correspondence: xy.wang@cau.edu.cn

Citation: Wang, X.; Song, R.; Clinchamps, M.; Dutheil, F. New Insights into High-Fat Diet with Chronic Diseases. *Nutrients* **2023**, *15*, 4031. https://doi.org/10.3390/nu15184031

Received: 7 September 2023
Accepted: 10 September 2023
Published: 18 September 2023

1. Introduction

Chronic diseases, encompassing conditions such as heart disease, cancer, and diabetes, represent a significant global health challenge and are the leading causes of mortality worldwide. Extensive research has been dedicated to investigating the relationship between dietary fat consumption and the risk of developing chronic diseases over the years. It is well-established that lifestyle modifications, particularly dietary choices, hold substantial potential for preventing and mitigating the risk of these chronic conditions. This Special Issue of *Nutrients*, entitled "High Fat Diet with Chronic Diseases", aims to contribute to our comprehension of the potential underlying mechanisms connecting high-fat diets to chronic diseases. Furthermore, this Special Issue places a strong emphasis on the development of effective therapeutic strategies for managing and combating these conditions. The studies in this issue are grouped into four categories: (i) obesity and metabolic disorders, (ii) pancreatic health and diabetes, (iii) skeletal and muscle health, and (iv) neurological and cognitive function. While these topics seem distinct, they share an underlying theme: the complex interplay between dietary factors, obesity, and their consequences on health.

2. Obesity and Metabolic Disorders

Several studies collectively provide a comprehensive exploration into the area of obesity and metabolism-related diseases. While each study explores distinct subjects, their collective findings reveal intricate underlying connections, with a notable emphasis on dietary supplements. The study authored by S. Heo et al. delves into the potential of *Cassia mimosoides* var. *nomame* Makino extract (EECM) as a means to combat obesity [1]. EECM demonstrates its efficacy by inhibiting adipogenesis and lipogenesis, primarily through the AMP-activated protein kinase pathway. C. Chou et al. scrutinize the influence of high fructose intake on lipolysis and reveal how valsartan and amlodipine can effectively mitigate lipolysis through PU.1 inhibition, underscoring the potential of these drugs in addressing fructose-induced obesity [2]. Y. Wang et al. explore the effects of coarse cereals on lipid metabolism and oxido-inflammatory responses [3]. Oats and tartary buckwheat, rich in polyphenols and dietary fiber, exhibit their potential in modulating gut microbiota, improving lipid metabolism, and reducing oxidative stress and inflammation in rats fed a high-fat diet. Additionally, the significance of probiotics in alleviating obesity is pivotal. The study by C. Wang et al. investigates the effects of fermented milk containing *Bifidobacterium animalis* subsp. *lactis* MN-Gup (MN-Gup), which could alleviate HFD-induced body weight gain, epididymal fat deposition, adipocyte hypertrophy, dyslipidemia and inflammation, which may play important roles in the mechanism underlying the alleviation of obesity [4]. Meanwhile, Y. Huo et al. found that *Bifidobacterium animalis* subsp. *lactis* A6 enhances adipose tissue fatty acid β-oxidation (FAO) to mitigate obesity development by increasing acetate levels and activating the GPR43-PPARα signaling pathway in mice [5].

In addition, two articles specifically delve into lipid metabolism in unique populations. The study by R. Lugarà et al. investigates the impact of a Western diet on gestating and lactating sows, revealing signs of lean metabolic syndrome characterized by disrupted cholesterol levels, decreased IGF1 levels, and indications of hepatic damage [6]. Spirulina supplementation in this context partially mitigates these effects, hinting at its potential to counteract some of the negative outcomes associated with a Western diet. R. Song et al. review lipid metabolism in the elderly, emphasizing how age-related changes in digestion, absorption, and lipid utilization contribute to excess fat accumulation [7]. This accumulation is linked to chronic lipid-related diseases in older individuals, and understanding altered lipid metabolism in these populations is crucial for developing targeted interventions to address age-related chronic diseases. In summary, these studies collectively contribute to our understanding of obesity and metabolic disorders by examining both dietary supplements and their potential intervention on fat metabolism. They provide valuable insights into potential therapeutic strategies and the intricate mechanisms underlying these conditions, spanning various populations.

3. Pancreatic Health and Diabetes

Three papers shed light on glucose homeostasis and diabetes. M.K. Brahma et al. study the impact of Nova1 or Bim deficiency in pancreatic β-cells on diabetes and obesity in mice [8]. The deletion of Nova1 or Bim did not affect glucose homeostasis or diabetes development in response to multiple low-dose streptozotocin (MLD-STZ)-induced β-cell dysfunction or high-fat diet-induced insulin resistance. Another study explored the effects of a long-term high-fat, high-fructose (HFHF) diet on diabetes development in rats. Rats on HFHF diets exhibited metabolic disorders, changes in pancreatic islet size, and increased insulin levels. Authors Y. Zhao et al. suggest that long-term HFHF diets and age-related structural and transcriptomic changes in pancreatic islets may contribute to type 2 diabetes development [9]. The review authored by Y. Qi et al. focuses on the role of gut microbiota in high-fat diet-induced diabetes [10]. It highlights the importance of diet and gut microbiome in the development of diabetes and suggests that modifying the gut microbial community through probiotic and prebiotic approaches could be a promising strategy for diabetes prevention. Together, these papers provide valuable insights into the complex relationship between glucose homeostasis, diet, and diabetes, offering potential directions for future research and therapeutic interventions.

4. Skeletal and Muscle Health

Two articles are related to diet-induced obesity and skeletal/muscle health. The study by E. Nebot et al. investigates how obesity affects bone health [11]. It establishes that obesity might lead to increased bone mineral density, yet also pose a significant risk for fractures. Interventions like caloric restriction and exercise effectively counter bone structure and mineral density changes and improve body composition by reducing body fat and increasing lean body mass. Y. Zou et al. employed a zebrafish model to explore the impact of a high-fat diet on muscle mitochondrial function [12]. Obesity was linked to reduced exercise capacity, decreased skeletal muscle fiber cross-sectional area, and elevated expression of atrophy-related markers. The study demonstrated that mitochondrial dysfunction contributes to muscle atrophy in obesity. These findings collectively emphasize the intricate connection between dietary factors, obesity, and their consequences on bone and muscle health.

5. Neurological and Cognitive Function

Two studies explore the connection between obesity and cognitive function. Y. Liu et al. found that high-calorie food-cues impaired food-related conflict control [13]. Participants displayed slower reaction times and reduced accuracy when dealing with high-calorie food images. EEG data exhibited a notable reduction in N2 amplitudes and a decline in theta power when exposed to high-calorie foods, which serves as an indicator of cognitive impairments. The study by H. Zhang et al. investigated cognitive decline in obese mice, highlighting the beneficial effects of swimming [14]. Obesity led to cognitive impairment, but an 8-week swimming regimen mitigated this decline by reducing inflammation, inhibiting the JNK/IRS-1/PI3K/Akt pathway, and activating the PGC-1α/BDNF pathway. Overall, these studies underscore the detrimental impact of high-calorie food cues on cognitive function and emphasize the potential cognitive benefits of exercise interventions in combating obesity-related cognitive decline.

6. Conclusions

This Special Issue of *Nutrients* delves into the intricate relationship between dietary fat intake and chronic diseases. Understanding these relationships is crucial for developing effective strategies to combat diet-induced chronic diseases and improve overall health. These papers emphasize the role of nutritional supplements in addressing obesity and metabolic disorders while highlighting the significance of tailored interventions in unique populations' lipid metabolism research. They include findings on glucose regulation and diabetes, encompassing studies on pancreatic β-cell and gut microbiota in diabetes development, and opening up promising pathways for future research and therapeutic interventions. Studies on obesity's effects on bone, muscle health, and cognitive function offer insights into interventions like exercise and caloric restriction. In summary, this Special Issue serves as a platform to enhance our understanding of the intricate interplay between dietary fat intake and chronic diseases, offering insights into potential avenues for prevention and treatment.

Author Contributions: Conceptualization, X.W.; writing—original draft preparation, X.W., R.S., M.C. and F.D.; writing—review and editing X.W. All authors have read and agreed to the published version of the manuscript.

Funding: This research was funded by the National Natural Science Foundation of China (Grant No. 32001676), and the 111 Project from the Education Ministry of China (No. B18053).

Conflicts of Interest: The authors declare no conflict of interest.

References

1. Heo, S.-W.; Chung, K.-S.; Yoon, Y.-S.; Kim, S.-Y.; Ahn, H.-S.; Shin, Y.-K.; Lee, S.-H.; Lee, K.-T. Standardized Ethanol Extract of Cassia Mimosoides Var. Nomame Makino Ameliorates Obesity via Regulation of Adipogenesis and Lipogenesis in 3T3-L1 Cells and High-Fat Diet-Induced Obese Mice. *Nutrients* **2023**, *15*, 613. [CrossRef]
2. Chou, C.-L.; Li, C.-H.; Fang, T.-C. Benefits of Valsartan and Amlodipine in Lipolysis through PU.1 Inhibition in Fructose-Induced Adiposity. *Nutrients* **2022**, *14*, 3759. [CrossRef]
3. Wang, Y.; Qi, W.; Guo, X.; Song, G.; Pang, S.; Fang, W.; Peng, Z. Effects of Oats, Tartary Buckwheat, and Foxtail Millet Supplementation on Lipid Metabolism, Oxido-Inflammatory Responses, Gut Microbiota, and Colonic SCFA Composition in High-Fat Diet Fed Rats. *Nutrients* **2022**, *14*, 2760. [CrossRef] [PubMed]
4. Wang, C.; Li, S.; Sun, E.; Xiao, R.; Wang, R.; Ren, Y.; He, J.; Zhang, Q.; Zhan, J. Effects of Fermented Milk Containing Bifidobacterium Animalis Subsp. Lactis MN-Gup (MN-Gup) and MN-Gup-Based Synbiotics on Obesity Induced by High Fat Diet in Rats. *Nutrients* **2022**, *14*, 2631. [CrossRef]
5. Huo, Y.; Zhao, G.; Li, J.; Wang, R.; Ren, F.; Li, Y.; Wang, X. Bifidobacterium Animalis Subsp. Lactis A6 Enhances Fatty Acid β-Oxidation of Adipose Tissue to Ameliorate the Development of Obesity in Mice. *Nutrients* **2022**, *14*, 598. [CrossRef]
6. Lugarà, R.; Renner, S.; Wolf, E.; Liesegang, A.; Bruckmaier, R.; Giller, K. Crossbred Sows Fed a Western Diet during Pre-Gestation, Gestation, Lactation, and Post-Lactation Periods Develop Signs of Lean Metabolic Syndrome That Are Partially Attenuated by Spirulina Supplementation. *Nutrients* **2022**, *14*, 3574. [CrossRef]
7. Song, R.; Hu, M.; Qin, X.; Qiu, L.; Wang, P.; Zhang, X.; Liu, R.; Wang, X. The Roles of Lipid Metabolism in the Pathogenesis of Chronic Diseases in the Elderly. *Nutrients* **2023**, *15*, 3433. [CrossRef]
8. Brahma, M.K.; Xiao, P.; Popa, M.; Negueruela, J.; Vandenbempt, V.; Demine, S.; Cardozo, A.K.; Gurzov, E.N. Noxa1 or Bim Deficiency in Pancreatic β-Cells Does Not Alter Multiple Low-Dose Streptozotocin-Induced Diabetes and Diet-Induced Obesity in Mice. *Nutrients* **2022**, *14*, 3866. [CrossRef] [PubMed]
9. Zhao, Y.; Wang, Q.-Y.; Zeng, L.-T.; Wang, J.-J.; Liu, Z.; Fan, G.-Q.; Li, J.; Cai, J.-P. Long-Term High-Fat High-Fructose Diet Induces Type 2 Diabetes in Rats through Oxidative Stress. *Nutrients* **2022**, *14*, 2181. [CrossRef] [PubMed]
10. Qi, Y.; Wang, X. The Role of Gut Microbiota in High-Fat-Diet-Induced Diabetes: Lessons from Animal Models and Humans. *Nutrients* **2023**, *15*, 922. [CrossRef] [PubMed]

11. Nebot, E.; Martínez, R.; Kapravelou, G.; Sánchez, C.; Llopis, J.; Aranda, P.; Porres, J.M.; López-Jurado, M.; Pietschmann, P. Combination of Caloric Restriction and a Mixed Training Protocol as an Effective Strategy to Counteract the Deleterious Effects in Trabecular Bone Microarchitecture Caused by a Diet-Induced Obesity in Sprague Dawley Rats. *Nutrients* **2022**, *14*, 3672. [CrossRef]
12. Zou, Y.-Y.; Chen, Z.-L.; Sun, C.-C.; Yang, D.; Zhou, Z.-Q.; Xiao, Q.; Peng, X.-Y.; Tang, C.-F. A High-Fat Diet Induces Muscle Mitochondrial Dysfunction and Impairs Swimming Capacity in Zebrafish: A New Model of Sarcopenic Obesity. *Nutrients* **2022**, *14*, 1975. [CrossRef]
13. Liu, Y.; Zhao, J.; Zhou, Y.; Yang, R.; Han, B.; Zhao, Y.; Pang, Y.; Yuan, H.; Chen, H. High-Calorie Food-Cues Impair Conflict Control: EEG Evidence from a Food-Related Stroop Task. *Nutrients* **2022**, *14*, 4593. [CrossRef]
14. Zhang, H.; Liang, J.-L.; Wu, Q.-Y.; Li, J.-X.; Liu, Y.; Wu, L.-W.; Huang, J.-L.; Wu, X.-W.; Wang, M.-H.; Chen, N. Swimming Suppresses Cognitive Decline of HFD-Induced Obese Mice through Reversing Hippocampal Inflammation, Insulin Resistance, and BDNF Level. *Nutrients* **2022**, *14*, 2432. [CrossRef]

nutrients

MDPI

Review

The Roles of Lipid Metabolism in the Pathogenesis of Chronic Diseases in the Elderly

Rui Song [1], Mengxiao Hu [1], Xiyu Qin [1], Lili Qiu [1], Pengjie Wang [2], Xiaoxu Zhang [2], Rong Liu [2] and Xiaoyu Wang [1,2,*]

[1] College of Food Science & Nutritional Engineering, China Agricultural University, Beijing 100083, China; songrui_123@cau.edu.cn (R.S.); humengxiao@cau.edu.cn (M.H.); qinxiyu@cau.edu.cn (X.Q.); qiulily@cau.edu.cn (L.Q.)

[2] Key Laboratory of Precision Nutrition and Food Quality, Department of Nutrition and Health, China Agricultural University, Beijing 100193, China; wpj1019@cau.edu.cn (P.W.); xiaoxuzhang0220@cau.edu.cn (X.Z.); liurong@cau.edu.cn (R.L.)

* Correspondence: xy.wang@cau.edu.cn

Abstract: Lipid metabolism plays crucial roles in cellular processes such as hormone synthesis, energy production, and fat storage. Older adults are at risk of the dysregulation of lipid metabolism, which is associated with progressive declines in the physiological function of various organs. With advancing age, digestion and absorption commonly change, thereby resulting in decreased nutrient uptake. However, in the elderly population, the accumulation of excess fat becomes more pronounced due to a decline in the body's capacity to utilize lipids effectively. This is characterized by enhanced adipocyte synthesis and reduced breakdown, along with diminished peripheral tissue utilization capacity. Excessive lipid accumulation in the body, which manifests as hyperlipidemia and accumulated visceral fat, is linked to several chronic lipid-related diseases, including cardiovascular disease, type 2 diabetes, obesity, and nonalcoholic fatty liver disease. This review provides a summary of the altered lipid metabolism during aging, including lipid digestion, absorption, anabolism, and catabolism, as well as their associations with age-related chronic diseases, which aids in developing nutritional interventions for older adults to prevent or alleviate age-related chronic diseases.

Keywords: elderly; lipid; digestion; absorption; metabolism; chronic diseases

Citation: Song, R.; Hu, M.; Qin, X.; Qiu, L.; Wang, P.; Zhang, X.; Liu, R.; Wang, X. The Roles of Lipid Metabolism in the Pathogenesis of Chronic Diseases in the Elderly. *Nutrients* **2023**, *15*, 3433. https://doi.org/10.3390/nu15153433

Academic Editor: Alan Stewart

Received: 10 July 2023
Revised: 26 July 2023
Accepted: 31 July 2023
Published: 3 August 2023

1. Introduction

Globally, the proportion of elderly people aged 65 years or over continues to increase. According to the United Nations statistics, in 2019, the elderly population aged 65 years or over accounted for 9% of the total population, and this proportion is projected to increase to 16% by 2050 and surpass 23% by 2100 [1]. The number of people aged 80 years or over is growing even faster as a consequence of increased global life expectancy. Age is a major risk factor for chronic diseases, and more than 50% of the elderly suffer from at least one kind of chronic disease. For example, cardiovascular disease (CVD) affects up to 70% of the elderly population, and diabetes affects about 20% of the elderly population [2–4]. The high cost of these chronic diseases brings major personal and public health burdens, which will even get worse in the future. Therefore, it is advisable to prevent the occurrence of these diseases as early as possible. These diseases are commonly accompanied by dyslipidemia and insulin resistance that result from lipid metabolism disorders, thus indicating that modulating lipid metabolism may be an effective way to prevent or alleviate these chronic diseases.

The dysregulation of lipid metabolism contributes to many age-related chronic diseases [5]. Lipid metabolism is a complex biochemical reaction, which refers to the process of the digestion, absorption, synthesis, and catabolism of lipids. As a result of aging, all gastrointestinal processes (such as movement, enzyme, hormone release, and so on) are altered, which in turn affects digestion and absorption, thereby leading to reduced nutrient uptake. Postprandial lipemia (PPL) is the phenomenon of elevated post-meal blood lipids, which typically decreases over time. Despite the reduced intestinal uptake, the excessive and prolonged PPL observed in the elderly suggests the sustained elevation of lipid concentrations after meals, thus indicating an imbalance in blood lipid circulation. The strict regulation of lipid metabolism in vivo after lipid uptake is crucial for maintaining lipid homeostasis. Changes in lipid synthesis and catabolism that occur with aging lead to abnormal lipid utilization in tissues. Blood lipid and other chronic disease precursors arise as a result of lipid metabolic disorders, which eventually develop into several chronic diseases, such as CVD, type 2 diabetes (T2D), nonalcoholic fatty liver disease (NAFLD), obesity, and so on [6]. Therefore, the disturbed lipid homeostasis in the elderly highlights the significance of the strict regulation of lipid metabolism in maintaining overall health.

This review covers the changes in lipid metabolism during aging, including lipid digestion and absorption, as well as the anabolism and catabolism capacity of triglycerides (TGs) and cholesterol and the pathogenesis of several lipid-related diseases in the elderly. Our aim is to gain insight into the lipid uptake and utilization patterns in the elderly to provide personalized nutritional recommendations for older adults, thereby preventing or alleviating lipid-related chronic diseases to promote healthy aging.

2. Disorder of Lipid Digestion and Absorption in the Elderly

Digestion and absorption play critical roles in whole-body lipid metabolism by providing lipid digestive products as metabolic substrates. Lipid digestion primarily occurs through digestive enzymes secreted by the pancreas, followed by absorption in the small intestine, which is the main pathway for exogenous lipids to enter the body. The excessive intake of dietary fats can lead to lipid accumulation in the body, which promotes the development of chronic lipid-related diseases. Age-related changes in lipid digestion may be attributed to degeneration of the digestive tract function, such as atrophied organs, reduced pancreatic secretion, insufficient lipase production, and decreased bile acid (BA) concentrations [7]. In addition, current studies have indicated that lipid absorption may decrease, while cholesterol absorption increases with age [8] (Figure 1). Therefore, lipid digestion and absorption processing may become a new target for dietary interventions to prevent dyslipidemia in the elderly and to treat lipid-related diseases.

2.1. Age-Related Changes in Lipid Digestion

Lipid digestion is an intricate process involving multiple stages. Lipid digestion begins in the mouth and continues in the stomach, with lingual lipase playing a minor role and gastric lipase contributing 10–30% to the overall lipid breakdown in adults [9]. Subsequently, the partially hydrolyzed lipids enter the duodenum as small lipid droplets (LDs). Since dietary lipids are hydrophobic, and lipase is hydrophilic, further emulsification takes place through the action of bile salts to increase the surface area for efficient digestion. These micelles can pass through the unstirred water layer, thereby allowing for lipid absorption to occur [10].

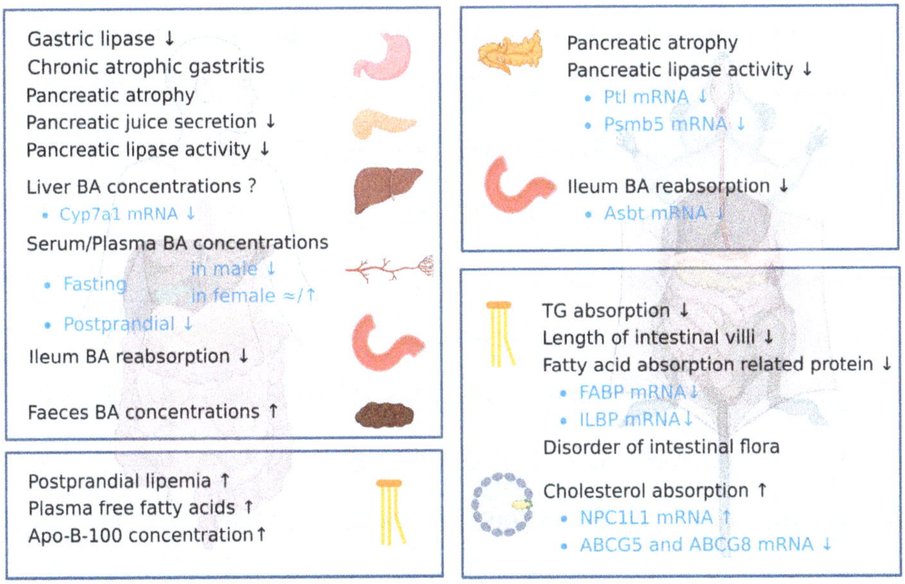

Figure 1. Age-related changes in lipid digestion and absorption. The image on the left depicts changes in the digestive tracts of elderly people, and the image on the right shows changes in the digestive tracts of aging rodents. Alterations in digestive function occur in the stomach, pancreas, liver, blood, intestines, and feces. Both TG and cholesterol absorption that occur in the intestine are described. ↑ means studies showing a significant increase; ↓ means studies showing a significant decrease; ≈ means studies showing no significant decrease; ? means this result has not been reported. Figure 1 was created using Biorender with permission for publication from Biorender.com (accessed on 9 July 2023).

2.1.1. Gastric Lipase Activity Decreases with Age

The contribution of the stomach to lipid digestion is reduced in comparison to younger individuals and is further diminished in elderly individuals with gastric atrophy or those with the chronic use of acid suppression medication. Previous studies have shown a significant negative correlation between age and gastric lipase activity; however, the mechanism behind this phenomenon has not been elucidated [11]. Gastric lipase activity is particularly stable at a low pH [12], but the high prevalence of chronic atrophic gastritis and the widespread use of proton pump inhibitor drugs in the elderly population could result in a decline in gastric acid secretion, which may affect gastric lipase activity [13].

2.1.2. Pancreatic Function Declines with Age Due to Decreased Pancreatic Lipase Expression

The pancreas is an important organ for digestion, and morphological changes occur with age. Magnetic resonance imaging (MRI) was used to scan the morphological characteristics of the pancreas in over two hundred healthy participants, which revealed that the pancreases of elderly subjects showed signs of atrophy. Moreover, the critical age range for pancreatic morphological changes was found to be 40 years old and 60 years old [14]. Consistently, MRI demonstrated increased pancreatic atrophy, lobulation, and steatosis in aged mice [15].

Along with pancreas atrophy, its function also declines. Numerous studies have demonstrated a positive correlation between physical aging and the decline in the exocrine function of the pancreas [14]. Specifically, the lower secretory flow rate of the pancreatic juice in humans indicated a significant reduction in the secretion ability with aging [16]. Moreover, the time point for changes in pancreatic exocrine function was found to be consistent with that of morphological changes, with degeneration occurring at 43 years of age [17].

The primary age-related change in pancreatic function is the decline of the pancreatic lipase activity. In healthy adults, pancreatic lipase is responsible for digesting approximately three-quarters of dietary TGs. However, its digestive efficiency decreases with aging [18]. The concentration and secretion volume of lipase showed a gradual decline in participants after the age of 30 [17]. Researchers collected duodenal fluid to evaluate the output and concentration of lipases after continuous intravenous infusions of irritants in elderly patients without pancreatic diseases. It was found that lipase concentrations declined significantly, by about 15%, in the older group compared to the younger group. Additionally, it was observed that lipase output levels decreased substantially by 45% [19]. Similar results were also observed in aging mice, in which the activity of the pancreatic lipase decreased significantly [20].

Pancreatic lipase activity may be reduced due to the accumulation of unfolded enzymes in the pancreas. The protein expressions of pancreatic lipases in elderly mice are consistent with those in younger mice [21]. As with other enzymes, pancreatic lipase is a linear polypeptide chain that must be folded into a unique 3D structure to perform its activity and physiological role [22]. Interestingly, it was found that the level of proteasome subunit beta type 5 (Psmb5), which facilitates the elimination of enzymes that have not been folded into specific structures, decreased significantly in 25-month-old mice compared to 3-month-old mice. Therefore, the accumulation of nonfunctional enzymes may affect the synthesis or secretion of effective lipases [21].

2.1.3. Aging Is Mainly Associated with a Decline in Bile Acid Levels Due to the Lower Ileum Reabsorption

BA plays a crucial role in facilitating the digestion and absorption of dietary lipids. This is due to the amphiphilic structure of BA, which allows it to function as an emulsifier [23]. Age-related changes in the composition of BA have been studied in both humans and animals. Although the changes varied depending on certain factors such as gender, BA levels showed a consistent decrease with age [10].

In the fasting state, the plasma concentrations of BA decreased with age in men, but they remained constant or showed a slight increase in women. Similarly, in the postprandial state, the responses of BAs were weaker in the elderly compared to young people. The serum BA levels in healthy young subjects increased rapidly after a meal, whereas elderly individuals showed a smaller increase in their serum BA levels after a meal [24]. For this reason, there is some indirect evidence to suggest that intestinal BA reabsorption becomes reduced in older adults. This hypothesis was further supported by the higher fecal concentrations of BAs in older adults, thereby indicating decreased reabsorption in the ileum [25]. Additionally, during the enterohepatic circulation in the elderly, a higher rate of deoxycholic acid (DCA) was found in the colon, which is a metabolic byproduct of cholic acid (CA) transformation [26]. This finding suggests a reduction in the reabsorption of BA in the ileum among older adults, thereby leading to an increased CA load in the colon.

As was consistent with human studies, the reabsorption of BA in the ileum of aging mice was similarly reduced, with about a 50% lower absorption of taurocholate than in younger mice. This was supported by the reduced gene expression of apical sodium-dependent bile acid transporter (Asbt), which is the key transporter regulating BA uptake in the intestinal apical membrane [27]. These observations suggest that the decreased expression of Asbt potentially acts as a rate-determining step in BA reabsorption.

2.2. Age-Related Changes in Lipid Absorption

After digestion, lipids are absorbed in the lower part of the duodenum and the upper part of the jejunum. The lipid hydrolysis products form micelles, which contain free fatty acids (FFA), monoacylglycerols, free cholesterol, and so on. These micelles could diffuse through the unstirred water layer and eventually release their lipid contents into the enterocytes. Typically, fatty acid translocase (CD36/FAT), fatty acid transporter 4 (FATP4), and membrane-associated fatty acid binding protein (FABP) transporters mediate TG absorption, while Niemann–Pick C1-like 1 (NPCIL1) mediates cholesterol absorption [28,29].

2.2.1. Triglyceride Absorption Decreases with Age

The absorption of dietary fat initiates an increase in the plasma TG concentration, and the resultant increase in chylomicron and very low-density lipoproteins (VLDL) in the blood in the postprandial state is known as PPL. Increased PPL has been typically observed in older adults [30,31]. Research projects have shown that PPL in older adults increases by about 130% compared to younger adults, which is associated with higher plasma FFA and apolipoprotein B-100 concentrations and a lower rate of oxidation of the ingested lipid [32]. By lowering dietary fat intake, PPL can be reduced by about 50%, and FFA can be reduced by about 190% in older adults [33]. However, it should be noted that, while PPL reflects absorption, its true significance is in the concentration of lipids in the bloodstream. Therefore, an increase in PPL in older adults may also be attributed to a decreased efficiency of lipid utilization.

Due to the significant time and cost required to validate studies in humans, the majority of aging data comes from shorter-lived rodents. The investigations on lipid absorption in aged mice have consistently reported decreased outcomes with humans. Researchers investigated changes in lipid absorption in 3- and 25-month-old mice and found that, after the oral administration of soybean oil, the increase in serum TG levels in older mice was significantly reduced [21]. Nevertheless, only a few studies have obtained similar results, and further research is needed to confirm this finding. There are several known mechanisms that could explain the decline of the absorption capacity in older individuals. Morphologically, changes in the small intestine may lead to a decline in nutrients absorption in aging mice, as the length of intestinal villi decreases with age [34]. Some researchers have also suggested that the decline in intestinal lipid intake may be associated with a reduction in the abundance of fatty acid (FA)-absorption-related proteins during aging. For example, studies have shown that the abundance of intestinal FABP and ileal lipid-binding protein (ILBP) decreases in aging rats [35]. Moreover, aging is often accompanied by intestinal flora disorders, and previous studies have indicated that the small intestinal microbiota could regulate the digestion and absorption of dietary fat in hosts [36,37].

In brief, the efficiency of dietary lipids absorption affects lipid homeostasis. However, the lipid absorption capacity in older adults remains unknown. As previously discussed, several known mechanisms may contribute to the decline in the absorption capacity in the elderly. However, further research is needed to determine the impact of aging on lipid absorption, including the absorption time and efficiency.

2.2.2. Cholesterol Absorption Increases with Age

Numerous investigations have indicated that the absorption of cholesterol will increase with age. Research has indicated that cholesterol absorption increased with age in mice [38]. In addition, a mouse model has shown that the expressions of the sterol efflux transporters adenosine triphosphate-binding cassette transporters G5 and G8 (ABCG5 and ABCG8) in the jejunum and ileum were down-regulated, and the expression of sterol efflux transporter NPC1L1 in duodenum and jejunum was up-regulated during aging [39]. The above changes may be related to the increased intestinal cholesterol absorption in aged mice. At present, the successful prescription drugs on the market treating hypercholesterolemia work by inhibiting NPC1L1 to reduce the cholesterol level in the blood. In addition, dietary phytosterol or stanol supplements could also treat hypercholesterolemia [40]. Further research is needed to investigate the efficacy and safety of these interventions, as well as to explore more effective interventions that may improve cholesterol homeostasis in aging populations.

3. Disorder of Lipid Anabolism and Catabolism in the Elderly

Lipid anabolism and catabolism are fundamental processes involved in the biosynthesis and degradation of FAs, TGs, cholesterol, and other lipid substances [41]. The liver is the most important organ for lipid synthesis and catabolism. However, lipids cannot be stored in hepatocytes, and adipose cells act as the main reservoir for fat storage. Lipids must be transported from where they are absorbed or synthesized to where they are used or stored. However, as lipids are insoluble in water, they have to bind to proteins to form lipoproteins before being transported. Thus, plasma lipoproteins are essential for lipid circulation throughout the body. It is important to note that complex changes occur in lipid metabolism during aging, which can contribute to the development of lipid-related diseases [42].

3.1. Disorder of Plasma Lipoprotein Metabolism during Aging

The dysregulation of the lipid metabolism in older adults is often reflected by changes in blood lipid levels, which are associated with the onset of various chronic diseases. Specifically, the levels of TGs, total cholesterol (TC), and low-density lipoprotein cholesterol (LDL-C) tend to increase, while high-density lipoprotein cholesterol (HDL-C) concentrations become relatively irregular during aging [43,44]. In humans and mice, plasma TG levels have been found to increase with age, thereby leading to its potential as a biomarker of aging due to its positive correlation with age [45–48]. By characterizing the metabolic changes in young and old mice, aging has been shown to be accompanied by an increase in the FFA levels [49]. In rats, plasma cholesterol levels also have an age-dependent increase that can be mitigated by regulating lipoprotein metabolism [50]. Regarding lipoprotein levels, both low-density lipoproteins (LDLs) and LDL-C have been reported to increase with age [51]. The determinant of plasma lipoprotein concentrations is affected by the rate at which lipoproteins are removed from the plasma. Additionally, the efficiency of lipoprotein plasma removal is reduced in the elderly [52]. High-density lipoproteins (HDLs) play a preventive role in atherosclerosis by exporting cholesterol and promoting cholesterol metabolism. However, changes happened in the composition of the HDLs during aging. HDLs isolated from elderly patients were reported to contain less cholesterol [53]. In addition, HDL function was impaired in older subjects, and its effectiveness in promoting cholesterol efflux and inhibiting LDL oxidation decreased [54]. Taken together, these data make a strong case for blood lipid and plasma lipoprotein levels being highly relevant to aging.

3.2. Age-Related Changes in Triglyceride Metabolism

TGs can be obtained by both exogenous pathways (dietary intake) and endogenous pathways (de novo synthesis). Exogenous TGs are absorbed in the intestine, formed as chylomicrons, and released into circulation. The liver plays a crucial role in TG metabolism, as it can take up and enzymatically hydrolyze chylomicrons, thereby converting them into residual components [55]. Endogenous TGs are mainly produced in the liver through glycolysis. Glycerol triphosphate reacts with esterified coenzyme A, thereby leading to the formation of TGs. Following their transportation by lipoproteins and utilization by extrahepatic tissues, excess TGs are stored in adipose cells. During starvation or sympathetic activation, adipose triglyceride lipase (ATGL) catalyzes the TGs in adipocytes to produce one FA and diacylglycerol (DAG). The DAG is then hydrolyzed by hormone-sensitive lipase (HSL) to generate another FA and monoacylglycerol (MAG). Finally, the MAG is hydrolyzed by monoglyceride lipase (MGL) to release glycerol and the last FA. The FAs produced during this process are used to provide energy through beta-oxidation (β-oxidation), which occurs in mitochondria and involves a series of enzymatic reactions that break down FAs into acyl-CoA and ketone bodies (acetoacetate, acetone, and β-hydroxybutyrate). These acyl-CoAs are further oxidized to produce TGs and ATP.

3.2.1. Fatty Acid Uptake and Triglyceride Synthesis Increase in Liver with Age

Studies have suggested that the liver lipid content in aged mice increases due to the internalization of remnant chylomicrons, which requires the absorption of circulating FAs by hepatic cells. FA uptake in 20-month-old rats resulted in a two-fold increase compared to 2-month-old rats, and this change was detected as early as in middle-aged rats (10-month-old) [56]. The FA transporter CD36 (CD36) on the hepatic cell membrane plays a central role in controlling the FA flux into the liver. A study showed that the index of CD36 in the liver biopsies of healthy elderly subjects increased significantly [57]. This finding was confirmed in older mice, with the gene and protein expression of CD36 being significantly increased in the aged group [57]. Additionally, the pregnane X receptor (PXR) induces liver CD36 mRNA expression, and PXR gene expression is positively associated with age, which may lead to increased FA uptake and liver TG accumulation with age [58,59].

Liver FAs can also be synthesized through de novo lipogenesis (DNL), and the genes responsible for liver DNL are upregulated during aging. The regulation of liver adipogenesis is primarily controlled by sterol regulatory element-binding protein-1c (SREBP-1c), which can increase the expression of other genes involved in adipogenesis, such as ATP citrate lyase (Acly), fatty acid synthase (Fasn), and stearoyl-CoA desaturase-1 (SCD-1). To identify the molecular mechanisms underlying increased FA synthesis in aged mice, researchers analyzed the expression of the hepatic lipogenesis genes, including SREBP-1c, Acly, Fasn, and SCD-1, and found that their expressions were significantly increased [46]. The expression of SREBP-1c is tightly regulated by nuclear receptor farnesoid X receptor (FXR) [60]. Studies have shown that the gene and protein expressions of FXR in the livers of older mice were decreased. The levels of the downstream target genes of FXR, such as small heterodimer partner (SHP) and bile salt export pump (Bsep), were also decreased [46]. These findings suggest that FXR regulates SREBP-1c in opposing directions and ultimately leads to increased TG synthesis with age.

Dietary patterns can impact the absorption and de novo synthesis of FAs in elderly individuals. The consumption of a high-fat diet (HFD) may lead to alterations in fat metabolism, thus further contributing to FA uptake and TG synthesis. Studies indicate that the gene and protein expressions of the CD36 in mice fed with a HFD were significantly higher than those in mice fed a normal diet, thereby suggesting that CD36-mediated liver FA uptake was enhanced [57]. Additionally, the expressions of SREBP-1c and acetyl-CoA carboxylase (ACC) genes were significantly increased in the liver of aged mice fed with a HFD, thereby activating the DNL pathway [61]. Conversely, the calorie restriction (CR) diet reduced the levels of CD36 and ACC protein expressions in elderly subjects fed a normal diet, thus equalizing their levels with those of the younger group [62].

3.2.2. Mitochondrial Dysfunction Leads to Impairment of the β-Oxidation

During aging, lipid synthesis increases and lipid catabolism decreases, thereby disrupting lipid homeostasis. Mitochondrial FA β-oxidation is the primary pathway for FA degradation, wherein it plays a crucial role in maintaining the energy balance [63]. Evidence has shown that liver β-oxidation becomes weaker in aged mice compared to young mice under HFD conditions, as confirmed by the decreased levels of the end-product β-hydroxybutyrate (BHBA) in the liver [61]. This decrease in FA oxidation was also observed in another study, which found that levels of BHBA were reduced in the plasma, thereby indicating a decline in the ketogenic capacity in aging mice [49]. Acylcarnitines transport FAs into the mitochondria and are essential for β-oxidation to occur. An increase in the plasma acylcarnitine concentration with age has been observed in healthy individuals, thus suggesting that mitochondrial function was impaired and β-oxidation declined [64].

Peroxisome proliferator-activated receptor α (PPARα) is a major regulator of the β-oxidation in all organs, and its dysfunction can contribute to age-related diseases. Studies have shown that the up-regulation of the advanced glycation end product receptor (RAGE) in the livers of aging mice leads to the down-regulation of PPARα and the decreased expression of downstream target genes regulated by PPARα, such as carnitine palmitoyltransferase 1a (CPT1a), carnitine palmitoyltransferase 1b (CPT1b), and acyl-CoA dehydrogenase medium chain (MCAD), which explains the weakened mitochondrial FA oxidation and the development of age-related fatty liver diseases [65]. Similarly, a notable down-regulation of the PPARα and its downstream target proteins, including CPT1a and acyl-CoA oxidase 1 (ACOX1), were observed in the kidneys of 24-month-old rats. This down-regulation was attributed to the regulatory effects of the miR-21 and ultimately contributed to the progression of renal fibrosis [66]. Furthermore, the down-regulation of the genes involved in FA transport and oxidation was observed in the heart during aging, thereby leading to cardiac dysfunction [66]. Therefore, decreased FA oxidation during aging contributes to the occurrence of a variety of age-related diseases.

Interestingly, increased FA oxidation has been observed in a long-lived mouse model, thus indicating the importance of this process for organism lifespan extension [67]. Building upon this discovery, investigations have explored the effects of various compounds, including resveratrol and astragaloside IV, which have the ability to enhance FA oxidation and positively influence lipid metabolism in aged mice [68,69]. These exciting findings not only shed light on the mechanism underlying healthy aging, but also hold promise for the development of novel interventions aimed at improving overall health and longevity.

3.3. *Age-Related Changes in Cholesterol Metabolism*

There are two sources of cholesterol in the body: exogenous cholesterol, which can be obtained from the diet via NPC1L1, and endogenous cholesterol, which is primarily synthesized in almost all tissues, with the liver being the most important organ [70]. The cholesterol biosynthetic pathway comprises nearly 30 enzymatic reactions that convert acetyl coenzyme A to cholesterol. However, cholesterol cannot be completely oxidized into carbon dioxide and water in the body. The primary route of the excretion of cholesterol is from the liver, where it is converted to BAs and excreted in the feces. The other route is through the ABCG5/G8 receptors, which remove cholesterol directly and efflux it into the gallbladder [71]. Cholesterol that enters the gut is metabolized by intestinal bacteria to form steroids, which are then excreted in the feces.

3.3.1. Endogenous Cholesterol Synthesis Increases during Aging

During aging, dietary cholesterol intake and endogenous cholesterol synthesis increase, which may contribute to the development of fatty liver disease. Studies have shown that hepatic cholesterol levels become elevated in senescence-accelerated-prone mice [72]. In a study on 28-month-old mice, the expressions of cholesterol-synthesis-related genes, such as glucose transporter 2 (GLUT2), glucokinase (GK), sterol regulatory element-binding protein-2 (SREBP-2), 3-hydroxy-3-methylglutaryl coenzyme A reductase (HMGCR), and 3-hydroxy-3-methylglutaryl-CoA synthase (HMGCS) were increased, and further cellular experiments showed that the reactive oxygen species (ROS) played an important role in this process [73]. In addition, other studies have found that the ROS leads to the complete activation of the 3-hydroxy-3-methylglutaryl coenzyme A (HMG-CoA) reductase, which is a key rate-limiting enzyme in cholesterol synthesis, thereby leading to the increase in cholesterol synthesis with age [74]. Furthermore, studies on aging rats have shown that the ROS can cause the dephosphorylation of the HMG-CoA reductase through the mitogen-activated protein kinase (p38/MAPK) pathway, thereby leading to its activation [75]. Since excessive cholesterol accumulation has harmful effects on the body, reducing cholesterol synthesis through interventions such as the mechanistic target of rapamycin (mTOR) and sirtuins is of great significance [71].

3.3.2. Bile Acid Enterohepatic Circulation and Lipoprotein Dynamics Changes with Age

Excess cholesterol, particularly LDL-C, is a significant risk factor for atherosclerotic CVD. To mitigate this risk, the body has developed multiple mechanisms to remove excess cholesterol from circulation, including both LDL receptor (LDLr)-dependent and LDLr-independent mechanisms. The LDLr-dependent mechanism involves the binding and internalization of LDL particles for lysosomal degradation, while the LDLr-independent mechanism involves the conversion of cholesterol into BA via cholesterol 7alpha-hydroxylase (CYP7A1), which is then followed by metabolism via the gut microbiota. BAs produced by the liver and gut are recycled and reused through enterohepatic circulation, which helps to conserve BA reserves and maintain cholesterol homeostasis.

As was previously mentioned, LDL-C levels increase with age, which is associated with a decrease in the LDL-C clearance. Normally, LDL-C and very low-density lipoprotein cholesterol (VLDL-C) are removed from circulation by hepatic LDLr to regulate cholesterol homeostasis [76]. However, studies have shown that hepatic LDLr expression decreases with age in humans, thus leading to a decrease in the LDL-C clearance from circulation [77]. This decrease has also been observed in elderly mice, as both the gene and protein expressions of the LDLrs decreased with age [78]. The reduction in LDLr expression may be mediated by the adenosine monophosphate-activated protein kinase (AMPK) and the downstream SREBP-2/proprotein convertase subtilisin kexin type 9 (PCSK9)/LDLr signaling pathway. AMPK levels increases with age, and they can regulate lipid metabolism by affecting the expression of the downstream target gene SREBP-2 [79]. PCSK9 is regulated by SREBP-2, which promotes the degradation of LDLrs on the surface of hepatic cells [80,81]. The gene expression of SREBP-2 and the blood levels of PCSK9 have been shown to increase with age [73,82,83]. Therefore, the activation of the AMPK/SREBP-2/PCSK9/LDLr signaling pathway may account for the age-related decrease in the LDLr and LDL-C clearance. Moreover, in LDLr knockout mice, the plasma LDL levels still increased with age, thereby indicating that other factors besides the reduction in LDLrs contribute to the imbalance in cholesterol metabolism throughout the body during aging [84].

Furthermore, BA synthesis decreases with age in humans. Previous studies on biopsied livers in the fasting state have shown reduced CYP7A1 expression in the elderly [85,86]. Additionally, aging is inversely correlated with Cyp7a1 mRNA levels [87]. The decrease in BA synthesis may reduce the utilization of cholesterol in the body, which could lead to an increase in the plasma cholesterol levels [88]. The age-related decreases in BA synthesis and their potential impact on cholesterol utilization are closely linked to the observed disturbances in the enterohepatic circulation of BAs during aging, which are evidenced by the reduced intestinal reabsorption and liver transport of the BAs. Aging was associated with the decreased mRNA expression of the intestinal BAs reabsorption transporter Asbt [27]. The mRNA expressions of the liver BA uptake transporters Na+/taurocholate cotransporting polypeptide (Ntcp) and organic anion transporting polypeptide 1b2 (Oatp1b2) were decreased during aging, whereas the expression of the BAs efflux transporter Bsep remained unchanged in aging mice [27,89]. Moreover, there were gender-dependent changes in the expression of these BA transporters during aging in mice, with Ntcp and Oatp1b2 mRNA expressions being reduced in aging female mice, but remaining constant in aging male mice [90]. These gender-specific differences in transporter expression may be regulated by changes in the estrogen levels in female mice after middle age.

4. Lipid-Related Chronic Diseases in the Elderly

Lipid metabolism disorder is one of the key pathogenic factors for the occurrence and development of a series of lipid-related chronic diseases. In general, lipid-related diseases include CVD, T2D, NAFLD, and obesity, which seriously threaten public health [91]. Given that changes in lipid metabolism occur in healthy older adults, it is important to note that these changes may contribute to pathological alterations. Therefore, understanding the role of lipid metabolism in the development of these diseases may provide new insights into their underlying mechanisms and facilitate the development of effective treatments and prevention for the elderly (Figure 2).

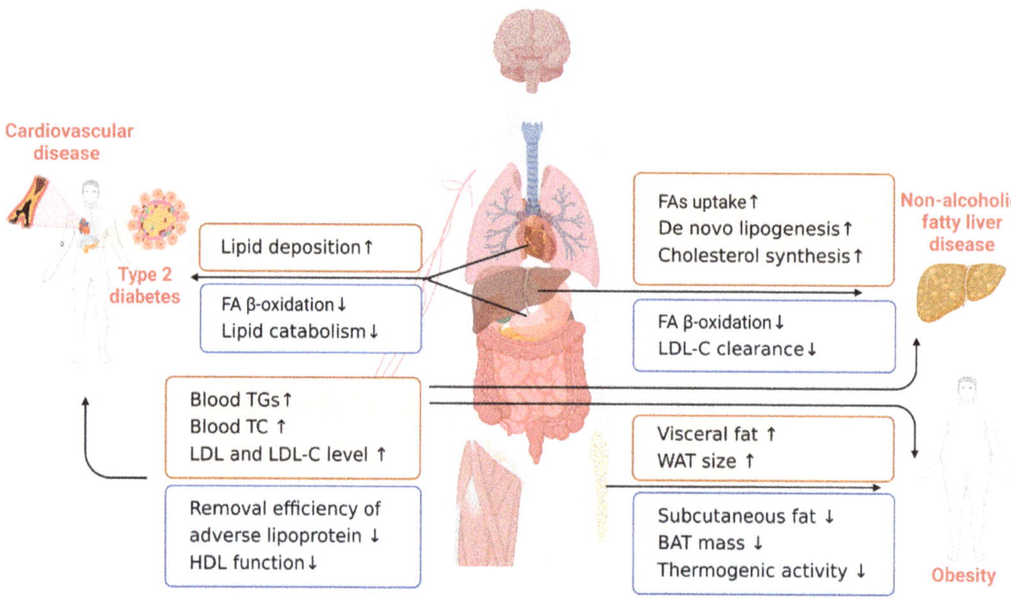

Figure 2. Key metabolic mechanisms of lipid dysregulation in the elderly and their associated chronic diseases. The body's lipid homeostasis is maintained through a combination of hepatic anabolism and catabolism, adipose tissue storage, peripheral tissue utilization, and subsequent balance in blood circulation. These endogenous lipid metabolic changes in the elderly include increased lipid production and accumulation (corresponding to the red box) and decreased lipid consumption and clearance (corresponding to the blue box), which result in age-related chronic diseases such as cardiovascular disease, type 2 diabetes, nonalcoholic fatty liver disease, and obesity. ↑ means studies showing a significant increase; ↓ means studies showing a significant decrease; FA—fatty acid; TG—triglyceride; TC—total cholesterol; LDL—low-density lipoprotein; LDL-C—low-density lipoprotein cholesterol; HDL—high-density lipoprotein; WAT—white adipose tissue; BAT—brown adipose tissue. Figure 2 was created using Biorender with permission for publication from Biorender.com (accessed on 9 July 2023).

4.1. *Cardiovascular Disease*

CVD is recognized as the primary cause of death in elderly individuals and is considered as a true disease of aging, with atherosclerosis being closely associated with lipid metabolism disorders [92,93]. Age-related disruptions in lipid metabolism result in the elevated levels of cholesterol, TGs, and LDLs in the bloodstream, thereby leading to lipid deposition beneath the inner lining of the blood vessels, which in turn forms atherosclerotic plaques. Evidence suggests that LDL-C, TC, and TG levels are positively correlated with the incidence of coronary heart disease, with the TC/HDL-C ratio being the strongest predictor of coronary heart disease in the elderly [2]. Thus, the continuous monitoring of these clinical markers is crucial during the aging process [92].

Furthermore, aging is associated with a greater tendency for lipid deposition in the heart and blood vessels. For instance, 25-month-old mice showed a significant accumulation of triacylglycerols in their heart tissues compared to 4-month-old mice [94]. Additionally, elderly mice exhibited higher rates of lipid deposition in their aortic arches than younger mice [95]. A study on mice has shown that the cluster of differentiation 73 (CD73), which is an exonuclease that catalyzes the conversion of adenosine monophosphate (AMP) to adenosine, can promote atherosclerosis on the vessel walls in aged mice by inhibiting lipid catabolism [96]. In addition, CD73 is an interesting molecule that could promote excess vascular lipid accumulation and accelerate atherosclerosis.

Preventing atherosclerosis requires a careful diet and exercise from a young age. High-TG and low-HDL-C levels at a young age are positively associated with accelerated epigenetic aging in midlife, thereby indicating that optimal lipid levels in early life may slow down epigenetic aging and delay the onset of diseases such as atherosclerosis [97]. Furthermore, older adults who engage in regular aerobic exercise have shown elevated HDL-C levels and decreased TC/HDL-C ratios, thereby suggesting that regular physical exercise can improve the lipid profile in older adults [43].

4.2. Type 2 Diabetes

T2D is a chronic metabolic disease that is prevalent among the elderly, wherein it is characterized by lipid metabolism disorders that lead to abnormal lipoprotein levels, which are often associated with diabetes progression. A study on Chinese older adults confirmed an independent inverse association between HDL-C levels and diabetes risk [98]. Additionally, the ratio of TG/HDL-C was positively correlated with T2D risk, and other lipid-related indicators have also been identified as risk factors for predicting T2D [99]. People with T2D have a significantly higher visceral adiposity index, which can serve as a predictor of T2D risk [100]. The triglyceride-to-glucose fasting index has also been considered as a supplementary indicator in diagnosing prediabetes in the elderly.

Studies have further shown that lipid metabolism disorders primarily affect glucose metabolism through insulin resistance, thereby leading to diabetes. Impaired lipid metabolism can contribute to the ectopic storage of adipocytes, thereby leading to insulin resistance occurrence or development [101]. Along with glucose intolerance, insulin resistance induces T2D during aging [102]. Research suggests that postprandial TG levels in T2D individuals are higher than in nondiabetic individuals due to insulin resistance in the intestinal epithelial cells. To improve the postprandial blood lipid and glucose indexes in T2D, researchers propose a focus on chylomicron metabolism [103].

4.3. Obesity

The increasing prevalence of obesity in the elderly population is a matter of concern [104]. Obesity can accelerate aging and contribute to the development of various age-related diseases, such as sarcopenia, T2D, CVD, and almost all lipid-related diseases, which pose a serious threat to the health of the elderly [105]. Research has indicated that the incidence of obesity in the elderly is associated with lipid metabolism disorders [106]. During the aging process, the activity of the sympathoadrenal system decreases, which may reduce lipid turnover rates, thereby leading to obesity in the elderly [107,108].

The distribution and function of adipose tissue change during the aging process. Older adults frequently experience an accumulation of visceral fat and a loss of subcutaneous fat, with an increase in white adipose tissue size and a decrease in brown adipose tissue function being the most noticeable changes [109]. Specifically, in older rats, adipocytes exhibit increased size, reduced DNL and lipolysis, and enhanced esterification [110]. Meanwhile, brown adipose tissue mass decreases, and its ability to produce heat weakens with aging, thereby leading to fat accumulation and the increased incidence of obesity in the elderly [111].

Adipose-derived stromal/stem cells (ASCs) play a critical role in energy balance maintenance, fat storage, and adipocyte homeostasis [112]. Studies have shown that the ability of ASCs to proliferate and differentiate decreases with age [113]. After transplanting the ASCs from young mice (2 months) into old mice (22 months), it was found that the plasticity of the stem cells, liver function, and lipid metabolism of the old mice improved, thereby indicating that ASCs had a positive effect on the lipid metabolism in aged animals and had the potential to be used as antiaging and antiobesity agents [114].

4.4. Nonalcoholic Fatty Liver Disease

NAFLD is a significant aging-related issue caused by the accumulation of excessive hepatocellular LDs, known as simple steatosis, which can progress to nonalcoholic steatohepatitis (NASH) and associated fibrosis [115]. Studies have indicated that NAFLD is prevalent in the elderly population in Asia, wherein it affects approximately 40% of this population and has the potential to develop into liver fibrosis with aging [116]. Similar observations have been reported in aging mice, wherein TG and cholesterol contents increased in the livers of aged mice [46,65,73].

As mentioned previously, studies have demonstrated that aging processes are likely to promote NAFLD through increased lipid synthesis and decreased lipid catabolism in the liver [46,65,73]. In addition, dietary fat intake significantly impacts the liver fat in aged mice. Under the condition of a HFD, the contents of neutral fat in the livers of aged mice increased compared with young mice. The contents of polyunsaturated FAs with liver-protective effects decreased, while the contents of saturated and monounsaturated FAs with lipotoxicity increased [61]. Thus, dietary restriction is protective against hepatocyte senescence and liver fat deposition in aged mice [117].

Furthermore, understanding the underlying mechanism and identifying effective therapeutic targets for age-related NAFLD is of great importance. Several potential targets have been identified in the context of age-related liver conditions. In the livers of aged mice, the levels of CDGSH iron–sulfur domain-containing protein 2 (Cisd2) decreased by approximately 50%, thus resulting in increased de novo hepatic fat synthesis and accumulation. Cisd2 is also a promising target for slowing down liver aging [118]. Additionally, previous experiments showed that, during the aging process, hepatic lipid deposition reduced in zinc finger gene 1 (JAZF1)-transgene mice, which was possibly due to the decreased gene expressions involved in adipose storage, including SREBP-1c, SCD-1, and fatty acid synthase (FAS) [119]. This suggests that JAZF1 may be another potential target, which could prevent lipid production.

5. Conclusions

Fundamentally, lipid metabolism is maintained through a delicate balance between lipid intestinal digestion and absorption, systemic synthesis, and catabolism. This review summarizes the adverse changes in lipid metabolism that occur during aging and their relationship with the development of age-related chronic diseases. The insights from this review have promising implications for developing nutrition and health guidelines regarding the distinct lipid metabolism needs of the aging population. The process of conducting a literature review has revealed inconsistencies in the age-related research findings, which are possibly due to increased individual differences in the aging process. Genetic variations, lifestyle choices, environmental exposures, and health conditions all influence aging outcomes. Addressing this issue requires rigorous, larger-scale prospective or interventional studies. In addition, utilizing more advanced approaches, such as single-cell sequencing, spatial transcriptomics, and proteomics could offer profound insights into the underlying mechanisms of age-related lipid metabolism disorders and may reduce research discrepancies.

Recent progress has shown that the aging process affects not only one aspect of lipid metabolism, but every regulatory component of it. Furthermore, lipid-related diseases are likely to be interconnected, with the occurrence of one disease increasing the risk of others. Further research is needed to explore the connections between these regulators and their interactions in causing age-related diseases. Accumulating evidence has shown that a nutritionally adequate diet is beneficial for all age-related diseases, and dietary intervention may be a new target for preventing excess fat deposition and for treating lipid-related diseases in the elderly. Overall, this review offers valuable insights into the prevention of nutrition-related chronic diseases in the elderly, which is beneficial for promoting a healthy lifespan.

Author Contributions: Conceptualization, R.S. and X.W.; writing—original draft preparation, R.S., M.H. and X.Q.; writing—review and editing, R.S., M.H., X.Q., L.Q., P.W. and X.W.; visualization, X.Z. and R.L.; supervision, X.W. All authors have read and agreed to the published version of the manuscript.

Funding: This research was funded by the National Natural Science Foundation of China (Grant No. 32001676), and the 111 Project from the Education Ministry of China (No. B18053).

Institutional Review Board Statement: Not applicable.

Informed Consent Statement: Not applicable.

Data Availability Statement: Not applicable.

Conflicts of Interest: The authors declare no conflict of interest.

References

1. United Nations, Department of Economic and Social Affairs, Population Division. World Population Prospects 2019: Highlights. 2019. ST/ESA/SER.A/423. Available online: https://population.un.org/wpp/publications/files/wpp2019_highlights.pdf (accessed on 15 June 2022).
2. Katsiki, N.; Kolovou, G.; Perez-Martinez, P.; Mikhailidis, D.P. Dyslipidaemia in the Elderly: To Treat or Not to Treat? *Expert Rev. Clin. Pharmacol.* **2018**, *11*, 259–278. [CrossRef]
3. Zemedikun, D.T.; Gray, L.J.; Khunti, K.; Davies, M.J.; Dhalwani, N.N. Patterns of Multimorbidity in Middle-Aged and Older Adults: An Analysis of the UK Biobank Data. *Mayo Clin. Proc.* **2018**, *93*, 857–866. [CrossRef] [PubMed]
4. Yan, Y.; Wu, T.; Zhang, M.; Li, C.; Liu, Q.; Li, F. Prevalence, Awareness and Control of Type 2 Diabetes Mellitus and Risk Factors in Chinese Elderly Population. *BMC Public Health* **2022**, *22*, 1382. [CrossRef] [PubMed]
5. Ames, B.N. Prolonging Healthy Aging: Longevity Vitamins and Proteins. *Proc. Natl. Acad. Sci. USA* **2018**, *115*, 10836–10844. [CrossRef]
6. Bergman, P.; Brighenti, S. Targeted Nutrition in Chronic Disease. *Nutrients* **2020**, *12*, 1682. [CrossRef]
7. Milan, A.M.; Cameron-Smith, D. Digestion and Postprandial Metabolism in the Elderly. *Adv. Food Nutr. Res.* **2015**, *76*, 79–124. [CrossRef]
8. Van der Wulp, M.Y.M.; Verkade, H.J.; Groen, A.K. Regulation of Cholesterol Homeostasis. *Mol. Cell Endocrinol.* **2013**, *368*, 1–16. [CrossRef] [PubMed]
9. Acevedo-Fani, A.; Singh, H. Biophysical Insights into Modulating Lipid Digestion in Food Emulsions. *Prog. Lipid Res.* **2022**, *85*, 101129. [CrossRef]
10. Perino, A.; Demagny, H.; Velazquez-Villegas, L.; Schoonjans, K. Molecular Physiology of Bile Acid Signaling in Health, Disease, and Aging. *Physiol. Rev.* **2021**, *101*, 683–731. [CrossRef]
11. Moreau, H.; Laugier, R.; Gargouri, Y.; Ferrato, F.; Verger, R. Human Preduodenal Lipase Is Entirely of Gastric Fundic Origin. *Gastroenterology* **1988**, *95*, 1221–1226. [CrossRef]
12. Sams, L.; Paume, J.; Giallo, J.; Carrière, F. Relevant PH and Lipase for in Vitro Models of Gastric Digestion. *Food Funct.* **2016**, *7*, 30–45. [CrossRef] [PubMed]
13. Feldman, M.; Cryer, B.; McArthur, K.; Huet, B.; Lee, E. Effects of Aging and Gastritis on Gastric Acid and Pepsin Secretion in Humans: A Prospective Study. *Gastroenterology* **1996**, *110*, 1043–1052. [CrossRef] [PubMed]
14. Wang, L.; Jia, H.; Lin, G.; Zheng, S. Magnetic Resonance Imaging Investigation of Age-Related Morphological Changes in the Pancreases of 226 Chinese. *Aging Med.* **2021**, *4*, 297–303. [CrossRef]
15. Jiang, Z.-E.; Jiang, C.; Chen, B.; Koh, C.S.; Yong, J.-H.; Park, D.-H.; Won, M.-H.; Lee, Y.-L. Age-Associated Changes in Pancreatic Exocrine Secretion of the Isolated Perfused Rat Pancreas. *Lab. Anim. Res.* **2013**, *29*, 19–26. [CrossRef]
16. Torigoe, T.; Ito, K.; Yamamoto, A.; Kanki, A.; Yasokawa, K.; Tamada, T.; Yoshida, K. Age-Related Change of the Secretory Flow of Pancreatic Juice in the Main Pancreatic Duct: Evaluation with Cine-Dynamic MRCP Using Spatially Selective Inversion Recovery Pulse. *AJR Am. J. Roentgenol.* **2014**, *202*, 1022–1026. [CrossRef] [PubMed]
17. Laugier, R.; Bernard, J.P.; Berthezene, P.; Dupuy, P. Changes in Pancreatic Exocrine Secretion with Age: Pancreatic Exocrine Secretion Does Decrease in the Elderly. *Digestion* **1991**, *50*, 202–211. [CrossRef] [PubMed]
18. Carriere, F.; Barrowman, J.A.; Verger, R.; Laugier, R. Secretion and Contribution to Lipolysis of Gastric and Pancreatic Lipases during a Test Meal in Humans. *Gastroenterology* **1993**, *105*, 876–888. [CrossRef]
19. Vellas, B.; Balas, D.; Moreau, J.; Bouisson, M.; Senegas-Balas, F.; Guidet, M.; Ribet, A. Exocrine Pancreatic Secretion in the Elderly. *Int. J. Pancreatol.* **1988**, *3*, 497–502. [CrossRef]
20. Yamamoto, K.; E, S.; Hatakeyama, Y.; Sakamoto, Y.; Tsuduki, T. High-Fat Diet Intake from Senescence Inhibits the Attenuation of Cell Functions and the Degeneration of Villi with Aging in the Small Intestine, and Inhibits the Attenuation of Lipid Absorption Ability in SAMP8 Mice. *J. Clin. Biochem. Nutr.* **2015**, *57*, 204–211. [CrossRef]
21. Yamamoto, K.; Kitano, Y.; Shuang, E.; Hatakeyama, Y.; Sakamoto, Y.; Honma, T.; Tsuduki, T. Decreased Lipid Absorption Due to Reduced Pancreatic Lipase Activity in Aging Male Mice. *Biogerontology* **2014**, *15*, 463–473. [CrossRef]

22. Kovacs, D.; Szabo, B.; Pancsa, R.; Tompa, P. Intrinsically Disordered Proteins Undergo and Assist Folding Transitions in the Proteome. *Arch. Biochem. Biophys.* **2013**, *531*, 80–89. [CrossRef] [PubMed]
23. De Aguiar Vallim, T.Q.; Tarling, E.J.; Edwards, P.A. Pleiotropic Roles of Bile Acids in Metabolism. *Cell Metab.* **2013**, *17*, 657–669. [CrossRef] [PubMed]
24. Humbert, L.; Rainteau, D.; Tuvignon, N.; Wolf, C.; Seksik, P.; Laugier, R.; Carrière, F. Postprandial Bile Acid Levels in Intestine and Plasma Reveal Altered Biliary Circulation in Chronic Pancreatitis Patients. *J. Lipid Res.* **2018**, *59*, 2202–2213. [CrossRef] [PubMed]
25. Nagengast, F.M.; van der Werf, S.D.; Lamers, H.L.; Hectors, M.P.; Buys, W.C.; van Tongeren, J.M. Influence of Age, Intestinal Transit Time, and Dietary Composition on Fecal Bile Acid Profiles in Healthy Subjects. *Dig. Dis. Sci.* **1988**, *33*, 673–678. [CrossRef]
26. Kullak-Ublick, G.A.; Paumgartner, G.; Berr, F. Long-Term Effects of Cholecystectomy on Bile Acid Metabolism. *Hepatology* **1995**, *21*, 41–45. [CrossRef] [PubMed]
27. Gao, T.; Feridooni, H.A.; Howlett, S.E.; Pelis, R.M. Influence of Age on Intestinal Bile Acid Transport in C57BL/6 Mice. *Pharmacol. Res. Perspect.* **2017**, *5*, e00287. [CrossRef]
28. Vergès, B. Intestinal Lipid Absorption and Transport in Type 2 Diabetes. *Diabetologia* **2022**, *65*, 1587–1600. [CrossRef]
29. Hussain, M.M. Intestinal Lipid Absorption and Lipoprotein Formation. *Curr. Opin. Lipidol.* **2014**, *25*, 200–206. [CrossRef]
30. Cohn, J.S.; McNamara, J.R.; Cohn, S.D.; Ordovas, J.M.; Schaefer, E.J. Postprandial Plasma Lipoprotein Changes in Human Subjects of Different Ages. *J. Lipid Res.* **1988**, *29*, 469–479. [CrossRef]
31. Issa, J.S.; Diament, J.; Forti, N. Postprandial Lipemia: Influence of Aging. *Arq. Bras. Cardiol.* **2005**, *85*, 15–19. [CrossRef]
32. Puga, G.M.; Meyer, C.; Everman, S.; Mandarino, L.J.; Katsanos, C.S. Postprandial Lipemia in the Elderly Involves Increased Incorporation of Ingested Fat in Plasma Free Fatty Acids and Small (Sf 20-400) Triglyceride-Rich Lipoproteins. *Am. J. Physiol. Endocrinol. Metab.* **2011**, *301*, E356–E361. [CrossRef]
33. Puga, G.M.; Meyer, C.; Mandarino, L.J.; Katsanos, C.S. Postprandial Spillover of Dietary Lipid into Plasma Is Increased with Moderate Amounts of Ingested Fat and Is Inversely Related to Adiposity in Healthy Older Men. *J. Nutr.* **2012**, *142*, 1806–1811. [CrossRef] [PubMed]
34. Gebert, N.; Cheng, C.-W.; Kirkpatrick, J.M.; Di Fraia, D.; Yun, J.; Schädel, P.; Pace, S.; Garside, G.B.; Werz, O.; Rudolph, K.L.; et al. Region-Specific Proteome Changes of the Intestinal Epithelium during Aging and Dietary Restriction. *Cell Rep.* **2020**, *31*, 107565. [CrossRef] [PubMed]
35. Woudstra, T.D.; Drozdowski, L.A.; Wild, G.E.; Clandinin, M.T.; Agellon, L.B.; Thomson, A.B.R. The Age-Related Decline in Intestinal Lipid Uptake Is Associated with a Reduced Abundance of Fatty Acid-Binding Protein. *Lipids* **2004**, *39*, 603–610. [CrossRef] [PubMed]
36. Biagi, E.; Franceschi, C.; Rampelli, S.; Severgnini, M.; Ostan, R.; Turroni, S.; Consolandi, C.; Quercia, S.; Scurti, M.; Monti, D.; et al. Gut Microbiota and Extreme Longevity. *Curr. Biol.* **2016**, *26*, 1480–1485. [CrossRef] [PubMed]
37. Martinez-Guryn, K.; Hubert, N.; Frazier, K.; Urlass, S.; Musch, M.W.; Ojeda, P.; Pierre, J.F.; Miyoshi, J.; Sontag, T.J.; Cham, C.M.; et al. Small Intestine Microbiota Regulate Host Digestive and Absorptive Adaptive Responses to Dietary Lipids. *Cell Host Microbe* **2018**, *23*, 458–469.e5. [CrossRef] [PubMed]
38. Wang, D.Q.-H. Aging Per Se Is an Independent Risk Factor for Cholesterol Gallstone Formation in Gallstone Susceptible Mice. *J. Lipid Res.* **2002**, *43*, 1950–1959. [CrossRef]
39. Duan, L.-P.; Wang, H.H.; Ohashi, A.; Wang, D.Q.-H. Role of Intestinal Sterol Transporters Abcg5, Abcg8, and Npc1l1 in Cholesterol Absorption in Mice: Gender and Age Effects. *Am. J. Physiol. Gastrointest. Liver Physiol.* **2006**, *290*, G269–G276. [CrossRef]
40. Wang, D.Q.-H. Regulation of Intestinal Cholesterol Absorption. *Annu. Rev. Physiol.* **2007**, *69*, 221–248. [CrossRef]
41. Schoeler, M.; Caesar, R. Dietary Lipids, Gut Microbiota and Lipid Metabolism. *Rev. Endocr. Metab. Disord.* **2019**, *20*, 461–472. [CrossRef]
42. Chung, K.W. Advances in Understanding of the Role of Lipid Metabolism in Aging. *Cells* **2021**, *10*, 880. [CrossRef] [PubMed]
43. Matthan, N.R.; Jalbert, S.M.; Lamon-Fava, S.; Dolnikowski, G.G.; Welty, F.K.; Barrett, H.R.; Schaefer, E.J.; Lichtenstein, A.H. TRL, IDL, and LDL Apolipoprotein B-100 and HDL Apolipoprotein A-I Kinetics as a Function of Age and Menopausal Status. *Arterioscler. Thromb. Vasc. Biol.* **2005**, *25*, 1691–1696. [CrossRef]
44. Feng, L.; Nian, S.; Tong, Z.; Zhu, Y.; Li, Y.; Zhang, C.; Bai, X.; Luo, X.; Wu, M.; Yan, Z. Age-Related Trends in Lipid Levels: A Large-Scale Cross-Sectional Study of the General Chinese Population. *BMJ Open* **2020**, *10*, e034226. [CrossRef] [PubMed]
45. Xia, X.; Chen, W.; McDermott, J.; Han, J.-D.J. Molecular and Phenotypic Biomarkers of Aging. *F1000Res* **2017**, *6*, 860. [CrossRef]
46. Xiong, X.; Wang, X.; Lu, Y.; Wang, E.; Zhang, Z.; Yang, J.; Zhang, H.; Li, X. Hepatic Steatosis Exacerbated by Endoplasmic Reticulum Stress-Mediated Downregulation of FXR in Aging Mice. *J. Hepatol.* **2014**, *60*, 847–854. [CrossRef] [PubMed]
47. Papsdorf, K.; Brunet, A. Linking Lipid Metabolism to Chromatin Regulation in Aging. *Trends Cell Biol.* **2019**, *29*, 97–116. [CrossRef]
48. Greenfield, M.S.; Kraemer, F.; Tobey, T.; Reaven, G. Effect of Age on Plasma Triglyceride Concentrations in Man. *Metabolism* **1980**, *29*, 1095–1099. [CrossRef]
49. Houtkooper, R.H.; Argmann, C.; Houten, S.M.; Cantó, C.; Jeninga, E.H.; Andreux, P.A.; Thomas, C.; Doenlen, R.; Schoonjans, K.; Auwerx, J. The Metabolic Footprint of Aging in Mice. *Sci. Rep.* **2011**, *1*, 134. [CrossRef]
50. Parini, P.; Angelin, B.; Rudling, M. Cholesterol and Lipoprotein Metabolism in Aging: Reversal of Hypercholesterolemia by Growth Hormone Treatment in Old Rats. *Arterioscler. Thromb. Vasc. Biol.* **1999**, *19*, 832–839. [CrossRef]
51. Ericsson, S.; Eriksson, M.; Vitols, S.; Einarsson, K.; Berglund, L.; Angelin, B. Influence of Age on the Metabolism of Plasma Low Density Lipoproteins in Healthy Males. *J. Clin. Investig.* **1991**, *87*, 591–596. [CrossRef]

52. Maranhão, R.C.; Pala, D.; Freitas, F.R. Lipoprotein Removal Mechanisms and Aging: Implications for the Cardiovascular Health of the Elderly. *Curr. Opin. Endocrinol. Diabetes Obes.* **2020**, *27*, 104–109. [CrossRef]
53. Mutharasan, R.K.; Thaxton, C.S.; Berry, J.; Daviglus, M.L.; Yuan, C.; Sun, J.; Ayers, C.; Lloyd-Jones, D.M.; Wilkins, J.T. HDL Efflux Capacity, HDL Particle Size, and High-Risk Carotid Atherosclerosis in a Cohort of Asymptomatic Older Adults: The Chicago Healthy Aging Study. *J. Lipid Res.* **2017**, *58*, 600–606. [CrossRef]
54. Berrougui, H.; Khalil, A. Age-Associated Decrease of High-Density Lipoprotein-Mediated Reverse Cholesterol Transport Activity. *Rejuvenation Res.* **2009**, *12*, 117–126. [CrossRef] [PubMed]
55. Alves-Bezerra, M.; Cohen, D.E. Triglyceride Metabolism in the Liver. *Compr. Physiol.* **2017**, *8*, 1–8. [CrossRef] [PubMed]
56. Murthy, V.K.; Oredipe, O.A.; Stiles, M.R.; Shipp, J.C. Increased Fatty Acid Uptake, a Factor in Increased Hepatic Triacylglycerol Synthesis in Aging Rats. *Mech. Ageing Dev.* **1986**, *37*, 49–54. [CrossRef]
57. Sheedfar, F.; Sung, M.M.; Aparicio-Vergara, M.; Kloosterhuis, N.J.; Miquilena-Colina, M.E.; Vargas-Castrillón, J.; Febbraio, M.; Jacobs, R.L.; de Bruin, A.; Vinciguerra, M.; et al. Increased Hepatic CD36 Expression with Age Is Associated with Enhanced Susceptibility to Nonalcoholic Fatty Liver Disease. *Aging* **2014**, *6*, 281–295. [CrossRef]
58. Xu, S.-F.; Hu, A.-L.; Xie, L.; Liu, J.-J.; Wu, Q.; Liu, J. Age-Associated Changes of Cytochrome P450 and Related Phase-2 Gene/Proteins in Livers of Rats. *PeerJ* **2019**, *7*, e7429. [CrossRef]
59. Zhou, J.; Febbraio, M.; Wada, T.; Zhai, Y.; Kuruba, R.; He, J.; Lee, J.H.; Khadem, S.; Ren, S.; Li, S.; et al. Hepatic Fatty Acid Transporter Cd36 Is a Common Target of LXR, PXR, and PPARgamma in Promoting Steatosis. *Gastroenterology* **2008**, *134*, 556–567. [CrossRef] [PubMed]
60. Watanabe, M.; Houten, S.M.; Wang, L.; Moschetta, A.; Mangelsdorf, D.J.; Heyman, R.A.; Moore, D.D.; Auwerx, J. Bile Acids Lower Triglyceride Levels via a Pathway Involving FXR, SHP, and SREBP-1c. *J. Clin. Investig.* **2004**, *113*, 1408–1418. [CrossRef]
61. Ishizuka, K.; Kon, K.; Lee-Okada, H.-C.; Arai, K.; Uchiyama, A.; Yamashina, S.; Yokomizo, T.; Ikejima, K. Aging Exacerbates High-Fat Diet-Induced Steatohepatitis through Alteration in Hepatic Lipid Metabolism in Mice. *J. Gastroenterol. Hepatol.* **2020**, *35*, 1437–1448. [CrossRef] [PubMed]
62. Teofilović, A.; Vratarić, M.; Veličković, N.; Vojnović Milutinović, D.; Mladenovic, A.; Prvulovic, M.; Djordjevic, A. Late-Onset Calorie Restriction Improves Lipid Metabolism and Aggravates Inflammation in the Liver of Old Wistar Rats. *Front. Nutr.* **2022**, *9*, 899255. [CrossRef] [PubMed]
63. Houten, S.M.; Violante, S.; Ventura, F.V.; Wanders, R.J.A. The Biochemistry and Physiology of Mitochondrial Fatty Acid β-Oxidation and Its Genetic Disorders. *Annu. Rev. Physiol.* **2016**, *78*, 23–44. [CrossRef] [PubMed]
64. Jarrell, Z.R.; Smith, M.R.; Hu, X.; Orr, M.; Liu, K.H.; Quyyumi, A.A.; Jones, D.P.; Go, Y.-M. Plasma Acylcarnitine Levels Increase with Healthy Aging. *Aging* **2020**, *12*, 13555–13570. [CrossRef] [PubMed]
65. Wan, J.; Wu, X.; Chen, H.; Xia, X.; Song, X.; Chen, S.; Lu, X.; Jin, J.; Su, Q.; Cai, D.; et al. Aging-Induced Aberrant RAGE/PPARα Axis Promotes Hepatic Steatosis via Dysfunctional Mitochondrial β Oxidation. *Aging Cell* **2020**, *19*, e13238. [CrossRef]
66. Chung, K.W.; Lee, E.K.; Lee, M.K.; Oh, G.T.; Yu, B.P.; Chung, H.Y. Impairment of PPARα and the Fatty Acid Oxidation Pathway Aggravates Renal Fibrosis during Aging. *J. Am. Soc. Nephrol.* **2018**, *29*, 1223–1237. [CrossRef] [PubMed]
67. Zhao, X.; Petrashen, A.P.; Sanders, J.A.; Peterson, A.L.; Sedivy, J.M. SLC1A5 Glutamine Transporter Is a Target of MYC and Mediates Reduced MTORC1 Signaling and Increased Fatty Acid Oxidation in Long-Lived Myc Hypomorphic Mice. *Aging Cell* **2019**, *18*, e12947. [CrossRef]
68. Gimeno-Mallench, L.; Mas-Bargues, C.; Inglés, M.; Olaso, G.; Borras, C.; Gambini, J.; Vina, J. Resveratrol Shifts Energy Metabolism to Increase Lipid Oxidation in Healthy Old Mice. *Biomed. Pharmacother.* **2019**, *118*, 109130. [CrossRef]
69. Luo, Z.; Wang, Y.; Xue, M.; Xia, F.; Zhu, L.; Li, Y.; Jia, D.; Chen, S.; Xu, G.; Lei, Y. Astragaloside IV Ameliorates Fat Metabolism in the Liver of Ageing Mice through Targeting Mitochondrial Activity. *J. Cell Mol. Med.* **2021**, *25*, 8863–8876. [CrossRef]
70. Luo, J.; Yang, H.; Song, B.-L. Mechanisms and Regulation of Cholesterol Homeostasis. *Nat. Rev. Mol. Cell Biol.* **2020**, *21*, 225–245. [CrossRef]
71. Morgan, A.E.; Mooney, K.M.; Wilkinson, S.J.; Pickles, N.A.; Mc Auley, M.T. Cholesterol Metabolism: A Review of How Ageing Disrupts the Biological Mechanisms Responsible for Its Regulation. *Ageing Res. Rev.* **2016**, *27*, 108–124. [CrossRef]
72. Kuhla, A.; Blei, T.; Jaster, R.; Vollmar, B. Aging Is Associated with a Shift of Fatty Metabolism toward Lipogenesis. *J. Gerontol. A Biol. Sci. Med. Sci.* **2011**, *66*, 1192–1200. [CrossRef] [PubMed]
73. Seo, E.; Kang, H.; Choi, H.; Choi, W.; Jun, H.-S. Reactive Oxygen Species-Induced Changes in Glucose and Lipid Metabolism Contribute to the Accumulation of Cholesterol in the Liver during Aging. *Aging Cell* **2019**, *18*, e12895. [CrossRef] [PubMed]
74. Mc Auley, M.T.; Mooney, K.M. LDL-C Levels in Older People: Cholesterol Homeostasis and the Free Radical Theory of Ageing Converge. *Med. Hypotheses* **2017**, *104*, 15–19. [CrossRef]
75. Pallottini, V.; Martini, C.; Cavallini, G.; Bergamini, E.; Mustard, K.J.; Hardie, D.G.; Trentalance, A. Age-Related HMG-CoA Reductase Deregulation Depends on ROS-Induced P38 Activation. *Mech. Ageing Dev.* **2007**, *128*, 688–695. [CrossRef]
76. Goldstein, J.L.; Brown, M.S. History of Discovery: The LDL Receptor. *Arterioscler. Thromb. Vasc. Biol.* **2009**, *29*, 431–438. [CrossRef] [PubMed]
77. Millar, J.S.; Lichtenstein, A.H.; Cuchel, M.; Dolnikowski, G.G.; Hachey, D.L.; Cohn, J.S.; Schaefer, E.J. Impact of Age on the Metabolism of VLDL, IDL, and LDL Apolipoprotein B-100 in Men. *J. Lipid Res.* **1995**, *36*, 1155–1167. [CrossRef] [PubMed]
78. Bose, C.; Bhuvaneswaran, C.; Udupa, K.B. Age-Related Alteration in Hepatic Acyl-CoA: Cholesterol Acyltransferase and Its Relation to LDL Receptor and MAPK. *Mech. Ageing Dev.* **2005**, *126*, 740–751. [CrossRef]

79. Li, X.; Hu, X.; Pan, T.; Dong, L.; Ding, L.; Wang, Z.; Song, R.; Wang, X.; Wang, N.; Zhang, Y.; et al. Kanglexin, a New Anthraquinone Compound, Attenuates Lipid Accumulation by Activating the AMPK/SREBP-2/PCSK9/LDLR Signalling Pathway. *Biomed. Pharmacother.* **2021**, *133*, 110802. [CrossRef]
80. Horton, J.D.; Cohen, J.C.; Hobbs, H.H. Molecular Biology of PCSK9: Its Role in LDL Metabolism. *Trends Biochem. Sci.* **2007**, *32*, 71–77. [CrossRef]
81. Chae, H.-S.; You, B.H.; Kim, D.-Y.; Lee, H.; Ko, H.W.; Ko, H.-J.; Choi, Y.H.; Choi, S.S.; Chin, Y.-W. Sauchinone Controls Hepatic Cholesterol Homeostasis by the Negative Regulation of PCSK9 Transcriptional Network. *Sci. Rep.* **2018**, *8*, 6737. [CrossRef]
82. Dubuc, G.; Tremblay, M.; Paré, G.; Jacques, H.; Hamelin, J.; Benjannet, S.; Boulet, L.; Genest, J.; Bernier, L.; Seidah, N.G.; et al. A New Method for Measurement of Total Plasma PCSK9: Clinical Applications. *J. Lipid Res.* **2010**, *51*, 140–149. [CrossRef] [PubMed]
83. Cui, Q.; Ju, X.; Yang, T.; Zhang, M.; Tang, W.; Chen, Q.; Hu, Y.; Haas, J.V.; Troutt, J.S.; Pickard, R.T.; et al. Serum PCSK9 Is Associated with Multiple Metabolic Factors in a Large Han Chinese Population. *Atherosclerosis* **2010**, *213*, 632–636. [CrossRef] [PubMed]
84. Du, W.; Wong, C.; Song, Y.; Shen, H.; Mori, D.; Rotllan, N.; Price, N.; Dobrian, A.D.; Meng, H.; Kleinstein, S.H.; et al. Age-Associated Vascular Inflammation Promotes Monocytosis during Atherogenesis. *Aging Cell* **2016**, *15*, 766–777. [CrossRef]
85. Einarsson, K.; Nilsell, K.; Leijd, B.; Angelin, B. Influence of Age on Secretion of Cholesterol and Synthesis of Bile Acids by the Liver. *N. Engl. J. Med.* **1985**, *313*, 277–282. [CrossRef] [PubMed]
86. Bertolotti, M.; Abate, N.; Bertolotti, S.; Loria, P.; Concari, M.; Messora, R.; Carubbi, F.; Pinetti, A.; Carulli, N. Effect of Aging on Cholesterol 7 Alpha-Hydroxylation in Humans. *J. Lipid Res.* **1993**, *34*, 1001–1007. [CrossRef]
87. Bertolotti, M.; Gabbi, C.; Anzivino, C.; Crestani, M.; Mitro, N.; Del Puppo, M.; Godio, C.; De Fabiani, E.; Macchioni, D.; Carulli, L.; et al. Age-Related Changes in Bile Acid Synthesis and Hepatic Nuclear Receptor Expression. *Eur. J. Clin. Investig.* **2007**, *37*, 501–508. [CrossRef]
88. Uchida, K.; Satoh, T.; Chikai, T.; Takase, H.; Nomura, Y.; Nakao, H.; Takeuchi, N. Influence of Cholesterol Feeding on Bile Acid Metabolism in Young and Aged Germ-Free Rats. *Jpn. J. Pharmacol.* **1996**, *71*, 113–118. [CrossRef]
89. Zhang, Y.-K.J.; Saupe, K.W.; Klaassen, C.D. Energy Restriction Does Not Compensate for the Reduced Expression of Hepatic Drug-Processing Genes in Mice with Aging. *Drug Metab. Dispos.* **2010**, *38*, 1122–1131. [CrossRef]
90. Fu, Z.D.; Csanaky, I.L.; Klaassen, C.D. Gender-Divergent Profile of Bile Acid Homeostasis during Aging of Mice. *PLoS ONE* **2012**, *7*, e32551. [CrossRef]
91. Li, Y.; Ma, Z.; Jiang, S.; Hu, W.; Li, T.; Di, S.; Wang, D.; Yang, Y. A Global Perspective on FOXO1 in Lipid Metabolism and Lipid-Related Diseases. *Prog. Lipid Res.* **2017**, *66*, 42–49. [CrossRef]
92. Castro, A.; Signini, É.F.; De Oliveira, J.M.; Di Medeiros Leal, M.C.B.; Rehder-Santos, P.; Millan-Mattos, J.C.; Minatel, V.; Pantoni, C.B.F.; Oliveira, R.V.; Catai, A.M.; et al. The Aging Process: A Metabolomics Perspective. *Molecules* **2022**, *27*, 8656. [CrossRef]
93. Donato, A.J.; Machin, D.R.; Lesniewski, L.A. Mechanisms of Dysfunction in the Aging Vasculature and Role in Age-Related Disease. *Circ. Res.* **2018**, *123*, 825–848. [CrossRef] [PubMed]
94. Eum, J.Y.; Lee, J.C.; Yi, S.S.; Kim, I.Y.; Seong, J.K.; Moon, M.H. Aging-Related Lipidomic Changes in Mouse Serum, Kidney, and Heart by Nanoflow Ultrahigh-Performance Liquid Chromatography-Tandem Mass Spectrometry. *J. Chromatogr. A* **2020**, *1618*, 460849. [CrossRef]
95. Pereira, T.M.; Nogueira, B.V.; Lima, L.C.; Porto, M.L.; Arruda, J.A.; Vasquez, E.C.; Meyrelles, S.S. Cardiac and Vascular Changes in Elderly Atherosclerotic Mice: The Influence of Gender. *Lipids Health Dis.* **2010**, *9*, 87. [CrossRef] [PubMed]
96. Sutton, N.R.; Bouïs, D.; Mann, K.M.; Rashid, I.M.; McCubbrey, A.L.; Hyman, M.C.; Goldstein, D.R.; Mei, A.; Pinsky, D.J. CD73 Promotes Age-Dependent Accretion of Atherosclerosis. *Arterioscler. Thromb. Vasc. Biol.* **2020**, *40*, 61–71. [CrossRef] [PubMed]
97. Gao, T.; Wilkins, J.T.; Zheng, Y.; Joyce, B.T.; Jacobs, D.R.; Schreiner, P.J.; Horvath, S.; Greenland, P.; Lloyd-Jones, D.; Hou, L. Plasma Lipid Profiles in Early Adulthood Are Associated with Epigenetic Aging in the Coronary Artery Risk Development in Young Adults (CARDIA) Study. *Clin. Epigenet.* **2022**, *14*, 16. [CrossRef]
98. Cao, X.; Tang, Z.; Zhang, J.; Li, H.; Singh, M.; Sun, F.; Li, X.; Li, C.; Wang, Y.; Guo, X.; et al. Association between High-Density Lipoprotein Cholesterol and Type 2 Diabetes Mellitus among Chinese: The Beijing Longitudinal Study of Aging. *Lipids Health Dis.* **2021**, *20*, 71. [CrossRef]
99. Zheng, D.; Li, H.; Ai, F.; Sun, F.; Singh, M.; Cao, X.; Jiang, J.; He, Y.; Tang, Z.; Guo, X. Association between the Triglyceride to High-Density Lipoprotein Cholesterol Ratio and the Risk of Type 2 Diabetes Mellitus among Chinese Elderly: The Beijing Longitudinal Study of Aging. *BMJ Open Diabetes Res. Care* **2020**, *8*, e000811. [CrossRef]
100. Ramírez-Vélez, R.; Pérez-Sousa, M.Á.; González-Ruíz, K.; Cano-Gutierrez, C.A.; Schmidt-RioValle, J.; Correa-Rodríguez, M.; Izquierdo, M.; Romero-García, J.A.; Campos-Rodríguez, A.Y.; Triana-Reina, H.R.; et al. Obesity- and Lipid-Related Parameters in the Identification of Older Adults with a High Risk of Prediabetes According to the American Diabetes Association: An Analysis of the 2015 Health, Well-Being, and Aging Study. *Nutrients* **2019**, *11*, 2654. [CrossRef]
101. Lambert, J.E.; Parks, E.J. Postprandial Metabolism of Meal Triglyceride in Humans. *Biochim. Biophys. Acta* **2012**, *1821*, 721–726. [CrossRef]
102. Liu, Z.; Jin, L.; Yang, J.-K.; Wang, B.; Wu, K.K.L.; Hallenborg, P.; Xu, A.; Cheng, K.K.Y. The Dysfunctional MDM2-P53 Axis in Adipocytes Contributes to Aging-Related Metabolic Complications by Induction of Lipodystrophy. *Diabetes* **2018**, *67*, 2397–2409. [CrossRef] [PubMed]

103. Søndergaard, E.; Johansen, R.F.; Jensen, M.D.; Nielsen, S. Postprandial VLDL-TG Metabolism in Type 2 Diabetes. *Metabolism* **2017**, *75*, 25–35. [CrossRef]
104. Batsis, J.A.; Villareal, D.T. Sarcopenic Obesity in Older Adults: Aetiology, Epidemiology and Treatment Strategies. *Nat. Rev. Endocrinol.* **2018**, *14*, 513–537. [CrossRef] [PubMed]
105. Tam, B.T.; Morais, J.A.; Santosa, S. Obesity and Ageing: Two Sides of the Same Coin. *Obes. Rev.* **2020**, *21*, e12991. [CrossRef]
106. Da Silva Alexandre, T.; Aubertin-Leheudre, M.; Carvalho, L.P.; de Oliveira Máximo, R.; Corona, L.P.; de Brito, T.R.P.; Nunes, D.P.; Santos, J.L.F.; de Oliveira Duarte, Y.A.; Lebrão, M.L. Dynapenic Obesity as an Associated Factor to Lipid and Glucose Metabolism Disorders and Metabolic Syndrome in Older Adults—Findings from SABE Study. *Clin. Nutr.* **2018**, *37*, 1360–1366. [CrossRef]
107. Arner, P.; Bernard, S.; Appelsved, L.; Fu, K.-Y.; Andersson, D.P.; Salehpour, M.; Thorell, A.; Rydén, M.; Spalding, K.L. Adipose Lipid Turnover and Long-Term Changes in Body Weight. *Nat. Med.* **2019**, *25*, 1385–1389. [CrossRef] [PubMed]
108. Guillermier, C.; Fazeli, P.K.; Kim, S.; Lun, M.; Zuflacht, J.P.; Milian, J.; Lee, H.; Francois-Saint-Cyr, H.; Horreard, F.; Larson, D.; et al. Imaging Mass Spectrometry Demonstrates Age-Related Decline in Human Adipose Plasticity. *JCI Insight* **2017**, *2*, e90349. [CrossRef] [PubMed]
109. Reyes-Farias, M.; Fos-Domenech, J.; Serra, D.; Herrero, L.; Sánchez-Infantes, D. White Adipose Tissue Dysfunction in Obesity and Aging. *Biochem. Pharmacol.* **2021**, *192*, 114723. [CrossRef]
110. Bonzón-Kulichenko, E.; Moltó, E.; Pintado, C.; Fernández, A.; Arribas, C.; Schwudke, D.; Gallardo, N.; Shevchenko, A.; Andrés, A. Changes in Visceral Adipose Tissue Plasma Membrane Lipid Composition in Old Rats Are Associated With Adipocyte Hypertrophy With Aging. *J. Gerontol. A Biol. Sci. Med. Sci.* **2018**, *73*, 1139–1146. [CrossRef]
111. Feng, J.; Xu, H.; Pan, F.; Hu, J.; Wu, Y.; Lin, N.; Zhang, X.; Ji, C.; Hu, Y.; Zhong, H.; et al. An Integrated Analysis of MRNA and LncRNA Expression Profiles Indicates Their Potential Contribution to Brown Fat Dysfunction with Aging. *Front. Endocrinol.* **2020**, *11*, 46. [CrossRef]
112. Silva, K.R.; Baptista, L.S. Adipose-Derived Stromal/Stem Cells from Different Adipose Depots in Obesity Development. *World J. Stem Cells* **2019**, *11*, 147–166. [CrossRef] [PubMed]
113. Kawagishi-Hotta, M.; Hasegawa, S.; Igarashi, T.; Yamada, T.; Takahashi, M.; Numata, S.; Kobayashi, T.; Iwata, Y.; Arima, M.; Yamamoto, N.; et al. Enhancement of Individual Differences in Proliferation and Differentiation Potentials of Aged Human Adipose-Derived Stem Cells. *Regen. Ther.* **2017**, *6*, 29–40. [CrossRef] [PubMed]
114. Wu, Q.; He, S.; Zhu, Y.; Pu, S.; Zhou, Z. Antiobesity Effects of Adipose-Derived Stromal/Stem Cells in a Naturally Aged Mouse Model. *Obesity* **2021**, *29*, 133–142. [CrossRef] [PubMed]
115. Carr, R.M.; Ahima, R.S. Pathophysiology of Lipid Droplet Proteins in Liver Diseases. *Exp. Cell Res.* **2016**, *340*, 187–192. [CrossRef] [PubMed]
116. Chen, T.-P.; Lai, M.; Lin, W.-Y.; Huang, K.-C.; Yang, K.-C. Metabolic Profiles and Fibrosis of Nonalcoholic Fatty Liver Disease in the Elderly: A Community-Based Study. *J. Gastroenterol. Hepatol.* **2020**, *35*, 1636–1643. [CrossRef]
117. Ogrodnik, M.; Miwa, S.; Tchkonia, T.; Tiniakos, D.; Wilson, C.L.; Lahat, A.; Day, C.P.; Burt, A.; Palmer, A.; Anstee, Q.M.; et al. Cellular Senescence Drives Age-Dependent Hepatic Steatosis. *Nat. Commun.* **2017**, *8*, 15691. [CrossRef]
118. Huang, Y.; Shen, Z.; Huang, C.; Lin, C.; Tsai, T. Cisd2 Slows down Liver Aging and Attenuates Age-related Metabolic Dysfunction in Male Mice. *Aging Cell* **2021**, *20*, e13523. [CrossRef]
119. Wei, Q.; Zhou, B.; Yang, G.; Hu, W.; Zhang, L.; Liu, R.; Li, M.; Wang, K.; Gu, H.F.; Guan, Y.; et al. JAZF1 Ameliorates Age and Diet-Associated Hepatic Steatosis through SREBP-1c -Dependent Mechanism. *Cell Death Dis.* **2018**, *9*, 859. [CrossRef]

nutrients

Review

The Role of Gut Microbiota in High-Fat-Diet-Induced Diabetes: Lessons from Animal Models and Humans

Yue Qi and Xiaofei Wang *

National Engineering Technology Research Center for Fruit and Vegetable Processing, Key Open Laboratory of Fruit and Vegetable Processing, Ministry of Agriculture and Rural Affairs, Beijing Key Laboratory of Food Non-thermal Processing, College of Food Science and Nutritional Engineering, China Agricultural University, Beijing 100083, China

* Correspondence: xiaofei.wang@cau.edu.cn

Abstract: The number of diabetes mellitus patients is increasing rapidly worldwide. Diet and nutrition are strongly believed to play a significant role in the development of diabetes mellitus. However, the specific dietary factors and detailed mechanisms of its development have not been clearly elucidated. Increasing evidence indicates the intestinal microbiota is becoming abundantly apparent in the progression and prevention of insulin resistance in diabetes. Differences in gut microbiota composition, particularly butyrate-producing bacteria, have been observed in preclinical animal models as well as human patients compared to healthy controls. Gut microbiota dysbiosis may disrupt intestinal barrier functions and alter host metabolic pathways, directly or indirectly relating to insulin resistance. In this article, we focus on dietary fat, diabetes, and gut microbiome characterization. The promising probiotic and prebiotic approaches to diabetes, by favorably modifying the composition of the gut microbial community, warrant further investigation through well-designed human clinical studies.

Keywords: high-fat diet; diabetes; gut microbiota; probiotic

Citation: Qi, Y.; Wang, X. The Role of Gut Microbiota in High-Fat-Diet-Induced Diabetes: Lessons from Animal Models and Humans. *Nutrients* **2023**, *15*, 922. https://doi.org/10.3390/nu15040922

Academic Editor: Manuel Sánchez Santos

Received: 30 December 2022
Revised: 2 February 2023
Accepted: 10 February 2023
Published: 12 February 2023

1. Introduction

Diabetes, the epidemic of the 21st century, has become one of the major threats to human health and has greatly increased the global burden of the disease [1,2]. The development of diabetes is associated with a number of factors, including excessive dietary intake, genetics, and a sedentary lifestyle [3]. Dietary composition is an important factor influencing the risk of developing diabetes [4], and the quantity and/or quality of dietary fat in diabetes have attracted considerable interest. Dietary fat, especially saturated fatty acid, has been considered to be an unhealthy dietary component due to its high energy density [5,6]. The excessive intake of dietary fat is thought to be associated with obesity and metabolic disorders [7], and the relationship between high-fat diets (HFDs) and diabetes has received extensive attention in past studies. Studies have confirmed that 59% of safflower oil can lead to insulin resistance in rats [8]. With HFD supplementation, beta cell senescence leads to a reduction in insulin release [9].

Gut microbes refer to microbiota present in the gastrointestinal tract and are associated with energy harvesting and storage and the metabolism of many metabolic functions, such as amino acids and carbohydrates [10,11]. Gut microbes are affected by diet, and when mice were shifted to a high-fat, high-sugar diet, the structure of the microbiota was altered within one day [12]. An HFD of 60% lard and soybean oil resulted in a decrease in *Bacteroidetes* and an increase in *Firmicutes* and *Proteobacteria* in mice [13]. Imbalances in gut microbes are associated with metabolic diseases such as obesity and diabetes through mechanisms such as increasing the amount of energy obtained from the diet, affecting fatty acid metabolism in the liver and adipose tissue, and increasing serum concentrations of branched-chain amino acids causing insulin resistance [14–16].

This review discusses possible metabolic dysregulation induced by an HFD, particularly the changes in diabetes and gut microbes. In order to mitigate the prevalence of diabetes caused by an HFD, appropriate animal models are selected to explore the cellular and molecular mechanisms between gut microbiome and diabetes. Moreover, this review highlights the anti-diabetic effects of dietary therapy, therapeutic interventions, and probiotics, as well as the mechanisms of their actions.

2. High-Fat-Diet-Induced Metabolic Dysfunction

2.1. Consumption of Dietary Fats Is Generally Increasing

Over thousands of years, diet consumption, in conjunction with other aspects of daily lifestyles such as exercise, has been associated with metabolic health. As one of the three important nutrients, dietary fats are mainly from edible oils, dairy products, meat, nuts, and other foods. They provide energy, act as a carrier of fat-soluble vitamins, and participate in the metabolism of cells and tissues as biologically active components [17,18]. Inadequate total dietary fat intake can easily lead to malnutrition. However, excessive dietary fat intake is also associated with nutrition-related diseases, including obesity, diabetes, heart disease, and cancer [19].

A typical Western diet contains different forms of fats, such as triglycerides, cholesterol, phospholipids, and long-chain fatty acids. Recently, people's dietary structure has gradually shifted towards a high-calorie diet with increasing dietary fat intake since the occurrence of economic development and industrialization. The Global Burden of Disease Nutrition and Chronicity Expert Group systematically assessed dietary consumption in 187 countries worldwide and showed that the global average intake of polyunsaturated fatty acids (PUFAs) was about 5%, and the average intake of saturated fatty acids (SFAs) was about 11% [20]. A study of 29 countries found that total trans fat (TFA) intake ranged from 0.3% to 4.2% of total energy intake (E%) in each country, with seven countries having trans-fat intakes above the WHO recommendation of 1% [21]. The Chinese National Nutrition Survey shows that between 1992 and 2002, the total fat intake of Chinese people rose from 22% to 29.8%, with the amount of energy obtained from animal food rising from 9.3% to 13.7% [22]. The excessive intake of dietary fat can lead to the development of various chronic diseases related to fat metabolism.

Both the US Public Health Dietary Recommendations and the UK Public Health Dietary Recommendations state that total fat consumption should be reduced to less than 35% of total energy intake, with the saturated fat intake being limited to less than 10% of daily calories [23]. A high-fat diet in humans refers to a calorie intake of 30–75% [24]. As can be seen from the literature, diets with different higher fatty acid compositions are considered to be HFDs. Stocks T et al. defined an HFD as an intake of 40–45% of energy derived from fat [25]. André J Tremblay defined an HFD in a cohort study as 37% fat intake, with saturated fat intake at 15%, monounsaturated fatty acids (MUFAs) at 12.7%, PUFA at 4.3%, and TFA at 3.5% [26]. Osterberg et al. regarded an HFD as 55% fat intake in their study, with saturated fat accounting for 25% of total energy intake [27]. Cameron J Holloway, in his study, chose an HFD as one in which 70% of the daily calorie intake is fat [28].

2.2. Excessive Dietary Fat Intake Exacerbated Metabolic Disorders

Excessive fat intake can lead to excess nutrients in the body and adversely cause systematic metabolic changes in blood plasma, liver, urine, and other organs, involving multiple metabolic pathways, including tricarboxylic acid cycle, glycolysis, lipogenesis, and gut microbiota functions together with the metabolisms of fatty acids, amino acids, choline, and others. These dynamic metabolic responses may result in the development and progression of HFD-induced metabolic disorders, including the dysbiosis of gut microbes [29] and the inflammation of peripheral tissues such as the central nervous system, liver, adipose tissue, and skeletal muscle [30,31].

Although there is still some controversy, a growing body of research points to the development of cardiovascular disease with excessive fat intake (especially SFA) [6,32]. An HFD fed to mice can lead to endothelial dysfunction by reducing the ability of vascular tissue to scavenge superoxide anions [33]. Compared to an HFD with unsaturated fatty acids such as olive oil, a 60%-lard diet reduces endothelial NO synthase activity, thereby affecting vascular homeostasis [34]. In a rabbit model, HFD induction led to early vascular injury through endothelial dysfunction and increased vascular reactivity [35]. A randomized controlled trial (RCT) has also shown that high SFA intake leads to increased plasma concentrations of medium and small LDL particles, increasing the risk of cardiovascular disease [36].

An association between an HFD and cognitive impairment and neurodegenerative diseases has also been found in human epidemiological studies [37]. SFA intake has been positively associated with Alzheimer's disease, dementia, mild cognitive impairment, and cognitive decline [38,39]. An HFD induces oxidative stress in the brain leading to cognitive impairment and enhances cerebral amyloid angiopathy promoting the development of Alzheimer's [40,41]. A high-fat palm oil diet for 16 months leads to amyloid deposits in the brains of mice, thought to be a marker of Alzheimer's disease [42]. Cognitive impairment due to an HFD may be related to oxidative stress. A 60%-lard diet increases brain inflammation in mice, significantly increasing the expression of the cytokines TNFα and IL-6, and the chemokine MCP-1, leading to impaired cognitive performance [40]. In rats, a diet of 40% fat for three months was found to impair learning and memory function, with more severe damage seen with a diet rich in SFA, lard, compared to a diet rich in unsaturated fatty acids, soybean oil [43].

It has been well documented in human and animal models that high-fat diets are associated with fatty liver disease. The HFD diet is widely used to induce hepatic steatosis or non-alcoholic fatty liver disease in experimental animals. The accumulation of triglycerides and cholesterol in the livers of rats fed 35% lard for 12 weeks occurs, which can lead to fatty liver degeneration [44]. Mice fed an HFD developed varying degrees of fatty liver disease [44]. Mice induced with an HFD for 16 weeks showed an obese and inflammatory phenotype, while the liver showed an increase in natural killer T cells and clusters of differentiation (CD)8+ T-cells, which play an important role in obesity-associated adipose tissue inflammation [45]. It has been suggested that total fat intake is positively associated with hepatic steatosis in overweight adolescents [46]. Lisis et al. confirmed that total fat intake was associated with non-alcoholic fatty liver in patients with hepatic steatosis [47].

There is a general consensus that a long-term HFD leads to diabetes, established in both animal and human experiments. Diabetes (defined as fasting blood glucose equal to or above 7 mmol/L) is a chronic metabolic disease caused by insulin abnormalities and manifested as an increase in blood glucose [44,48]. Diabetes can lead to a variety of complications, including retinopathy, nephropathy, peripheral neuropathy, cardiovascular and cerebrovascular complications, arteriopathy of the lower limbs, and hypertension [49]. According to the classification proposed by the American Diabetes Association and adopted by the WHO [45], there are four types of diabetes: type 1 diabetes, type 2 diabetes, gestational diabetes, and other special types of diabetes such as neonatal diabetes, etc. [46]. According to the 10th edition of the Diabetes Atlas published by the International Diabetes Federation [47], the global prevalence of diabetes among people aged 20–79 years was predicted to be about 10.5% (536.6 million people) in 2021, rising to 12.2% (783.2 million people) in 2045. The 2017 Global Burden of Disease Study states that diabetes is the leading cause of diet-related death and disability, second only to cardiovascular disease and cancer [50]. The pathogenesis of type 2 diabetes is complex, and the causes of diabetes have not yet been fully explored. Diabetes is associated with a number of factors, including lifestyle, genetic, and environmental factors, with diet playing an important role in the pathogenesis of type 2 diabetes. Excessive energy intake is thought to be a major cause of the type 2 diabetes epidemic [51]. In a nurses' health study, higher dietary intake of TFA was associated with increased diabetes [52]. A prospective cohort study noted that whole grain intake was negatively associated with type 2 diabetes [53]. With a long-term HFD, a lack of exercise, genetics, and aging, the body becomes metabolically disturbed, and the balance of blood glucose in the body is disturbed, causing an increase in blood glucose and leading to the development of type 2 diabetes. An HFD is currently one of the main methods of inducing diabetes in rodent models, and HFD-induced diabetes in rodents is associated with weight gain, hyperglycemia, insulin resistance, hyperinsulinemia, and accumulation of lipids [50]. An HFD can lead to hyperglycemia, insulin resistance, and damage to pancreatic beta cells by affecting glucose and lipid metabolism in the metabolic organs [51].

3. HFD-Induced Diabetes in Animal Models and Humans

An HFD leads to fat accumulation and increased blood sugar, causing insulin resistance and beta cell damage, causing diabetes in humans [30]. The ethical aspects of human research have necessitated the development of animal models of diabetes. Animal models of diabetes can better explore the pathogenesis of diabetes and help to reveal the pathogenesis of diabetes. Currently, common models of diabetes include rodents, non-human primate models, large animals, and non-mammalian models [54]. Within these models, HFD induction is a common approach, and common symptoms of HFD-induced diabetes in animal models include weight gain, hyperinsulinemia, and disruption of glucose homeostasis [55].

3.1. HFD-Induced Diabetes in Human Intervention Studies

The strongest evidence about the relationship between diet and the progression of disease comes from RCTs. Most of the current studies on the relationship between an HFD and diabetes are cohort studies. An HFD can lead to high type 2 diabetes by affecting glucose and lipid metabolism, which in turn can impair the function of major metabolic organs [56], including adipose tissue, pancreas, and liver. Table 1 summarizes the relationship between dietary fat and diabetes in human studies.

Adipose tissue is a loose connective tissue consisting of cells filled with lipids [57]. As an important organ involved in energy homeostasis, adipose tissue produces various bioactive substances, such as adipocytokines and fatty acids, which play a key role in the development of diabetes [58]. An HFD also has an effect on gene expression in adipose tissue; an RCT of patients with metabolic syndrome found that a high saturated fat diet increased the expression of lipolytic genes, which may be associated with impaired insulin sensitivity [59].

The pancreatic beta cells can maintain blood glucose stability by secreting insulin to promote glucose uptake by peripheral tissues [60]. Type 2 diabetes eventually develops when pancreatic beta cells do not secrete enough insulin to meet the demands of insulin resistance. The decrease in beta cell mass in type 2 diabetics is due to beta cell apoptosis [61]. In pre-diabetes, blood glucose can still be maintained at normal levels due to the compensatory response of the beta cells [62]. As oxidative stress and inflammatory responses proceed in later stages, the compensatory mechanisms of the beta cells are continuously compromised, eventually leading to the development of type 2 diabetes [63].

Under normal physiological conditions, hepatic glucose production is regulated by a combination of insulin and glucagon, with glucagon inducing hepatic glucose production and insulin inhibiting it [64]. As there is insulin resistance in diabetes, the inability of insulin to suppress liver glucose production leads to hyperglycemia [65]. An HFD can lead to fat accumulation in the liver, causing insulin resistance and thus disrupting blood glucose homeostasis. An HFD has been shown to significantly increase liver fat levels in 56% of obese women [66].

Table 1. High-fat diet and human diabetes intervention studies.

Diet	Participants	Duration	Findings	References
		Randomized controlled intervention trials (RCTs)		
50 E % carbohydrate, 20 E % protein, 5 E% PUFAs • SFA: 20 E% SFAs, 5 E% MUFAs • cis-MUFA: 20 E% cis-MUFAs, 5 E% SFAs • trans MUFA: 20 E% trans-MUFAs, 5 E% SFAs	Obese type 2 diabetes patients aged 42–58 (*N* = 16)	6 weeks	No difference in postprandial glucose and serum lipids; increased serum insulin and C-peptide for SAT and trans MUFA diets	[67]
45 E% carbohydrate, 15 E% protein • Saturated fat diet (butter and margarine) • Monounsaturated fatty acid diets (oleic acid)	Healthy people aged 30–65 (*N* = 162)	3 months	Insulin sensitivity was significantly impaired for SAT diet, while there was no difference for MUFA diet	[68]
• Control group: regular diet • Intervention group: carbohydrate >50 E%, fat <30 E%	Overweight people aged >40 with glucose tolerance (7.8–11.1) mmol/l (*N* = 102)	3.1 years	55% reduction in the incidence of diabetes in the intervention group	[69]
		Cohort		
Fat intake (total, SFA, MUFA, and PUFA)	Healthy people aged 40–69 (*N* = 1173)	2 years	Total fat is negatively associated with insulin sensitivity	[70]
Fat intake (SFA, MUFA, PUFA, TFA, long-chain omega-3 PUFA, and animal and vegetable fat)	Healthy women aged 45–50 (*N* = 35,988)	11 years	Diabetes incidence is negatively associated with vegetable fats	[71]
Fat intake (total fat, SAT, MUFA-oleic acid, PUFA-linoleic acid)	Healthy men aged 40–75 (*N* = 42,504)	12 years	Total fat and SAT intake are associated with a higher risk of type 2 diabetes	[72]
Foods high in fat (vegetable oils, butter, margarine, nuts and seeds, and cakes and biscuits)	European Prospective Investigation into Cancer (*N* = 340,234)	9 years	Margarine consumption is positively associated with diabetes risk	[73]
Fat intake (SFA, MUFA, PUFA, TFA, animal fats, vegetable fats, marine omega-3 fatty acids, non-marine omega-3 fatty acids, and omega-6 linoleic acid (18:2n-6))	The people who were free of diabetes but were at high cardiovascular risk were aged 55–80 (*N* = 3349)	4.3 years	SAT and animal fats (cheese and butter) are associated with a higher risk of diabetes	[74]
Fat intake (SFA, MUFA, and PUFA)	Healthy women aged 45–50 (*N* = 8370)	6 years	Intake of MUFA, total n-3 PUFA, α-linolenic acid, and n-6 PUFA were positively associated with the incidence of diabetes	[75]
Total fat, SFA, MUFA, PUFA, and TFA	Healthy women aged 45–50 (*N* = 84,204)	14 years	TFA intake was positively associated with the risk of diabetes, while PUFA intake was negatively associated with the direction of diabetes	[76]

Table 1. *Cont.*

Diet	Participants	Duration	Findings	References
Type of fat and amount of fat: oils and margarine used during cooking and at the table	Healthy women aged 30–55 (*N* = 83,648)	32 years	Higher intakes of linoleic acid are associated with a lower risk of type 2 diabetes	[77]
	Healthy women aged 25–44 (*N* = 88,610)	22 years		
	Healthy men aged 40–75 (*N* = 41,771)	26 years		
Consumption of nuts and peanut butter (monounsaturated and polyunsaturated fatty acids)	Healthy women aged 35–49 (*N* = 83,818)	16 years	Women who ate nuts or peanut butter at least five times a week had a lower risk of developing diabetes	[78]

3.2. HFD-Induced Diabetes in Animal Models

Intervention human studies have necessitated the development of animal models of diabetes, including mice, rats, *Drosophila*, zebrafish, and so on, to better explore the pathogenesis of diabetes and help to reveal the pathogenesis of diabetes. An HFD affects the major insulin-sensitive tissues of the animal model, including adipose tissue, the pancreas, and the liver. Rodent models are the most widely used animal models of diabetes and have been well-studied. Large animal models have physiological conditions more similar to those of humans, including physiological and pathological features. Primate models are very similar to humans but are more expensive and have a longer life cycle. Non-mammalian models, such as fruit flies and zebrafish, have a variety of advantages, such as short growth cycles, simple husbandry, low cost, and high reproductive capacity [54,79]. Table 2 summaries the animal models induced by an HFD, including fat type and amount, duration, animal species, and symptoms

Adipose tissue has an important role in the development of diabetes, and adipose tissue is a major site for storing gluconeogenic substrates and energy [80]. An HFD can alter the expression of genes in adipose tissue, down-regulating genes encoding lipid metabolizing enzymes or markers of lipid differentiation, and increasing genes encoding markers of inflammation [81]. The B-cell activating factor (BAFF) is a tumor necrosis factor (TNF) ligand family protein that is a key factor in the development of poor glucose tolerance [82]. Mice fed an HFD had significantly increased BAFF in visceral adipose tissue and serum [83]. The pro-inflammatory cytokine TNF-α is associated with insulin resistance [84], and an HFD of both lard and soybean oil can increase TNF-α expression levels in adipose tissue [85].

The pancreas is a key site for regulating the secretion of insulin and glucagon, and an HFD can have an impact on the pancreas, leading to the development of diabetes. Several studies have pointed out the mechanism by which an HFD can enhance the compensation of pancreatic β-cells. For example, Jonatan Ahrén et al. found an increase in β-cell volume and β-cell numbers after feeding mice with a 60% lard diet for three months [86]. Kanno et al. found that the compensatory mechanism of islet cells in those on an HFD resulted mainly from increased levels of insulin translation [87]. Ribeiro et al. found that an HFD induced islet hypertrophy and a compensatory morpho-functional shift in pancreatic β-cells [88]. However, it has also been suggested that an HFD can directly lead to the degeneration of islet cell function. The levels of glucose transporters (GLUT)2 and glucokinase mRNA in rat pancreas were significantly reduced after 10 weeks of HFD feeding, and an HFD can reduce insulin secretion by impairing signal transduction in pancreatic β-cells [89]. In ZDF rats fed an HFD for a long period of time, the pancreas developed fat accumulation, which may have led to pancreatic fibrosis, acinar cell damage, and pancreatic stellate cell activation [90,91].

The liver is the main site of carbohydrate and fat utilization and plays an important role in controlling glucose intake, fat metabolism, and energy balance [92]. Liver fat accumulation is associated with insulin resistance, and excess fat intake leads to increased levels of free fatty acids and increased triglyceride deposition in the liver [93].

Table 2. Animal model of high-fat-diet-induced diabetes mellitus.

High-Fat Diet	Duration	Mode	Findings	References
335 g/kg corn oil and lard	11 weeks	Japanese fancy mouse 1	Impaired glucose tolerance, hyperglycemia, hyperinsulinemia, and obesity	[94]
58% lard	12 months	C57BL/6J mice	Weight increase, circulating insulin increase, and impaired glucose tolerance	[55]
42% lard 42% olive	12 weeks	Male Wistar rats	Obesity and insulin resistance	[95]
43% fat	Different ages	Nile rat	Hyperinsulinemia, high blood glucose, insulin resistance, abdominal adiposity, and impaired glucose clearance	[96]
20% coconut oil	14 days	*Drosophila*	Induced insulin resistance, elevated triglyceride and circulating glucose, and elevated expression of glass bottom boat (a *Drosophila* homolog of mammalian transforming growth factor-β)	[97]
30% fat vegetable shortening and beef tallow	8 weeks	Guinea pigs	Impaired glucose tolerance, β-cell hyperplasia, compensatory hyperinsulinemia, and dyslipidemia with hepatocellular steatosis	[98]
80% fat (lard)	7 weeks	Dogs	Decreased insulin sensitivity	[99]
8% trans fatty acids	6 years	African green monkeys	Increased intra-abdominal fat deposition, hyperinsulinemia, elevated fructosamine, and reduced muscle AKT (protein kinase) phosphorylation	[100]
Six feeds/day (11% fat)	8 weeks	Zebrafish	Increased blood glucose, impaired glucose tolerance, and insulin resistance	[101]

3.3. *Gut Microbiota Dysbiosis in HFD-Induced Diabetes*

From a physiological point of view, one of the most important links between an HFD and diabetes is the gut microbiota–host axis, as well as the factors released from intestinal metabolites, mediating bidirectional communication between the intestines and the host. Specific intestinal flora community profiles have been suggested to promote type 2 diabetes. Type 2 diabetic patients have dysbiosis of gut microbes with a reduced abundance of butyrate-producing bacteria (including *Eubacterium rectale*, *Roseburia intestinalis*, and *Roseburia inulinivorans*) and an increased abundance of pathogens (such as *Clostridium ramosum*, *Clostridium symbiosum*, *Eggerthella lenta*, and *Escherichia coli*) [102]. Emerging evidence demonstrates that changes in the ratios between gut microbiota, such as the ratios of *Bacteroides* and *Firmicutes*, are associated with the development of type 2 diabetes [103].

Recent studies have also found that type 2 diabetes caused by an HFD may be associated with gut microbes. Diet is a major factor influencing the composition and function of gut microbes. Animal experiments have shown that an HFD affects gut microbes. Compared to the normal diet, mice fed a high-fat diet were more susceptible to diabetes, which may be associated with a reduction in *Bifidobacteria* [104]. In mice on a 45% HFD, there was a decrease in *Bacteroidetes* and an increase in *Firmicutes* and *Proteobacteria* [13]. The HFD reduced the mice's *Akkermansia*, *Coprococcus*, and *Ruminococcus*, increased *Odoribacter* and *Parabacteroides*, and led to a reduction in short-chain fatty acids (SCFAs) [105]. SCFAs can alleviate diet-induced insulin resistance, and a reduction in SCFAs may lead to type 2 diabetes [106]. An HFD can affect the production of immunoglobulin A, a key regulator of glucose homeostasis, an immune-derived molecule in the gut [107]. Table 3 below summaries the gut microbiota of animals with high-fat-diet-induced diabetes.

Table 3. Effect of an HFD on gut microbes in diabetic animal models.

Mode	Mode	High-Fat Diet	Duration	Sample	Impact on Microbiota	References
Mice	C57BL/6	72% fat (corn oil and lard)	3 months	Ileum, caecum, and colon	Decrease *Bacteroidetes*, *Proteobacteria* Increase *Firmicutes*, *Deferribacteres*, *Lachnospiraceae*	[108]
		60% fat	13 weeks	Caecum	Decrease *Bacteroidetes* Increase *Proteobacteria*	[109]
		60% fat (soybean oil and lard)	8 weeks	Fecal	Decrease *Bacteroidetes* Increase *Firmicutes*, *Deferribacteres*, *Actinobacteria*	[110]
		60% fat	16 weeks	Fecal	Decrease *Actinobacteria* Increase *Proteobacteria*, the ratio of *Bacteroidetes* to *Firmicutes*	[111]
		60% fat +STZ	5 weeks	Fecal	Increased ratio of *Firmicutes* to *Bacteroidetes* Decrease *Rikenellaceae* Increase *Ruminococcaceae* and *Erysipelotrichaceae*	[112]
		60% fat (soybean oil and lard)	18 weeks	Fecal	Decrease *Akkermansia* Increase *Muribaculaceae* and *Eubacterium*	[113]
		60% fat (soybean oil and lard) +STZ	11 weeks	Fecal	Decrease *Bacteroides* Increase *Firmicutes*	[114]
		60% fat (soybean oil and lard) +STZ	6 weeks	Fecal	Increase the ratio of *Firmicutes/Bacteroidetes* Decrease *Akkermansia*, *Muribaculaceae*, *Bacteroides*, *Fusobacterium*, and *Dubosiella* Increase *Colidextribacter* and *Helicobacter*	[115]
	C57BL/6J	60% fat (soybean oil and lard)	8 weeks	Fecal	Decrease *Bacteroidetes* Increase *Firmicutes*, *Proteobacteria*, *Deferribacteres*	[116]
		60% fat (soybean oil and lard)	8 weeks	Cecal	Decrease *Bacteroidetes* Increase *Firmicutes*	[117]
		41% fat	15 weeks	Fecal	Decrease *Akkermansia*, *Coprococcus*, and *Ruminococcus* Increase *Odoribacter* and *Parabacteroides*	[105]
		60% fat +STZ	12 weeks	Fecal	Decrease *Bacteroidetes* Increase *Firmicutes*	[118]
		60% fat (soybean oil and lard)	17 weeks	Fecal	Decrease *Actinobacteria*	[119]
		72% fat (corn oil and lard)	4 weeks	Cecal	Decrease *Lactobacillus* spp., *Bifidobacterium* spp., and *Bacteroides-Prevotella* spp.	[120]
		60% fat (soybean oil and lard)	12 weeks	Fecal	Increased ratio of *Firmicutes* to *Bacteroidetes* Decrease *Bacteroidetes* Increase *Proteobacteria*, *Firmicutes*	[121]
		45% fat (lard)	8 weeks	Fecal	Decrease *Bacteroidetes* and *Actinobacteria* Increase *Firmicutes*	[122]

Table 3. *Cont.*

Mode	Mode	High-Fat Diet	Duration	Sample	Impact on Microbiota	References
		60% fat (soybean oil and lard) +STZ	7 weeks	Fecal	Decrease *Verrucomicrobia* Increase *Saccharibacteria*	[123]
		45% fat (soybean oil and lard)	14 weeks	Fecal	Decrease *Akkermansia* Increase the ratio of *Firmicutes* and *Bacteroidetes*	[124]
	BALB/c	40% fat +STZ	8 weeks	Fecal	Decrease *Firmicutes, Proteobacteria,* and *Actinobacteria* Increase *Bacteroidetes, Actinobacteria*	[125]
	Swiss	55% fat	12 weeks	Fecal	Decrease *Firmicutes, Actinobacteria* Increase *Bacteroidetes*	[126]
Rats	Wistar rats	58% fat +STZ	12 weeks	Fecal	Decrease *Lactobacillus* spp. Increase *Bifidobacterium* spp.	[127]
		60% fat	6 months	Fecal	Decrease *Actinobacteria, Proteobacteria,* and *Bacteroidetes* Increase *Firmicutes*	[128]
	Sprague Dawley rats	60% fat (soybean oil and lard)	10 weeks	Fecal	Decrease *Bacteroides/Prevotella* Increase *Firmicutes, Bifidobacterium* spp., *Enterobacteriaceae,* and *C. leptum.*	[129]
		10% lard + normal diet	12 weeks	Fecal	Decrease *Firmicutes* Increase *Bacteroidetes, Proteobacteria*	[130]
		60% fat (soybean oil and lard)	4 weeks	Fecal	Increase the ratio of *Firmicutes* to *Bacteroidetes*	[131]
		7% lard + normal diet	9 weeks	Fecal	Decrease *Proteobacteria* Increase *Firmicutes*	[132]
		High-fat diet (lard)	12 weeks	Fecal	Decrease *Actinobacteria, Proteobacteria* Increase *Firmicutes*	[133]
		High-fat diet (soybean oil and lard)	15 weeks	Colonic	Decrease *Clostridium* and *Faecalibacterium* Increase *Bacteroides, Butyricoccus, Parabacteroides, Rikenella, Bifidobacterium, Allobaculum, Dehalobacterium, Lactobacillus, Oscillospira, Ruminococcus,* and *Desulfovibrio*	[134]
		45% fat (soybean oil)	24 weeks	Fecal and cecal	Decrease *Bacteroidetes* Increase *Firmicutes*	[135]

4. Measurements to Treat Diabetes

With aging and urbanization, the number of people with diabetes is increasing, and the prevalence of diabetes continues to rise [136]. Therefore, the prevention and treatment of diabetes is now an important issue for people. Current treatments for diabetes include medication, dietary interventions, and physical activity [137]. In recent years, with intensive research into gut microbes and diabetes, probiotics may be a new way to treat diabetes.

4.1. Therapeutic Interventions for Diabetes

When dietary interventions are not feasible, medication may be considered as a strategy to prevent the development of type 2 diabetics. The chemical drugs used can be divided into biguanides, sulfonylureas, thiazolidinediones, glucosidase inhibitors, etc., according to the mechanism of action and chemical structure [138,139]. Metformin, a biguanide, is the most widely used oral hypoglycemic drug, with the advantages of safety and effectiveness, cardiovascular protection, and low cost, and is recommended as a first-line drug by the

American Diabetes Association and the European Diabetes Association [140]. Metformin acts primarily on the liver and can improve hyperglycemia by inhibiting hepatic glucose production [141]. Metformin can act by inhibiting mitochondrial respiratory chain complex I, increasing the AMP/ATP ratio and activating AMPK-activated protein kinase [142,143]. The therapeutic effect of metformin may be related to its effect on gut microbiota, confirmed in animal models and clinical studies. Metformin decreases *Bacteroides fragilis* in type 2 diabetics and also increases the abundance of the mucin-degrading bacterium *Akkermansia* in HFD mice [144,145]. Sulphonylureas are a class of drugs that promote insulin secretion and can act by binding to the SUR subunit of the ATP-sensitive potassium channel in pancreatic cells [146]. However, the effects of sulfonylureas on gut microbiota have still not been well studied. Thiazolidinediones improve insulin sensitivity, and rosiglitazone and pioglitazone are representatives of these drugs. Thiazolidinediones are agonists of peroxisome proliferator receptor gamma (PPARγ), which can enhance insulin target tissues (muscle, fat, liver) and accelerate glucose utilization by activating PPARγ to promote the expression of genes related to glucose transport and lipid metabolism [147,148]. The relative abundance of *Proteobacteria* decreased after the treatment of HFD mice with pioglitazone [149]. Thiazolidinediones may cause a variety of side effects, such as heart failure, cardiovascular death, edema, and fractures [150]. Oral α- glucosidase inhibitors can improve hyperglycemia by delaying the breakdown of carbohydrates into glucose. Currently, the three clinically approved glucosidase inhibitors include acarbose, voglibose, and miglitol [151]. A double-blind RCT of acarbose altered gut microbiota in prediabetic patients, decreasing *Ruminococcaceae* and *Lachnospiraceae* and increasing *Lactobacillaceae*, *Ruminococcaceae*, and *Veillonellaceae* [152].

4.2. Dietary Interventions to Alleviate Diabetes

Most health organizations point to dietary interventions as a powerful treatment for diabetes, with controlled diets improving insulin sensitivity and reducing the risk of diabetes and its complications [153]. The American College of Lifestyle Medicine believes that diabetes can be treated with dietary interventions that use whole food, plant-based eating patterns, and increase the intake of unrefined plant foods in the daily diet while eliminating or minimizing the intake of animal foods and refined foods, and with moderate exercise in life [154]. The impact of dietary interventions on diabetes includes effects through indirect weight loss and direct consumption of a variety of nutrients with health benefits [155]. Being overweight is considered to be one of the important factors associated with the risk of diabetes [156,157]. Dietary interventions can reduce weight and improve diabetes by reducing the intake of fat.

As carbohydrate catabolism causes blood glucose to rise, reducing carbohydrate intake in the daily diet can be a good treatment for type 2 diabetes [158]. A low-carbohydrate diet, as defined by the American Diabetes Association, is 130 g/day or less than 26% of total daily energy intake [159]. The traditional Mediterranean diet of minimally processed whole grains has also been shown to have significant benefits for diabetes [160].

Specific types of dietary fat may affect diabetes. The KANWU study found that replacing a diet with monounsaturated fatty acids (23%E for MUFA, 8%E for SFA, 6%E for PUFA) over saturated fatty acids (17%E for SFA, 14%E for MUFA, 6%E for PUFA) improved insulin sensitivity at a total fat intake below the median (37E%) [68]. Another study also confirmed that a diet rich in MUFA could improve central fat distribution and insulin resistance [161]. The Iowa Women's Study found a reduced risk of diabetes when saturated fatty acids were replaced with unsaturated fatty acids [71], and a study by Summers et al. also noted that replacing saturated fatty acids with unsaturated fatty acids improved insulin sensitivity and abdominal fat accumulation [162]. n-3 PUFA improves high-fat-diet-induced insulin resistance. n-3 PUFA in fish oil improves insulinemia, lipid metabolism, and glucose metabolism in insulin-resistant rats [163]. Different dietary fat types may influence the affinity of insulin receptors by affecting the fatty acid composition of cell membranes [164]. It has also been suggested that dietary fat can modulate the expression of genes involved in lipid metabolisms that affect diabetes, such as fatty acid transport proteins and fatty acid synthases [165]. Increased inflammation may lead to insulin resistance, and PUFA acid intake may improve inflammation, e.g., n-3 PUFA may inhibit Toll-like receptors on the cell surface and reduce the production of inflammatory cytokines [166,167].

Other than the chemical drugs currently used to treat diabetes, natural plant foods such as fruits and vegetables, which are rich in nutrients such as antioxidants and polyphenols, can improve adipokines and oxidative stress, significantly improving beta cell function and insulin sensitivity [168,169]. Table 4 below summaries the potential mechanisms of natural products in food for the treatment of diabetes.

Table 4. Potential mechanisms of natural products in foods in animal models of diabetes.

Natural Products	Model	Potential Mechanisms	References
hesperidin	male C57BL/KsJ-*db/db* mice	↑ hepatic glucokinase activity, glycogen concentration, plasma insulin, C-peptide, and leptin ↓ hepatic glucose-6-phosphatase and phosphoenolpyruvate carboxykinase	[170]
cyanidin 3-glucoside	HFD-induced obese rat and *db/db* mice	↑insulin sensitivity, phosphorylation of forkhead box O1 ↓inflammatory cytokines, hepatic triglyceride, c-Jun N-terminal kinase activation	[171]
quercetin	*db/db* mice	↑insulin, triglyceride, glycogen, the ratio of B-cell lymphoma-2/Bcl2-Associated X ↓the activation of caspase-3, -9, -12	[172]
kaempferol	HFD-fed C57BL/6 male mice	↑AKT and hexokinase activity ↓pyruvate carboxylase and glucose-6 phosphatase activity	[173]
ferulic acid	C57BL/KsJ *db/db* mice	↑plasma insulin, hepatic glycogen synthesis, and glucokinase activity ↓total cholesterol and low-density lipoprotein cholesterol	[174]
resveratrol	streptozotocin-nicotinamide-induced diabetic rats	↑insulin ↓blood glucose, glycosylated hemoglobin, TNF-α, IL-1β, IL-6, NF-κB p65 unit, nitric oxide, superoxide dismutase, catalase, glutathione peroxidase, and glutathione-S-transferase	[175]
genistein	streptozotocin-induced diabetic mice	↑insulin, protein expression of cyclin D1, islet β-cell proliferation, survival, and mass	[176]
anthocyanins	HFD-fed Zucker rats	↑adipose and skeletal muscle PPAR activity ↓triglycerides, abdominal fat mass, insulin resistance	[177]
conophylline	streptozotocin-treated and Goto-Kakizaki rats	↑ insulin, β-cell differentiation	[178]
berberine	streptozotocin-induced rats	↑insulin sensitivity, insulin receptor mRNA, protein kinase C activity	[179]
ginsenosides	HFD-fed C57BL/6J mice	↑glucose uptake ↓ TNF-α-induced activation of MAPK and NF-κB signaling pathway	[180]

↑: Increased gene expression, increased content in the body, improved insulin sensitivity. ↓: Gene expression decreases, content decreases.

4.3. Potential Probiotics Help with Diabetes

Probiotics refer to the beneficial microbiota which inhabit the gut and have a variety of health functions [181]. The development of diabetes is closely linked to gut microbes, and therefore the regulation of gut microbes through probiotics could be a new approach to treating diabetes. The hypoglycemic effect of probiotics has been confirmed in both in vitro and in vivo experiments. Zhu et al. found that most of the Lactobacillus species inhibited dipeptidyl peptidase IV and α-glucosidase by cell-free excretory supernatants and cell-free extracts prepared from 21 Lactobacillus species [182]. Several studies have shown that probiotics can lower blood sugar to varying degrees in diabetic animals, such as *Lactobacillus plantarum* [183], *Lactobacillus casei* [184], *Lactobacillus rhamnosus* [185], and *Clostridium butyricum* [186]. A meta-analysis indicated that probiotics significantly lowered hemoglobin A1c, fasting blood glucose, fasting insulin, triglycerides, and total cholesterol, and improved the symptoms of diabetes [187]. High endotoxemia was demonstrated in high-fat-fed mice. Supplementation with oligofructose to increase the number of gut *Bifidobacteria* revealed that endotoxemia was negatively correlated with *Bifidobacteria* and also improved glucose tolerance and insulin secretion with increased *Bifidobacteria* [188].

The mechanisms by which probiotics improve diabetes include direct effects on the gut microbiota, anti-inflammatory and immunomodulatory effects, reduction in oxidative stress, and involvement in glucose homeostasis [181,189,190]. Gut microbial dysbiosis in diabetic patients leads to increased gut permeability and increased concentrations of bacterial endotoxins such as lipopolysaccharides, inducing inflammation and, ultimately, systemic insulin resistance [191]. Probiotics such as *Lactobacillus paracasei* can restore the expression of the tight junction protein in the colon, thereby reducing serum lipopolysaccharide and inflammatory cytokine levels [192]. Diabetes leads to increased systemic oxidative stress, and the intake of beneficial bacteria can significantly improve fasting blood glucose and the antioxidant status in diabetics [193,194]. *Bifidobacterium lactis* improves glucose uptake and GLU4 translocation through the insulin signaling pathway AKT and insulin receptor substrate-1, increases the expression of GLUT4 and insulin-sensitivity-related genes, and regulates glucose metabolism [195].

5. Conclusions

A long-term HFD, especially the excessive intake of saturated fats, could have a variety of adverse effects on human body health and even lead to chronic diseases, including diabetes. Animal models of diabetes can better explore the pathogenesis of diabetes and help to reveal the pathogenesis of diabetes. Daily diet can have a direct impact on the composition and function of host gut microbiota. Excessive fat intake can lead to the imbalances of gut microbiota, including changes in the ratio of *Bacteroidetes* and *Firmicutes*, a decrease in butyrate-producing bacteria, and an increase in the abundance of pathogens. A gut microbiota imbalance may further disturb host metabolism, such as decreased amounts of SCFA and immunoglobulin A, ultimately leading to diabetes. Within animal models, HFD-induced diabetes is accompanied by weight gain, hyperinsulinemia, and the disruption of glucose homeostasis. At present, the treatment of diabetes includes dietary interventions and medication. The causal relationship between gut microbiota and diabetes and its underlying mechanisms are still not fully elucidated, and further research is needed. In the near future, as research into the mechanisms of diabetes and gut microbes intensifies, probiotics may become a new method of treatment for diabetes.

Author Contributions: Writing—original draft preparation, Y.Q.; writing—review and editing, X.W.; supervision, X.W.; funding acquisition, X.W. All authors have read and agreed to the published version of the manuscript.

Funding: This article was funded by Project 32200387 of the National Natural Science Foundation of China.

Institutional Review Board Statement: Not applicable.

Informed Consent Statement: Not applicable.

Data Availability Statement: Not applicable.

Conflicts of Interest: The authors declare no conflict of interest.

References

1. Global Burden of Disease Study Collaborators 2013; Vos, T.; Allen, C.; Arora, M.; Barber, R.M.; Bhutta, Z.A.; Brown, A.; Liang, X.; Kawashima, T.; Coggeshall, M.; et al. Global, regional, and national incidence, prevalence, and years lived with disability for 301 acute and chronic diseases and injuries in 188 countries, 1990–2013: A systematic analysis for the Global Burden of Disease Study 2013. *Lancet* **2015**, *386*, 743–800.
2. Must, A. The Disease Burden Associated With Overweight and Obesity. *JAMA* **1999**, *282*, 1523–1529. [CrossRef] [PubMed]
3. Bai, Z.; Huang, X.; Wu, G.; Ye, H.; Huang, W.; Nie, Q.; Chen, H.; Yin, J.; Chen, Y.; Nie, S. Polysaccharides from red kidney bean alleviating hyperglycemia and hyperlipidemia in type 2 diabetic rats via gut microbiota and lipid metabolic modulation. *Food Chem.* **2023**, *404*, 134598. [CrossRef] [PubMed]
4. Risérus, U.; Willett, W.C.; Hu, F.B. Dietary fats and prevention of type 2 diabetes. *Prog. Lipid Res.* **2009**, *48*, 44–51. [CrossRef]
5. Drewnowski, A.; Darmon, N. The economics of obesity: Dietary energy density and energy cost. *Am. J. Clin. Nutr.* **2005**, *82*, 265S–273S. [CrossRef]
6. Wali, J.A.; Jarzebska, N.; Raubenheimer, D.; Simpson, S.J.; Rodionov, R.N.; O'Sullivan, J.F. Cardio-Metabolic Effects of High-Fat Diets and Their Underlying Mechanisms—A Narrative Review. *Nutrients* **2020**, *12*, 1505. [CrossRef]
7. Hariri, N.; Thibault, L. High-fat diet-induced obesity in animal models. *Nutr. Res. Rev.* **2010**, *23*, 270–299. [CrossRef]
8. Oakes, N.D.; Cooney, G.J.; Camilleri, S.; Chisholm, D.J.; Kraegen, E.W. Mechanisms of Liver and Muscle Insulin Resistance Induced by Chronic High-Fat Feeding. *Diabetes* **1997**, *46*, 1768–1774. [CrossRef]
9. Sone, H.; Kagawa, Y. Pancreatic beta cell senescence contributes to the pathogenesis of type 2 diabetes in high-fat diet-induced diabetic mice. *Diabetologia* **2005**, *48*, 58–67. [CrossRef]
10. Gill, S.R.; Pop, M.; DeBoy, R.T.; Eckburg, P.B.; Turnbaugh, P.J.; Samuel, B.S.; Gordon, J.I.; Relman, D.A.; Fraser-Liggett, C.M.; Nelson, K.E. Metagenomic Analysis of the Human Distal Gut Microbiome. *Science* **2006**, *312*, 1355–1359. [CrossRef]
11. Kurokawa, K.; Itoh, T.; Kuwahara, T.; Oshima, K.; Toh, H.; Toyoda, A.; Takami, H.; Morita, H.; Sharma, V.K.; Srivastava, T.P.; et al. Comparative Metagenomics Revealed Commonly Enriched Gene Sets in Human Gut Microbiomes. *DNA Res.* **2007**, *14*, 169–181. [CrossRef]
12. Turnbaugh, P.J.; Ridaura, V.K.; Faith, J.J.; Rey, F.E.; Knight, R.; Gordon, J.I. The Effect of Diet on the Human Gut Microbiome: A Metagenomic Analysis in Humanized Gnotobiotic Mice. *Sci. Transl. Med.* **2009**, *1*, 6ra14. [CrossRef] [PubMed]
13. Hildebrandt, M.A.; Hoffmann, C.; Sherrill–Mix, S.A.; Keilbaugh, S.A.; Hamady, M.; Chen, Y.-Y.; Knight, R.; Ahima, R.S.; Bushman, F.; Wu, G.D. High-Fat Diet Determines the Composition of the Murine Gut Microbiome Independently of Obesity. *Gastroenterology* **2009**, *137*, 1716–1724.E2. [CrossRef] [PubMed]
14. Pedersen, H.K.; Gudmundsdottir, V.; Nielsen, H.B.; Hyotylainen, T.; Nielsen, T.; Jensen, B.A.H.; Forslund, K.; Hildebrand, F.; Prifti, E.; Falony, G.; et al. Human gut microbes impact host serum metabolome and insulin sensitivity. *Nature* **2016**, *535*, 376–381. [CrossRef]
15. Turnbaugh, P.J.; Ley, R.E.; Mahowald, M.A.; Magrini, V.; Mardis, E.R.; Gordon, J.I. An Obesity-Associated Gut Microbiome with Increased Capacity for Energy Harvest. *Nature* **2006**, *444*, 1027–1031. [CrossRef]
16. Kimura, I.; Ozawa, K.; Inoue, D.; Imamura, T.; Kimura, K.; Maeda, T.; Terasawa, K.; Kashihara, D.; Hirano, K.; Tani, T.; et al. The gut microbiota suppresses insulin-mediated fat accumulation via the short-chain fatty acid receptor GPR43. *Nat. Commun.* **2013**, *4*, 1829. [CrossRef]
17. Calder, P.C. Functional Roles of Fatty Acids and Their Effects on Human Health. *J. Parenter. Enter. Nutr.* **2015**, *39*, 18S–32S. [CrossRef] [PubMed]
18. Machate, D.J.; Figueiredo, P.S.; Marcelino, G.; Guimarães, R.D.C.A.; Hiane, P.A.; Bogo, D.; Pinheiro, V.A.Z.; de Oliveira, L.C.S.; Pott, A. Fatty Acid Diets: Regulation of Gut Microbiota Composition and Obesity and Its Related Metabolic Dysbiosis. *Int. J. Mol. Sci.* **2020**, *21*, 4093. [CrossRef]
19. Rolls, B.J.; Shide, D.J. The Influence of Dietary Fat on Food Intake and Body Weight. *Nutr. Rev.* **2009**, *50*, 283–290. [CrossRef]
20. Imamura, F.; Micha, R.; Khatibzadeh, S.; Fahimi, S.; Shi, P.; Powles, J.; Mozaffarian, D. Global Burden of Diseases Nutrition and Chronic Diseases Expert Group. Dietary quality among men and women in 187 countries in 1990 and 2010: A systematic assessment. *Lancet Glob. Health* **2015**, *3*, e132–e142. [CrossRef]
21. Wanders, A.J.; Zock, P.L.; Brouwer, I.A. Trans Fat Intake and Its Dietary Sources in General Populations Worldwide: A Systematic Review. *Nutrients* **2017**, *9*, 840. [CrossRef] [PubMed]
22. Wang, Y.; Mi, J.; Shan, X.-Y.; Wang, Q.J.; Ge, K.-Y. Is China facing an obesity epidemic and the consequences? The trends in obesity and chronic disease in China. *Int. J. Obes.* **2007**, *31*, 177–188. [CrossRef]
23. Harcombe, Z. US dietary guidelines: Is saturated fat a nutrient of concern? *Br. J. Sports Med.* **2018**, *53*, 1393–1396. [CrossRef] [PubMed]
24. Hohos, N.M.; Skaznik-Wikiel, M.E. High-Fat Diet and Female Fertility. *Endocrinology* **2017**, *158*, 2407–2419. [CrossRef] [PubMed]
25. Stocks, T.; Taylor, M.A.; Ängquist, L.; MacDonald, I.A.; Arner, P.; Holst, C.; Oppert, J.-M.; Martinez, J.; Rössner, S.; Polak, J.; et al. Change in proportional protein intake in a 10-week energy-restricted low- or high-fat diet, in relation to changes in body size and metabolic factors. *Obes. Facts* **2013**, *6*, 217–227. [CrossRef]
26. Tremblay, A.J.; Lamarche, B.; Guay, V.; Charest, A.; Lemelin, V.; Couture, P. Short-term, high-fat diet increases the expression of key intestinal genes involved in lipoprotein metabolism in healthy men. *Am. J. Clin. Nutr.* **2013**, *98*, 32–41. [CrossRef]

27. Osterberg, K.L.; Boutagy, N.E.; McMillan, R.P.; Stevens, J.R.; Frisard, M.I.; Kavanaugh, J.W.; Davy, B.M.; Davy, K.P.; Hulver, M.W. Probiotic supplementation attenuates increases in body mass and fat mass during high-fat diet in healthy young adults. *Obesity* **2015**, *23*, 2364–2370. [CrossRef] [PubMed]
28. Holloway, C.J.; E Cochlin, L.; Emmanuel, Y.; Murray, A.; Codreanu, I.; Edwards, L.M.; Szmigielski, C.; Tyler, D.J.; Knight, N.S.; Saxby, B.K.; et al. A high-fat diet impairs cardiac high-energy phosphate metabolism and cognitive function in healthy human subjects. *Am. J. Clin. Nutr.* **2011**, *93*, 748–755. [CrossRef] [PubMed]
29. Murphy, E.A.; Velazquez, K.T.; Herbert, K.M. Influence of high-fat diet on gut microbiota. *Curr. Opin. Clin. Nutr. Metab. Care* **2015**, *18*, 515–520. [CrossRef]
30. Duan, Y.; Zeng, L.; Zheng, C.; Song, B.; Li, F.; Kong, X.; Xu, K. Inflammatory Links Between High Fat Diets and Diseases. *Front. Immunol.* **2018**, *9*, 2649. [CrossRef] [PubMed]
31. Liu, Y.; Palanivel, R.; Rai, E.; Park, M.; Gabor, T.V.; Scheid, M.P.; Xu, A.; Sweeney, G. Adiponectin Stimulates Autophagy and Reduces Oxidative Stress to Enhance Insulin Sensitivity During High-Fat Diet Feeding in Mice. *Diabetes* **2014**, *64*, 36–48. [CrossRef]
32. Eckel, R.H.; Jakicic, J.M.; Ard, J.D.; de Jesus, J.M.; Miller, N.H.; Hubbard, V.S.; Lee, I.-M.; Lichtenstein, A.H.; Loria, C.M.; Millen, B.E.; et al. 2013 AHA/ACC Guideline on Lifestyle Management to Reduce Cardiovascular Risk. *J. Am. Coll. Cardiol.* **2014**, *63*, 2960–2984. [CrossRef]
33. Gentile, D.; Fornai, M.; Pellegrini, C.; Colucci, R.; Benvenuti, L.; Duranti, E.; Masi, S.; Carpi, S.; Nieri, P.; Nericcio, A.; et al. Luteolin Prevents Cardiometabolic Alterations and Vascular Dysfunction in Mice With HFD-Induced Obesity. *Front. Pharmacol.* **2018**, *9*, 1094. [CrossRef]
34. Martins, M.; Catta-Preta, M.; Mandarim-De-Lacerda, C.; Águila, M.; Brunini, T.; Mendes-Ribeiro, A. High fat diets modulate nitric oxide biosynthesis and antioxidant defence in red blood cells from C57BL/6 mice. *Arch. Biochem. Biophys.* **2010**, *499*, 56–61. [CrossRef]
35. Alarcon, G.; Roco, J.; Medina, M.; Medina, A.; Peral, M.; Jerez, S. High fat diet-induced metabolically obese and normal weight rabbit model shows early vascular dysfunction: Mechanisms involved. *Int. J. Obes.* **2018**, *42*, 1535–1543. [CrossRef]
36. Chiu, S.; Williams, P.T.; Krauss, R.M. Effects of a very high saturated fat diet on LDL particles in adults with atherogenic dyslipidemia: A randomized controlled trial. *PLoS ONE* **2017**, *12*, e0170664. [CrossRef]
37. Yeh, T.-S.; Yuan, C.; Ascherio, A.; Rosner, B.A.; Blacker, D.; Willett, W.C. Long-term intake of total energy and fat in relation to subjective cognitive decline. *Eur. J. Epidemiol.* **2021**, *37*, 133–146. [CrossRef]
38. Laitinen, M.; Ngandu, T.; Rovio, S.; Helkala, E.-L.; Uusitalo, U.; Viitanen, M.; Nissinen, A.; Tuomilehto, J.; Soininen, H.; Kivipelto, M. Fat Intake at Midlife and Risk of Dementia and Alzheimer's Disease: A Population-Based Study. *Dement. Geriatr. Cogn. Disord.* **2006**, *22*, 99–107. [CrossRef] [PubMed]
39. Eskelinen, M.H.; Ngandu, T.; Helkala, E.; Tuomilehto, J.; Nissinen, A.; Soininen, H.; Kivipelto, M. Fat intake at midlife and cognitive impairment later in life: A population-based CAIDE study. *Int. J. Geriatr. Psychiatry* **2008**, *23*, 741–747. [CrossRef]
40. Pistell, P.J.; Morrison, C.; Gupta, S.; Knight, A.G.; Keller, J.; Ingram, D.K.; Bruce-Keller, A.J. Cognitive impairment following high fat diet consumption is associated with brain inflammation. *J. Neuroimmunol.* **2010**, *219*, 25–32. [CrossRef] [PubMed]
41. Lin, B.; Hasegawa, Y.; Takane, K.; Koibuchi, N.; Cao, C.; Kim-Mitsuyama, S. High-Fat-Diet Intake Enhances Cerebral Amyloid Angiopathy and Cognitive Impairment in a Mouse Model of Alzheimer's Disease, Independently of Metabolic Disorders. *J. Am. Hear. Assoc.* **2016**, *5*, e003154. [CrossRef] [PubMed]
42. Busquets, O.; Ettcheto, M.; Pallàs, M.; Beas-Zarate, C.; Verdaguer, E.; Auladell, C.; Folch, J.; Camins, A. Long-term exposition to a high fat diet favors the appearance of β-amyloid depositions in the brain of C57BL/6J mice. A potential model of sporadic Alzheimer's disease. *Mech. Ageing Dev.* **2017**, *162*, 38–45. [CrossRef] [PubMed]
43. Greenwood, C.E.; Winocur, G. Learning and memory impairment in rats fed a high saturated fat diet. *Behav. Neural Biol.* **1990**, *53*, 74–87. [CrossRef]
44. Razaz, J.M.; Rahmani, J.; Varkaneh, H.K.; Thompson, J.; Clark, C.; Abdulazeem, H.M. The health effects of medical nutrition therapy by dietitians in patients with diabetes: A systematic review and meta-analysis. *Prim. Care Diabetes* **2019**, *13*, 399–408. [CrossRef] [PubMed]
45. Adler, A.; Bennett, P.; Chair, S.C.; Gregg, E.; Narayan, K.V.; Schmidt, M.I.; Sobngwi, E.; Tajima, N.; Tandon, N.; Unwin, N.; et al. Reprint of: Classification of diabetes mellitus. *Diabetes Res. Clin. Pract.* **2021**, 108972. [CrossRef] [PubMed]
46. DeFronzo, R.A.; Ferrannini, E.; Groop, L.; Henry, R.R.; Herman, W.H.; Holst, J.J.; Hu, F.B.; Kahn, C.R.; Raz, I.; Shulman, G.I.; et al. Type 2 diabetes mellitus. *Nat. Rev. Dis. Prim.* **2015**, *1*, 15019. [CrossRef]
47. Sun, H.; Saeedi, P.; Karuranga, S.; Pinkepank, M.; Ogurtsova, K.; Duncan, B.B.; Stein, C.; Basit, A.; Chan, J.C.; Mbanya, J.C.; et al. IDF Diabetes Atlas: Global, regional and country-level diabetes prevalence estimates for 2021 and projections for 2045. *Diabetes Res. Clin. Pract.* **2021**, *183*, 109119. [CrossRef]
48. Alberti, K.G.; Zimmet, P.Z. Definition, diagnosis and classification of diabetes mellitus and its complications. Part 1: Diagnosis and classification of diabetes mellitus provisional report of a WHO consultation. *Diabet. Med.* **1998**, *15*, 539–553. [CrossRef]
49. American Diabetes Association. Diagnosis and Classification of Diabetes Mellitus. *Diabetes Care* **2013**, *36* (Suppl. 1), S67–S74. [CrossRef]
50. GBD 2017 Diet Collaborators. Health effects of dietary risks in 195 countries, 1990–2017: A systematic analysis for the Global Burden of Disease Study 2017. *Lancet* **2019**, *393*, 1958–1972. [CrossRef]
51. Hu, F.B. Globalization of Diabetes. *Diabetes Care* **2011**, *34*, 1249–1257. [CrossRef]

52. Hu, F.B.; Manson, J.E.; Stampfer, M.J.; Colditz, G.; Liu, S.; Solomon, C.G.; Willett, W.C. Diet, Lifestyle, and the Risk of Type 2 Diabetes Mellitus in Women. *N. Engl. J. Med.* **2001**, *345*, 790–797. [CrossRef]
53. De Munter, J.S.L.; Hu, F.B.; Spiegelman, D.; Franz, M.; van Dam, R.M. Whole Grain, Bran, and Germ Intake and Risk of Type 2 Diabetes: A Prospective Cohort Study and Systematic Review. *PLoS Med.* **2007**, *4*, e261. [CrossRef] [PubMed]
54. Kleinert, M.; Clemmensen, C.; Hofmann, S.M.; Moore, M.C.; Renner, S.; Woods, S.C.; Huypens, P.; Beckers, J.; de Angelis, M.H.; Schürmann, A.; et al. Animal models of obesity and diabetes mellitus. *Nat. Rev. Endocrinol.* **2018**, *14*, 140–162. [CrossRef]
55. Winzell, M.S.; Ahrén, B. The High-Fat Diet–Fed Mouse. *Diabetes* **2004**, *53*, S215–S219. [CrossRef]
56. Prasad, M.; Rajagopal, P.; Devarajan, N.; Veeraraghavan, V.P.; Palanisamy, C.P.; Cui, B.; Patil, S.; Jayaraman, S. A comprehensive review on high -fat diet-induced diabetes mellitus: An epigenetic view. *J. Nutr. Biochem.* **2022**, *107*, 109037. [CrossRef] [PubMed]
57. Ahima, R.S.; Flier, J.S. Adipose Tissue as an Endocrine Organ. *Trends Endocrinol. Metab.* **2000**, *11*, 327–332. [CrossRef]
58. Tilg, H.; Moschen, A.R. Inflammatory Mechanisms in the Regulation of Insulin Resistance. *Mol. Med.* **2008**, *14*, 222–231. [CrossRef]
59. Camargo, A.; Meneses, M.E.; Pérez-Martínez, P.; Delgado-Lista, J.; Rangel-Zúñiga, O.A.; Marín, C.; Almadén, Y.; Yubero-Serrano, E.M.; González-Guardia, L.; Fuentes, F.; et al. Dietary fat modifies lipid metabolism in the adipose tissue of metabolic syndrome patients. *Genes Nutr.* **2014**, *9*, 1–9. [CrossRef]
60. Rorsman, P.; Braun, M. Regulation of Insulin Secretion in Human Pancreatic Islets. *Annu. Rev. Physiol.* **2013**, *75*, 155–179. [CrossRef] [PubMed]
61. Butler, A.E.; Janson, J.; Bonner-Weir, S.; Ritzel, R.; Rizza, R.A.; Butler, P.C. Beta-Cell Deficit and Increased beta-Cell Apoptosis in Humans With Type 2 Diabetes. *Diabetes* **2003**, *52*, 102–110. [CrossRef]
62. Prentki, M.; Nolan, C.J. Islet beta cell failure in type 2 diabetes. *J. Clin. Investig.* **2006**, *116*, 1802–1812. [CrossRef] [PubMed]
63. Lv, C.; Sun, Y.; Zhang, Z.Y.; Aboelela, Z.; Qiu, X.; Meng, Z.-X. β-cell dynamics in type 2 diabetes and in dietary and exercise interventions. *J. Mol. Cell Biol.* **2022**, *14*, mjac046. [CrossRef] [PubMed]
64. Titchenell, P.M.; Lazar, M.A.; Birnbaum, M.J. Unraveling the Regulation of Hepatic Metabolism by Insulin. *Trends Endocrinol. Metab.* **2017**, *28*, 497–505. [CrossRef]
65. Rizza, R.A. Pathogenesis of Fasting and Postprandial Hyperglycemia in Type 2 Diabetes: Implications for Therapy. *Diabetes* **2010**, *59*, 2697–2707. [CrossRef]
66. Westerbacka, J.; Lammi, K.; Häkkinen, A.-M.; Rissanen, A.; Salminen, I.; Aro, A.; Yki-Järvinen, H. Dietary Fat Content Modifies Liver Fat in Overweight Nondiabetic Subjects. *J. Clin. Endocrinol. Metab.* **2005**, *90*, 2804–2809. [CrossRef]
67. Christiansen, E.; Schnider, S.; Palmvig, B.; Tauber-Lassen, E.; Pedersen, O. Intake of a Diet High in *Trans* Monounsaturated Fatty Acids or Saturated Fatty Acids: Effects on postprandial insulinemia and glycemia in obese patients with NIDDM. *Diabetes Care* **1997**, *20*, 881–887. [CrossRef] [PubMed]
68. Vessby, B.; Uusitupa, M.; Hermansen, K.; Riccardi, G.; Rivellese, A.A.; Tapsell, L.C.; Nälsén, C.; Berglund, L.; Louheranta, A.; Rasmussen, B.M.; et al. Substituting dietary saturated for monounsaturated fat impairs insulin sensitivity in healthy men and women: The KANWU study. *Diabetologia* **2001**, *44*, 312–319. [CrossRef] [PubMed]
69. Penn, L.; White, M.; Oldroyd, J.; Walker, M.; Alberti, K.G.M.; Mathers, J.C. Prevention of type 2 diabetes in adults with impaired glucose tolerance: The European Diabetes Prevention RCT in Newcastle upon Tyne, UK. *BMC Public Health* **2009**, *9*, 342. [CrossRef]
70. Mayer-Davis, E.J.; Monaco, J.H.; Hoen, H.M.; Carmichael, S.; Vitolins, M.Z.; Rewers, M.J.; Haffner, S.M.; Ayad, M.F.; Bergman, R.N.; Karter, A.J. Dietary fat and insulin sensitivity in a triethnic population: The role of obesity. The Insulin Resistance Atherosclerosis Study (IRAS). *Am. J. Clin. Nutr.* **1997**, *65*, 79–87. [CrossRef]
71. Meyer, K.A.; Kushi, L.H.; Jacobs, D.R.; Folsom, A.R. Dietary Fat and Incidence of Type 2 Diabetes in Older Iowa Women. *Diabetes Care* **2001**, *24*, 1528–1535. [CrossRef]
72. van Dam, R.M.; Willett, W.C.; Rimm, E.B.; Stampfer, M.J.; Hu, F.B. Dietary Fat and Meat Intake in Relation to Risk of Type 2 Diabetes in Men. *Diabetes Care* **2002**, *25*, 417–424. [CrossRef] [PubMed]
73. Buijsse, B.; Boeing, H.; Drogan, D.; Schulze, M.B.; Feskens, E.J.; Amiano, P.; Barricarte, A.; Clavel-Chapelon, F.; de Lauzon-Guillain, B.; Fagherazzi, G.; et al. Consumption of fatty foods and incident type 2 diabetes in populations from eight European countries. *Eur. J. Clin. Nutr.* **2014**, *69*, 455–461. [CrossRef]
74. Guasch-Ferré, M.; Becerra-Tomás, N.; Ruiz-Canela, M.; Corella, D.; Schröder, H.; Estruch, R.; Ros, E.; Arós, F.; Gómez-Gracia, E.; Fiol, M.; et al. Total and subtypes of dietary fat intake and risk of type 2 diabetes mellitus in the Prevención con Dieta Mediterránea (PREDIMED) study. *Am. J. Clin. Nutr.* **2017**, *105*, 723–735. [CrossRef]
75. Alhazmi, A.; Stojanovski, E.; McEvoy, M.; Garg, M.L. Macronutrient intake and type 2 diabetes risk in middle-aged Australian women. Results from the Australian Longitudinal Study on Women's Health. *Public Health Nutr.* **2013**, *17*, 1587–1594. [CrossRef]
76. Salmerón, J.; Hu, F.B.; E Manson, J.; Stampfer, M.J.; A Colditz, G.; Rimm, E.B.; Willett, W.C. Dietary fat intake and risk of type 2 diabetes in women. *Am. J. Clin. Nutr.* **2001**, *73*, 1019–1026. [CrossRef] [PubMed]
77. Zong, G.; Liu, G.; Willett, W.C.; Wanders, A.J.; Alssema, M.; Zock, P.L.; Hu, F.B.; Sun, Q. Associations Between Linoleic Acid Intake and Incident Type 2 Diabetes Among U.S. Men and Women. *Diabetes Care* **2019**, *42*, 1406–1413. [CrossRef] [PubMed]
78. Jiang, R. Nut and Peanut Butter Consumption and Risk of Type 2 Diabetes in Women. *JAMA* **2002**, *288*, 2554–2560. [CrossRef]
79. Renner, S.; Dobenecker, B.; Blutke, A.; Zöls, S.; Wanke, R.; Ritzmann, M.; Wolf, E. Comparative aspects of rodent and nonrodent animal models for mechanistic and translational diabetes research. *Theriogenology* **2016**, *86*, 406–421. [CrossRef]
80. Erion, D.M.; Park, H.-J.; Lee, H.-Y. The role of lipids in the pathogenesis and treatment of type 2 diabetes and associated co-morbidities. *BMB Rep.* **2016**, *49*, 139–148. [CrossRef] [PubMed]

81. Moraes, R.; Blondet, A.; Birkenkamp-Demtroeder, K.; Tirard, J.; Orntoft, T.F.; Gertler, A.; Durand, P.; Naville, D.; Bégeot, M. Study of the Alteration of Gene Expression in Adipose Tissue of Diet-Induced Obese Mice by Microarray and Reverse Transcription-Polymerase Chain Reaction Analyses. *Endocrinology* **2003**, *144*, 4773–4782. [CrossRef] [PubMed]
82. Kim, B.; Do, M.-S.; Hyun, C.-K. B-cell-activating factor deficiency attenuates high-fat diet-induced glucose intolerance by potentiating adipose tissue function. *Biochem. Biophys. Res. Commun.* **2015**, *464*, 1171–1177. [CrossRef]
83. Hamada, M.; Abe, M.; Miyake, T.; Kawasaki, K.; Tada, F.; Furukawa, S.; Matsuura, B.; Hiasa, Y.; Onji, M. B Cell-Activating Factor Controls the Production of Adipokines and Induces Insulin Resistance. *Obesity* **2011**, *19*, 1915–1922. [CrossRef] [PubMed]
84. Hotamisligil, G.S.; Shargill, N.S.; Spiegelman, B.M. Adipose Expression of Tumor Necrosis Factor-α: Direct Role in Obesity-Linked Insulin Resistance. *Science* **1993**, *259*, 87–91. [CrossRef] [PubMed]
85. Wang, X.; Cheng, M.; Zhao, M.; Ge, A.; Guo, F.; Zhang, M.; Yang, Y.; Liu, L.; Yang, N. Differential effects of high-fat-diet rich in lard oil or soybean oil on osteopontin expression and inflammation of adipose tissue in diet-induced obese rats. *Eur. J. Nutr.* **2012**, *52*, 1181–1189. [CrossRef] [PubMed]
86. Ahrén, J.; Ahrén, B.; Wierup, N. Increased β-cell volume in mice fed a high-fat diet: A dynamic study over 12 months. *Islets* **2010**, *2*, 353–356. [CrossRef] [PubMed]
87. Kanno, A.; Asahara, S.-I.; Masuda, K.; Matsuda, T.; Kimura-Koyanagi, M.; Seino, S.; Ogawa, W.; Kido, Y. Compensatory hyperinsulinemia in high-fat diet-induced obese mice is associated with enhanced insulin translation in islets. *Biochem. Biophys. Res. Commun.* **2015**, *458*, 681–686. [CrossRef] [PubMed]
88. Ribeiro, R.A.; Santos-Silva, J.C.; Vettorazzi, J.F.; Cotrim, B.B.; Mobiolli, D.D.M.; Boschero, A.C.; Carneiro, E.M. Taurine supplementation prevents morpho-physiological alterations in high-fat diet mice pancreatic β-cells. *Amino Acids* **2012**, *43*, 1791–1801. [CrossRef]
89. Kim, Y.; Iwashita, S.; Tamura, T.; Tokuyama, K.; Suzuki, M. Effect of High-Fat Diet on the Gene Expression of Pancreatic GLUT2 and Glucokinase in Rats. *Biochem. Biophys. Res. Commun.* **1995**, *208*, 1092–1098. [CrossRef]
90. Matsuda, A.; Makino, N.; Tozawa, T.; Shirahata, N.; Honda, T.; Ikeda, Y.; Sato, H.; Ito, M.; Kakizaki, Y.; Akamatsu, M.; et al. Pancreatic Fat Accumulation, Fibrosis, and Acinar Cell Injury in the Zucker Diabetic Fatty Rat Fed a Chronic High-Fat Diet. *Pancreas* **2014**, *43*, 735–743. [CrossRef] [PubMed]
91. Yan, M.-X.; Ren, H.-B.; Kou, Y.; Meng, M.; Li, Y.-Q. Involvement of Nuclear Factor Kappa B in High-Fat Diet-Related Pancreatic Fibrosis in Rats. *Gut Liver* **2012**, *6*, 381–387. [CrossRef] [PubMed]
92. Langhans, W. Role of the liver in the control of glucose-lipid utilization and body weight. *Curr. Opin. Clin. Nutr. Metab. Care* **2003**, *6*, 449–455. [CrossRef] [PubMed]
93. Kusunoki, M.; Tsutsumi, K.; Hara, T.; Ogawa, H.; Nakamura, T.; Miyata, T.; Sakakibara, F.; Fukuzawa, Y.; Suga, T.; Kakumu, S.; et al. Correlation between lipid and glycogen contents in liver and insulin resistance in high-fat[ndash]fed rats treated with the lipoprotein lipase activator NO-1886. *Metabolism* **2002**, *51*, 792–795. [CrossRef]
94. Kobayashi, M.; Ohno, T.; Tsuchiya, T.; Horio, F. Characterization of diabetes-related traits in MSM and JF1 mice on high-fat diet. *J. Nutr. Biochem.* **2004**, *15*, 614–621. [CrossRef]
95. Buettner, R.; Parhofer, K.G.; Woenckhaus, M.; Wrede, C.E.; Kunz-Schughart, L.A.; Schölmerich, J.; Bollheimer, L.C. Defining high-fat-diet rat models: Metabolic and molecular effects of different fat types. *J. Mol. Endocrinol.* **2006**, *36*, 485–501. [CrossRef] [PubMed]
96. Chaabo, F.; Pronczuk, A.; Maslova, E.; Hayes, K.C. Nutritional correlates and dynamics of diabetes in the Nile rat (Arvicanthis niloticus): A novel model for diet-induced type 2 diabetes and the metabolic syndrome. *Nutr. Metab.* **2010**, *7*, 29. [CrossRef] [PubMed]
97. Hong, S.-H.; Kang, M.; Lee, K.-S.; Yu, K. High fat diet-induced TGF-β/Gbb signaling provokes insulin resistance through the tribbles expression. *Sci. Rep.* **2016**, *6*, 30265. [CrossRef]
98. Podell, B.K.; Ackart, D.F.; Richardson, M.A.; DiLisio, J.E.; Pulford, B.; Basaraba, R.J. A model of type 2 diabetes in the guinea pig using sequential diet-induced glucose intolerance and streptozotocin treatment. *Dis. Model. Mech.* **2017**, *10*, 151–162. [CrossRef]
99. Kaiyala, K.J.; Prigeon, R.L.; Kahn, S.E.; Woods, S.C.; Porte, D.; Schwartz, M.W. Reduced β-cell function contributes to impaired glucose tolerance in dogs made obese by high-fat feeding. *Am. J. Physiol. Metab.* **1999**, *277*, E659–E667. [CrossRef] [PubMed]
100. Kavanagh, K.; Jones, K.L.; Sawyer, J.; Kelley, K.; Carr, J.; Wagner, J.D.; Rudel, L.L. Trans Fat Diet Induces Abdominal Obesity and Changes in Insulin Sensitivity in Monkeys*. *Obesity* **2007**, *15*, 1675–1684. [CrossRef]
101. Zang, L.; Shimada, Y.; Nishimura, N. Development of a Novel Zebrafish Model for Type 2 Diabetes Mellitus. *Sci. Rep.* **2017**, *7*, 1461. [CrossRef] [PubMed]
102. Qin, J.; Li, Y.; Cai, Z.; Li, S.; Zhu, J.; Zhang, F.; Liang, S.; Zhang, W.; Guan, Y.; Shen, D.; et al. A metagenome-wide association study of gut microbiota in type 2 diabetes. *Nature* **2012**, *490*, 55–60. [CrossRef] [PubMed]
103. Larsen, N.; Vogensen, F.K.; Van Den Berg, F.W.J.; Nielsen, D.S.; Andreasen, A.S.; Pedersen, B.K.; Al-Soud, W.A.; Sørensen, S.J.; Hansen, L.H.; Jakobsen, M. Gut Microbiota in Human Adults with Type 2 Diabetes Differs from Non-Diabetic Adults. *PLoS ONE* **2010**, *5*, e9085. [CrossRef] [PubMed]
104. Shao, T.; Yu, Q.; Zhu, T.; Liu, A.; Gao, X.; Long, X.; Liu, Z. Inulin from Jerusalem artichoke tubers alleviates hyperglycaemia in high-fat-diet-induced diabetes mice through the intestinal microflora improvement. *Br. J. Nutr.* **2019**, *123*, 308–318. [CrossRef]
105. Ju, M.; Liu, Y.; Li, M.; Cheng, M.; Zhang, Y.; Deng, G.; Kang, X.; Liu, H. Baicalin improves intestinal microecology and abnormal metabolism induced by high-fat diet. *Eur. J. Pharmacol.* **2019**, *857*, 172457. [CrossRef] [PubMed]

106. Lin, H.V.; Frassetto, A.; Kowalik, E.J., Jr.; Nawrocki, A.R.; Lu, M.M.; Kosinski, J.R.; Hubert, J.A.; Szeto, D.; Yao, X.; Forrest, G.; et al. Butyrate and Propionate Protect against Diet-Induced Obesity and Regulate Gut Hormones via Free Fatty Acid Receptor 3-Independent Mechanisms. *PLoS ONE* **2012**, *7*, e35240. [CrossRef] [PubMed]
107. Luck, H.; Khan, S.; Kim, J.H.; Copeland, J.K.; Revelo, X.S.; Tsai, S.; Chakraborty, M.; Cheng, K.; Chan, Y.T.; Nøhr, M.K.; et al. Gut-associated IgA+ immune cells regulate obesity-related insulin resistance. *Nat. Commun.* **2019**, *10*, 1–17. [CrossRef] [PubMed]
108. Serino, M.; Luche, E.; Gres, S.; Baylac, A.; Bergé, M.; Cenac, C.; Waget, A.; Klopp, P.; Iacovoni, J.; Klopp, C.; et al. Metabolic adaptation to a high-fat diet is associated with a change in the gut microbiota. *Gut* **2011**, *61*, 543–553. [CrossRef] [PubMed]
109. Yin, R.; Xue, Y.; Hu, J.; Hu, X.; Shen, Q. The effects of diet and streptozotocin on metabolism and gut microbiota in a type 2 diabetes mellitus mouse model. *Food Agric. Immunol.* **2020**, *31*, 723–739. [CrossRef]
110. Wang, H.; Tang, W.; Zhang, P.; Zhang, Z.; He, J.; Zhu, D.; Bi, Y. Modulation of gut microbiota contributes to effects of intensive insulin therapy on intestinal morphological alteration in high-fat-diet-treated mice. *Acta Diabetol.* **2019**, *57*, 455–467. [CrossRef]
111. Wang, H.-Y.; Guo, L.-X.; Hu, W.-H.; Peng, Z.-T.; Wang, C.; Chen, Z.-C.; Liu, E.Y.; Dong, T.T.; Wang, T.-J.; Tsim, K.W. Polysaccharide from tuberous roots of Ophiopogon japonicus regulates gut microbiota and its metabolites during alleviation of high-fat diet-induced type-2 diabetes in mice. *J. Funct. Foods* **2019**, *63*, 103593. [CrossRef]
112. Liu, S.; Qin, P.; Wang, J. High-Fat Diet Alters the Intestinal Microbiota in Streptozotocin-Induced Type 2 Diabetic Mice. *Microorganisms* **2019**, *7*, 176. [CrossRef]
113. Wang, D.; Liu, J.; Zhong, L.; Ding, L.; Zhang, Q.; Yu, M.; Li, M.; Xiao, X. Potential benefits of metformin and pioglitazone combination therapy via gut microbiota and metabolites in high-fat diet-fed mice. *Front. Pharmacol.* **2022**, *13*, 1004617. [CrossRef]
114. Yang, T.; Zhou, W.; Xu, W.; Ran, L.; Yan, Y.; Lu, L.; Mi, J.; Zeng, X.; Cao, Y. Modulation of gut microbiota and hypoglycemic/hypolipidemic activity of flavonoids from the fruits of *Lycium barbarum* on high-fat diet/streptozotocin-induced type 2 diabetic mice. *Food Funct.* **2022**, *13*, 11169–11184. [CrossRef]
115. Yin, W.; Zhang, S.-Q.; Pang, W.-L.; Chen, X.-J.; Wen, J.; Hou, J.; Wang, C.; Song, L.-Y.; Qiu, Z.-M.; Liang, P.-T.; et al. Tang-Ping-San Decoction Remodel Intestinal Flora and Barrier to Ameliorate Type 2 Diabetes Mellitus in Rodent Model. *Diabetes Metab. Syndr. Obes. Targets Ther.* **2022**, *15*, 2563–2581. [CrossRef]
116. Wu, R.; Zhao, D.; An, R.; Wang, Z.; Li, Y.; Shi, B.; Ni, Q. Linggui Zhugan Formula Improves Glucose and Lipid Levels and Alters Gut Microbiota in High-Fat Diet-Induced Diabetic Mice. *Front. Physiol.* **2019**, *10*, 918. [CrossRef]
117. Ma, S.; Tian, S.; Sun, J.; Pang, X.; Hu, Q.; Li, X.; Lu, Y. Broccoli microgreens have hypoglycemic effect by improving blood lipid and inflammatory factors while modulating gut microbiota in mice with type 2 diabetes. *J. Food Biochem.* **2022**, *46*, e14145. [CrossRef] [PubMed]
118. Yan, J.; Li, J.; Xue, Q.; Xie, S.; Jiang, J.; Li, P.; Du, B. Bacillus sp. DU-106 ameliorates type 2 diabetes by modulating gut microbiota in high-fat-fed and streptozotocin-induced mice. *J. Appl. Microbiol.* **2022**, *133*, 3126–3138. [CrossRef] [PubMed]
119. Zandani, G.; Anavi-Cohen, S.; Tsybina-Shimshilashvili, N.; Sela, N.; Nyska, A.; Madar, Z. Broccoli Florets Supplementation Improves Insulin Sensitivity and Alters Gut Microbiome Population—A Steatosis Mice Model Induced by High-Fat Diet. *Front. Nutr.* **2021**, *8*, 680241. [CrossRef] [PubMed]
120. Cani, P.D.; Bibiloni, R.; Knauf, C.; Waget, A.; Neyrinck, A.M.; Delzenne, N.M.; Burcelin, R. Changes in Gut Microbiota Control Metabolic Endotoxemia-Induced Inflammation in High-Fat Diet-Induced Obesity and Diabetes in Mice. *Diabetes* **2008**, *57*, 1470–1481. [CrossRef] [PubMed]
121. Nakamura, A.; Yokoyama, Y.; Tanaka, K.; Benegiamo, G.; Hirayama, A.; Zhu, Q.; Kitamura, N.; Sugizaki, T.; Morimoto, K.; Itoh, H.; et al. Asperuloside Improves Obesity and Type 2 Diabetes through Modulation of Gut Microbiota and Metabolic Signaling. *Iscience* **2020**, *23*, 101522. [CrossRef]
122. Nyavor, Y.; Estill, R.; Edwards, H.; Ogden, H.; Heideman, K.; Starks, K.; Miller, C.; May, G.; Flesch, L.; McMillan, J.; et al. Intestinal nerve cell injury occurs prior to insulin resistance in female mice ingesting a high-fat diet. *Cell Tissue Res.* **2019**, *376*, 325–340. [CrossRef]
123. Chen, Y.; Zhu, L.; Hu, W.; Wang, Y.; Wen, X.; Yang, J. Simiao Wan modulates the gut microbiota and bile acid metabolism during improving type 2 diabetes mellitus in mice. *Phytomedicine* **2022**, *104*, 154264. [CrossRef]
124. Song, H.; Chu, Q.; Yan, F.; Yang, Y.; Han, W.; Zheng, X. Red pitaya betacyanins protects from diet-induced obesity, liver steatosis and insulin resistance in association with modulation of gut microbiota in mice. *J. Gastroenterol. Hepatol.* **2016**, *31*, 1462–1469. [CrossRef]
125. Rehman, A.U.; Siddiqui, N.Z.; Farooqui, N.A.; Alam, G.; Gul, A.; Ahmad, B.; Asim, M.; Khan, A.I.; Xin, Y.; Zexu, W.; et al. Morchella esculenta mushroom polysaccharide attenuates diabetes and modulates intestinal permeability and gut microbiota in a type 2 diabetic mice model. *Front. Nutr.* **2022**, *9*, 984695. [CrossRef] [PubMed]
126. Bagarolli, R.A.; Tobar, N.; Oliveira, A.G.; Araújo, T.G.; Carvalho, B.M.; Rocha, G.Z.; Vecina, J.F.; Calisto, K.; Guadagnini, D.; Prada, P.D.O.; et al. Probiotics modulate gut microbiota and improve insulin sensitivity in DIO mice. *J. Nutr. Biochem.* **2017**, *50*, 16–25. [CrossRef]
127. Khat-Udomkiri, N.; Toejing, P.; Sirilun, S.; Chaiyasut, C.; Lailerd, N. Antihyperglycemic effect of rice husk derived xylooligosaccharides in high-fat diet and low-dose streptozotocin-induced type 2 diabetic rat model. *Food Sci. Nutr.* **2019**, *8*, 428–444. [CrossRef]

128. Huh, Y.-J.; Seo, J.-Y.; Nam, J.; Yang, J.; McDowell, A.; Kim, Y.-K.; Lee, J.-H. Bariatric/Metabolic Surgery Induces Noticeable Changes of Microbiota and Their Secreting Extracellular Vesicle Composition in the Gut. *Obes. Surg.* **2019**, *29*, 2470–2484. [CrossRef]
129. Cowan, T.E.; Palmnäs, M.S.; Yang, J.; Bomhof, M.R.; Ardell, K.L.; Reimer, R.A.; Vogel, H.J.; Shearer, J. Chronic coffee consumption in the diet-induced obese rat: Impact on gut microbiota and serum metabolomics. *J. Nutr. Biochem.* **2014**, *25*, 489–495. [CrossRef] [PubMed]
130. Zeng, Z.; Guo, X.; Zhang, J.; Yuan, Q.; Chen, S. *Lactobacillus paracasei* modulates the gut microbiota and improves inflammation in type 2 diabetic rats. *Food Funct.* **2021**, *12*, 6809–6820. [CrossRef] [PubMed]
131. Wang, K.; Wang, Y.; Chen, S.; Gu, J.; Ni, Y. Insoluble and Soluble Dietary Fibers from Kiwifruit (*Actinidia deliciosa*) Modify Gut Microbiota to Alleviate High-Fat Diet and Streptozotocin-Induced TYPE 2 Diabetes in Rats. *Nutrients* **2022**, *14*, 3369. [CrossRef] [PubMed]
132. Zhu, Y.; Bai, J.; Zhang, Y.; Xiao, X.; Dong, Y. Effects of bitter melon (*Momordica charantia* L.) on the gut microbiota in high fat diet and low dose streptozocin-induced rats. *Int. J. Food Sci. Nutr.* **2016**, *67*, 686–695. [CrossRef]
133. Peng, M.; Wang, L.; Su, H.; Zhang, L.; Yang, Y.; Sun, L.; Wu, Y.; Ran, L.; Liu, S.; Yin, M.; et al. Ginsenoside Rg1 improved diabetes through regulating the intestinal microbiota in high-fat diet and streptozotocin-induced type 2 diabetes rats. *J. Food Biochem.* **2022**, *46*, e14321. [CrossRef] [PubMed]
134. Zhu, Y.; Dong, L.; Huang, L.; Shi, Z.; Dong, J.; Yao, Y.; Shen, R. Effects of oat β-glucan, oat resistant starch, and the whole oat flour on insulin resistance, inflammation, and gut microbiota in high-fat-diet-induced type 2 diabetic rats. *J. Funct. Foods* **2020**, *69*, 103939. [CrossRef]
135. Hereu, M.; Ramos-Romero, S.; Busquets, C.; Atienza, L.; Amézqueta, S.; Miralles-Pérez, B.; Nogués, M.R.; Méndez, L.; Medina, I.; Torres, J.L. Effects of combined d-fagomine and omega-3 PUFAs on gut microbiota subpopulations and diabetes risk factors in rats fed a high-fat diet. *Sci. Rep.* **2019**, *9*, 16628. [CrossRef] [PubMed]
136. Khan, M.A.B.; Hashim, M.J.; King, J.K.; Govender, R.D.; Mustafa, H.; Al Kaabi, J. Epidemiology of Type 2 Diabetes—Global Burden of Disease and Forecasted Trends. *J. Epidemiol. Glob. Health* **2020**, *10*, 107–111. [CrossRef]
137. Nyenwe, E.A.; Jerkins, T.W.; Umpierrez, G.E.; Kitabchi, A.E. Management of type 2 diabetes: Evolving strategies for the treatment of patients with type 2 diabetes. *Metabolism* **2011**, *60*, 1–23. [CrossRef]
138. Chaudhury, A.; Duvoor, C.; Reddy Dendi, V.S.; Kraleti, S.; Chada, A.; Ravilla, R.; Marco, A.; Shekhawat, N.S.; Montales, M.T.; Kuriakose, K.; et al. Clinical Review of Antidiabetic Drugs: Implications for Type 2 Diabetes Mellitus Management. *Front. Endocrinol.* **2017**, *8*, 6. [CrossRef]
139. Thulé, P.M. Mechanisms of current therapies for diabetes mellitus type 2. *Adv. Physiol. Educ.* **2012**, *36*, 275–283. [CrossRef] [PubMed]
140. Inzucchi, S.E.; Bergenstal, R.M.; Buse, J.B.; Diamant, M.; Ferrannini, E.; Nauck, M.; Peters, A.L.; Tsapas, A.; Wender, R.; Matthews, D.R. Management of Hyperglycemia in Type 2 Diabetes, 2015: A Patient-Centered Approach: Update to a Position Statement of the American Diabetes Association and the European Association for the Study of Diabetes. *Diabetes Care* **2014**, *38*, 140–149. [CrossRef]
141. Pryor, R.; Cabreiro, F. Repurposing metformin: An old drug with new tricks in its binding pockets. *Biochem. J.* **2015**, *471*, 307–322. [CrossRef]
142. Viollet, B.; Guigas, B.; Garcia, N.S.; Leclerc, J.; Foretz, M.; Andreelli, F. Cellular and molecular mechanisms of metformin: An overview. *Clin. Sci.* **2012**, *122*, 253–270. [CrossRef] [PubMed]
143. Fujita, Y.; Inagaki, N. Metformin: New Preparations and Nonglycemic Benefits. *Curr. Diabetes Rep.* **2017**, *17*, 5. [CrossRef] [PubMed]
144. Shin, N.R.; Lee, J.C.; Lee, H.Y.; Kim, M.S.; Whon, T.W.; Lee, M.S.; Bae, J.W. An increase in the *Akkermansia* spp. population induced by metformin treatment improves glucose homeostasis in diet-induced obese mice. *Gut* **2014**, *63*, 727–735. [CrossRef]
145. Sun, L.; Xie, C.; Wang, G.; Wu, Y.; Wu, Q.; Wang, X.; Liu, J.; Deng, Y.; Xia, J.; Chen, B.; et al. Gut microbiota and intestinal FXR mediate the clinical benefits of metformin. *Nat. Med.* **2018**, *24*, 1919–1929. [CrossRef] [PubMed]
146. Proks, P.; Reimann, F.; Green, N.; Gribble, F.; Ashcroft, F. Sulfonylurea Stimulation of Insulin Secretion. *Diabetes* **2002**, *51*, S368–S376. [CrossRef] [PubMed]
147. Colca, J.R. The TZD insulin sensitizer clue provides a new route into diabetes drug discovery. *Expert Opin. Drug Discov.* **2015**, *10*, 1259–1270. [CrossRef] [PubMed]
148. Thangavel, N.; Al Bratty, M.; Javed, S.A.; Ahsan, W.; Alhazmi, H.A. Targeting Peroxisome Proliferator-Activated Receptors Using Thiazolidinediones: Strategy for Design of Novel Antidiabetic Drugs. *Int. J. Med. Chem.* **2017**, *2017*, 1–20. [CrossRef] [PubMed]
149. Bai, J.; Zhu, Y.; Dong, Y. Response of gut microbiota and inflammatory status to bitter melon (*Momordica charantia* L.) in high fat diet induced obese rats. *J. Ethnopharmacol.* **2016**, *194*, 717–726. [CrossRef]
150. Xu, B.; Xing, A.; Li, S. The forgotten type 2 diabetes mellitus medicine: Rosiglitazone. *Diabetol. Int.* **2021**, *13*, 49–65. [CrossRef] [PubMed]
151. Klochkov, V.G.; Bezsonova, E.N.; Dubar, M.; Melekhina, D.D.; Temnov, V.V.; Zaryanova, E.V.; Lozinskaya, N.A.; Babkov, D.A.; Spasov, A.A. Towards multi-target antidiabetic agents: In vitro and in vivo evaluation of 3,5-disubstituted indolin-2-one derivatives as novel α-glucosidase inhibitors. *Bioorg. Med. Chem. Lett.* **2021**, *55*, 128449. [CrossRef]

152. Zhang, X.; Fang, Z.; Zhang, C.; Xia, H.; Jie, Z.; Han, X.; Chen, Y.; Ji, L. Effects of Acarbose on the Gut Microbiota of Prediabetic Patients: A Randomized, Double-blind, Controlled Crossover Trial. *Diabetes Ther.* **2017**, *8*, 293–307. [CrossRef]
153. Mann, J. Nutrition Recommendations for the Treatment and Prevention of Type 2 Diabetes and the Metabolic Syndrome: An Evidenced-Based Review. *Nutr. Rev.* **2006**, *64*, 422–427. [CrossRef]
154. Rosenfeld, R.M.; Kelly, J.H.; Agarwal, M.; Aspry, K.; Barnett, T.; Davis, B.C.; Fields, D.; Gaillard, T.; Gulati, M.; Guthrie, G.E.; et al. Dietary Interventions to Treat Type 2 Diabetes in Adults with a Goal of Remission: An Expert Consensus Statement from the American College of Lifestyle Medicine. *Am. J. Lifestyle Med.* **2022**, *16*, 342–362. [CrossRef]
155. Schröder, H. Protective mechanisms of the Mediterranean diet in obesity and type 2 diabetes. *J. Nutr. Biochem.* **2007**, *18*, 149–160. [CrossRef]
156. Chan, J.M.; Rimm, E.B.; A Colditz, G.; Stampfer, M.J.; Willett, W.C. Obesity, Fat Distribution, and Weight Gain as Risk Factors for Clinical Diabetes in Men. *Diabetes Care* **1994**, *17*, 961–969. [CrossRef]
157. Colditz, G.; Willett, W.C.; Rotnitzky, A.; Manson, J.E. Weight Gain as a Risk Factor for Clinical Diabetes Mellitus in Women. *Ann. Intern. Med.* **1995**, *122*, 481–486. [CrossRef]
158. Feinman, R.D.; Pogozelski, W.K.; Astrup, A.; Bernstein, R.K.; Fine, E.J.; Westman, E.C.; Accurso, A.; Frassetto, L.; Gower, B.A.; McFarlane, S.I.; et al. Dietary carbohydrate restriction as the first approach in diabetes management: Critical review and evidence base. *Nutrition* **2015**, *31*, 1–13. [CrossRef] [PubMed]
159. Accurso, A.; Bernstein, R.K.; Dahlqvist, A.; Draznin, B.; Feinman, R.D.; Fine, E.J.; Gleed, A.; Jacobs, D.B.; Larson, G.; Lustig, R.H.; et al. Dietary carbohydrate restriction in type 2 diabetes mellitus and metabolic syndrome: Time for a critical appraisal. *Nutr. Metab.* **2008**, *5*, 9. [CrossRef] [PubMed]
160. Tosti, V.; Bertozzi, B.; Fontana, L. Health Benefits of the Mediterranean Diet: Metabolic and Molecular Mechanisms. *J. Gerontol. Ser. A* **2018**, *73*, 318–326. [CrossRef] [PubMed]
161. Paniagua, J.; de la Sacristana, A.G.; Romero, I.; Vidal-Puig, A.; Latre, J.; Sanchez, E.; Perez-Martinez, P.; Lopez-Miranda, J.; Perez-Jimenez, F. Monounsaturated Fat–Rich Diet Prevents Central Body Fat Distribution and Decreases Postprandial Adiponectin Expression Induced by a Carbohydrate-Rich Diet in Insulin-Resistant Subjects. *Diabetes Care* **2007**, *30*, 1717–1723. [CrossRef]
162. Summers, L.K.M.; Fielding, B.A.; Bradshaw, H.A.; Ilic, V.; Beysen, C.; Clark, M.L.; Moore, N.R.; Frayn, K.N. Substituting dietary saturated fat with polyunsaturated fat changes abdominal fat distribution and improves insulin sensitivity. *Diabetologia* **2002**, *45*, 369–377. [CrossRef]
163. Luo, J.; Rizkalla, S.W.; Boillot, J.; Alamowitch, C.; Chaib, H.; Bruzzo, F.; Desplanque, N.; Dalix, A.M.; Durand, G.; Slama, G. Dietary (n-3) polyunsaturated fatty acids improve adipocyte insulin action and glucose metabolism in insulin-resistant rats: Relation to membrane fatty acids. *J. Nutr.* **1996**, *126*, 1951–1958.
164. Ginsberg, B.H.; Brown, T.J.; Simon, I.; A Spector, A. Effect of the Membrane Lipid Environment on the Properties of Insulin Receptors. *Diabetes* **1981**, *30*, 773–780. [CrossRef] [PubMed]
165. Pégorier, J.-P.; Le May, C.; Girard, J. Control of Gene Expression by Fatty Acids. *J. Nutr.* **2004**, *134*, S2444–S2449. [CrossRef]
166. Lee, J.Y.; Zhao, L.; Youn, H.S.; Weatherill, A.R.; Tapping, R.; Feng, L.; Lee, W.H.; Fitzgerald, K.A.; Hwang, D.H. Saturated Fatty Acid Activates but Polyunsaturated Fatty Acid Inhibits Toll-like Receptor 2 Dimerized with Toll-like Receptor 6 or 1. *J. Biol. Chem.* **2004**, *279*, 16971–16979. [CrossRef]
167. Baynes, H.W.; Mideksa, S.; Ambachew, S. The role of polyunsaturated fatty acids (n-3 PUFAs) on the pancreatic β-cells and insulin action. *Adipocyte* **2018**, *7*, 1–7. [CrossRef]
168. Kahleova, H.; Tura, A.; Hill, M.; Holubkov, R.; Barnard, N.D. A Plant-Based Dietary Intervention Improves Beta-Cell Function and Insulin Resistance in Overweight Adults: A 16-Week Randomized Clinical Trial. *Nutrients* **2018**, *10*, 189. [CrossRef] [PubMed]
169. Kahleova, H.; Matoulek, M.; Malinska, H.; Oliyarnik, O.; Kazdova, L.; Neskudla, T.; Skoch, A.; Hajek, M.; Hill, M.; Kahle, M.; et al. Vegetarian diet improves insulin resistance and oxidative stress markers more than conventional diet in subjects with Type 2 diabetes. *Diabet. Med.* **2011**, *28*, 549–559. [CrossRef] [PubMed]
170. Jung, U.J.; Choi, M.-S.; Lee, M.-K.; Jeong, K.-S. The Hypoglycemic Effects of Hesperidin and Naringin Are Partly Mediated by Hepatic Glucose-Regulating Enzymes in C57BL/KsJ-db/db Mice. *J. Nutr.* **2004**, *134*, 2499–2503. [CrossRef] [PubMed]
171. Guo, H.; Xia, M.; Zou, T.; Ling, W.; Zhong, R.; Zhang, W. Cyanidin 3-glucoside attenuates obesity-associated insulin resistance and hepatic steatosis in high-fat diet-fed and db/db mice via the transcription factor FoxO1. *J. Nutr. Biochem.* **2012**, *23*, 349–360. [CrossRef]
172. Zhuang, M.; Qiu, H.; Li, P.; Hu, L.; Wang, Y.; Rao, L. Islet protection and amelioration of type 2 diabetes mellitus by treatment with quercetin from the flowers of Edgeworthia gardneri. *Drug Des. Dev. Ther.* **2018**, *12*, 955–966. [CrossRef]
173. Alkhalidy, H.; Moore, W.; Wang, A.; Luo, J.; McMillan, R.P.; Wang, Y.; Zhen, W.; Hulver, M.W.; Liu, D. Kaempferol ameliorates hyperglycemia through suppressing hepatic gluconeogenesis and enhancing hepatic insulin sensitivity in diet-induced obese mice. *J. Nutr. Biochem.* **2018**, *58*, 90–101. [CrossRef] [PubMed]
174. Jung, E.H.; Kim, S.R.; Hwang, I.K.; Ha, T.Y. Hypoglycemic Effects of a Phenolic Acid Fraction of Rice Bran and Ferulic Acid in C57BL/KsJ-*db/db* Mice. *J. Agric. Food Chem.* **2007**, *55*, 9800–9804. [CrossRef]
175. Palsamy, P.; Subramanian, S. Ameliorative potential of resveratrol on proinflammatory cytokines, hyperglycemia mediated oxidative stress, and pancreatic β-cell dysfunction in streptozotocin-nicotinamide-induced diabetic rats. *J. Cell. Physiol.* **2010**, *224*, 423–432. [CrossRef]

176. Fu, Z.; Zhang, W.; Zhen, W.; Lum, H.; Nadler, J.L.; Bassaganya-Riera, J.; Jia, Z.; Wang, Y.; Misra, H.; Liu, D. Genistein Induces Pancreatic β-Cell Proliferation through Activation of Multiple Signaling Pathways and Prevents Insulin-Deficient Diabetes in Mice. *Endocrinology* **2010**, *151*, 3026–3037. [CrossRef]
177. Seymour, E.M.; Tanone, I.I.; Urcuyo-Llanes, D.E.; Lewis, S.K.; Kirakosyan, A.; Kondoleon, M.G.; Kaufman, P.B.; Bolling, S.F. Blueberry Intake Alters Skeletal Muscle and Adipose Tissue Peroxisome Proliferator-Activated Receptor Activity and Reduces Insulin Resistance in Obese Rats. *J. Med. Food* **2011**, *14*, 1511–1518. [CrossRef]
178. Fujii, M.; Takei, I.; Umezawa, K. Antidiabetic effect of orally administered conophylline-containing plant extract on streptozotocin-treated and Goto-Kakizaki rats. *Biomed. Pharmacother.* **2009**, *63*, 710–716. [CrossRef] [PubMed]
179. Kong, W.-J.; Zhang, H.; Song, D.-Q.; Xue, R.; Zhao, W.; Wei, J.; Wang, Y.-M.; Shan, N.; Zhou, Z.-X.; Yang, P.; et al. Berberine reduces insulin resistance through protein kinase C–dependent up-regulation of insulin receptor expression. *Metabolism* **2009**, *58*, 109–119. [CrossRef] [PubMed]
180. Dai, S.; Hong, Y.; Xu, J.; Lin, Y.; Si, Q.; Gu, X. Ginsenoside Rb2 promotes glucose metabolism and attenuates fat accumulation via AKT-dependent mechanisms. *Biomed. Pharmacother.* **2018**, *100*, 93–100. [CrossRef] [PubMed]
181. Sun, Z.; Sun, X.; Li, J.; Li, Z.; Hu, Q.; Li, L.; Hao, X.; Song, M.; Li, C. Using probiotics for type 2 diabetes mellitus intervention: Advances, questions, and potential. *Crit. Rev. Food Sci. Nutr.* **2019**, *60*, 670–683. [CrossRef] [PubMed]
182. Zeng, Z.; Luo, J.; Zuo, F.; Zhang, Y.; Ma, H.; Chen, S. Screening for potential novel probiotic Lactobacillus strains based on high dipeptidyl peptidase IV and α-glucosidase inhibitory activity. *J. Funct. Foods* **2015**, *20*, 486–495. [CrossRef]
183. Li, C.; Ding, Q.; Nie, S.-P.; Zhang, Y.-S.; Xiong, T.; Xie, M.-Y. Carrot Juice Fermented with *Lactobacillus plantarum* NCU116 Ameliorates Type 2 Diabetes in Rats. *J. Agric. Food Chem.* **2014**, *62*, 11884–11891. [CrossRef] [PubMed]
184. Chen, P.; Zhang, Q.; Dang, H.; Liu, X.; Tian, F.; Zhao, J.; Chen, Y.; Zhang, H.; Chen, W. Antidiabetic effect of Lactobacillus casei CCFM0412 on mice with type 2 diabetes induced by a high-fat diet and streptozotocin. *Nutrition* **2014**, *30*, 1061–1068. [CrossRef]
185. Chen, P.; Zhang, Q.; Dang, H.; Liu, X.; Tian, F.; Zhao, J.; Chen, Y.; Zhang, H.; Chen, W. Oral administration of Lactobacillus rhamnosus CCFM0528 improves glucose tolerance and cytokine secretion in high-fat-fed, streptozotocin-induced type 2 diabetic mice. *J. Funct. Foods* **2014**, *10*, 318–326. [CrossRef]
186. Jia, L.; Li, D.; Feng, N.; Shamoon, M.; Sun, Z.; Ding, L.; Zhang, H.; Chen, W.; Sun, J.; Chen, Y.Q. Anti-diabetic Effects of Clostridium butyricum CGMCC0313.1 through Promoting the Growth of Gut Butyrate-producing Bacteria in Type 2 Diabetic Mice. *Sci. Rep.* **2017**, *7*, 1–15. [CrossRef] [PubMed]
187. Kocsis, T.; Molnár, B.; Németh, D.; Hegyi, P.; Szakács, Z.; Bálint, A.; Garami, A.; Soós, A.; Márta, K.; Solymár, M. Probiotics have beneficial metabolic effects in patients with type 2 diabetes mellitus: A meta-analysis of randomized clinical trials. *Sci. Rep.* **2020**, *10*, 1–14. [CrossRef] [PubMed]
188. Cani, P.D.; Neyrinck, A.M.; Fava, F.; Knauf, C.; Burcelin, R.G.; Tuohy, K.M.; Gibson, G.R.; Delzenne, N.M. Selective increases of bifidobacteria in gut microflora improve high-fat-diet-induced diabetes in mice through a mechanism associated with endotoxaemia. *Diabetologia* **2007**, *50*, 2374–2383. [CrossRef]
189. Miraghajani, M.; Dehsoukhteh, S.S.; Rafie, N.; Hamedani, S.G.; Sabihi, S.; Ghiasvand, R. Potential mechanisms linking probiotics to diabetes: A narrative review of the literature. *Sao Paulo Med J.* **2017**, *135*, 169–178. [CrossRef]
190. Wang, F.; Jiang, H.; Shi, K.; Ren, Y.; Zhang, P.; Cheng, S. Gut bacterial translocation is associated with microinflammation in end-stage renal disease patients. *Nephrol. Carlton Vic.* **2012**, *17*, 733–738. [CrossRef] [PubMed]
191. Cani, P.D.; Amar, J.; Iglesias, M.A.; Poggi, M.; Knauf, C.; Bastelica, D.; Neyrinck, A.M.; Fava, F.; Tuohy, K.M.; Chabo, C.; et al. Metabolic endotoxemia initiates obesity and insulin resistance. *Diabetes* **2007**, *56*, 1761–1772. [CrossRef] [PubMed]
192. Hung, S.-C.; Tseng, W.-T.; Pan, T.-M. Lactobacillus paracasei subsp. paracasei NTU 101 ameliorates impaired glucose tolerance induced by a high-fat, high-fructose diet in Sprague-Dawley rats. *J. Funct. Foods* **2016**, *24*, 472–481. [CrossRef]
193. Rains, J.L.; Jain, S.K. Oxidative stress, insulin signaling, and diabetes. *Free Radic. Biol. Med.* **2011**, *50*, 567–575. [CrossRef] [PubMed]
194. Ejtahed, H.S.; Mohtadi-Nia, J.; Homayouni-Rad, A.; Niafar, M.; Asghari-Jafarabadi, M.; Mofid, V. Probiotic yogurt improves antioxidant status in type 2 diabetic patients. *Nutrition* **2012**, *28*, 539–543. [CrossRef]
195. Kim, S.-H.; Huh, C.-S.; Choi, I.-D.; Jeong, J.-W.; Ku, H.-K.; Ra, J.-H.; Kim, T.-Y.; Kim, G.-B.; Sim, J.-H.; Ahn, Y.-T. The anti-diabetic activity of *Bifidobacterium lactis* HY8101 in vitro and in vivo. *J. Appl. Microbiol.* **2014**, *117*, 834–845. [CrossRef] [PubMed]

nutrients

MDPI

Article

Standardized Ethanol Extract of *Cassia mimosoides* var. *nomame* Makino Ameliorates Obesity via Regulation of Adipogenesis and Lipogenesis in 3T3-L1 Cells and High-Fat Diet-Induced Obese Mice

So-Won Heo [1,2], Kyung-Sook Chung [1], Young-Seo Yoon [1,2], Soo-Yeon Kim [1,3], Hye-Shin Ahn [4], Yu-Kyong Shin [4], Sun-Hee Lee [4,*] and Kyung-Tae Lee [1,2,*]

1 Department of Pharmaceutical Biochemistry, College of Pharmacy, Kyung Hee University, 26 Kyungheedae-ro, Seoul 02447, Republic of Korea
2 Department of Biomedical and Pharmaceutical Science, Graduate School, Kyung Hee University, 26 Kyungheedae-ro, Seoul 02447, Republic of Korea
3 Department of Fundamental Pharmaceutical Science, Graduate School, Kyung Hee University, 26 Kyungheedae-ro, Seoul 02447, Republic of Korea
4 Department of New Material Development, COSMAXBIO, Seongnam 13486, Republic of Korea
* Correspondence: bt-shlee@cosmax.com (S.-H.L.); ktlee@khu.ac.kr (K.-T.L.); Tel.: +82-31-8018-0390 (S.-H.L.); +82-2-961-0860 (K.-T.L.)

Abstract: Obesity is a major cause of conditions such as type 2 diabetes and non-alcoholic fatty liver disease, posing a threat to public health worldwide. Here, we analyzed the anti-obesity effects of a standardized ethanol extract of *Cassia mimosoides* var. *nomame* Makino (EECM) in vitro and in vivo. Treatment of 3T3-L1 adipocytes with EECM suppressed adipogenesis and lipogenesis via the AMP-activated protein kinase pathway by downregulating the expression levels of CCAAT/enhancer-binding protein-alpha, peroxisome proliferator-activated receptor (PPAR)-γ, sterol regulatory element-binding protein-1, and fatty acid synthase and upregulating the acetyl-CoA carboxylase. EECM inhibited mitotic clonal expansion during early adipocyte differentiation. Oral administration of EECM for 10 weeks significantly alleviated body weight gain and body fat accumulation in high-fat diet (HFD)-fed mice. EECM mitigated adipogenesis and lipid accumulation in white adipose and liver tissues of HFD-induced obese mice. It regulated the levels of adipogenic hormones including insulin, leptin, and adipokine in the blood plasma. In brown adipose tissue, EECM induced the expression of thermogenic factors such as uncoupling protein-1, PPAR-α, PPARγ co-activator-1α, sirtuin 1, and cytochrome c oxidase IV. EECM restored the gut microbiome composition at the phylum level and alleviated dysbiosis. Therefore, EECM may be used as a promising therapeutic agent for the prevention of obesity.

Keywords: *Cassia mimosoides*; high-fat diet; AMPK; adipogenesis; lipogenesis; white adipose tissue; brown adipose tissue; microbiota

Citation: Heo, S.-W.; Chung, K.-S.; Yoon, Y.-S.; Kim, S.-Y.; Ahn, H.-S.; Shin, Y.-K.; Lee, S.-H.; Lee, K.-T. Standardized Ethanol Extract of *Cassia mimosoides* var. *nomame* Makino Ameliorates Obesity via Regulation of Adipogenesis and Lipogenesis in 3T3-L1 Cells and High-Fat Diet-Induced Obese Mice. *Nutrients* **2023**, *15*, 613. https://doi.org/10.3390/nu15030613

Academic Editor: Susanna Iossa

Received: 29 December 2022
Revised: 20 January 2023
Accepted: 22 January 2023
Published: 25 January 2023

1. Introduction

Incidence of obesity and overweight is very high worldwide, and recent statistics estimate that more than 1.9 billion adults are overweight worldwide, of which more than 650 million adults develop obesity. As the prevalence of obesity increases, the risk of various diseases and complications such as type 2 diabetes (T2D), non-alcoholic fatty liver disease (NAFLD), hypertension, cardiovascular disease (CVD), insulin resistance, and some types of cancers also increases [1–3]. Obesity ultimately results in hypertrophy and hyperplasia of adipocytes and adipogenesis via an imbalance between energy intake and expenditure [4]. Regulation of adipogenesis may be an effective therapeutic strategy for treating obesity. Therefore, it is important to identify potential adipogenesis-regulating anti-obesity drugs.

Mitotic clonal expansion (MCE) is an essential process during the early differentiation of adipocytes. After the proliferation of adipocytes, MCE is initiated, which induces intracellular lipid accumulation, resulting in an adipocyte phenotype [5]. Adipogenesis and lipogenesis are modulated mainly by AMP-activated protein kinase (AMPK), a major regulator of the cellular energy balance pathway [6]. AMPK activity suppresses the expression of crucial adipogenic transcription factors, including CCAAT/enhancer-binding protein alpha (C/EBPα), peroxisome proliferator-activated receptor gamma (PPARγ), and sterol regulatory element-binding protein (SREBP)-1 [7]. In addition, AMPK regulates the expression of lipolysis and lipogenesis proteins such as acetyl-CoA carboxylase (ACC) and fatty acid synthase (FAS) [8]. Therefore, inhibition of adipogenic and lipogenic signaling of adipocytes via the AMPK pathway may potentially alleviate obesity.

Body weight gain is often induced by white adipose tissue (WAT), which stores nutrient-derived energy [9]. In contrast, brown adipose tissue (BAT) plays an important role in energy expenditure via thermogenesis and is composed of more mitochondria than WAT [10]. Thermogenesis in BAT is induced by the upregulation of mitochondrial factors, such as uncoupling protein-1 (UCP-1), cytochrome c oxidase (COX)-IV, and thermogenic proteins, such as PPARγ co-activator-1α (PGC-1α), sirtuin 1 (SIRT1), and PPARα. Interestingly, the activation of thermogenesis in BAT appears to inhibit body weight gain [11] and diabetes status [12].

In human intestine, the gut microbiota consists of various bacterial phyla such as Firmicutes, Bacteroidetes, Proteobacteria, Actinobacteria, and Verrucomicrobia. Moreover, Firmicutes and Bacteroidetes, two dominant phyla in the gut microbiota, constitute more than 90% of the total bacterial community [13]. The Firmicutes-to-Bacteroidetes (F/B) ratio has a critical effect on the maintenance of homeostasis in the intestinal environment. Interestingly, a rapid change in the F/B ratio can lead to intestinal microbial imbalance, called dysbiosis, which correlates with obesity [14]. Therefore, it is important to develop novel anti-obesity drugs and evaluate their effects on the gut microbiome.

Various synthetic drugs exert anti-obesity effects, but their application is limited by certain side effects, such as diarrhea, CVD, and dizziness. Therefore, many studies have focused on developing anti-obesity drugs based on natural products with relatively low side effects. *Cassia mimosoides* var. *nomame* Makino grows in most regions of Korea and is also distributed in Japan and China. It is used in herbal tea, and its fruit is used for medicinal purposes. Previous studies have reported that the 10% methanol extract of *C. mimosoides* var. *nomame* Makino attenuates myocardial injury [15] and brain damage [16] by inhibiting apoptosis. However, the anti-obesity effects of *C. mimosoides* var. *nomame* Makino and its molecular signaling pathway have not yet been reported. Therefore, we investigated the effects of the 30% ethanol extract of *C. mimosoides* var. *nomame* Makino (EECM) on obesity and clarified the underlying mechanisms in 3T3-L1 cells and high-fat diet (HFD)-induced obese animal models.

2. Materials and Methods

2.1. EECM Preparation, Identification, and Standardization

The leaves of *C. mimosoides* var. *nomame* Makino were collected from Youngwolgun (Gangwon-do, Republic of Korea). *C. mimosoides* var. *nomame* Makino was identified by miDNA Genome Research Institute (Gunsan, Jeollabuk-do, Republic of Korea). A voucher specimen (COS2007) was stored and extracted according to the instruction of COSMAX BIO (Seongnam, Gyeonggi-do, Republic of Korea). After 30% ethanol extraction, the yield of 30% ethanol extract of *C. mimosoides* var. *nomame* Makino (EECM) was 19.66% (dried leaves of *C. mimosoides* var. *nomame* Makino; 1.0 kg, extract residue; 196.6 g). Ultra-performance liquid chromatography (UPLC) analysis was performed with an Agilent 6545 Q-TOF LC/MS spectrometer (Agilent Technologies, Santa Clara, CA, USA). The chromatographic separation was achieved on EclipsePlus C18 (1.8 µm, 50 × 2.1 mm, Agilent Technologies) and column temperature at 35 °C. The mobile phase consisted of acetonitrile (solvent A) and 0.1% formic acid in water (solvent B). The gradient condition of the mobile phase was 0–10 min, 10% to 100%; 10–15 min, 100% to 100%; 15–20 min, 100% to 10%, as a percentage of solvent A. The flow rate was set at 0.7 mL/min, and the injection volume was 1.0 μL. The monitored wavelength of the UV detector was 210, 254, and 315 nm. The conditions of MS analysis in the positive or negative ion mode were as follows: mass range: 100–1700 m/z; acquisition rate: 1 spectra/s; acquisition time: 100 ms/spectrum; gas temperature: 320 °C; drying gas: 8 L/min; nebulizer: 35 psi; sheath gas temperature: 350 °C; sheath gas flow: 11 L/min; Vcap: 3500 V; MS fragmentor: 100 V. The gradient mobile phases was set up as follows: solvent A: 100% acetonitrile and solvent B: distilled water with a gradient elution as follows: 0 to 40 min, 10% to 30%; 40 to 50 min, 30% to 100%; 50 to 55 min, 100% to 100%; 55 to 60 min, 100% to 10% as a percent of solvent A at a flow rate of 1.0 mL/min.

2.2. *Differentiation of Adipocytes*

The cell culture of 3T3-L1 mouse preadipocyte and differentiation was performed as in our previous report [7]. During the differentiation of 3T3-L1 mouse preadipocyte, the media were changed to DM containing 1 μM insulin with or without EECM (6.25, 12.5, or 25 μg/mL) every 2 days and processed repeatedly for 7 days.

2.3. *Cell Viability*

Cells were cultured at 1 × 10^5 cells/mL into 24-well plates with GM for 3 days. Then, the medium was changed to DM including MDI with or without EECM (6.25, 12.5, or 25 μg/mL) and the treatment of EECM until 7 days. To examine cell viability of the EECM, the tetrazolium compound [3-(4, 5-dimethylthiazol-2-yl)-5-(3-carboxymethoxyphenyl)-2-(4-sulfophenyl)-2H-tetrazolium, inner salt; MTS] solution (Promega Corporation, Madison, WI, USA) was treated with 10% of the media volume in each well and then incubated for 2 h. After MTS treatment, the measurement of absorbance was performed at 490 nm.

2.4. *Oil Red O Staining*

After the differentiation of 3T3-L1 cells, the cells were then fixed with 4% formaldehyde in PBS for 40 min at 25 °C. 0.2% Oil Red O in isopropanol was diluted with DW and filtered using filter paper. The fixed cells were dried completely and then stained with Oil Red O solution for 1 h in dark conditions. The stained intracellular lipid droplets were observed with a microscope (OLYMPUS CORPORATION, Tokyo, Japan). In addition, the intracellular lipid droplets were liquified with 100% isopropanol and quantified by measuring the absorbance at 510 nm.

2.5. *Western Blot Analysis*

Protein lysates were prepared by PRO-PREP™ solution (Intron Biotechnology, Seoul, Korea) from cells or tissues. The lysate was reacted on ice, centrifuged at 13,000× *g* rpm at 4 °C, and the protein-contained supernatant was obtained. After the quantification of proteins (30 μg) using the Bradford assay, proteins were separated by SDS-PAGE and then transferred onto the PVDF membrane. The PVDF membranes were treated with the primary antibodies in skim milk solution for 18 h at 4 °C. The immunoblotted membranes were reacted with the secondary antibodies in skim milk solution at 25 °C and then detected by our previous report [7].

2.6. *MCE Determination*

For the analysis of MCE, propidium iodide (PI) staining was implemented. The 3T3-L1 cells were cultured with or without EECM treatment for 24 h and harvested. Subsequently, the cells were resuspended in PBS and fixed for 18 h in 70% cold ethanol at −20 °C. After fixation, the cells were stained with 100 μg/mL PI staining buffer containing RNase for 30 min in dark conditions. PI stained cells were analyzed by our previous report [17].

2.7. HFD-Induced Obese Animals Experiments

C57BL/6J mice (male, 5–6 weeks, 20–22 g) were purchased from Orient Bio Inc. (Seongnam, Republic of Korea) and cared for under standard conditions (light–dark cycle: 12 h, temperature: 22 ± 1 °C, humidity 40–60%) for 7 days. Mice were classified into 5 groups (n = 8/group) to inspect the anti-obesity effect of EECM: (i) control (CON) group, (ii) 60% HFD-induced obese group, (iii) HFD + orlistat administration group, (iv) HFD + EECM administration group (EECM 100 mg/kg, p.o.), and (v) HFD + EECM administration group (EECM 300 mg/kg, p.o.). The CON group was fed a normal diet, and the other group was fed a 60% HFD to induce obesity for 10 weeks. Simultaneously with the diet feeding, the CON and HFD groups were administered with vehicle, and orlistat and EECM groups were administered orlistat (20 mg/kg, p.o.) and EECM (100 or 300 mg/kg, p.o.), respectively. All administration of samples was daily provided to experimental animals. The body weight was measured 2 days per week. At the end of the experiment, sacrifice was performed surgically, and obtained tissues were immediately stored at −80 °C.

2.8. Body Fat Composition Analysis

The body fat composition of experimental animals was determined using dual-energy X-ray absorptiometry (DEXA) (InAlyzer, Medikors, Seongnam, Republic of Korea). The lean tissue, fat tissue, and differentiated tissue from lean to fat were exposed as blue, red, and green/yellow, respectively.

2.9. Hematoxylin and Eosin (H&E) Staining Analysis

White adipose tissue (WAT) and liver were obtained after mice sacrifice and then fixed with 4% formaldehyde in PBS buffer for 24 h. Fixed tissues were sliced and embedded into a paraffin block. H&E staining was performed, and then, stained tissue was observed with a light microscope optically, and the adipocyte size was measured by using an instrument of the microscope system (cell Sense standard ver.1.9, OLYMPUS CORPORATION, Tokyo, Japan).

2.10. Biochemical Examination of Blood Plasma

At the sacrifice, the blood of mice was collected from a vein and placed in heparin-coated tubes. The preparation and biochemical examination of plasma were followed by the guideline of T&P Bio Co. (Gwangju, Republic of Korea).

2.11. Genomic DNA Extraction and Microbiome Taxonomic Profiling (MTP)

Total genomic DNA (gDNA) from stool was extracted with the QIAamp® Fast DNA Stool Mini Kit (Qiagen, Hilden, Germany) as the instruction of manufacture. Then, the bacterial 16S ribosomal RNA V3-V4 region was amplified by PCR. The amplified product was performed purification by magnetic bead reaction and detected by electrophoresis on agarose gel (1.5%). Whole products were pooled at a constant concentration to produce a library. The library was quantified at 2 nM using KAPA Library Quantification Kit Illumina® platforms (Roche, Basel, Switzerland), and the sequence was analyzed using the Illumina iSeq 100 sequencing system (Illumina, San Diego, CA, USA) finally.

2.12. Statistical Analysis

Statistical values were calculated by the Prism Graph Pad 5.0 program (GraphPad Software Inc., San Diego, CA, USA). All data were expressed as mean ± SD for in vitro experiments and SEM for in vivo experiments. The analysis was performed by Tukey's range test. p-values of less than 0.05 were considered statistically significant.

3. Results

3.1. Identification of Quercitrin and Standardization of EECM via UPLC–High-Resolution Mass Spectrometry (HR/MS)

The UPLC chromatogram of EECM at 254 nm is shown in Figure 1A. By comparing the retention time, MS data, and UV/vis spectra with those presented in previous literature reports, the major peak at a retention time of 4.033 min [M + H] ions (m/z 499) was identified as quercitrin (Figure 1A,B). Qualitative analysis confirmed quercitrin by comparing the HPLC retention time data with the analytical standard of quercitrin (phytolab, GmbH & Co., Germany; Figure 1C,D).

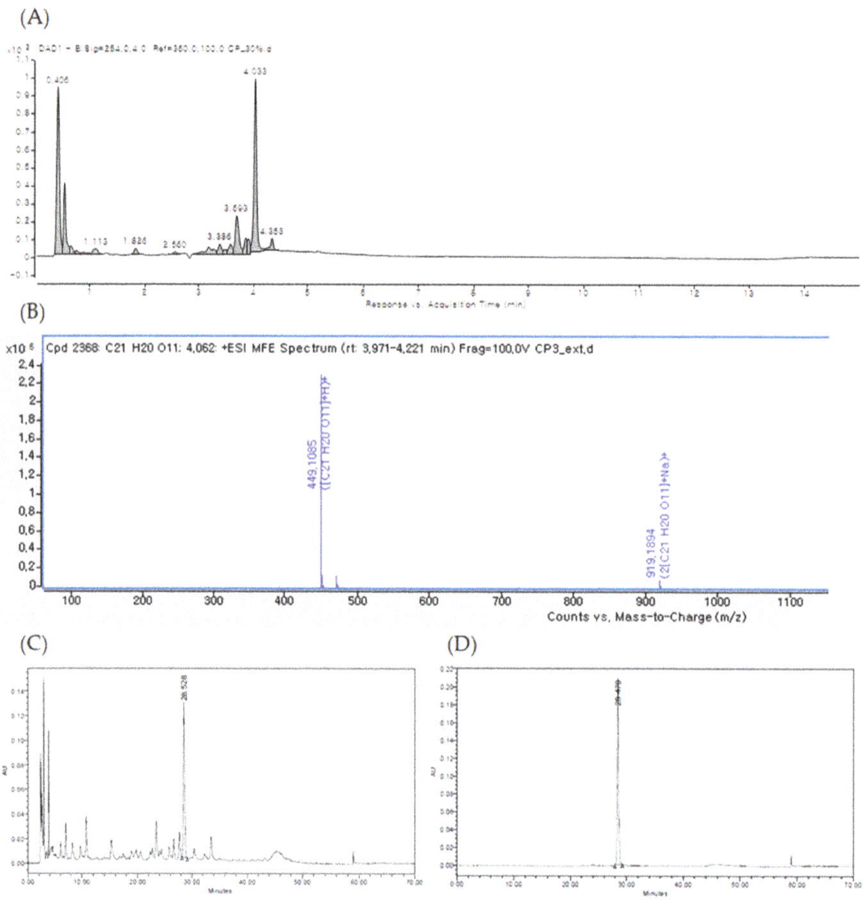

Figure 1. (**A**) UPLC of the ethanol extract of *C. mimosoides* var. *nomame* Makino (EECM) detected at 254 nm. (**B**) High-resolution mass spectrometry (HR/MS) spectrum of [M + H] ions (m/z 499) of quercitrin (Rt: 4.033 min). High-performance liquid chromatography (HPLC) profiles of (**C**) EECM and (**D**) standard quercitrin.

3.2. EECM Inhibits Differentiation and Regulates Lipogenesis and Adipogenesis via AMPK Pathway in 3T3-L1 Preadipocytes

To investigate the effects of EECM on 3T3-L1 preadipocytes, we first performed the MTS assay at various concentrations (Figure S1). Cell viability was shown to be less than 80% with EECM concentration of 50 or 100 µg/mL, so we used 6.25 to 25 µg/mL of EECM for further experiments. After differentiation of 3T3-L1 cells with EECM, cell viability was shown to be more than 80% at all concentrations up to 25 μg/mL. Thus, we performed the following experiments with EECM concentrations of 6.25–25 μg/mL. We assessed the effect of EECM on intracellular lipid accumulation via Oil Red O staining. EECM visually and quantitatively inhibited lipid accumulation during investigation of the differentiated 3T3-L1 cells (Figure 2A,B). Next, we examined the regulation of EECM on the protein expression levels of phosphorylated AMPKα, lipogenic enzymes including ACC and FAS, and adipogenic transcription factors including C/EBPα, PPARγ, and SREBP-1. Phosphorylation of AMPKα (Thr172) decreased with MDI treatment and was restored by EECM (Figure 2C). Moreover, EECM increased the expression levels of phospho-ACC (Ser79) and pre-SREBP-1 and reduced the expression of total ACC, FAS, C/EBPα, and PPARγ in the differentiated cells (Figure 2D,E). Our data revealed that EECM disturbs lipogenesis and adipogenesis during adipocyte differentiation by regulating the AMPK pathway.

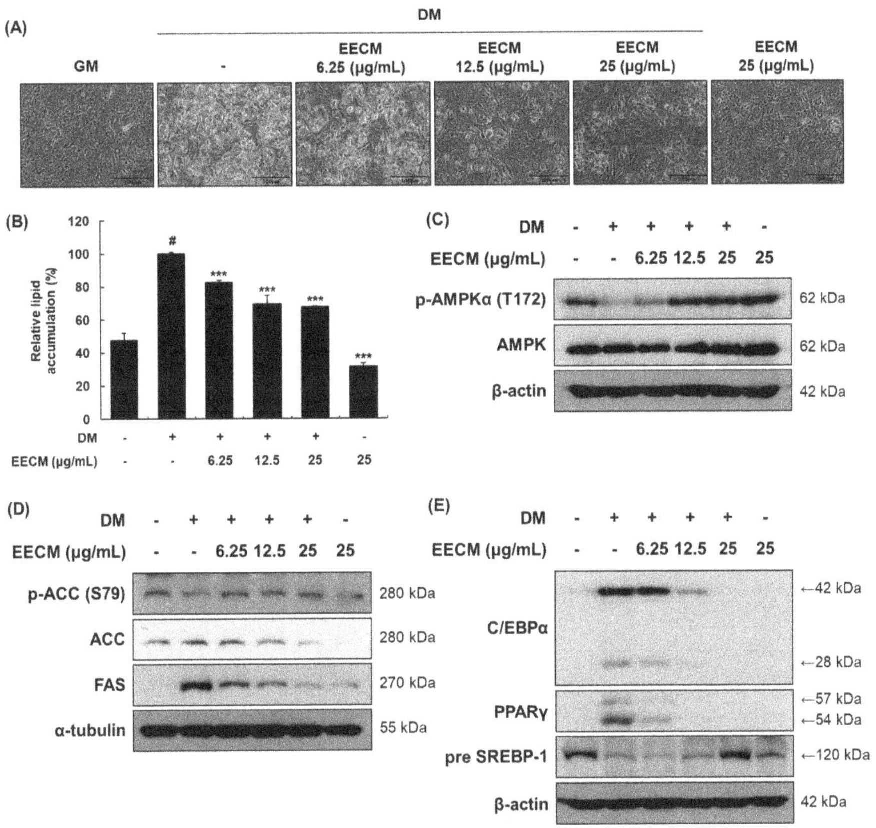

Figure 2. Inhibitory effects of EECM on intracellular lipid accumulation, adipogenesis, and lipogenesis in 3T3-L1 cells. The 3T3-L1 cells were cultured in the growth medium (GM) for four days. After the cells reached confluency, they were cultured with or without methylisobutylxanthine, dexamethasone, and insulin (MDI) and various concentrations of EECM. (**A**) Relative microscopic images of Oil-Red-O-stained 3T3-L1 cells. (**B**) Quantitative lipid accumulation ratio. After differentiation, total protein was extracted and detected by via Western blotting analysis. Expression levels of (**C**) AMPKα pathway proteins and (**D**) lipogenic and (**E**) adipogenic transcription factors. Values are represented as the mean ± standard deviation (SD). # $p < 0.05$ vs. the GM group; *** $p < 0.001$ vs. the differentiation medium (DM) group.

3.3. EECM Suppresses MCE during Early Differentiation of 3T3-L1 Preadipocytes

When the cells are differentiated by MDI treatment, MCE progresses to the early stage. To demonstrate the inhibitory effect of EECM on MCE, we conducted cell cycle analysis after propidium iodide (PI) staining using flow cytometry. In Figure 3A, which shows 3T3-L1 preadipocytes differentiation with MDI, the G0/G1 phase ratio decreased considerably (from 71.1 to 47.5%), and the G2/M phase ratio increased accordingly (from 19.9 to 41.6%). The changes in this ratio suggested that the cell differentiation was initiated by MDI treatment, and MCE was successfully induced. In contrast, EECM induced an increase in the G0/G1 phase ratio compared to GM treatment (from 47.5 to 48.3, 54.6, and 57.3%, respectively) while reducing the G2/M phase ratio (from 41.6 to 41.5, 34.6, and 29.7%, respectively) in a concentration-dependent manner. Additionally, we investigated the protein expression levels of cell-cycle-related factors following EECM treatment in preadipocytes via Western blotting analysis. Expression levels of p21, a cell cycle inhibitor, decreased but showed a tendency to increase after EECM treatment. In contrast, the expression levels of cyclin D1, cyclin E1, cyclin B1, CDK6, and CDK2, indicating cell-cycle-related regulators, were increased after MDI treatment compared to those observed after GM treatment, but treatment with EECM, especially at high concentrations (25 μg/mL), markedly decreased the expression levels during the differentiation of adipocytes (Figure 3B).

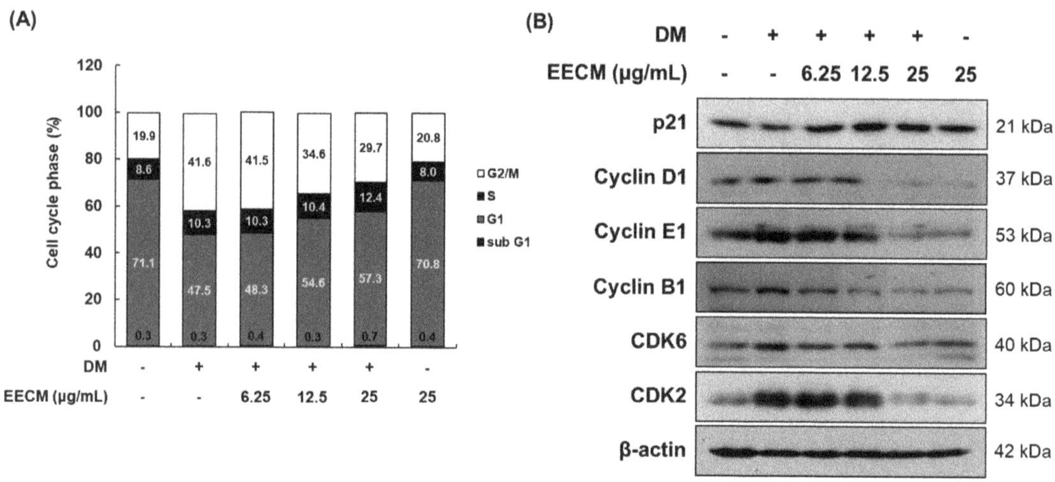

Figure 3. Regulatory effect of EECM on the MCE in 3T3-L1 cells. Differentiated cells were treated with EECM (6.25, 12.5, and 25 μg/mL) for 24 h. (**A**) Cell cycle phase ratio and (**B**) cell-cycle-related proteins expression were measured via propidium iodide (PI) staining and Western blotting, respectively.

3.4. EECM Obstructs the Body Weight Gain of HFD-Induced Obese Mice

We next investigated whether EECM alleviated body weight gain and body fat accumulation in the tissues of HFD-induced obese mice. Body weight was considerably increased in the HFD group compared to the normal diet (CON) group during the 10-week administration period, whereas EECM administration significantly reduced the body weight gain from three weeks (Figure 4A). In addition, HFD-induced final body weight (HFD group: 42.24 ± 1.94 g vs. EECM 100 mg/kg group: 35.94 ± 0.83 g, $p < 0.001$; EECM 300 mg/kg group: 35.53 ± 1.06 g, $p < 0.001$) and weight gain (HFD group: 18.64 ± 0.44 g vs. EECM 100 mg/kg group: 12.99 ± 0.88 g, $p < 0.001$; EECM 300 mg/kg group: 12.91 ± 0.99 g, $p < 0.001$) were significantly attenuated by EECM treatment (Figure 4B,C).

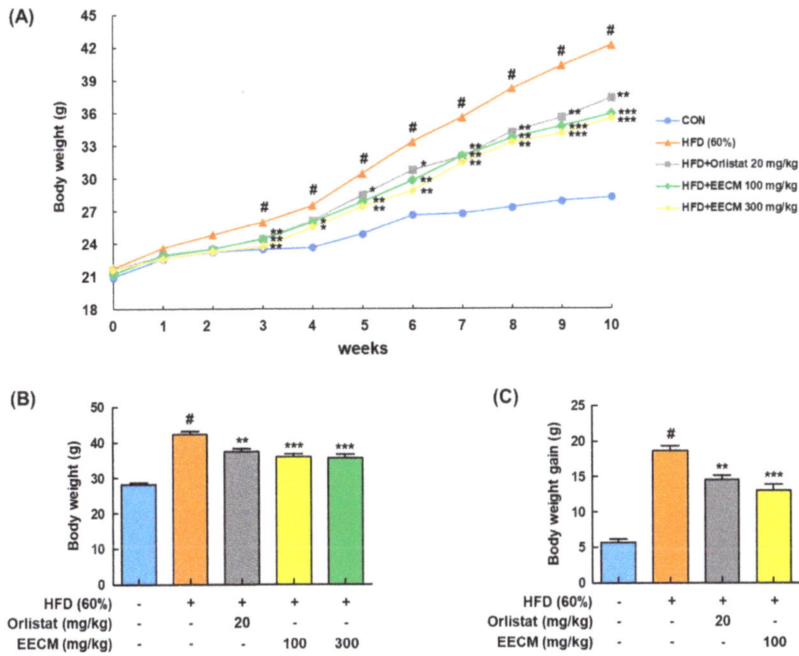

Figure 4. Repressive effect of EECM on body weight gain in HFD (60%)-induced obese mice. Mice of EECM treatment were provided the indicated dose of EECM (100 and 300 mg/kg) daily during the experiment. (**A**) Body weight changes, (**B**) total body weight at the endpoint of the experiment, and (**C**) total body weight gain for 10 weeks. Values are represented as the mean ± SEM of eight mice per group. # $p < 0.05$ vs. the control (CON) group; * $p < 0.05$, ** $p < 0.01$, and *** $p < 0.001$ vs. the HFD (60%)-induced obese group.

3.5. EECM Changes the Fat Composition in HFD-Induced Obese Mice

For the identification of fat composition by EECM administration, we analyzed the body composition using DEXA in HFD-induced obese mice. As shown in Figure 5A,B, body fat accumulation was significantly induced in the HFD group compared to the CON group (CON group: 5.81 ± 0.68 g vs. HFD group: 18.38 ± 0.86 g, $p < 0.05$). On the contrary, EECM inhibited fat accumulation similarly to in the orlistat group, representing a positive control, and effectively improved the total body fat weight (HFD group: 18.38 ± 0.86 g vs. orlistat group: 13.97 ± 1.33 g, $p < 0.05$; EECM 100 mg/kg group: 12.20 ± 1.55 g, $p < 0.05$; EECM 300 mg/kg group: 10.73 ± 0.63 g, $p < 0.01$). In addition, EECM reduced the fat pad weight gain of subcutaneous, mesenteric, gonadal, and perirenal fat compared to that in the HFD group in a dose-dependent manner (Figure 5C–F), indicating that the daily supplementation of EECM can suppress body weight gain via the regulation of body fat composition.

3.6. EECM Alleviates the Production of Insulin, Adipokines, and Lipid Parameters in Blood Plasma of HFD-Induced Obese Mice

Next, to analyze the effects of EECM on the production of insulin and adipokines in plasma, including leptin and adiponectin, and lipid profiles in HFD-induced obese mice were investigated. HFD group had significantly higher levels of insulin and leptin than the CON group, whereas administration of EECM reduced those levels similarly to the orlistat group (Figure 6A,B). Levels of adiponectin decreased in the HFD group compared to those in the CON group but were significantly recovered by EECM treatment (Figure 6C). As shown in Table 1, EECM administration significantly ameliorated HFD-induced TC, TG, and LDL levels but had no significant effects on LDL and HDL levels.

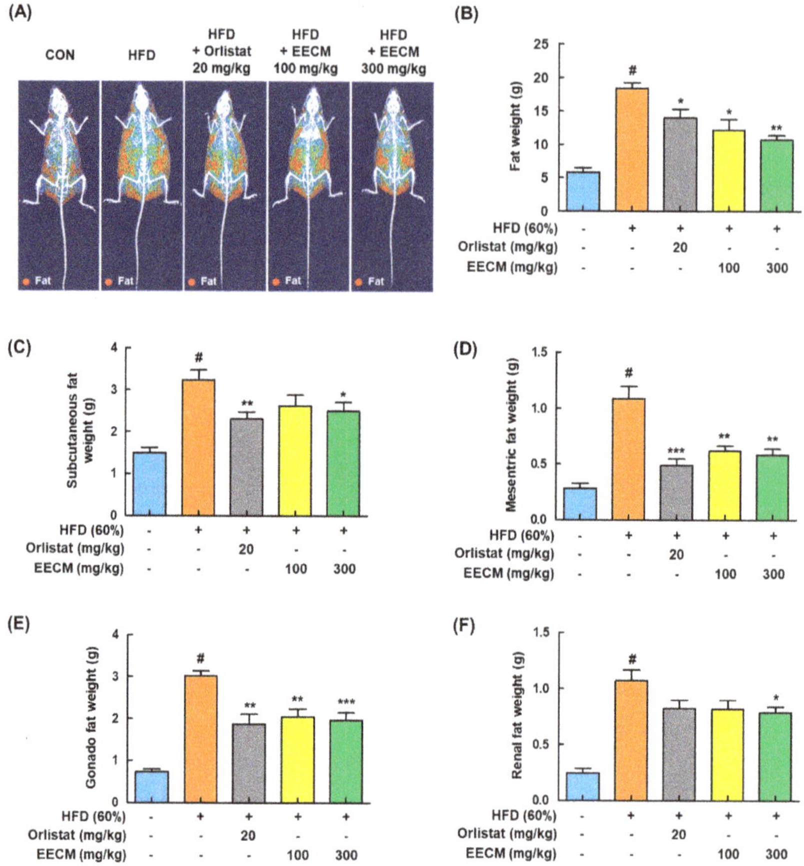

Figure 5. Regulatory effects of EECM on body composition and fat mass in HFD (60%)-induced obese mice. (**A**) Radiography images of the fat tissues are presented. (**B**) Fat weight in each group was measured via DEXA analysis. Fat was obtained from each tissue after the sacrifice of mice, and the weight was measured. (**C–F**) Subcutaneous fat, mesenteric fat, gonadal fat, and perirenal fat weights. Values are represented as the mean ± SEM (DEXA analysis; $n = 3$/group, fat in tissue weight: $n = 8$/group). # $p < 0.05$ vs. the CON group; * $p < 0.05$, ** $p < 0.01$, and *** $p < 0.001$ vs. the HFD (60%)-induced obese group.

Table 1. Effects of the ethanol extract of *Cassia mimosoides* var. *nomame* Makino (EECM) on the lipid parameters in HFD (60%)-induced obese mice.

Lipid Profiles	CON	HFD	Orlistat (20 mg/kg)	EECM (100 mg/kg)	EECM (300 mg/kg)
TC (mg/dL)	104.88 ± 5.08	157.00 ± 3.14 #	130.25 ± 13.19 **	145.00 ± 6.54	142.38 ± 4.04 *
TG (mg/dL)	55.50 ± 4.18	112.25 ± 5.87 #	67.75 ± 8.35 **	70.00 ± 9.14 **	63.00 ± 9.58 ***
LDL (mg/dL)	10.13 ± 0.81	15.00 ± 0.50 #	11.88 ± 0.91 *	12.63 ± 0.53	12.63 ± 0.46
HDL (mg/dL)	85.75 ± 2.10	92.88 ± 4.97	113.13 ± 9.27	102.25 ± 3.22	100.50 ± 2.05

Values are represented as the mean ± standard error of the mean (SEM) ($n = 8$/group). # $p < 0.05$ vs. the CON group; * $p < 0.05$, ** $p < 0.01$, and *** $p < 0.001$ vs. the HFD (60%)-induced obese group.

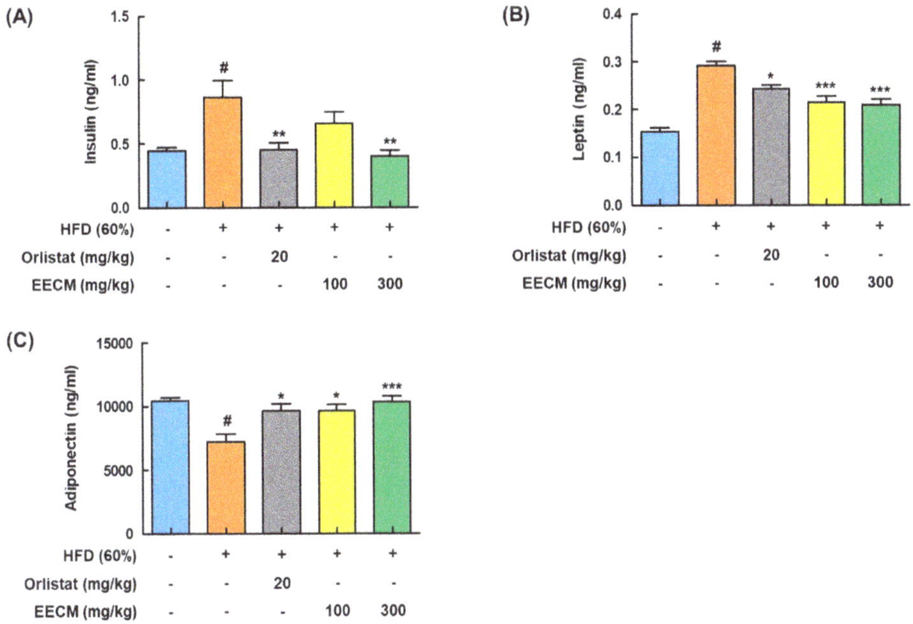

Figure 6. Regulatory effect of EECM on the production of adipocyte-released hormones in HFD (60%)-induced obese mice. Plasma levels of (**A**) insulin, (**B**) leptin, and (**C**) adiponectin were detected using ELISA kit. Values are represented as the mean ± SEM (n = 8/group). # $p < 0.05$ vs. the CON group; * $p < 0.05$, ** $p < 0.01$, and *** $p < 0.001$ vs. the HFD (60%)-induced obese group.

3.7. EECM Prevents Hypertrophy of Lipid Droplets and Adipogenesis in Subcutaneous Fat of HFD-Induced Obese Mice

We compared the histological construction of subcutaneous fat pads in each group using H&E staining. In the HFD group, lipid droplets of adipocytes were extremely enlarged in size compared to those in the CON group (CON group: 40.68 ± 1.78 μM vs. HFD group: 136.81 ± 4.50 μM, $p < 0.001$); conversely, EECM administration abated the magnification of adipocytes according to the increase in concentration (HFD group: 136.81 ± 4.50 μM vs. EECM 100 mg/kg group: 48.13 ± 2.67 μM, $p < 0.001$; EECM 300 mg/kg group: 41.70 ± 3.49 μM, $p < 0.001$; Figure 7A,B). Based on histological analysis, we verified the protein expression levels of adipogenesis-related proteins such as AMPKα, lipogenic enzymes, and adipogenic transcription factors. Phosphorylation of AMPKα (Thr172) was downregulated by HFD but restored by EECM administration (Figure 7C). Consistent with the in vitro data, EECM also upregulated the HFD-induced ACC phosphorylation (Ser79) (Figure 7D). Moreover, adipogenic factors including PPARγ, C/EBPα, and mature-SREBP-1 were expressed at higher levels in the HFD group than in the CON group, but EECM treatment lowered these levels (Figure 7E). Our data indicate the inhibitory effect of EECM on adipocyte hypertrophy is due to the regulation of adipogenesis-related proteins in the subcutaneous fat tissues of HFD-induced obese mice.

3.8. EECM Ameliorates Lipid Accumulation in Liver Tissue of HFD-Induced Obese Mice

To verify the repressive effect of EECM on lipid accumulation leading to fatty liver, we performed H&E staining of the liver tissues of the animal model. As shown in Figure 8A, HFD induced lipid accumulation in the liver tissues, whereas EECM treatment restored this change remarkably and was similar to the positive control, the orlistat group. In addition, EECM significantly suppressed the weight gain of liver tissue in HFD-induced obese mice (HFD group: 1.59 ± 0.04 g vs. EECM 100 mg/kg group: 1.18 ± 0.03 g, $p < 0.001$;

EECM 300 mg/kg group: 1.11 ± 0.06 g, $p < 0.001$; Figure 8B). In line with the results for subcutaneous fat, the levels of phospho-AMPKα (Thr172), phospho-ACC (Ser79), pre-SREBP-1, PPARγ, and C/EBPα were also downregulated following EECM administration (Figure 8C–E). These results indicate that EECM prevents liver steatosis by regulating the AMPK pathway as well as lipogenesis and adipogenesis in HFD-induced obese mice.

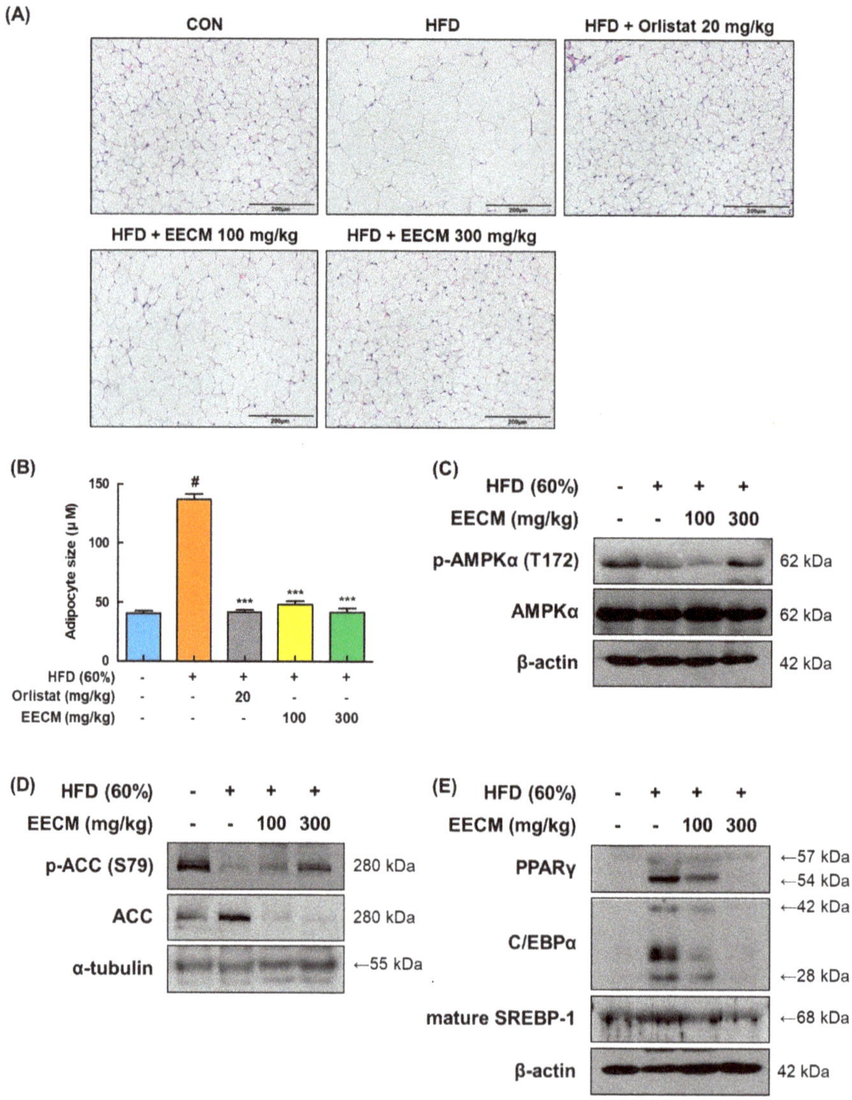

Figure 7. Effects of EECM on adipocyte size, adipogenesis, and lipogenesis in the subcutaneous fat tissues of HFD (60%)-induced obese mice. (**A**) Histological analysis of subcutaneous fat tissue via H&E staining at ×20 magnification. (**B**) Representative diameter of adipocytes in the tissue. Expression levels of (**C**) p-AMPKα, (**D**) lipogenic proteins, and (**E**) adipogenic transcription factors were determined via Western blotting analysis. Values are represented as the mean ± SEM (n = 8/group). # $p < 0.05$ vs. the CON group; *** $p < 0.001$ vs. the HFD (60%)-induced obese group.

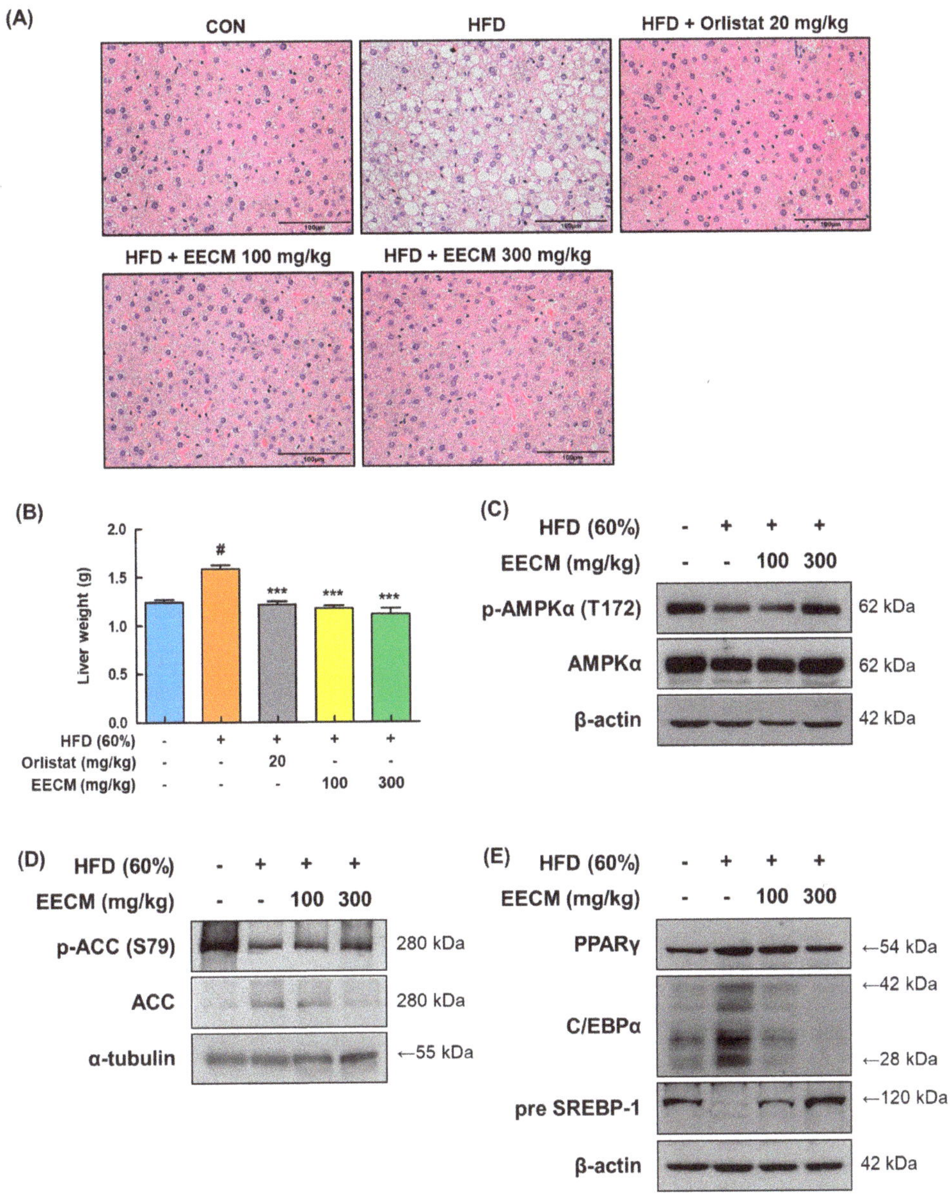

Figure 8. Effect of EECM on lipid accumulation in the liver tissues of HFD (60%)-induced obese mice. (**A**) The images of H&E stained the liver tissues at ×40 magnification. (**B**) Comparison of liver weights of each group. Expression levels of (**C**) p-AMPKα, (**D**) lipogenic proteins, and (**E**) adipogenic transcription factors were determined via Western blotting analysis. Values are represented as the mean ± SEM (n = 8/group). # $p < 0.05$ vs. the CON group; *** $p < 0.001$ vs. the HFD (60%)-induced obese group.

3.9. EECM Upregulates Thermogenic-Mediated Proteins Levels in Brown Adipose Tissue (BAT) of HFD-Induced Obese Mice

Mitochondrial thermogenesis in BAT provides protection against fat accumulation owing to its energy-dissipating activity [18]. After the mice were sacrificed, EECM administration slightly reduced the BAT weight compared to that in the HFD group (Figure 9A). Expression levels of major thermogenic-related factors, including PGC-1α, PPARα, SIRT1, and UCP-1, were markedly upregulated after EECM treatment compared to those in the HFD group. Additionally, EECM enhanced the protein levels of COX-IV and ATP synthase in the mitochondria (Figure 9B). These data imply that EECM-induced thermogenesis results in increased energy expenditure in the BAT of HFD-fed mice.

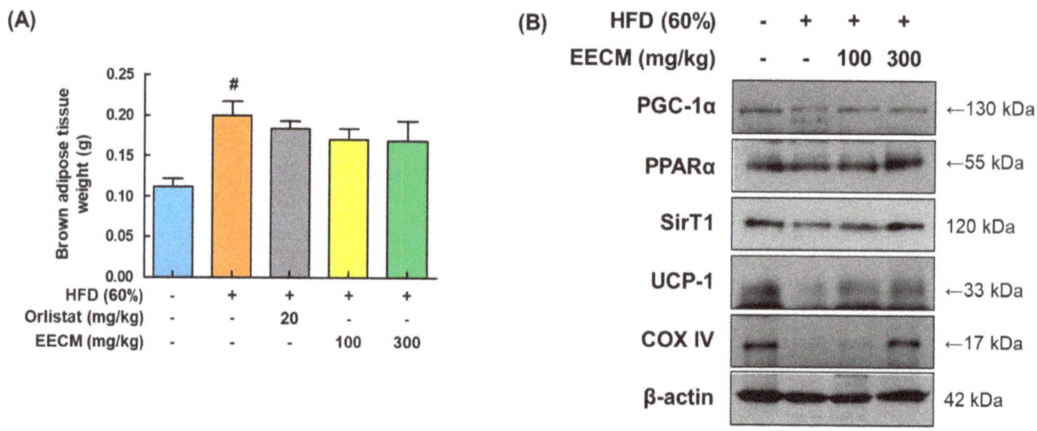

Figure 9. Stimulating effects of EECM on thermogenesis in the BAT. (**A**) Weights of BAT in each group. (**B**) Expression levels of thermogenesis-related proteins PGC-1α, PPARα, SIRT1, UCP-1, and COX-IV were determined via Western blotting analysis. Values are represented as the mean ± SEM (n = 8/group). $^{\#}$ $p < 0.05$ vs. the CON group.

3.10. EECM Regulates the Gut Microbiota Composition and Alleviates the Dysbiosis in HFD-Induced Obese Mice

Finally, we evaluated the effects of EECM on the composition and colonization of gut microbiota in an animal model. To assess the variation in the phylogenetic relationship of microbial clusters between each group, we confirmed the principal coordinate analysis plot based on β-diversity. In the plot, the HFD group data were formed by shifting from the CON group community, and the groups treated with EECM were clustered away from the HFD group (Figure 10A). Next, we analyzed the phylum-level composition of the microbiome in each group (Figure 10B–E). HFD feeding increased the relative F/B ratio and decreased the relative ratio of Bacteroidetes, whereas EECM administration inversely regulated the ratio of each microbiota at the phylum level. Dysbiosis is frequently observed in obese individuals, and a large change in the F/B ratio is considered to be an indicator of dysbiosis [14]. Here, the F/B ratio significantly increased in the HFD group compared to that in the CON group, but EECM moderated this ratio compared to that in the HFD group (Figure 10F). These results indicate that EECM regulates the shift in gut microbiome composition and prevents dysbiosis in HFD-induced obese mice.

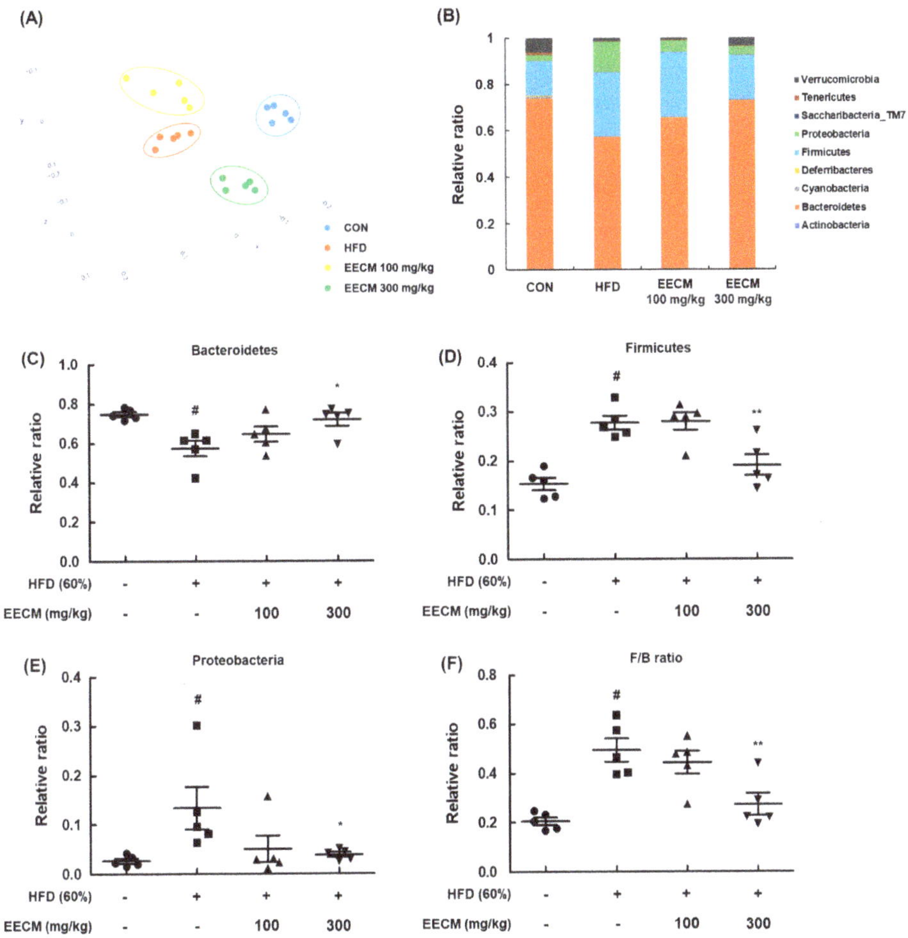

Figure 10. Regulatory effects of EECM on microbiota composition and colonization. (**A**) β-diversity between groups was analyzed using the principal coordinate analysis (PCoA) plot. (**B**) Ratio of phylum level composition ratio of each group. (**C–F**) Relative ratio of Bacteroidetes, Firmicutes, and Proteobacteria and the Firmicutes/Bacteroidetes (F/B) ratio. Values are represented as the mean ± SEM of five mice per group. # $p < 0.05$ vs. the CON group; * $p < 0.05$ and ** $p < 0.01$ vs. the HFD (60%)-induced obese group.

4. Discussion

Being overweight and obese are major risk factors for metabolic syndromes, CVD, T2D, and some cancer types [19]. According to the World Health Organization, the number of obese people nearly tripled between 1975 and 2016 [20]. Although there are various anti-obesity drugs, their limited effects and possible side effects demand a constant need to find novel and safe compounds for anti-obesity treatment. Natural-product-derived anti-obesity therapeutics are less toxic than synthetic drugs [21], and some drugs for obesity complications are derived from natural product materials [22]. *C. mimosoides* var. *nomame* Makino, a plant that grows naturally in Korea, Japan, and China, has the potential to be developed as a functional food material due to its scavenging of reactive oxygen species [15]. However, the methanol extract was not used because it is well-known that the toxicity of methanol is serious and cannot be used as an extraction solvent for development in health functional foods. Therefore, methanol extraction was excluded and ethanol used instead, as this experiment was designed to evaluate an efficacy study for the further development of health functional food. In the present study, EECM treatment not only resulted in a remarkable reduction in the number and diameter of 3T3-L1 preadipocytes during differentiation but also a decrease in body weight gain and fat weight in HFD-induced obese mice. Interestingly, as EECM administration did not affect the levels of GPT, GOT, and BUN in the blood plasma compared to those in the control or HFD groups (Figure S2), we believe that EECM does not cause considerable damage to liver and kidney function. Our data revealed EECM as a natural product with great potential for anti-obesity treatment. We further examined its target signaling activity in vitro and in vivo.

Insulin is a hormone secreted by pancreatic β cells, and the regulation of insulin clearance in the plasma is weak in obese subjects and patients with T2D [23]. Insulin concentration in the plasma of overweight individuals has been shown to correlate with plasma leptin levels [24]. Leptin, an adipokine secreted by adipocytes, is mainly associated with obesity, and the degree of leptin resistance in the blood plasma of most obese individuals is low [25]. Among the major adipokines, adiponectin plays an important role in obesity and influences weight loss; therefore, it has the potential to treat insulin resistance and obesity [26,27]. In this study, EECM administration downregulated the plasma levels of insulin and ameliorated the secretion of adipokines, including leptin and adiponectin. Furthermore, EECM administration decreased the plasma levels of lipid parameters including TC, TG, and LDL although it did not significantly affect the HDL levels. Therefore, our findings revealed that EECM mitigated dyslipidemia in an HFD-induced obese mouse model via the regulation of lipid profiles.

The early period of adipocyte differentiation involves complex signaling including adipogenesis, lipogenesis, and cell cycle regulation [28]. PPARγ and C/EBPα are the main regulators of adipogenesis that are involved in adipocyte differentiation [29]. PPARγ expression is induced during the proliferation and differentiation of adipocytes, and it promotes intracellular lipid accumulation [30]. In addition, the cross-regulation between PPARγ and C/EBPα is important for the transcriptional regulation of adipogenesis [31]. SREBP-1 is present as an ectopic protein of the pre-form (inactive), which binds SREBP cleavage-activating protein to the complex in the endoplasmic reticulum (ER) membrane, and the mature form (active), which is cleaved and released from the ER membrane [32]. Mature-SREBP-1 enters the nucleus and sustains lipogenesis in adipocytes at the transcriptional level [33,34]. Phosphorylation (Thr 172) of AMPK inhibits adipogenesis, fatty acid storage, and lipogenesis by increasing the activity of related factors, such as PPARγ, C/EBPα, and SREBP-1 [35]. Activated AMPK is associated with the regulation of lipogenic enzymes including ACC, a major regulatory factor in sterol and lipid synthesis [36], and FAS [37]. ACC catalyzes the conversion of acetyl-CoA to malonyl-CoA following the phosphorylation of ACC (inactive form) by AMPK, and the production of malonyl-CoA is reduced, eventually resulting in fatty acid oxidation [38]. In the present study, EECM exerted anti-obesity effects in vitro and in vivo by suppressing the expression of PPARγ, C/EBPα, SREBP-1, and FAS and p-ACC expression via the AMPK pathway.

During early adipocyte differentiation, MCE is tightly regulated by cell cycle regulators and differentiation-related factors prior to terminal differentiation [39]. The proliferation of adipocytes is stopped, and the cells enter the MCE process by treatment with MDI, a differentiation inducer [40,41]. Our results showed that treatment with EECM increased the protein expression levels of cyclin D1, cyclin E1, cyclin B1, CDK6, and CDK2 in MDI-differentiated 3T3-L1 pre-adipocytes. Moreover, EECM treatment restored cell cycle progression from the G2/M phase. In addition, the protein expression of p21, a CDK inhibitor that regulates G1/S phase progression [42], was increased by EECM treatment although the recovery level was weak. Therefore, EECM suppressed MCE during the early differentiation of preadipocytes.

Among primary adipose tissues, BAT plays a significant role in energy expenditure and thermogenesis [10]. Energy dissipation occurs in mitochondria and leads to ATP generation, which further increases heat production [43]. UCP-1, the major thermogenic regulator of the mitochondrial inner membrane, causes heat generation by uncoupling respiration from ATP production [44]. In addition, the mitochondrial biomarker, COX IV, is a major regulator of oxidative phosphorylation that mediates the final step of the electron transfer chain in mitochondria [43]. PGC-1α induces the expression of thermogenic genes, including the *PPAR* family and *UCP-1*, via an adaptive process for energy dissipation, resulting in the lipolysis of cellular TG [45,46]. AMPK increases cellular NAD^+ levels and activates SIRT-1 (NAD^+-dependent deacetylase), resulting in PGC-1α activation and upregulation of adiponectin expression [47]. Therefore, hepatic lipid accumulation can be mitigated via the AMPK/SIRT1 pathway [48]. In this study, we found that the administration of EECM increased the protein expression of thermogenesis-related factors, including PGC-1α, PPARα, UCP-1, and COX IV, and the activation of AMPKα, which induced the expression of SIRT1. Based on these results, we believe that the modulating effect of EECM on thermogenesis in BAT may affect lipolysis in HFD-induced obese mice.

Several studies have reported that the composition and colonization of the gut microbiome are related to various dietary and pathophysiological changes such as obesity [49–52]. In general, the gut microbiome in the human intestine has a dominant F/B ratio (approximately 90%) [53]. Proteobacteria is often reported in diseases such as obesity, NAFLD, and metabolic syndrome [7]. Previous studies reported that the proportion of Firmicutes was higher than that of Bacteroidetes in the obese colon, whereas the opposite was observed in the lean state [17,54]. Accordingly, the F/B ratio is considered as one of the measures to analyze obesity and dysbiosis [14]. In this study, we observed that the composition of the microbiome was altered in HFD-induced obese mice compared to that in the normal-diet group, and it was recovered by EECM administration. Moreover, EECM also reduced the F/B ratio compared to that in the HFD group, indicating its anti-obesity effect.

5. Conclusions

In summary, our results revealed that EECM ameliorates obesity in vitro and in vivo. Treatment with EECM suppressed MCE and differentiation of 3T3-L1 preadipocytes and downregulated the protein expression levels of adipogenesis-related markers. In the subcutaneous fat and liver of HFD-induced obese mice, oral administration of EECM significantly decreased the body weight gain and lipid accumulation and suppressed the protein expression of adipogenic factors. In addition, EECM activated thermogenesis in BAT by upregulating the expression levels of thermogenic proteins. EECM also restored gut microbiota dysbiosis and blood levels of insulin, leptin, and adiponectin in HFD-induced obese mice. Taken together, our findings confirm the anti-obesity effects of EECM and highlight its potential to be used as a functional food ingredient from natural products.

Supplementary Materials: The following supporting information can be downloaded at: https://www.mdpi.com/article/10.3390/nu15030613/s1, Figure S1: Cytotoxicity of EECM in 3T3-L1; Figure S2: Hepatoxicity and nephrotoxicity of EECM in the blood plasma of HFD-induced obese mice; Figure S3: Representative area of adipocytes in the tissue; Figures S4–S8: Quantitative values for Western blot in Figures 2, 3, 7, 8 and 9.

Author Contributions: Conceptualization, S.-W.H., K.-S.C., S.-H.L. and K.-T.L.; data curation, S.-W.H. and H.-S.A.; formal analysis, S.-W.H. and H.-S.A.; funding acquisition, S.-H.L. and K.-T.L.; investigation, S.-W.H., K.-S.C., H.-S.A. and Y.-K.S.; methodology, S.-W.H., Y.-S.Y., S.-Y.K. and H.-S.A.; project administration, S.-H.L. and K.-T.L.; resources, H.-S.A., Y.-K.S. and S.-H.L.; supervision, K.-S.C. and K.-T.L.; validation, S.-W.H., Y.-S.Y., S.-Y.K. and H.-S.A.; visualization, S.-W.H.; writing—original draft, S.-W.H., K.-S.C., H.-S.A., Y.-K.S., S.-H.L. and K.-T.L.; writing—review and editing, S.-W.H., K.-S.C. and K.-T.L. All authors will be informed about each step of manuscript processing including submission, revision, revision reminder, etc., via emails from our system or assigned Assistant Editor. All authors have read and agreed to the published version of the manuscript.

Funding: This work was supported by "Food Functionality Evaluation program" under the Ministry of Agriculture, Food, and Rural Affairs and partly by the Korea Food Research Institute (G0220300-01).

Institutional Review Board Statement: All procedures were executed with the Guidelines for Animal Care and Use of Kyung Hee University, and the animal model experiment was approved by the experimentation Committee of Kyung Hee University (KHSASP-22-210, Approval date: 13 May 2022).

Informed Consent Statement: Not applicable.

Data Availability Statement: Not applicable.

Conflicts of Interest: The authors declare no conflict of interest.

References

1. Haslam, D.W.; James, W.P. Obesity. *Lancet* **2005**, *366*, 1197–1209. [CrossRef]
2. Geum, N.G.; Son, H.J.; Yeo, J.H.; Yu, J.H.; Choi, M.Y.; Lee, J.W.; Baek, J.K.; Jeong, J.B. Anti-obesity activity of Heracleum moellendorffii root extracts in 3T3-L1 adipocytes. *Food Sci. Nutr.* **2021**, *9*, 5939–5945. [CrossRef] [PubMed]
3. Kopelman, P.G. Obesity as a medical problem. *Nature* **2000**, *404*, 635–643. [CrossRef] [PubMed]
4. Jakab, J.; Miskic, B.; Miksic, S.; Juranic, B.; Cosic, V.; Schwarz, D.; Vcev, A. Adipogenesis as a Potential Anti-Obesity Target: A Review of Pharmacological Treatment and Natural Products. *Diabetes Metab. Syndr. Obes.* **2021**, *14*, 67–83. [CrossRef] [PubMed]
5. Tang, Q.Q.; Otto, T.C.; Lane, M.D. Mitotic clonal expansion: A synchronous process required for adipogenesis. *Proc. Natl. Acad. Sci. USA* **2003**, *100*, 44–49. [CrossRef] [PubMed]
6. Carling, D. AMPK signalling in health and disease. *Curr. Opin. Cell Biol.* **2017**, *45*, 31–37. [CrossRef]
7. Han, H.S.; Lee, H.H.; Gil, H.S.; Chung, K.S.; Kim, J.K.; Kim, D.H.; Yoon, J.; Chung, E.K.; Lee, J.K.; Yang, W.M.; et al. Standardized hot water extract from the leaves of *Hydrangea serrata* (Thunb.) Ser. alleviates obesity via the AMPK pathway and modulation of the gut microbiota composition in high fat diet-induced obese mice. *Food Funct.* **2021**, *12*, 2672–2685. [CrossRef]
8. Fang, K.; Wu, F.; Chen, G.; Dong, H.; Li, J.; Zhao, Y.; Xu, L.; Zou, X.; Lu, F. Diosgenin ameliorates palmitic acid-induced lipid accumulation via AMPK/ACC/CPT-1A and SREBP-1c/FAS signaling pathways in LO2 cells. *BMC Complement. Altern. Med.* **2019**, *19*, 255. [CrossRef]
9. Fenzl, A.; Kiefer, F.W. Brown adipose tissue and thermogenesis. *Horm. Mol. Biol. Clin. Investig.* **2014**, *19*, 25–37. [CrossRef]
10. Gesta, S.; Tseng, Y.H.; Kahn, C.R. Developmental origin of fat: Tracking obesity to its source. *Cell* **2007**, *131*, 242–256. [CrossRef]
11. Vijgen, G.H.; Bouvy, N.D.; Teule, G.J.; Brans, B.; Schrauwen, P.; van Marken Lichtenbelt, W.D. Brown adipose tissue in morbidly obese subjects. *PLoS ONE* **2011**, *6*, e17247. [CrossRef] [PubMed]
12. Ouellet, V.; Routhier-Labadie, A.; Bellemare, W.; Lakhal-Chaieb, L.; Turcotte, E.; Carpentier, A.C.; Richard, D. Outdoor temperature, age, sex, body mass index, and diabetic status determine the prevalence, mass, and glucose-uptake activity of 18F-FDG-detected BAT in humans. *J. Clin. Endocrinol. Metab.* **2011**, *96*, 192–199. [CrossRef]
13. Magne, F.; Gotteland, M.; Gauthier, L.; Zazueta, A.; Pesoa, S.; Navarrete, P.; Balamurugan, R. The Firmicutes/Bacteroidetes Ratio: A Relevant Marker of Gut Dysbiosis in Obese Patients? *Nutrients* **2020**, *12*, 1474. [CrossRef]
14. Stojanov, S.; Berlec, A.; Strukelj, B. The Influence of Probiotics on the Firmicutes/Bacteroidetes Ratio in the Treatment of Obesity and Inflammatory Bowel disease. *Microorganisms* **2020**, *8*, 1715. [CrossRef] [PubMed]
15. Lim, S.H.; Lee, J. Methanol Extract of Cassia mimosoides var. nomame Attenuates Myocardial Injury by Inhibition of Apoptosis in a Rat Model of Ischemia-Reperfusion. *Prev. Nutr. Food Sci.* **2012**, *17*, 177–183. [CrossRef] [PubMed]
16. Kim, K.-H.; Lee, J.-W. Methanol Extract of *Cassia mimosoides* var. *nomame* and Its Ethyl Acetate Fraction Attenuate Brain Damage by Inhibition of Apoptosis in a Rat Model of Ischemia-Reperfusion. *J. Food Sci. Nutr.* **2010**, *15*, 255–261. [CrossRef]
17. Lee, S.Y.; Chung, K.S.; Son, S.R.; Lee, S.Y.; Jang, D.S.; Lee, J.K.; Kim, H.J.; Na, C.S.; Lee, S.H.; Lee, K.T. A Botanical Mixture Consisting of Inula japonica and Potentilla chinensis Relieves Obesity via the AMPK Signaling Pathway in 3T3-L1 Adipocytes and HFD-Fed Obese Mice. *Nutrients* **2022**, *14*, 3685. [CrossRef] [PubMed]
18. Saito, M.; Matsushita, M.; Yoneshiro, T.; Okamatsu-Ogura, Y. Brown Adipose Tissue, Diet-Induced Thermogenesis, and Thermogenic Food Ingredients: From Mice to Men. *Front. Endocrinol. (Lausanne)* **2020**, *11*, 222. [CrossRef]
19. Matsuda, M.; Shimomura, I. Increased oxidative stress in obesity: Implications for metabolic syndrome, diabetes, hypertension, dyslipidemia, atherosclerosis, and cancer. *Obes. Res. Clin. Pract.* **2013**, *7*, e330–e341. [CrossRef]
20. Nam, G.E.; Kim, Y.H.; Han, K.; Jung, J.H.; Rhee, E.J.; Lee, S.S.; Kim, D.J.; Lee, K.W.; Lee, W.Y.; Korean Society for the Study of Obesity. Obesity Fact Sheet in Korea, 2019: Prevalence of Obesity and Abdominal Obesity from 2009 to 2018 and Social Factors. *J. Obes. Metab. Syndr.* **2020**, *29*, 124–132. [CrossRef]
21. Osadebe, P.O.; Odoh, E.U.; Uzor, P.F. Natural Products as Potential Sources of Antidiabetic Drugs. *J. Pharm. Res. Int.* **2014**, *4*, 2075–2095. [CrossRef]
22. Rena, G.; Hardie, D.G.; Pearson, E.R. The mechanisms of action of metformin. *Diabetologia* **2017**, *60*, 1577–1585. [CrossRef]
23. Koh, H.E.; Cao, C.; Mittendorfer, B. Insulin Clearance in Obesity and Type 2 Diabetes. *Int. J. Mol. Sci.* **2022**, *23*, 596. [CrossRef]
24. Havel, P.J.; Kasim-Karakas, S.; Mueller, W.; Johnson, P.R.; Gingerich, R.L.; Stern, J.S. Relationship of plasma leptin to plasma insulin and adiposity in normal weight and overweight women: Effects of dietary fat content and sustained weight loss. *J. Clin. Endocrinol. Metab.* **1996**, *81*, 4406–4413. [CrossRef]

25. Considine, R.V.; Sinha, M.K.; Heiman, M.L.; Kriauciunas, A.; Stephens, T.W.; Nyce, M.R.; Ohannesian, J.P.; Marco, C.C.; McKee, L.J.; Bauer, T.L.; et al. Serum immunoreactive-leptin concentrations in normal-weight and obese humans. *N. Engl. J. Med.* **1996**, *334*, 292–295. [CrossRef] [PubMed]
26. Achari, A.E.; Jain, S.K. Adiponectin, a Therapeutic Target for Obesity, Diabetes, and Endothelial Dysfunction. *Int. J. Mol. Sci.* **2017**, *18*, 1321. [CrossRef] [PubMed]
27. Kawano, J.; Arora, R. The role of adiponectin in obesity, diabetes, and cardiovascular disease. *J. Cardiometab. Syndr.* **2009**, *4*, 44–49. [CrossRef] [PubMed]
28. Farmer, S.R. Transcriptional control of adipocyte formation. *Cell Metab.* **2006**, *4*, 263–273. [CrossRef]
29. Lefterova, M.I.; Lazar, M.A. New developments in adipogenesis. *Trends Endocrinol. Metab.* **2009**, *20*, 107–114. [CrossRef]
30. Moseti, D.; Regassa, A.; Kim, W.K. Molecular Regulation of Adipogenesis and Potential Anti-Adipogenic Bioactive Molecules. *Int. J. Mol. Sci.* **2016**, *17*, 124. [CrossRef]
31. Wu, Z.; Rosen, E.D.; Brun, R.; Hauser, S.; Adelmant, G.; Troy, A.E.; McKeon, C.; Darlington, G.J.; Spiegelman, B.M. Cross-regulation of C/EBP alpha and PPAR gamma controls the transcriptional pathway of adipogenesis and insulin sensitivity. *Mol. Cell* **1999**, *3*, 151–158. [CrossRef]
32. Xiao, X.; Song, B.L. SREBP: A novel therapeutic target. *Acta Biochim. Biophys. Sin. (Shanghai)* **2013**, *45*, 2–10. [CrossRef] [PubMed]
33. Sekiya, M.; Yahagi, N.; Matsuzaka, T.; Takeuchi, Y.; Nakagawa, Y.; Takahashi, H.; Okazaki, H.; Iizuka, Y.; Ohashi, K.; Gotoda, T.; et al. SREBP-1-independent regulation of lipogenic gene expression in adipocytes. *J. Lipid Res.* **2007**, *48*, 1581–1591. [CrossRef] [PubMed]
34. Crewe, C.; Zhu, Y.; Paschoal, V.A.; Joffin, N.; Ghaben, A.L.; Gordillo, R.; Oh, D.Y.; Liang, G.; Horton, J.D.; Scherer, P.E. SREBP-regulated adipocyte lipogenesis is dependent on substrate availability and redox modulation of mTORC1. *JCI Insight* **2019**, *5*, e129397. [CrossRef] [PubMed]
35. Chen, S.; Li, Z.; Li, W.; Shan, Z.; Zhu, W. Resveratrol inhibits cell differentiation in 3T3-L1 adipocytes via activation of AMPK. *Can. J. Physiol. Pharmacol.* **2011**, *89*, 793–799. [CrossRef] [PubMed]
36. Hardie, D.G.; Carling, D. The AMP-activated protein kinase–fuel gauge of the mammalian cell? *Eur. J. Biochem.* **1997**, *246*, 259–273. [CrossRef]
37. Student, A.K.; Hsu, R.Y.; Lane, M.D. Induction of fatty acid synthetase synthesis in differentiating 3T3-L1 preadipocytes. *J. Biol. Chem.* **1980**, *255*, 4745–4750. [CrossRef]
38. Kahn, B.B.; Alquier, T.; Carling, D.; Hardie, D.G. AMP-activated protein kinase: Ancient energy gauge provides clues to modern understanding of metabolism. *Cell Metab.* **2005**, *1*, 15–25. [CrossRef]
39. Fajas, L. Adipogenesis: A cross-talk between cell proliferation and cell differentiation. *Ann. Med.* **2003**, *35*, 79–85. [CrossRef]
40. Tang, Q.Q.; Lane, M.D. Adipogenesis: From stem cell to adipocyte. *Annu. Rev. Biochem.* **2012**, *81*, 715–736. [CrossRef]
41. Han, H.S.; Chung, K.S.; Shin, Y.K.; Lee, S.H.; Lee, K.T. Standardized *Hydrangea serrata* (Thunb.) Ser. Extract Ameliorates Obesity in db/db Mice. *Nutrients* **2021**, *13*, 3624. [CrossRef] [PubMed]
42. Oh, J.M.; Chun, S. Ginsenoside CK Inhibits the Early Stage of Adipogenesis via the AMPK, MAPK, and AKT Signaling Pathways. *Antioxidants* **2022**, *11*, 1890. [CrossRef] [PubMed]
43. Qu, Y.; Chen, S.; Zhou, L.; Chen, M.; Li, L.; Ni, Y.; Sun, J. The different effects of intramuscularly-injected lactate on white and brown adipose tissue in vivo. *Mol. Biol. Rep.* **2022**, *49*, 8507–8516. [CrossRef] [PubMed]
44. Klingenberg, M.; Huang, S.G. Structure and function of the uncoupling protein from brown adipose tissue. *Biochim. Biophys. Acta* **1999**, *1415*, 271–296. [CrossRef]
45. Puigserver, P.; Wu, Z.; Park, C.W.; Graves, R.; Wright, M.; Spiegelman, B.M. A cold-inducible coactivator of nuclear receptors linked to adaptive thermogenesis. *Cell* **1998**, *92*, 829–839. [CrossRef]
46. Haemmerle, G.; Moustafa, T.; Woelkart, G.; Buttner, S.; Schmidt, A.; van de Weijer, T.; Hesselink, M.; Jaeger, D.; Kienesberger, P.C.; Zierler, K.; et al. ATGL-mediated fat catabolism regulates cardiac mitochondrial function via PPAR-alpha and PGC-1. *Nat. Med.* **2011**, *17*, 1076–1085. [CrossRef]
47. You, M.; Rogers, C.Q. Adiponectin: A key adipokine in alcoholic fatty liver. *Exp. Biol. Med. (Maywood)* **2009**, *234*, 850–859. [CrossRef]
48. Chen, X.Y.; Cai, C.Z.; Yu, M.L.; Feng, Z.M.; Zhang, Y.W.; Liu, P.H.; Zeng, H.; Yu, C.H. LB100 ameliorates nonalcoholic fatty liver disease via the AMPK/Sirt1 pathway. *World J. Gastroenterol.* **2019**, *25*, 6607–6618. [CrossRef]
49. Abenavoli, L.; Scarpellini, E.; Colica, C.; Boccuto, L.; Salehi, B.; Sharifi-Rad, J.; Aiello, V.; Romano, B.; De Lorenzo, A.; Izzo, A.A.; et al. Gut Microbiota and Obesity: A Role for Probiotics. *Nutrients* **2019**, *11*, 2690. [CrossRef]
50. Castaner, O.; Goday, A.; Park, Y.M.; Lee, S.H.; Magkos, F.; Shiow, S.T.E.; Schroder, H. The Gut Microbiome Profile in Obesity: A Systematic Review. *Int. J. Endocrinol.* **2018**, *2018*, 4095789. [CrossRef]
51. Liu, B.N.; Liu, X.T.; Liang, Z.H.; Wang, J.H. Gut microbiota in obesity. *World J. Gastroenterol.* **2021**, *27*, 3837–3850. [CrossRef] [PubMed]
52. Cuevas-Sierra, A.; Ramos-Lopez, O.; Riezu-Boj, J.I.; Milagro, F.I.; Martinez, J.A. Diet, Gut Microbiota, and Obesity: Links with Host Genetics and Epigenetics and Potential Applications. *Adv. Nutr.* **2019**, *10*, S17–S30. [CrossRef] [PubMed]

53. Human Microbiome Project, C. Structure, function and diversity of the healthy human microbiome. *Nature* **2012**, *486*, 207–214. [CrossRef] [PubMed]
54. Pedersen, R.; Ingerslev, H.C.; Sturek, M.; Alloosh, M.; Cirera, S.; Christoffersen, B.O.; Moesgaard, S.G.; Larsen, N.; Boye, M. Characterisation of gut microbiota in Ossabaw and Gottingen minipigs as models of obesity and metabolic syndrome. *PLoS ONE* **2013**, *8*, e56612. [CrossRef]

nutrients

MDPI

Article

Benefits of Valsartan and Amlodipine in Lipolysis through PU.1 Inhibition in Fructose-Induced Adiposity

Chu-Lin Chou [1,2,3,4,†], Ching-Hao Li [5,6,†] and Te-Chao Fang [1,2,7,8,*]

[1] Division of Nephrology, Department of Internal Medicine, School of Medicine, College of Medicine, Taipei Medical University, Taipei 110, Taiwan
[2] TMU Research Center of Urology and Kidney, Taipei Medical University, Taipei 110, Taiwan
[3] Division of Nephrology, Department of Internal Medicine, Shuang Ho Hospital, Taipei Medical University, New Taipei City 235, Taiwan
[4] Division of Nephrology, Department of Internal Medicine, Hsin Kuo Min Hospital, Taipei Medical University, Taoyuan City 320, Taiwan
[5] Department of Physiology, School of Medicine, College of Medicine, Taipei Medical University, Taipei 110, Taiwan
[6] Graduate Institute of Medical Sciences, College of Medicine, Taipei Medical University, Taipei 110, Taiwan
[7] Graduate Institute of Clinical Medicine, College of Medicine, Taipei Medical University, Taipei 110, Taiwan
[8] Division of Nephrology, Department of Internal Medicine, Taipei Medical University Hospital, Taipei Medical University, Taipei 110, Taiwan
* Correspondence: fangtechao@gmail.com; Tel.: +886-2-2737-2181
† These authors contributed equally to this work.

Abstract: High fructose intake has been implicated in obesity and metabolic syndrome, which are related to increased cardiovascular mortality. However, few studies have experimentally examined the role of renin–angiotensin system blockers and calcium channel blockers (CCB) in obesity. We investigated the effects of valsartan (an angiotensin II receptor blocker) and amlodipine (a CCB) on lipolysis through the potential mechanism of PU.1 inhibition. We observed that high fructose concentrations significantly increased adipose size and triglyceride, monoacylglycerol lipase, adipose triglyceride lipase, and stearoyl-CoA desaturase-1 (SCD1), activating transcription factor 3 and PU.1 levels in adipocytes in vitro. Subsequently, PU.1 inhibitor treatment was able to reduce triglyceride, SCD1, and PU.1 levels. In addition, elevated levels of triglyceride and PU.1, stimulated by a high fructose concentration, decreased with valsartan and amlodipine treatment. Overall, these findings suggest that high fructose concentrations cause triacylglycerol storage in adipocytes through PU.1-mediated activation. Furthermore, valsartan and amlodipine treatment reduced triacylglycerol storage in adipocytes by inhibiting PU.1 activation in high fructose concentrations in vitro. Thus, the benefits of valsartan and amlodipine in lipolysis may be through PU.1 inhibition in fructose-induced adiposity, and PU.1 inhibition might have a potential therapeutic role in lipolysis in fructose-induced obesity.

Keywords: activating transcription factor 3; adiposity; calcium channel blocker; fructose; PU.1; renin–angiotensin system blocker

Citation: Chou, C.-L.; Li, C.-H.; Fang, T.-C. Benefits of Valsartan and Amlodipine in Lipolysis through PU.1 Inhibition in Fructose-Induced Adiposity. *Nutrients* **2022**, *14*, 3759. https://doi.org/10.3390/nu14183759

Academic Editor: Jean-François Landrier

Received: 1 August 2022
Accepted: 6 September 2022
Published: 12 September 2022

1. Introduction

Fructose intake has been linked to an increase in the epidemiological risk of metabolic syndrome [1,2]. Moreover, metabolic syndrome is epidemiologically associated with central obesity, hypertension, hyperglycemia, and hypertriglyceridemia, leading to an increase in cardiovascular disease and stroke [3,4]. In rodent models, a high-fructose (60% fructose) diet can also aggravate the activity of the renin–angiotensin system (RAS), causing insulin resistance associated with metabolic syndrome [1,2], as verified in our previous studies [3,4]. In addition, high fructose flux and metabolism in the liver lead to a rise in hepatic triglyceride accumulation [5]. Increased fructose consumption could be particularly harmful in patients with impaired glucose and lipid metabolism and increased proinflammatory cytokine expression because of the metabolic properties involved [6,7]. Thus, excessive fructose consumption has become a crucial public health concern globally, particularly in relation to obesity, nonalcoholic fatty liver disease, and type 2 diabetes [8,9].

PU.1, encoded by the Spi1 gene, is a transcription factor with multiple functions during normal and leukemogenic hematopoiesis [10]. During these processes, cellular specification is driven by primary lineage determinants, such as the transcription factor PU.1 [10]. Emerging evidence shows that PU.1 inhibition may be a potential therapeutic agent for fat degradation and its possible mechanism may be as follows [11–14]. For example, PU.1 expression in adipocytes increases significantly in genetic obesity (agouti A^{vy}) [11], high-fat-diet-fed obese mice [12], and high glucose stimulation in vivo [11], which leads to insulin resistance and inflammation through the upregulation of inflammatory cytokines (TNFα, IL-1β, and IL-6) and NADPH oxidase activity [11]. Moreover, Lin et al. showed that PU.1 expression was increased only in visceral but not subcutaneous adipose tissues of obese mice, and the adipocytes were responsible for this increase in PU.1 expression [11]. In addition, activating transcription factor 3 (ATF3) functions as a transcriptional activator or repressor [15,16] and is a hub of the cellular adaptive response network, leading to induction under various inflammatory conditions. In our previous study, ATF3 deficiency in mice results in higher serum levels of triglycerides, glucose, insulin, inflammatory cytokines (ICAM-1 and TNF-α), and increasing visceral adiposity, similar to metabolic syndrome [17]. A study on *Drosophila melanogaster* was shown that ATF3 deficiency leads to chronic inflammation and metabolic disturbances [18].

The risk and prevalence of hypertriglyceridemia and intrahepatic lipid accumulation with increased fructose consumption has already been reported [8]. Postprandial hypertriglyceridemia and adiposity from excessive fructose consumption are risk factors for cardiovascular disease [8,19]. Fructose absorbed from the intestine is metabolized in the liver, where it causes lipogenesis and glycolysis, leading to triglyceride and glucose production [8,20].

Postprandial hypertriglyceridemia, a state caused by metabolic disruption resulting from excessive fructose consumption, may increase visceral adipose deposition via lipogenesis. Stearoyl-CoA desaturase-1 (SCD1) is required for the biosynthesis of triglyceride during lipogenesis [21]. Meanwhile, free fatty acids are also mobilized from adipose tissue triglycerides by the action of adipose triglyceride lipase (ATGL), hormone-sensitive lipase, and monoacylglycerol lipase (MGL), which are involved in adipose tissue lipolysis, including the hydrolysis process of triglyceride, fatty acids, and glycerol [22–24].

Visceral adiposity contributes to hepatic triglyceride accumulation and hepatic insulin resistance by increasing the portal delivery of free fatty acids to the liver [25]. By contrast, after fasting or the high energy demands of physical activity, lipolysis is vital for supplying fatty acids and glycerol as energy to tissues [23,26]. In addition to providing energy, intermediate and end products in the adipose lipolysis process are able to regulate the metabolic process in nonadipose tissue [27]. Changes in the lipolysis process have frequently been associated with lipodystrophy, hyperlipidemia, insulin resistance, type 2 diabetes, depression, and cancer [23,27]. Furthermore, PU.1 reportedly expresses itself in adipocytes as an insulin resistance and inflammation factor, which increases significantly with high glucose stimulation in vivo [11]. However, few studies have focused on PU.1 in lipolysis.

Our previous studies with fructose-fed rodents with metabolic syndrome revealed that RAS blockers could ameliorate hypertriglyceridemia [3] and visceral adiposity [28], as well as improve aortic endothelial functions and reduce oxidative stress [4]. In addition, as previously reported, calcium channel blockers (CCB) attenuate white adipose tissue dysfunction and increase adipocyte differentiation in type 2 diabetic KK-A(y) mice [29]. However, few studies have experimentally examined the role of RAS blockers and CCB in adiposity. Therefore, we investigated the effects of RAS blockers and CCB on lipolysis through the potential mechanism of PU.1 inhibition.

2. Materials and Methods

2.1. Cell Culture and Adipocyte Differentiation

Mouse 3T3-L1 was purchased from American Type Culture Collection (ATCC, Manassas, VA, USA). The cells were routinely nourished in low-glucose Dulbecco's modified Eagle's medium, supplemented with 10% (*v*/*v*) fetal bovine serum, 1% penicillin/streptomycin mixture, and glutamate, at 37 °C in a humidified atmosphere containing 5% CO_2. The complete medium was refreshed every 2 days.

For adipocyte differentiation, 3T3-L1 cells were seeded at 4×10^4 per cm^2 (See the Supplemental Figure S1). Two days later (70–80% confluency), the complete medium was replaced with differentiation medium A (MA: complete medium supplemented with 50 mg/L insulin, 0.5 mM of 3-isobutyl-1-methylxanthine, and 1 μM of dexamethasone) for 2 days (denoted as Day 0). The cells were then incubated with differentiation medium B (MB: complete medium supplemented with 50 mg/L insulin). The MB was refreshed every 2 days. In general, the adipocyte differentiation could be identified from Day 3, and an apparent complete differentiation can be achieved between Day 7 and Day 10 [30]. In the control treatment, 3T3-L1 cells were differentiated with MB alone, whereas in the testing groups, 3T3-L1 cells were incubated with MB in the presence of fructose at various concentrations (1, 2, and 4 mg/mL). Cells in control and testing groups were collected after the same incubation period, and samples were collected at Day 7 of the experimental protocol. The RAS/or CCB blocker or the PU1 inhibitor was supplied in MB medium throughout the differentiation period.

2.2. Oil Red O Staining

Preadipocyte differentiation was imaged at Day 3 and Day 7. In addition, the accumulation of lipid droplets was quantified by Oil Red O (ORO) staining. Briefly, the differentiated adipocytes were washed twice with phosphate-buffered saline and fixed for 1 h in 10% formalin. The cells were then immersed briefly in 60% isopropanol and air dried. After a 30-min incubation with the ORO working solution, the cells were washed several times in distilled water to remove the unbinding dye. The accumulation of lipid droplets was determined by microscopy. For quantitative measurements, the ORO staining was eluted with 100% isopropanol (*v*/*v*), and the optical absorbance was measured at 595 nm [30], which was based on the quantitative method of the previous studies [30–32].

2.3. Western Blotting

Whole-protein lysate of differentiated adipocytes was extracted using a RIPA lysis buffer (containing a protease and phosphatase inhibitor cocktail). Western blotting was performed as described in [33]. Briefly, 40 μg of whole lysate was pipetted into a 12% separation gel and then subjected to constant voltage (90 V) SDS-PAGE. After separation, the proteins inside the gels were transferred to polyvinylidene fluoride (PVDF) membranes with a constant current (400 mA). The PVDF membranes were then washed with tris-buffered saline (with 0.05% Tween-20) containing 5% bovine serum albumin. Subsequently, the membranes were covered with primary and horseradish peroxidase (HRP)-conjugated secondary antibodies, respectively. The primary antibodies used in this study at the following dilutions in phosphate-buffered saline with Tween 20 (PBST): anti-activating transcription factor 3, 1/1000 (ATF3; GTX30069, GeneTex, Irvine, CA, USA), anti-PU.1, 1/1000 (GTX101581, GeneTex), and anti-β-actin, 1/10,000 (Sigma-Aldrich, St. Louis, MO, USA). Enhanced chemiluminescence was used for the visualization of the blotting bands, which were captured using a BioSpectrum AC imaging system (UVP, Upland, CA, USA). The intensity of the blotting bands was quantified using Gel-Pro analyzer software (version 4.0, Media Cybernetics, Rockville, MD, USA).

2.4. Reverse Transcription-Quantitative Polymerase Chain Reaction Assay

Total RNA was extracted using the Trizol reagent (Thermo Fisher Scientific, Waltham, MA, USA) according to the manufacturer's protocol. cDNA was synthesized using the

MMLV RT-Script kit (BioGenesis, Taipei, Taiwan), and a quantitative polymerase chain reaction (qPCR) was performed using OmicsGreen qPCR Master Mix with ROX dye (Omicsbio, New Taipei City, Taiwan) according to the manufacturer's protocol. Three-step PCR cycling was conducted using LightCycler Nano software (Roche Molecular Systems, Almere, The Netherlands) with primer sequences [34] and the following settings: 95 °C for 2 min, followed by 40 cycles at 95 °C for 20 s, 60 °C for 20 s, and 72 °C for 40 s, and then a final extension step of 95 °C for 15 s, 60 °C for 1 min, 95 °C for 15 s, and 60 °C for 15 s. Glyceraldehyde-3-phosphate dehydrogenase was used as an internal control for normalization. Gene expression was calculated using the $2^{-\Delta\Delta Cq}$ method [35].

2.5. *Hematoxylin and Eosin Staining and the Quantification of Lipid Droplet Size*

Fat tissues of mice fed with normal diet and 8-week high-fructose diet were obtained from our previous study (Protocol Number: LAC-2016-0093), and all experimental procedures were approved by the Institutional Animal Care and Use Committee of Taipei Medical University (Protocol Number: LAC-2016-0093) and were in strict accordance with the recommendations of the Guide for the Care and Use of Laboratory Animals set by the National Institutes of Health [17]. Fat tissues were embedded in paraffin and then they were cut into 0.5-µm-thick slices. In brief, control group, comprising wild-type mice that were fed a standard chow diet for 8 weeks; and fructose group, comprising wild-type mice that were fed a high-fructose (60% fructose) diet for 8 weeks. To investigate the histological changes, hematoxylin and eosin (HE; Bio-Check Laboratories, New Taipei City, Taiwan) staining was conducted in accordance with the manufacturer's protocol. HE-stained slices of fat tissue were observed under a microscope (Olympus IX70, Tokyo, Japan) equipped with ViewPoint Virtual Slide Viewing Software, (PreciPoint, Freising, Germany), and the size of the lipid droplets was manually traced using the software's scaling tool. The average lipid droplet diameter was calculated on the basis of >50 lipid droplets per image [33].

2.6. *Immunohistochemistry Staining*

To observe the expression pattern of ATF3 and PU.1, the paraffin-embedded slices of fat tissue were deparaffined, antigen retrieval was performed, and the antigens were incubated with the antibodies specific to ATF3 (GTX02578, GeneTex) at a dilution of 1/200 and PU.1 (DF13270, Affinity Biosciences, OH, USA) at a dilution of 1/50, respectively. Following routine procedures regarding secondary antibody incubation, washing, and HRP-DAB (3,3′-diaminobenzidine)-based chromogenicity, the chromogenic reactivity was detected using the light microscope, as described previously [33]. Images were acquired, and the immunoreactivity was quantified using ImageJ 1.51 software (NIH, Bethesda, MD, USA).

2.7. *Statistical Analyses*

All the results are expressed as mean ± standard deviation (SD). A one-way analysis of variance (ANOVA), followed by the Newman–Keuls post hoc test or one-way repeated-measures ANOVA, was performed for group comparisons. Additionally, a two-way ANOVA (first factor: treatment group; second factor: time period), with the two drug treatments as the factors, was performed to compare the groups. A Student t test was performed for the unpaired data when appropriate. p values significant at the <0.05, <0.01, and <0.001 levels are reported.

3. Results

3.1. *Effects of Different Fructose Concentrations on Adipocyte Size and Triglyceride, ATF3, PU.1, MGL, ATGL, and SCD1 Levels in Adipocytes In Vitro*

We examined the effects of high fructose concentrations on lipogenesis and lipolysis in vitro treated by DB2313 (PU.1 inhibitor), amlodipine (a CCB), valsartan (an angiotensin II receptor blocker, ARB), and untreated. Figure 1 presents the effects of different fructose concentrations on triglyceride levels by measuring ORO staining and adipose size in adipocytes in vitro. Through this method, we discovered that an increased fructose concentration stimulated adipocytes, causing adipose enlargement and triglyceride accumulation. Figure 2 indicates the effects of different fructose concentrations on ATF3 and PU.1 protein levels, ATF3, PU.1, MGL, ATGL, and SCD1 mRNA levels, and the immunohistochemistry (IHC) intensity of ATF3 and PU.1 in adipocytes in vitro. An increased fructose concentration gradually raised ATF3 and PU.1 protein levels and of PU.1 mRNA levels but reduced ATF3 mRNA levels. Moreover, an increased fructose concentration of 4 mg/mL augmented SCD1, ATGL, and MGL mRNA levels. In IHC staining, a 4 mg/mL fructose concentration enhanced ATF3 and PU.1 expression in adipocytes in vitro.

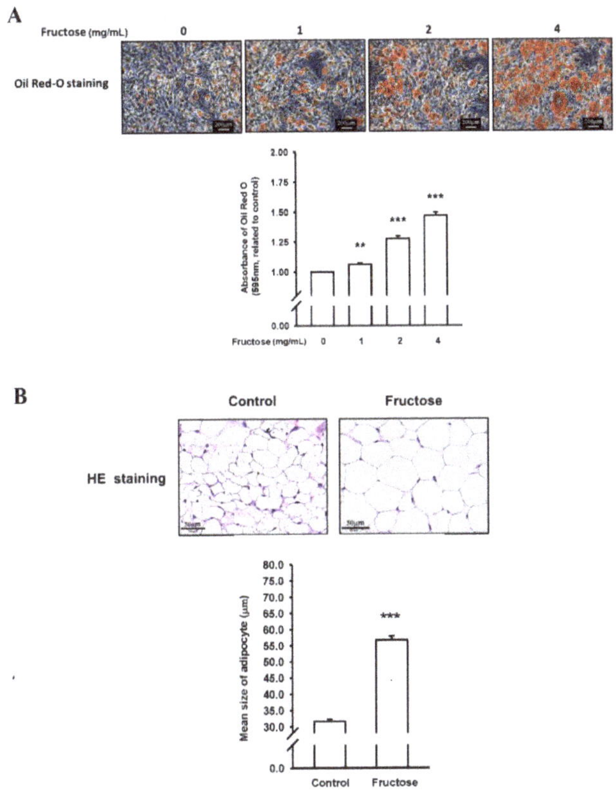

Figure 1. Effects of different fructose concentrations on (**A**) triglyceride levels by measuring Oil Red O (ORO) staining in adipocytes in vitro and (**B**) adipose size obtained from our previous study [17]. N = 6, the results from 6 independent experiments in cell culture experiments and from 6 mice in tissue used study. Values are presented as mean ± standard deviation (SD). **, and *** denote $p < 0.01$, and <0.001 vs. the control groups at Day 7 of the experimental protocol, respectively.

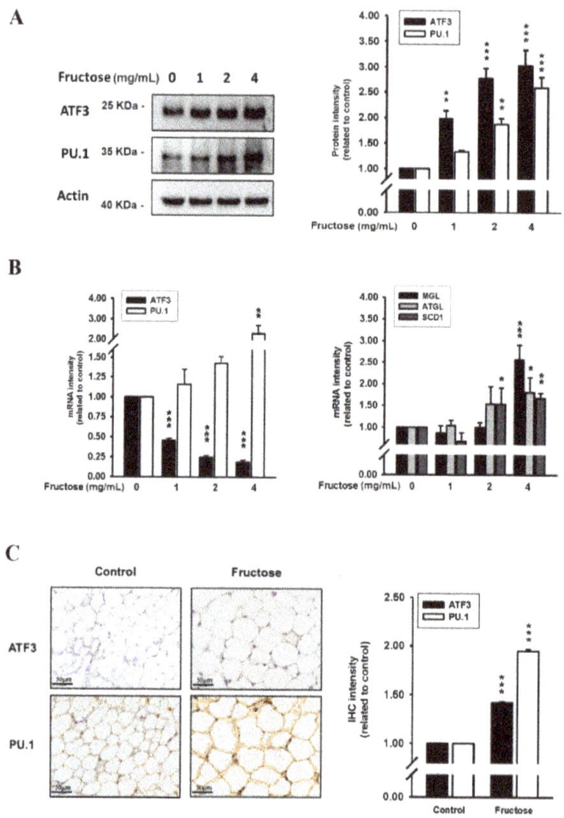

Figure 2. Effects of different fructose concentrations on (**A**) ATF3 and PU.1 protein levels in adipocytes in vitro, (**B**) ATF3, PU.1, MGL, ATGL, and SCD1 mRNA levels in adipocytes in vitro, and (**C**) the immunohistochemistry intensity of ATF3 and PU.1 obtained from our previous study [17]. N = 6, the results from 6 independent experiments in cell culture experiments and from 6 mice in tissue used study. Values are presented as mean ± SD. *, **, and *** denote $p < 0.05$, <0.01, and <0.001 vs. the control groups at Day 7 of the experimental protocol, respectively.

3.2. *Effects of a PU.1 Inhibitor on Adipocyte Triglyceride, ATF3, PU.1, MGL, ATGL, and SCD1 in a 4 mg/mL Fructose Concentration In Vitro*

The effects of the fructose concentration (4 mg/mL) with and without a PU.1 inhibitor (DB2313, 5 nM) on triglyceride levels by measuring ORO staining, ATF3, and PU.1 protein levels, and ATF3, PU.1, MGL, ATGL, and SCD1 mRNA levels in adipocytes in vitro are illustrated in Figure 3. Triglyceride, ATF3 protein, and PU.1 protein levels increased as a result of fructose stimulation and subsequently decreased after treatment with the PU.1 inhibitor. Moreover, PU.1 mRNA levels increased with a relatively high fructose concentration of 4 mg/mL and then decreased following PU.1 inhibitor treatment; however, the ATF3 mRNA levels were inversely increased by the PU.1 inhibitor. The levels of MGL, ATGL, and SCD1 induced by the 4 mg/mL fructose concentration were significantly higher than in the control without fructose. The PU.1 inhibitor increased MGL and ATGL mRNA levels and reduced SCD1 mRNA levels. Supplemental Figure S2 showed the protein level of ATGL was induced, whereas the SCD1 was decreased, during adipogenesis in responded to the treatment of fructose. Basically, the protein level of ATGL was coincided with its mRNA pattern. However, the translational level of SCD1 was obviously repressed upon the maturation of adipocyte. The PU.1 inhibitor reduced SCD1 protein level, but unaltered the ATGL protein level.

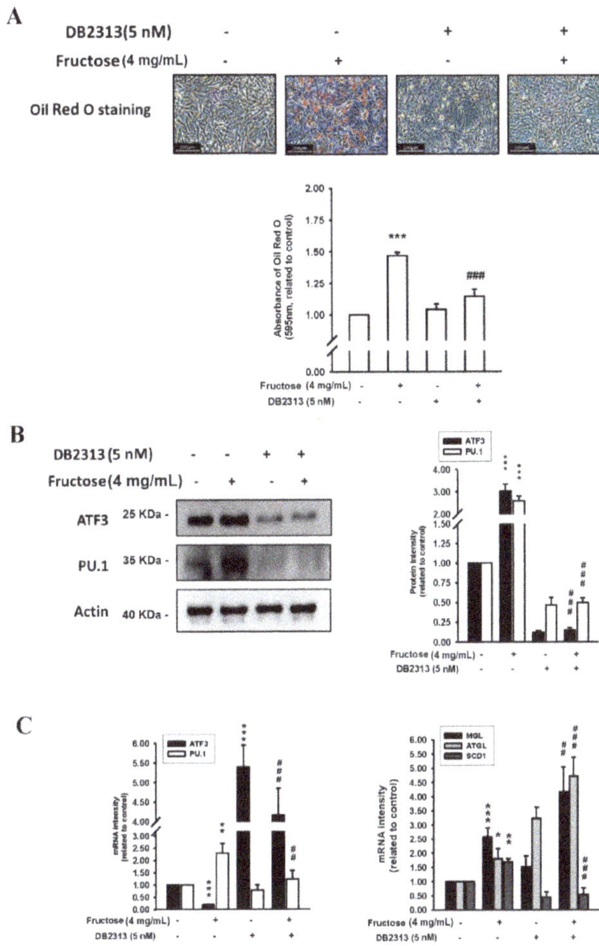

Figure 3. Effects of fructose concentration (4 mg/mL) with and without a PU.1 inhibitor (DB2313, 5 nM) on (**A**) triglyceride levels by measuring ORO staining, (**B**) ATF3 and PU.1 protein levels, and (**C**) ATF3, PU.1, MGL, ATGL, and SCD1 mRNA levels in adipocytes in vitro. N = 6, the results from 6 independent experiments in cell culture experiments. Values are presented as mean ± SD. *, **, and *** denote $p < 0.05$, <0.01, and <0.001 vs. the control groups at Day 7 of the experimental protocol, respectively. ##, and ### denote p <0.01, and <0.001 vs. the 4 mg/mL fructose group at Day 7 of the experimental protocol, respectively.

3.3. Effects of ARB and CCB on Adipocyte Triglyceride, ATF3, PU.1, MGL, ATGL, and SCD1 Levels in a 4 mg/mL Fructose Concentration In Vitro

Figure 4 depicts the effects of a 4 mg/mL M fructose concentration with and without amlodipine (10 μM) and valsartan (10 μM) on triglyceride levels by measuring ORO staining, ATF3 and PU.1 protein levels, and ATF3, PU.1, MGL, ATGL, and SCD1 mRNA levels in adipocytes in vitro. The triglyceride levels increased by the high fructose concentration were reduced by the amlodipine and valsartan treatment. The high fructose concentration also increased ATF3 and PU.1 protein levels; subsequently, only the PU.1

protein levels decreased as a result of the valsartan and amlodipine treatment, although the reduction caused by valsartan was nonsignificant. Moreover, the valsartan and amlodipine treatment significantly reduced the PU.1 mRNA levels. The increased MGL, ATGL, and SCD1 mRNA levels as a result of the 4 mg/mL fructose concentration remained unaffected by the valsartan treatment. Additionally, although the valsartan treatment resulted in a decreasing trend in SCD1 mRNA levels, the reduction was nonsignificant. Supplemental Figure S2 showed Co-incubation with valsartan (10 μM) and amlodipine (10 μM) sustained the high-expressed ATGL protein level and the repression of SCD1 protein translation.

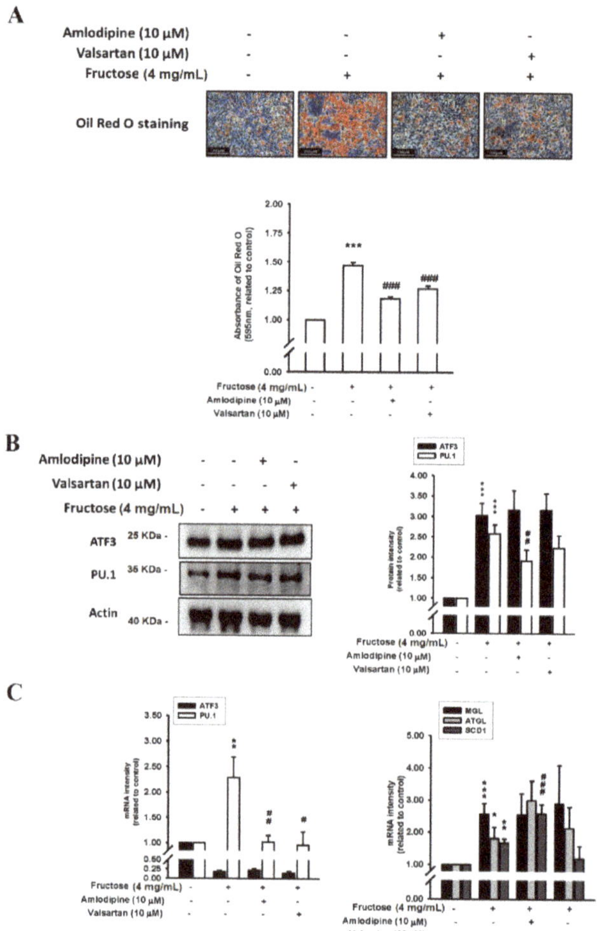

Figure 4. Effects of fructose concentration (4 mg/mL) with and without amlodipine (10 μM) and valsartan (10 μM) on (**A**) triglyceride levels by measuring ORO staining, (**B**) ATF3 and PU.1 protein levels, and (**C**) ATF3, PU.1, MGL, ATGL, and SCD1 mRNA levels in adipocytes in vitro. N = 6. the results from 6 independent experiments in cell culture experiments. Values are presented as mean ± SD. *, **, and *** denote $p < 0.05$, <0.01, and <0.001 vs. the control groups at Day 7 of the experimental protocol, respectively. #, ##, and ### denote $p < 0.05$, <0.01, and <0.001 vs. the 4 mg/mL fructose group at Day 7 of the experimental protocol, respectively.

4. Discussion

The key findings of our study are summarized in this section. First, high fructose concentrations increased adipocyte size and triglyceride content in adipocytes in vitro.

Second, the high fructose concentration significantly increased MGL, ATGL, and SCD1 levels; subsequently, SCD1 levels decreased following PU.1 inhibitor treatment in vitro. Third, the high fructose concentration increased adipose triglyceride, ATF3, and PU.1 levels, and triglyceride and PU.1 levels decreased after PU.1 inhibitor treatment. Fourth, increased triglyceride and PU.1 levels in adipocytes treated with high fructose concentrations were reduced with the amlodipine and valsartan treatment. Overall, these findings suggest that high fructose concentrations cause lipogenesis through PU.1 activation, and amlodipine and valsartan treatment activate lipolysis through a potential mechanism that inhibits PU.1 activation.

High fructose intake contributes to the development of obesity [8]. Fructose is metabolized in the intestine and transported into the liver, where approximately two-thirds of the fructose intake can be metabolized; the remaining intake is metabolized by other tissues, including adipose tissue expressed as the fructose transporter GLUT5 [36]. Furthermore, postprandial hypertriglyceridemia and visceral adipose deposition are metabolic disturbances resulting from excessive fructose consumption. Our study revealed that fructose stimulation increased MGL, ATGL, and SCD1 levels, and that SCD1 could subsequently be reduced by the PU.1 inhibitor. In addition, through the action of ATGL and MGL in the adipose tissue, adipose tissue triglycerides were metabolized into free fatty acids, which are involved in adipose tissue lipolysis.

In our study, the data revealed that fructose stimulation increased the size of adipose tissue and the lipid droplets, as well as increased ATF3 and PU.1 protein levels, which were subsequently lowered by the PU.1 inhibitor. Ruan et al. demonstrated the inhibition of PU.1 in the adipogenesis of ovine primary preadipocytes by limiting CCAAT-enhancer-binding protein-β [13]. In adipocyte-specific PU.1-knockout mice, peroxisome proliferator–activated receptor-gamma (PPARγ) is more active when PU.1 expression is reduced in adipocytes [12], indicating that increased PU.1 modifies the adipocyte PPARγ in adipocytes. PPARγ is highly expressed in adipose tissue, and it can be activated to alter the adipocyte phenotype and upregulate genes involved in fatty acid metabolism and triglyceride storage [14]. For example, PPARγ-activating ligands improve the function of fatty tissues, such as the expression of adiponectin and the reduction in nonesterified fatty acid intake in plasma [14]. This evidence suggests that fructose-stimulated PU.1 may be a mediator that contributes in part to the regulation of adipogenesis in adipose tissue. Thus, PU.1 could be a potential therapeutic agent for fat degradation through improving oxidative stress and regulating PPARγ-mediated metabolic processes.

Whether ATF3 affects adipose lipid processing is still being researched. In this in vitro study, we evaluated the association between ATF3 and lipolysis in adipose cells with and without high fructose stimulation. Our previous study demonstrated that the loss of ATF3 in mice was accompanied by elevated serum levels of glucose, insulin, triglycerides, inflammatory cytokines (TNF-α and ICAM-1), and larger visceral adiposity compared with wild-type mice [17]. This resembled metabolic syndrome and indicated that ATF3 could be involved in regulating lipogenic properties. That study further observed that high fructose stimulated an increased protein response in ATF3 and PU.1 and enhanced lipid droplet sizes with cumulative MGL, ATGL, and SCD1 levels, which were subsequently reduced by using the PU.1 inhibitor. Furthermore, Suganami et al.'s study provided evidence that ATF3, induced in obese adipose tissue, acts as a transcriptional repressor of the negative feedback mechanism that attenuates saturated fatty acid/toll-like receptor 4 signaling and macrophage activation in obese adipose tissue [37]. Moreover, our colleagues found that a synthetic ATF3 inducer, ST32da, promoted ATF3 expression to downregulate adipokine genes, increasing white adipose tissue browning and reducing lipogenesis in ATF3-knockout mice with a high-fat diet [34].

In this study, we found that there is a discrepancy between ATF3 protein and mRNA levels. As mRNA is eventually translated into protein, it is usually assumed that there is some sort of correlation between the levels of mRNA and protein. However, poor correlations between the levels of mRNA and protein are common, and this could be achieved by many factors. For example, the half-lives of mRNA and protein may differ. In addition, the modification of mRNA (e.g., methylation) might change the translational efficiency of mRNA. Additionally, the increased protein level may down-regulate mRNA synthesis and up-regulate mRNA degradation. For example, Zhu and his colleagues found the ATF3 protein in colorectal cancer patients was statistically up-regulated, but the ATF3 mRNA level was without significant change [38], suggesting a discrepancy between ATF3 protein and mRNA levels. The overexpressing ATF3 (by plasmid) inhibited adipocyte differentiation of 3T3-L1 cells [39,40]. However, a synthetic ATF3 inducer ST32da not only promoted ATF3 expression, but also induced adipocyte browning, suggesting the need for ATF3 during glucose metabolism and lipogenesis [34]. Moreover, ATF3 can have opposite effects on different types of cells. The functions of ATF3 might depend on its transcriptional milieu, thus the levels of protein and mRNA of ATF3 were investigated in this study.

Our study is the first to reveal that amlodipine and valsartan treatment could reduce fat droplets and PU.1 protein levels in adipocytes. Previous studies have reported the beneficial role of amlodipine and valsartan in improving lipoprotein oxidation, increasing superoxide dismutase activity and improving the effects of atherosclerosis in experimental and human studies [41–49]. In particular, local angiotensin II plays a key role in promoting the adipogenic differentiation of mesenchymal stem cells from human fat tissue through type 2 angiotensin receptors [50]. Furthermore, the activation of RAS by angiotensin II treatment increases endoplasmic reticulum stress and causes the inflammation of adipocytes through angiotensin II receptor type 1 [51]. ARBs have therefore been reported to have antioxidant and anti-inflammatory effects and result in improved fat-droplet size, as well as the differentiation and dysregulation of adipocytokines in culture adipocytes and mice [52–55]. In the future, the association between PU.1 inhibitors and RAS-induced paracrine effects merits investigation. In addition, nifedipine, a CCB, attenuates white adipose tissue dysfunction in type 2 diabetic KK-A(y) mice, demonstrating that nifedipine increases adipocyte numbers and the expression of PPARγ and adipocyte fatty acid-binding proteins related to adipocyte differentiation [29]. Furthermore, in essential hypertensive patients, Harano et al. observed that the responses of ketone bodies during insulin sensitivity tests at 30 min improved after amlodipine treatment, which reflects the effect of insulin on lipolysis in adipose tissue and hepatic fatty acid oxidation [44]. As for the role of ARBs and CCBs in adipocyte PU.1 inhibition, further studies are required to identify the mechanism in detail.

Our study demonstrated that high fructose stimulation increased adipose size and triglyceride, MGL, ATGL, SCD1, ATF3, and PU.1 levels in adipocytes in vitro; subsequently, PU.1 inhibitor treatment reduced triglyceride, SCD1, and PU.1. In addition, we revealed that increased triglyceride, and PU.1 levels in adipocytes stimulated by relatively high fructose concentrations were reduced by treatment with amlodipine and valsartan. Amlodipine increased fructose-induced SCD1 mRNA levels, which may be because amlodipine partially inhibited PU.1 and also had the other signaling pathways on lipogenesis independent of PU.1 signaling. These results indicate that high fructose concentrations cause triacylglycerol storage in adipocyte droplets through PU.1 activation. Furthermore, amlodipine and valsartan treatment can improve triacylglycerol storage in adipocyte droplets by inhibiting PU.1 activation. Thus, the benefits of valsartan and amlodipine in lipolysis may be through PU.1 inhibition in fructose-induced adiposity, and PU.1 inhibition might have a potential therapeutic role in lipolysis in fructose-induced obesity.

Supplementary Materials: The following supporting information can be downloaded at: https://www.mdpi.com/article/10.3390/nu14183759/s1, Figure S1: Experimental protocol for cell cultures. Figure S2: Effects PU.1 inhibitor, valsartan, and amlodipine on the protein levels of ATGL, SCD1 in adipocytes in vitro.

Author Contributions: Conceptualization, C.-H.L. and T.-C.F.; methodology, C.-H.L.; software, C.-H.L.; validation, C.-L.C., C.-H.L. and T.-C.F.; formal analysis, C.-H.L.; investigation, C.-H.L. and T.-C.F.; resources, C.-H.L. and T.-C.F.; data curation, C.-H.L. and T.-C.F.; writing—original draft preparation, C.-L.C.; writing—review and editing, C.-L.C. and T.-C.F.; visualization, C.-L.C.; supervision, C.-H.L. and T.-C.F.; project administration, C.-H.L. and T.-C.F.; funding acquisition, C.-H.L. and T.-C.F. All authors have read and agreed to the published version of the manuscript.

Funding: This study was financed by grant DP2-111-21121-01-O-07-02 and 110TMU-TMUH-17 from Taipei Medical University, and Taipei Medical University Hospital, Taipei Medical University, respectively.

Institutional Review Board Statement: The study was conducted according to the guidelines of the Declaration of Helsinki and approved by the Research Ethics Board of Taipei Medical University. All experimental procedures were approved by the Institutional Animal Care and Use Committee of Taipei Medical University (Protocol Number: LAC-2016-0093) and were in strict accordance with the recommendations of the Guide for the Care and Use of Laboratory Animals set by the National Institutes of Health.

Informed Consent Statement: Not applicable.

Data Availability Statement: Data is contained within the article.

Conflicts of Interest: The authors declare no conflict of interest.

References

1. Hwang, I.S.; Ho, H.; Hoffman, B.B.; Reaven, G.M. Fructose-induced insulin resistance and hypertension in rats. *Hypertension* **1987**, *10*, 512–516. [CrossRef] [PubMed]
2. Catena, C.; Cavarape, A.; Novello, M.; Giacchetti, G.; Sechi, L.A. Insulin receptors and renal sodium handling in hypertensive fructose-fed rats. *Kidney Int.* **2003**, *64*, 2163–2171. [CrossRef] [PubMed]
3. Chou, C.L.; Lai, Y.H.; Lin, T.Y.; Lee, T.J.; Fang, T.C. Aliskiren prevents and ameliorates metabolic syndrome in fructose-fed rats. *Arch. Med. Sci.* **2011**, *7*, 882–888. [CrossRef] [PubMed]
4. Chou, C.L.; Pang, C.Y.; Lee, T.J.; Fang, T.C. Direct renin inhibitor prevents and ameliorates insulin resistance, aortic endothelial dysfunction and vascular remodeling in fructose-fed hypertensive rats. *Hypertens. Res.* **2013**, *36*, 123–128. [CrossRef]
5. Huang, D.; Dhawan, T.; Young, S.; Yong, W.H.; Boros, L.G.; Heaney, A.P. Fructose impairs glucose-induced hepatic triglyceride synthesis. *Lipids Health Dis.* **2011**, *10*, 20. [CrossRef]
6. Basciano, H.; Federico, L.; Adeli, K. Fructose, insulin resistance, and metabolic dyslipidemia. *Nutr. Metab.* **2005**, *2*, 5. [CrossRef]
7. Rutledge, A.C.; Adeli, K. Fructose and the metabolic syndrome: Pathophysiology and molecular mechanisms. *Nutr. Rev.* **2007**, *65*, S13–S23. [CrossRef]
8. Taskinen, M.R.; Packard, C.J.; Boren, J. Dietary Fructose and the Metabolic Syndrome. *Nutrients* **2019**, *11*, 1987. [CrossRef]
9. Merino, B.; Fernandez-Diaz, C.M.; Cozar-Castellano, I.; Perdomo, G. Intestinal Fructose and Glucose Metabolism in Health and Disease. *Nutrients* **2019**, *12*, 94. [CrossRef]
10. Burda, P.; Laslo, P.; Stopka, T. The role of PU.1 and GATA-1 transcription factors during normal and leukemogenic hematopoiesis. *Leukemia* **2010**, *24*, 1249–1257. [CrossRef]
11. Lin, L.; Pang, W.; Chen, K.; Wang, F.; Gengler, J.; Sun, Y.; Tong, Q. Adipocyte expression of PU.1 transcription factor causes insulin resistance through upregulation of inflammatory cytokine gene expression and ROS production. *Am. J. Physiol. Endocrinol. Metab.* **2012**, *302*, E1550–E1559. [CrossRef] [PubMed]
12. Lackey, D.E.; Reis, F.C.G.; Isaac, R.; Zapata, R.C.; El Ouarrat, D.; Lee, Y.S.; Bandyopadhyay, G.; Ofrecio, J.M.; Oh, D.Y.; Osborn, O. Adipocyte PU.1 knockout promotes insulin sensitivity in HFD-fed obese mice. *Sci. Rep.* **2019**, *9*, 14779. [CrossRef] [PubMed]
13. Ruan, C.; Li, X.; Hu, J.; Zhang, Y.; Zhao, X. MITF and PU.1 inhibit adipogenesis of ovine primary preadipocytes by restraining C/EBPbeta. *Cell. Mol. Biol. Lett.* **2017**, *22*, 2. [CrossRef] [PubMed]
14. Sharma, A.M.; Staels, B. Review: Peroxisome proliferator-activated receptor gamma and adipose tissue–understanding obesity-related changes in regulation of lipid and glucose metabolism. *J. Clin. Endocrinol. Metab.* **2007**, *92*, 386–395. [CrossRef]
15. Thompson, M.R.; Xu, D.; Williams, B.R. ATF3 transcription factor and its emerging roles in immunity and cancer. *J. Mol. Med.* **2009**, *87*, 1053–1060. [CrossRef]
16. Hai, T.; Wolfgang, C.D.; Marsee, D.K.; Allen, A.E.; Sivaprasad, U. ATF3 and stress responses. *Gene Expr.* **1999**, *7*, 321–335.
17. Chou, C.L.; Li, C.H.; Lin, H.; Liao, M.H.; Wu, C.C.; Chen, J.S.; Sue, Y.M.; Fang, T.C. Role of activating transcription factor 3 in fructose-induced metabolic syndrome in mice. *Hypertens. Res.* **2018**, *41*, 589–597. [CrossRef]
18. Rynes, J.; Donohoe, C.D.; Frommolt, P.; Brodesser, S.; Jindra, M.; Uhlirova, M. Activating transcription factor 3 regulates immune and metabolic homeostasis. *Mol. Cell. Biol.* **2012**, *32*, 3949–3962. [CrossRef]
19. Lambadiari, V.; Korakas, E.; Tsimihodimos, V. The Impact of Dietary Glycemic Index and Glycemic Load on Postprandial Lipid Kinetics, Dyslipidemia and Cardiovascular Risk. *Nutrients* **2020**, *12*, 2204. [CrossRef]

20. Hannou, S.A.; Haslam, D.E.; McKeown, N.M.; Herman, M.A. Fructose metabolism and metabolic disease. *J. Clin. Investig.* **2018**, *128*, 545–555. [CrossRef]
21. Paton, C.M.; Ntambi, J.M. Biochemical and physiological function of stearoyl-CoA desaturase. *Am. J. Physiol. Endocrinol. Metab.* **2009**, *297*, E28–E37. [CrossRef] [PubMed]
22. Lafontan, M.; Langin, D. Lipolysis and lipid mobilization in human adipose tissue. *Prog. Lipid Res.* **2009**, *48*, 275–297. [PubMed]
23. Ahmadian, M.; Wang, Y.; Sul, H.S. Lipolysis in adipocytes. *Int. J. Biochem. Cell Biol.* **2010**, *42*, 555–559. [CrossRef]
24. Large, V.; Peroni, O.; Letexier, D.; Ray, H.; Beylot, M. Metabolism of lipids in human white adipocyte. *Diabetes Metab.* **2004**, *30*, 294–309. [CrossRef]
25. Stanhope, K.L.; Havel, P.J. Fructose consumption: Potential mechanisms for its effects to increase visceral adiposity and induce dyslipidemia and insulin resistance. *Curr. Opin. Lipidol.* **2008**, *19*, 16–24. [CrossRef]
26. Guyenet, S.J.; Schwartz, M.W. Clinical review: Regulation of food intake, energy balance, and body fat mass: Implications for the pathogenesis and treatment of obesity. *J. Clin. Endocrinol. Metab.* **2012**, *97*, 745–755. [CrossRef] [PubMed]
27. Luo, L.; Liu, M. Adipose tissue in control of metabolism. *J. Endocrinol.* **2016**, *231*, R77–R99. [CrossRef]
28. Chou, C.L.; Lin, H.; Chen, J.S.; Fang, T.C. Renin inhibition improves metabolic syndrome, and reduces angiotensin II levels and oxidative stress in visceral fat tissues in fructose-fed rats. *PLoS ONE* **2017**, *12*, e0180712. [CrossRef]
29. Iwai, M.; Kanno, H.; Inaba, S.; Senba, I.; Sone, H.; Nakaoka, H.; Horiuchi, M. Nifedipine, a calcium-channel blocker, attenuated glucose intolerance and white adipose tissue dysfunction in type 2 diabetic KK-A(y) mice. *Am. J. Hypertens.* **2011**, *24*, 169–174. [CrossRef]
30. Zebisch, K.; Voigt, V.; Wabitsch, M.; Brandsch, M. Protocol for effective differentiation of 3T3-L1 cells to adipocytes. *Anal. Biochem.* **2012**, *425*, 88–90. [CrossRef]
31. Biltz, N.K.; Meyer, G.A. A novel method for the quantification of fatty infiltration in skeletal muscle. *Skelet. Muscle* **2017**, *7*, 1. [CrossRef] [PubMed]
32. Kraus, N.A.; Ehebauer, F.; Zapp, B.; Rudolphi, B.; Kraus, B.J.; Kraus, D. Quantitative assessment of adipocyte differentiation in cell culture. *Adipocyte* **2016**, *5*, 351–358. [CrossRef]
33. Chen, T.S.; Lai, P.F.; Kuo, C.H.; Day, C.H.; Chen, R.J.; Ho, T.J.; Yeh, Y.L.; Mahalakshmi, B.; Padmaviswanadha, V.; Kuo, W.W.; et al. Resveratrol enhances therapeutic effect on pancreatic regeneration in diabetes mellitus rats receiving autologous transplantation of adipose-derived stem cells. *Chin. J. Physiol.* **2020**, *63*, 122–127. [PubMed]
34. Cheng, C.F.; Ku, H.C.; Cheng, J.J.; Chao, S.W.; Li, H.F.; Lai, P.F.; Chang, C.C.; Don, M.J.; Chen, H.H.; Lin, H. Adipocyte browning and resistance to obesity in mice is induced by expression of ATF3. *Commun. Biol.* **2019**, *2*, 389. [CrossRef]
35. Hsu, C.N.; Jen, C.Y.; Chen, Y.H.; Peng, S.Y.; Wu, S.C.; Yao, C.L. Glucocorticoid transiently upregulates mitochondrial biogenesis in the osteoblast. *Chin. J. Physiol.* **2020**, *63*, 286–293. [PubMed]
36. Zwarts, I.; van Zutphen, T.; Kruit, J.K.; Liu, W.; Oosterveer, M.H.; Verkade, H.J.; Uhlenhaut, N.H.; Jonker, J.W. Identification of the fructose transporter GLUT5 (SLC2A5) as a novel target of nuclear receptor LXR. *Sci. Rep.* **2019**, *9*, 9299. [CrossRef]
37. Suganami, T.; Yuan, X.; Shimoda, Y.; Uchio-Yamada, K.; Nakagawa, N.; Shirakawa, I.; Usami, T.; Tsukahara, T.; Nakayama, K.; Miyamoto, Y.; et al. Activating transcription factor 3 constitutes a negative feedback mechanism that attenuates saturated Fatty acid/toll-like receptor 4 signaling and macrophage activation in obese adipose tissue. *Circ. Res.* **2009**, *105*, 25–32. [CrossRef]
38. Zhu, H.; Liu, M.; Zhang, N.; Pan, H.; Lin, G.; Li, N.; Wang, L.; Yang, H.; Yan, K.; Gong, F. Serum and Adipose Tissue mRNA Levels of ATF3 and FNDC5/Irisin in Colorectal Cancer Patients With or Without Obesity. *Front. Physiol.* **2018**, *9*, 1125. [CrossRef]
39. Jang, M.K.; Kim, C.H.; Seong, J.K.; Jung, M.H. ATF3 inhibits adipocyte differentiation of 3T3-L1 cells. *Biochem. Biophys. Res. Commun.* **2012**, *421*, 38–43. [CrossRef]
40. Seong, S.; Kim, J.H.; Kim, K.; Kim, I.; Koh, J.T.; Kim, N. Alternative regulatory mechanism for the maintenance of bone homeostasis via STAT5-mediated regulation of the differentiation of BMSCs into adipocytes. *Exp. Mol. Med.* **2021**, *53*, 848–863. [CrossRef]
41. Turgan, N.; Habif, S.; Kabaroglu, C.G.; Mutaf, I.; Ozmen, D.; Bayindir, O.; Uysal, A. Effects of the calcium channel blocker amlodipine on serum and aortic cholesterol, lipid peroxidation, antioxidant status and aortic histology in cholesterol-fed rabbits. *J. Biomed. Sci.* **2003**, *10*, 65–72. [CrossRef] [PubMed]
42. Rashidi, B.; Mohammadi, M.; Mirzaei, F.; Badalzadeh, R.; Reisi, P. Amlodipine treatment decreases plasma and carotid artery tissue levels of endothelin-1 in atherosclerotic rabbits. *Pathophysiology* **2011**, *18*, 137–142. [CrossRef]
43. Yao, K.; Ina, Y.; Nagashima, K.; Ohmori, K.; Ohno, T. Antioxidant effects of calcium antagonists in rat brain homogenates. *Biol. Pharm. Bull.* **2000**, *23*, 766–769. [CrossRef] [PubMed]
44. Harano, Y.; Kageyama, A.; Hirose, J.; Asakura, Y.; Yokota, T.; Ikebuchi, M.; Suzuki, M.; Omae, T. Improvement of insulin sensitivity for glucose metabolism with the long-acting Ca-channel blocker amlodipine in essential hypertensive subjects. *Metabolism* **1995**, *44*, 315–319. [CrossRef]
45. Cole, B.K.; Keller, S.R.; Wu, R.; Carter, J.D.; Nadler, J.L.; Nunemaker, C.S. Valsartan protects pancreatic islets and adipose tissue from the inflammatory and metabolic consequences of a high-fat diet in mice. *Hypertension* **2010**, *55*, 715–721. [CrossRef]
46. Tomono, Y.; Iwai, M.; Inaba, S.; Mogi, M.; Horiuchi, M. Blockade of AT1 receptor improves adipocyte differentiation in atherosclerotic and diabetic models. *Am. J. Hypertens.* **2008**, *21*, 206–212. [CrossRef]
47. Whaley-Connell, A.; Govindarajan, G.; Habibi, J.; Hayden, M.R.; Cooper, S.A.; Wei, Y.; Ma, L.; Qazi, M.; Link, D.; Karuparthi, P.R.; et al. Angiotensin II-mediated oxidative stress promotes myocardial tissue remodeling in the transgenic (mRen2) 27 Ren2 rat. *Am. J. Physiol. Endocrinol. Metab.* **2007**, *293*, E355–E363. [CrossRef]

48. Saisho, Y.; Komiya, N.; Hirose, H. Effect of valsartan, an angiotensin II receptor blocker, on markers of oxidation and glycation in Japanese type 2 diabetic subjects: Blood pressure-independent effect of valsartan. *Diabetes Res. Clin. Pract.* **2006**, *74*, 201–203. [CrossRef]
49. Hussein, O.; Shneider, J.; Rosenblat, M.; Aviram, M. Valsartan therapy has additive anti-oxidative effect to that of fluvastatin therapy against low-density lipoprotein oxidation: Studies in hypercholesterolemic and hypertensive patients. *J. Cardiovasc. Pharmacol.* **2002**, *40*, 28–34. [CrossRef]
50. Sysoeva, V.Y.; Ageeva, L.V.; Tyurin-Kuzmin, P.A.; Sharonov, G.V.; Dyikanov, D.T.; Kalinina, N.I.; Tkachuk, V.A. Local angiotensin II promotes adipogenic differentiation of human adipose tissue mesenchymal stem cells through type 2 angiotensin receptor. *Stem Cell Res.* **2017**, *25*, 115–122. [CrossRef]
51. Menikdiwela, K.R.; Ramalingam, L.; Allen, L.; Scoggin, S.; Kalupahana, N.S.; Moustaid-Moussa, N. Angiotensin II Increases Endoplasmic Reticulum Stress in Adipose Tissue and Adipocytes. *Sci. Rep.* **2019**, *9*, 8481. [CrossRef] [PubMed]
52. Kurata, A.; Nishizawa, H.; Kihara, S.; Maeda, N.; Sonoda, M.; Okada, T.; Ohashi, K.; Hibuse, T.; Fujita, K.; Yasui, A.; et al. Blockade of Angiotensin II type-1 receptor reduces oxidative stress in adipose tissue and ameliorates adipocytokine dysregulation. *Kidney Int.* **2006**, *70*, 1717–1724. [CrossRef] [PubMed]
53. Munoz, M.C.; Giani, J.F.; Dominici, F.P.; Turyn, D.; Toblli, J.E. Long-term treatment with an angiotensin II receptor blocker decreases adipocyte size and improves insulin signaling in obese Zucker rats. *J. Hypertens.* **2009**, *27*, 2409–2420. [CrossRef]
54. Takemori, K.; Inoue, T.; Ito, H. Effects of angiotensin II type 1 receptor blocker and adiponectin on adipocyte dysfunction in stroke-prone spontaneously hypertensive rats. *Lipids Health Dis.* **2013**, *12*, 108. [CrossRef] [PubMed]
55. Furuhashi, M.; Ura, N.; Takizawa, H.; Yoshida, D.; Moniwa, N.; Murakami, H.; Higashiura, K.; Shimamoto, K. Blockade of the renin-angiotensin system decreases adipocyte size with improvement in insulin sensitivity. *J. Hypertens.* **2004**, *22*, 1977–1982. [CrossRef]

nutrients

MDPI

Article

Crossbred Sows Fed a Western Diet during Pre-Gestation, Gestation, Lactation, and Post-Lactation Periods Develop Signs of Lean Metabolic Syndrome That Are Partially Attenuated by Spirulina Supplementation

Rosamaria Lugarà [1], Simone Renner [2,3], Eckhard Wolf [2,3], Annette Liesegang [4], Rupert Bruckmaier [5] and Katrin Giller [1,*]

1 Animal Nutrition, ETH Zurich, Eschikon 27, 8315 Lindau, Switzerland
2 German Center for Diabetes Research (DZD), Ingolstaedter Landstrasse 1, 85764 Neuherberg, Germany
3 Molecular Animal Breeding and Biotechnology, Department of Veterinary Sciences, Ludwig-Maximilian University Munich, Gene Center, Feodor-Lynen-Strasse 25, 81377 Munich, Germany
4 Animal Nutrition, Vetsuisse Faculty, University of Zurich, Winterthurerstrasse 270, 8057 Zurich, Switzerland
5 Veterinary Physiology, Vetsuisse Faculty, University of Bern, Bremgartenstrasse 109a, 3001 Bern, Switzerland
* Correspondence: katrin.giller@usys.ethz.ch; Tel.: +41-52-3549209

Abstract: Excessive dietary intake of fats and sugars ("Western diet", WD) is one of the leading causes of obesity. The consumption of the microalga *Arthrospira platensis* (spirulina, Sp) is increasing due to its presumed health benefits. Both WD and Sp are also consumed by pregnant and breastfeeding women. This study investigated if gestating and lactating domestic pigs are an appropriate model for WD-induced metabolic disturbances similar to those observed in humans and if Sp supplementation may attenuate any of these adverse effects. Pigs were fed a WD high in fat, sugars, and cholesterol or a control diet. Half of the animals per diet group were supplemented with 20 g Sp per day. The WD did not increase body weight or adipose tissue accumulation but led to metabolic impairments such as higher cholesterol concentration in plasma, lower IGF1 plasma levels, and signs of hepatic damage compared to the control group. Spirulina supplementation could not reduce all the metabolic impairments observed in WD-fed animals. These findings indicate limited suitability of gestating and lactating domestic pigs as a model for WD but a certain potential of low-dose Sp supplementation to partially attenuate negative WD effects.

Keywords: dietary fat; dietary sugar; microalgae; gene expression; liver steatosis; skeletal muscle; insulin-like growth factor; pig model

Citation: Lugarà, R.; Renner, S.; Wolf, E.; Liesegang, A.; Bruckmaier, R.; Giller, K. Crossbred Sows Fed a Western Diet during Pre-Gestation, Gestation, Lactation, and Post-Lactation Periods Develop Signs of Lean Metabolic Syndrome That Are Partially Attenuated by Spirulina Supplementation. *Nutrients* **2022**, *14*, 3574. https://doi.org/10.3390/nu14173574

Academic Editors: Xiaoyu Wang and Frédéric Dutheil

Received: 22 July 2022
Accepted: 26 August 2022
Published: 30 August 2022

1. Introduction

Excessive energy intake in the form of a "Western diet" (WD), often combined with a sedentary lifestyle, is considered the leading cause of the obesity epidemic, which affects approximately 30% of the adult population in industrialized countries today [1]. Chronic excessive intake of dietary saturated fat, sugars, and cholesterol may lead to an alteration in the activity of transcription factors that results in a dysregulation of gene expression favoring intra-organ lipid accumulation and, in the long term, other metabolic disturbances [2–4]. Even before the onset of the metabolic syndrome, including visible changes in body weight (BW) and body composition, shifts in metabolic and inflammatory blood parameters point towards an unfavorable metabolic state, as observed in rodents and primates [5–7].

Excessive energy intake and obesity are also increasingly prevalent in pregnant and breastfeeding women [8]. Since pregnancy and lactation are generally characterized by complex changes in metabolism as well as by intense immune activity [9,10], unhealthy diets and obesity during these periods additionally strain the organism [11].

The filamentous blue-green microalga *Arthrospira platensis* (also known as spirulina, Sp) is rich in bioactive phytochemicals, such as essential fatty acids (FA), vitamins, polyphenols, carotenoids, tocopherols, and phycocyanin [12]. These are presumed to contribute to the many beneficial health effects reported for Sp, particularly its antioxidant, anti-inflammatory, hypolipemic, and hepato-protective effects [13,14]. In vivo studies showed that Sp exerts these protective effects by modulating the expression levels and activities of transcription factors, thus regulating downstream transcription of target genes [15,16]. Regarding the potential beneficial health effects, the utilization of Sp as a supplement in human nutrition, including pregnant and breastfeeding women, is becoming increasingly popular [17]. Still, to the best of our knowledge, the potential metabolic effects of Sp intake during pregnancy and lactation have not yet been investigated.

Because of ethical issues that would arise from studies in pregnant and lactating women, different animal models have been employed, with rodents representing the most frequently used model for human nutrition and metabolism [18–20]. However, studies in rodents present some limitations that prevent the translation of research results to humans [21]. The pig (*Sus scrofa*) can be considered an interesting alternative to rodent models and might help bridge the gap between classical animal models and actual human physiology. Pigs share many anatomical, physiological, genomic and metabolic similarities with humans [22–24]. Minipigs have been mostly used to develop novel porcine models for metabolic disorders, but they are far more sensitive to an obesogenic diet than humans, because of a genetic predisposition for obesity [18]. Domestic pigs do not have this genetic predisposition and might thus be biologically closer to humans than the minipigs. Studies in pregnant and lactating pigs may shed new light on translational research investigating the metabolic effects of maternal WD.

Therefore, in the present study, we investigated if a diet rich in saturated FA (SFA), sucrose, fructose, and cholesterol in comparison to a commercial pig diet may induce a detrimental, obesity-related phenotype in gestating and lactating domestic pigs and if low-dose Sp supplementation may attenuate these potential adverse effects. So far, no study has evaluated if low-dose Sp supplementation proportional per unit of BW to recommendations in human nutrition (max. 8 g/day) to gestating and lactating pigs fed a WD might effectively attenuate adverse WD effects. We hypothesized that (i) feeding a WD to domestic pigs induces an obesity-like metabolic phenotype similar to that observed in humans; (ii) the effects of the WD differ during pre-gestation, gestation, lactation, and post-lactation periods; and (iii) low-dose supplementation of Sp may attenuate or even prevent the unfavorable phenotype observed in pigs fed a WD.

2. Materials and Methods

2.1. Animals, Housing, and Experimental Design

The experiment was approved by the Cantonal Veterinary Office of Zurich, Switzerland (license number ZH157/18), and performed in accordance with the Swiss legislation on animal rights and welfare. Lugarà et al. [25] provide a detailed description of the experiment. In brief, 24 female Large White × Landrace pigs (initial BW: 119.3 ± 8.2 kg, mean ± standard deviation (SD); average age: 5.6 ± 0.8 months; same sire; obtained from as few litters as possible to reduce the genetic diversity also from the maternal side) were housed in the Metabolic Center of the AgroVet-Strickhof research station (Lindau, Switzerland). Despite years of intense selection for leaner meat, when compared to other breeds such as Pietrain, such crossbreeds still possess genetic traits that allow for an increased intramuscular fat accumulation [26] and may thus also be susceptible to a WD. Due to limited housing capacity, the experiment was conducted in two identical runs (12 sows each, spring to winter in 2019 and 2020, respectively). The experimental design and schedule are outlined in Figure 1. Animals were housed in pairs at 17–22 °C and 60% relative humidity and fed ad libitum via automatic feeding stations (Pig Performance Tester, Nedap, Groenlo, The Netherlands), which recorded individual daily feed intake. Water and compressed straw were accessible ad libitum. After 6–8 weeks of pre-gestational feeding, animals were cycle synchronized via 18 days of oral application of Regumate® (MSD, New Animal Health, Wellington, New Zealand). Artificial insemination (AI) was performed with fresh sperm from the same boar (PREMO® Large White, Suisag, Sempach, Switzerland) for all animals. Animals were controlled for pregnancy on days 18 (heat detection) and 25 (ultrasonography) post AI. Animals not pregnant on day 25 were excluded from the experiment as further described below. From day 44 of gestation until weaning (30.4 ± 0.8 days after farrowing), animals were kept in individual farrowing pens. For weaning, the sows were moved back in pairs to the initial pens where they stayed for another 5 weeks until being slaughtered.

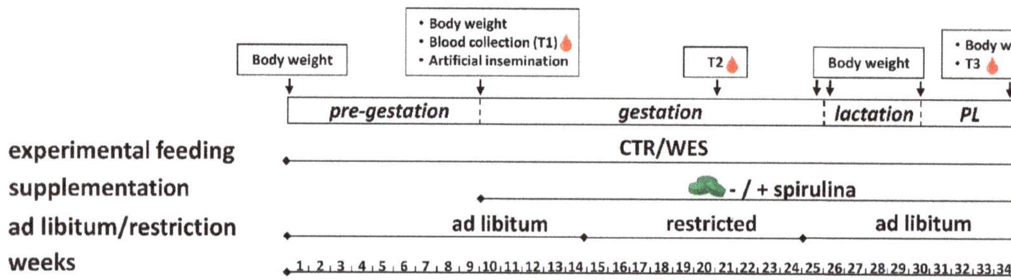

Figure 1. Schematic representation of the feeding experiment performed on female crossbred pigs in two runs (2019 and 2020) from spring to winter (34 weeks in total). Experimental sows were fed a control (CTR) or Western (WES) diet from the pre-gestation period to the post-lactation (PL) period for a total of 34 weeks. At artificial insemination, spirulina supplementation was started for half of both diet groups. Body weight was recorded at the start and end of each experimental period. Blood was collected at three time points (T1, T2, T3).

2.2. Experimental Diets

All diets used in this experiment were produced by Weinlandmuehle Truellikon (Glanzmann AG, Truellikon, Switzerland). For an initial adaption period of 2 weeks, all animals were offered a commercial compound feed designed to cover the basic nutritional requirements for gestating pigs. This diet was also used as the control diet (CTR_G) during the whole pre-gestation, gestation, and post-lactation periods (Table 1). After the adaption period, animals were assigned to two dietary groups (n = 12 per group). Initially, sisters sharing the same mother were randomly distributed into the two groups. The remaining animals were randomly assigned, but groups were balanced for a similar average BW. All animals were fed ad libitum either the control diet (CTR group) or a WD (WES group) for an initial pre-gestation period of about 9 weeks before AI (Table 1, Figure 1). The WES_G diet contained 150 g/kg freshly fed fully hydrogenated palm oil (ALIkon, Erbo Spraytech AG, Buetzberg, Switzerland), 200 g/kg saccharose (crystalline sugar, Schweizer Zucker AG, Aarberg, Switzerland), 150 g/kg fructose (Fructofin C, Danisco, Kotka, Finland), and 2 g/kg cholesterol (cholesterol from sheep wool ≥92.5%, Sigma-Aldrich, Buchs, Switzerland). As intended, the WES feed contained substantially more ether extract (+450%), total sugars (+570%), and cholesterol (+1532%) and ultimately provided a higher gross energy content (+3.8 MJ/kg DM) than the CTR feed. In contrast, the CTR feed provided a higher starch content than the WES feed (+82%). The fiber content was slightly higher in the WES than in the CTR feed (8.4 vs. 6.5 g/kg DM). In terms of FA, the WES feed contained a higher proportion of total SFA (+262%) than the CTR feed. Contrarily, the CTR feed contained higher proportions of unsaturated FA. In spirulina, the major FAs were C16:0, C18:3 n–6 (γ-linolenic acid, GLA), LA, C16:1 n–7, and C18:0 (in descending order).

Table 1. Ingredients and composition of experimental diets.

	Gestation Diet		Lactation Diet		
	CTR_G	WES_G	CTR_L	WES_L	Spirulina
Ingredients (g/kg as fed)					
Hydrogenated palm oil	-	150	-	150	-
Saccharose	-	200	-	200	-
D-Fructose	-	150	-	150	-
Cholesterol	-	2.0	-	2.0	-
Corn germ	590	298	488	200	-
Wheat	300	-	300	-	-
Soybean meal	45.0	130	145	217	-
Lignocellulose †	26.6	29.0	26.6	29.0	-
Monocalcium phosphate	13.0	20.0	11.2	20.0	-
Calcium carbonate	11.0	6.00	11.6	6.00	-
NaCl	6.2	6.0	6.2	6.0	-
Vitamin and mineral premix ‡	5.0	5.0	5.0	5.0	-
L-Lysine	2.3	2.5	3.6	5.4	-
DL-Methionine	-	1.0	0.7	1.9	-
DL-Tryptophan	0.2	0.4	0.5	1.0	-
L-Threonine	0.7	1.3	1.8	3.4	-
Valine	-	-	-	2.2	-
Chemical composition (g/kg dry matter (DM) if not stated otherwise)					
DM (g/kg fresh matter)	893	936	898	934	935
Total ash	47.8	47.1	54.9	54.3	90.2
Crude protein	110	93	150	162	638
Ether extract	26	158	29	143	47
Starch	462	93	408	74	51
Total sugars	46	359	52	326	<0.5
Crude fiber	6.7	8.7	6.2	8.1	n.a.
Cholesterol	0.03	0.48	0.02	0.45	n.a.
Gross energy (MJ/kg DM)	13.4	17.2	13.0	16.8	n.a.
Fatty acids (FA) composition (g/100 g total FA) §					
C12:0	0.14	0.64	0.30	0.08	0.03
C14:0	0.15	0.91	0.12	0.11	0.17
C16:0	15.9	30.7	12.3	5.9	42.9
iso C16:0	0.06	0.00	0.06	0.00	1.93
C16:1 n–7	0.15	0.01	0.12	0.03	5.44
C16:1*x*	0.34	0.00	0.04	0.00	0.05
C17:1	0.03	0.00	0.04	0.03	0.33
C18:0	4.4	56.4	14.8	81.4	3.4
C18:1 *cis*-9	26.0	3.3	21.9	2.8	1.6
C18:1 *cis*-11	0.88	0.11	0.67	0.12	0.73
C18:2 n–6 (LA)	47.6	5.0	45.1	5.0	16.5
C18:3 n–3 (ALA)	1.93	0.34	2.28	0.68	0.43
C18:3 n–6 (GLA)	0.0	0.0	0.0	0.0	23.2
C20:0	0.49	0.89	0.64	2.00	0.09
C20:1 n–9	0.42	0.04	0.34	0.06	0.12
C20:2 n–6	0.11	0.01	0.05	0.00	0.30
C20:3 n–6	0.00	0.00	0.00	0.00	0.38
C22:0	0.22	0.35	0.35	1.15	0.00
∑ Saturated FA	22.0	90.3	29.1	91.1	49.2
∑ Monounsaturated FA	28.0	4.3	23.3	3.1	8.9
∑ Polyunsaturated FA	50.0	5.40	47.6	5.75	41.9
∑n–6	47.7	5.0	45.2	5.0	40.4
∑n–3	1.9	0.3	2.3	0.7	0.4
n–6/n–3	25.1	14.7	19.6	7.4	101

CTR: Control; FA: fatty acids; WES: Western. n.a.: not analyzed. † Jeluvet Lignocellulose (Jelu-Werk J. Ehrler GmbH & Co. KG, Rosenberg, Germany). ‡ Naco Premix for breeding pigs (Vital AG, Oberentfelden, Switzerland), contains (per kg): lysine, 0.30 g; methionine, 0.10 g; Ca, 174 g; Fe (iron sulphate monohydrate E1), 16 g; native Mg, 8 g; Cu (copper sulphate pentahydrate E4), 3 g; Se (sodium selenite E 8), 0.05 g; Zn (zinc oxide 3b603), 18 g; iodine (calcium iodate anhydride 3b202), 0.2 g; vitamin D_3, 0.009 g; vitamin A, 0.72 g; vitamin E, 10 g; mineral oil, 10 g. § Only fatty acids (FA) with a proportion >0.3 g/100 g total FA in at least one of the feed items are displayed.

Feed allowance was 2 and 3 kg/day from day 40 to day 90 and from day 91 to day 105 of gestation, respectively. These restrictions were implemented to prevent problems at parturition that occur more frequently when sows are too heavy. One week before the expected parturition (day 107 of gestation), the diets of both groups were switched to lactation-type diets (CTR_L, WES_L) (Table 1, Figure 1). The lactation diets were fed ad libitum until weaning.

Two days after the final Regumate® administration and thus 3 days before AI, daily manual supplementation with 20 g of spirulina (Sp) in tablets (IGV Planttech GmbH, Nuthetal, Germany) was started for half of the animals in both dietary groups (−: without Sp; +: with Sp). This ultimately resulted in four experimental groups, namely CTR−, CTR+, WES−, and WES+. The Sp supplementation was continued until slaughter.

2.3. *Data and Sample Collection during the Feeding Experiment*

Throughout the experiment, individual daily feed intake was determined on 2 days per week, either from data registered by the feeding stations in the pens or by recording the amounts of supply and leftovers in the farrowing pens. The BW was measured using an animal scale (Mettler-Toledo, Greifensee, Switzerland) at the start and end of each experimental period. Backfat thickness at the level of the last rib was monitored using a portable ultrasound instrument (Carometec, Frontmatec Kolding, Kolding, Denmark) at the same time points. Feed samples were collected at the start of each run and then every 8 weeks throughout the experiment for compositional analyses. One week before expected farrowing, total feces were collected over a period of 24 h to analyze fecal fat excretion. All samples were stored at −20 °C until further analyses.

2.4. *Organ Collection at Slaughter*

After 16 h of overnight fasting, the sows were transported within 30–45 min to the abattoir at Tierspital (University of Zurich, Switzerland). They were slaughtered on six different dates in random order by exsanguination after electrical stunning. Liver and perirenal visceral adipose tissue (VAT) were collected warm and weighed. Samples of liver and abdominal muscle (*Obliquus externus abdominis*) were snap-frozen in liquid nitrogen and stored at −80 °C until further analyses. Additional liver and muscle samples were vacuum-sealed in plastic bags and stored at −20 °C for FA analysis. Liver, muscle, and VAT samples were fixed in 4% paraformaldehyde prepared in phosphate buffered saline (both: Sigma-Aldrich) at 4 °C for 48 h.

2.5. *Isolation and Quality Assessment of Liver and Muscle RNA*

Total hepatic RNA was isolated after homogenization (MagNA Lyser Green Beads and MagNA Lyser Instrument (Roche, Basel Switzerland), 7000× *g*, 30 s) using the AllPrep DNA/RNA Mini kit (Qiagen, Hilden, Germany) according to the manufacturer's instructions. To isolate total RNA from muscle tissue, samples were homogenized using TRIzol™ Reagent (Thermo-Fisher, CA, USA) and the MagNA Lyser Instrument (6500× *g*, 25 s). After homogenization, chloroform (Sigma-Aldrich) was added, and samples were centrifuged (12,000× *g*, 10 min, 4 °C). The upper phase was mixed with 100% ethanol, and the RNA purification was performed using the RNeasy Mini Kit (Qiagen) according to the manufacturer's instructions. The quantity of isolated RNA was estimated using a NanoDrop One spectrophotometer (Thermo Fisher, Waltham, MA, USA).

2.6. *Transcriptome Analyses*

Library preparation and RNA sequencing for liver and muscle tissue was performed by the Functional Genomics Center Zurich (Switzerland). For timing reasons, RNA sequencing was performed only for animals from the first run group, while qPCR was later used to validate the results in all animals (see next paragraph). Analysis of RNA integrity using a Fragment Analyzer System (Agilent Technologies, Wilmington, NC, USA) resulted in RQN numbers ≥8.6 for all samples. An RNA input of 1 µg was used for library preparation using the TruSeq Stranded mRNA Library Prep kit (Illumina, San Diego, CA, USA) following the manufacturer's instructions. Quality-controlled libraries were sequenced in a single multiplex on a NovaSeq instrument (Illumina) in 100 bp single-end read mode. Overall, the sequencing depth was about 25 million reads per sample. Read alignment was executed using STAR [27] and the porcine genome (Sscrofa11.1) as reference. Raw read counts were calculated with the featureCounts option of the Rsubread package [28]. The analyses of differentially abundant transcripts (DATs) were performed in R (version 4.1.2, R Core Team, 2020) using the DESeq2 package (version 1.30.0, Michael Love, Boston, MA, USA). The analysis generated the p-values corresponding to a differential expression test and the log2 Fold change (log2FC). Genes were considered differentially expressed at $p < 0.05$ and considering $-1.5 > FC > 1.5$. Comparative analyses were performed for WES− sows vs. CTR− sows and both Sp-supplemented groups vs their respective non-supplemented groups. Differentially expressed genes were subjected to functional analyses using the Ingenuity Pathway Analysis (IPA) software (version 76765844, Qiagen, Hilden, Germany, www.qiagen.com/ingenuity (accessed on 10 January 2021)). The overlap p-value was calculated using Fisher's exact test, and the z-score provides a prediction of the directional effect.

2.7. Validation of Transcriptome Data via qPCR

Reverse transcription was performed using 1 µg of hepatic RNA of all experimental animals with the GoScript Reverse Transcription System (Promega, Madison, WI, USA) according to the manufacturer's instructions. Quantitative real-time polymerase chain reaction was performed using the KAPA Sybr Fast Mix (Kapa Biosystems, Wilmington, NC, USA) according to the manufacturer's instructions on a CFX384 Real-Time PCR Detection System (Bio-Rad, Munich, Germany) with a 2-step amplification program (1 s at 95 °C, 30 s at 60 °C, 39 cycles). Primers (Table 2) were ordered at Microsynth (Balgach, Switzerland). Six reference genes were tested (ubiquitin B (UBB), actin B (ACTB), H3.3 histone A (H3F3A), hypoxanthine phosphoribosyltransferase 1 (HPRT1), tyrosine 3-monooxygenase/tryptophan 5-monooxygenase activation protein zeta (YWHAZ), succinate dehydrogenase complex flavoprotein subunit A (SDHA)). The RefFinder web tool was used to determine the most stable combination of reference genes for normalization of target gene expression [29]. The three most stable reference genes (H3F3A, YWHAZ, and UBB) were selected using the recommended comprehensive ranking, and the geometrical mean of their Cqs was used to calculate the ΔCq value of each target gene. Results are shown as log2 fold change, where the fold change was calculated as $2^{-\Delta\Delta Cq}$.

Table 2. List of primer sequences.

Gene	Forward Primer (5′→3′)	Reverse Primer (5′→3′)	Amplicon Length	Accession Number
ACTB	GATCTGGCACCACACCTTCT	AGAGACAGCACAGCCTGGAT	174	NM_173979.3
CYP1A2	GCCCAGCCCTACTCTGCAA	CCAGGAGATGGCTGTGGTAA	250	XM_005666124.3
CRP	TGAACACAGGCTCTCACATCC	CAAGCCAGACACTTGAATGCC	70	XM_003355107.4
GSTA1	CAGGACACCCAGGACCAATC	GTCTCAGGTACATTCCGGGAG	202	NM_214389.2
GSTA4	GCTCGGAGTGGACCCAGAAAA	TTCGGGTCTGCACCAACTTC	243	NM_001243379.1
H3F3A	AGGAGGTCTCTATACCATGGCTC	GAGCAATTTCCCGCACCAGA	245	NM_213930.1
HPRT1	TGCTGAGGATTTGGAGAAGG	CAACAGGTCGGCAAAGAACT	154	NM_001034035.2
IGF1	TGGTGGACGCTCTTCAGTTCG	ACAGTACATCTCCAGCCTCCTC	155	NM_214256.1
LIPG	CGAAACTCAGTTCCTCTGCTCT	TGGCTGTTGCATTGAAGCCA	247	XM_013992843.2
PCK1	GGGCATCATCTTCGGAGGG	AGTTGTAGCCGAAGAACGGC	182	XM_005673043.3
PTDGS	AAGAACTACGCCCTGCTCCA	ATGGCCAGGTCCTGAGAGT	231	NM_214228.1
SDHA	GCAGAACCTGATGCTTTGTG	CGTAGGAGAGCGTGTGCTT	185	NM_174178.2
UBB	CATTGTTGGCGGTTTCGCT	TTGACCTGTGAGTGAAGGCA	85	NM_001105309.1
YWHAZ	ATTGGGTCTGGCCCTTAACT	GCGTGCTGTCTTTGTATGACTC	146	XM_021088756.1

2.8. Compositional Analyses of Diets, Liver, Skeletal Muscle, and Feces

In all samples analyzed, compositional analyses were performed with similar methods. For analyses of dry matter (DM) and total ash (method 942.05; AOAC International, 1995), a thermo-gravimetric device (TGA 701, Leco Corporation, St. Joseph, MI, USA) was used. Using a C-N analyzer (Leco-CN 2000, Leco Corporation), nitrogen (N) content was determined (method 968.06; AOAC International, 1995), and crude protein was calculated as 6.25 × N. Ether extract was assessed by Soxhlet extraction using petroleum ether as a solvent (B-811, Büchi, Flawil, Switzerland; AOAC index no. 963.15). The gross energy was measured using a bomb calorimeter (Calorimeter C7000, IKA-Werke GmbH & Co. KG, Staufen, Germany). The cholesterol was extracted as described by Madzlan [30] and measured with a gas chromatograph (HP 6890, Agilent) equipped with a J&W Ultra1 column (25 m × 0.32 mm, Agilent Technologies). Total sugars (glucose, fructose, sucrose, lactose, maltose, xylose, and galactose) and starch content in the diets were analyzed by Eurofins Scientific AG (Schönenwerd, Switzerland) using, respectively, an HPAEC-PAD technique and an enzymatic assay (ISO 15914).

Hepatic total lipids were extracted as previously described by Lan et al. [31]. Briefly, 50 mg of frozen liver tissue was weighed, placed in ice-cold isopropanol, homogenized using a MagNA Lyser Instrument (Roche) at 7000× *g* for 30 s, and incubated for 10 min at room temperature while shaking. The samples were centrifuged at 1107× *g* for 10 min. Triglyceride and total cholesterol contents were analyzed in the supernatant using enzymatic reaction assays (Fluitest® TG and Fluitest® Chol, both from Analyticon, Lichtenfels, Germany) according to the manufacturer's instructions. Glycogen content was measured in both liver and muscle using the Glycogen Assay Kit (Fluorimetric) (Cell BioLabs, San Diego, CA, USA) following the manufacturer's instructions.

2.9. Histological Analyses of Liver, Skeletal Muscle, and Visceral Adipose Tissue

Histological analyses were performed at the Laboratory for Animal Model Pathology (University of Zurich). Samples were trimmed and embedded in paraffin wax. Consecutive sections (3–5 μm) were prepared, mounted on glass slides, and stained in hematoxylin and eosin (Sigma-Aldrich), according to the manufacturer's instructions. Adipocyte area was measured from microphotographs (Zeiss Axioskop 2, Zeiss, Jena, Germany) using the software ImageJ (v1.53e, National Institute of Health, Bethesda, Annapolis, MD, USA).

2.10. Blood Collection

During physical restraint with a snare, blood samples were collected at the end of the pre-gestation period (T1) and about 1 month before the expected parturition (T2) after 12 h of overnight fasting via jugular venipuncture into serum and plasma tubes (Z; clot activator, and K3-EDTA, S-Monovette, Sarstedt AG&Co. KG, Nuembrecht, Germany). At slaughter (T3), blood was collected during exsanguination into serum and plasma tubes. Tubes were centrifuged at 1500× *g* for 10 min at 4 °C. Serum and plasma supernatants were collected and stored at −80 °C until analyses.

2.11. Serum and Plasma Assays

Triglyceride and total cholesterol concentrations were determined in plasma by enzymatic reagent kits following the manufacturer's instructions (Fluitest® TG and Fluitest® Chol). Glucose was measured in serum samples using Multi-purpose kits—Glucose GOD FS (Dyasys, Diagnostic Systems GmbH, Holzheim, Germany). Hepatic biomarkers were analyzed in serum samples. In particular, alanine transaminase (ALT) and aspartate transaminase (AST) were analyzed using ALAT (GPT) FS and ASAT (GOT) FS (both from Dyasys), respectively. Concentrations of bilirubin, creatinine, gamma-glutamyl transferase (GGT), and glutamate dehydrogenase (GLDH) were analyzed using Bilirubin Auto Total FS (Dyasis), Creatinine FS (Dyasys), Gamma-GT FS (Szasz mod. /IFCC stand.) (Dyasys), and GLDH FS DGKC (Dyasys). Concentrations of glucose, creatinine, AST, GGT, GLDH, bilirubin, and ALT were measured using an automated analyzer (Cobas Miras Plus instrument, Roche Diagnostics AG, Rotkreuz, Switzerland). The insulin-like growth factor 1 (IGF1) was measured using a radioimmunoassay as described in detail by Vicari et al. [32]. Plasma insulin levels were measured using an ELISA kit (Mercodia, Uppsala, Sweden). The HOMA-IR index (Homeostasis model assessment of insulin resistance) evaluating the insulin resistance [33] was calculated from values obtained for fasting glucose (mmol/mL) and insulin (μIU/mL) using the formula: (Insulin × Glucose)/22.5. Total antioxidant capacity (TAC) and the ferric reducing antioxidant power (FRAP) were measured in serum using the OxiSelect Total Antioxidant Capacity Assay kit (Cell BioLab, San Diego, CA, USA) and the OxiSelect Ferric Reducing Antioxidant Power Assay kit (Cell BioLabs) according to the manufacturer's instructions.

2.12. Fatty Acid Analyses in Diets and Tissues

The FA in diets, liver, and abdominal muscle were analyzed as described in detail by Wolf et al. [34]. Briefly, from diets and tissues, FA were extracted with hexane:isopropanol in a ratio of 3:2 (*v/v*) using an accelerated solvent extractor (ASE 200, Dionex Corporation, Sunnyvale, CA, USA). The C11:0 (Fluka Chemie, Buchs, Switzerland) was used as internal standard. Methylation to FA methyl esters (FAME) was performed according to IUPAC (method 2.301). A gas chromatograph (HP 6890, Agilent Technologies) equipped with a CP7421 column (200 m × 0.25 mm, Varian, Lake Forest, IL, USA) was used to separate the FAME. Sunflower oil was used to produce a standard curve. Peak identification was further confirmed using chromatograms from Collomb and Buehler [35].

2.13. Statistical Analyses

Due to the experimental setup, sample sizes differed between the different experimental periods. The pre-gestation period with two experimental diets resulted in n = 12 per group. After forming the four experimental subgroups, three animals had to be excluded after the pre-gestation period because of not becoming pregnant, and one animal (WES+ group) had to be excluded because of miscarriage, resulting in group sizes: CTR− n = 5, CTR+ n = 6, WES− n = 5, and WES+ n = 4 for the gestation period. During the lactation period, one CTR+ sow experienced excessive weight loss and had to be euthanized, thus CTR+ was reduced to n = 5 in the lactation and post-lactation period as well as at slaughter. A CTR− animal was excluded due to agalactia from analyses during the lactation period (thus CTR− n = 4 in the lactation period). One of the non-pregnant WES+ animals was re-introduced into the experiment for the post-lactation period and slaughter (thus n = 5 for all experimental groups). All statistical analyses were performed using R [36] (version 4.1.2, R Core Team, Vienna, Austria). The packages lme4 and lmerTest were used to create linear mixed effect models, considering diet, spirulina supplementation, and their nested interaction as fixed effects. Animal, year of experiment, and mother were included as random effects. Normal distribution was confirmed by graphically inspecting the residuals and by using the Shapiro-Wilk test. When the fixed effects met a statistical significance, the normalized datasets were submitted to pairwise comparisons among least squares means using the emmeans package (v1.6.0, Russell V. Length, Iowa, USA). Differences were considered significant at $p < 0.05$ and a trend at $0.05 < p < 0.10$.

3. Results

3.1. Dietary Intake and Growth Performance

Feed intake did not significantly differ during any of the experimental periods between WES and CTR animals (Figure 2A). Therefore, WES animals had a significantly higher energy intake during the pre-gestation, gestation, and lactation periods (Figure 2B). The fat and sugar intakes in WES animals were significantly higher compared to CTR animals by about 5–6-fold and 6–7-fold, respectively, during all experimental periods (Figure 2C,D). In turn, the CTR animals had a significantly higher starch intake of about 5–6-fold during all experimental periods (Figure 2E). The supplementation of Sp did not significantly affect the feed intake and thus also did not affect the intake of energy, fat, sugar, and starch (Figure 2A–E).

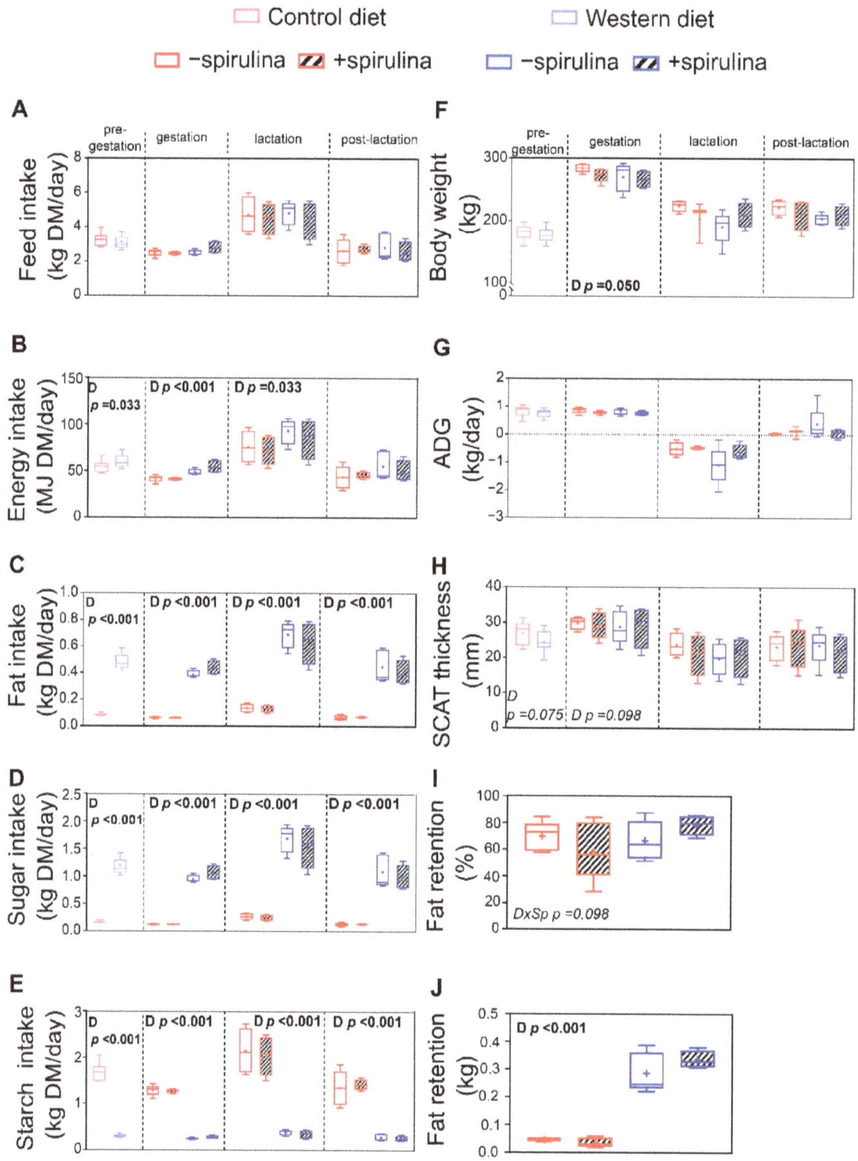

Figure 2. Effects of diet and spirulina supplementation on feed intake and growth performance of the sows during pre-gestation, gestation, lactation, and post-lactation, as well as body fat retention determined one week before expected farrowing. Experimental sows received either a control (CTR) or Western (WES) diet throughout all four experimental periods, while spirulina supplementation was started for half of CTR/WES groups at the end of the pre-gestation period. Differences in (**A**) feed, (**B**) energy, (**C**) fat, (**D**) sugar, and (**E**) starch intake during the four experimental periods; (**F**) body weight (BW) of experimental animals at the end of each period; (**G**) average daily gain (ADG) of experimental animals during each experimental period; (**H**) subcutaneous adipose tissue (SCAT) thickness at the end of each experimental period; (**I**) relative fat retention and (**J**) absolute fat retention one week before expected farrowing calculated from fat intake and fecal fat excretion during 24 h. Sp: spirulina; D: diet. Data are presented in boxplots with Spear style whiskers (min to max).

With the exception of a higher BW in CTR compared to WES sows shortly before farrowing (279 vs. 269 ± 3.4 kg), the animals' BW did not significantly differ between the dietary groups during the entire experiment (Figure 2F). Both absolute BW gain (data not shown) and average daily gain (ADG) were not significantly affected during any of the four experimental periods (Figure 2G). The Sp supplementation did not affect animal growth during any of the experimental periods (Figure 2F,G). The subcutaneous adipose tissue (SCAT) thickness did not significantly differ at any time point (Figure 2H). The absolute fat retention was about seven-fold higher in WES compared to CTR animals (0.30 vs. 0.04 ± 0.031 kg), while it was not significantly affected by Sp supplementation (Figure 2J).

3.2. *Differential Impact of WES Diet on Hepatic and Skeletal Muscle Transcriptome*

RNA sequencing of liver samples from WES− and CTR− sows revealed 446 differentially abundant transcripts (DATs), among them 183 with lower and 263 with higher abundance in the WES− compared to the CTR− group (Figure 3A). Biological functions related to glucose metabolism (glucose metabolism disorders, quantity of carbohydrates) and weight gain were predicted to be inhibited, while pathways related to lipid metabolism (FA metabolism, lipid oxidation, and quantity of protein-lipid complex in blood) were predicted to be activated (Figure 3B). Moreover, predicted activation of pathways related to hepatic inflammation and oxidative stress (inflammation of organ, chronic inflammatory disorder, quantity of reactive oxygen species) and liver damage (liver lesion, hepatic steatosis) was observed in WES− compared to CTR− sows. Particularly, analyses on regulator effects predicted inhibition of IGF1 signaling, activating the necrotic pathway (Figure 3C). In muscle samples, 729 DATs (116 with lower and 613 with higher abundance in the WES− group) were revealed (Figure 3D). In contrast to the liver, biological functions related to glucose metabolism were predicted to be activated in WES− compared to CTR− sows in the skeletal muscle (diabetes mellitus, metabolism of polysaccharides and carbohydrates), while differences in lipid metabolism were only indicated by a predicted activation of FA uptake (Figure 3E). Biological functions related to immune cell functionality (proliferation and cell death of immune cells, leukocyte migration, immune response of cells) were predicted to be activated in the muscle of WES− compared to CTR−animals. Upstream analyses showed predicted activation of both pro- and anti-inflammatory cytokines, such as tumor necrosis factor (TNF), interleukin 6 (IL-6), and interleukin 10 (IL-10), as well as a predicted inhibition of the aryl hydrocarbon receptor (AHR), thus suggesting a regulatory effect of the WES diet on inflammatory pathways also in the muscle tissue (Figure 3F).

3.3. *Changes in Gene Expression Influenced by Spirulina*

Spirulina supplementation altered the hepatic gene expression profile both in sows on the control diet (442 DATs; 188 with higher and 254 with lower abundance in CTR+ vs. CTR− samples) and on WD (524 DATs; 249 with higher and 275 with lower abundance in WES+ vs. WES− samples). Notably, only 15 and 25 transcripts were commonly up- and downregulated by Sp supplementation in both diet groups (Figure 4A). Dietary Sp resulted in some similarly regulated pathways in both diet groups (Figure 4B). Indeed, Sp supplementation may have inhibited biological functions such as metabolism of carbohydrates and FA, concentrations of cholesterol, D-glucose, and triacylglycerol, insulin resistance, inflammatory response, and apoptosis in livers of both diet groups (Figure 4B). On the other hand, Sp supplementation may have induced a diet-dependent regulation of the necrotic pathway that was predicted to be inhibited in CTR+ compared to CTR− but activated in WES+ compared to WES− sows. Likewise, biological functions related to the quantity of carbohydrates, insulin, and steroids in blood as well as size of body and lipid synthesis were predicted to be differentially regulated, showing activation in livers of CTR+ compared to CTR− but inhibition in WES+ compared to WES− animals.

In the muscle samples from sows fed the control diet, Sp supplementation altered the transcript abundance of 531 genes (397 up- and 134 down-regulated in CTR+ vs. CTR- samples). Muscle samples from WES+ sows revealed 262 DATs (172 up- and 90 down-regulated compared with WES− samples). The two diet groups had only 10 DATs in common—seven with increased and three with decreased abundances after Sp supplementation (Figure 4C). Among the differentially regulated biological functions, only weight gain was predicted to be similarly regulated in the skeletal muscle of both diet groups, being inhibited in Sp-supplemented compared to non-supplemented animals (Figure 4D). Biological functions such as apoptosis, cell movement of leukocytes, cellular infiltration by leukocytes, diabetes mellitus, and growth of connective tissue were predicted activated in CTR+ compared to CTR− sows but inhibited in WES+ compared to WES− sows. In contrast, inflammation was predicted to be inhibited in CTR+ compared to CTR− sows and activated in WES+ compared to WES− sows.

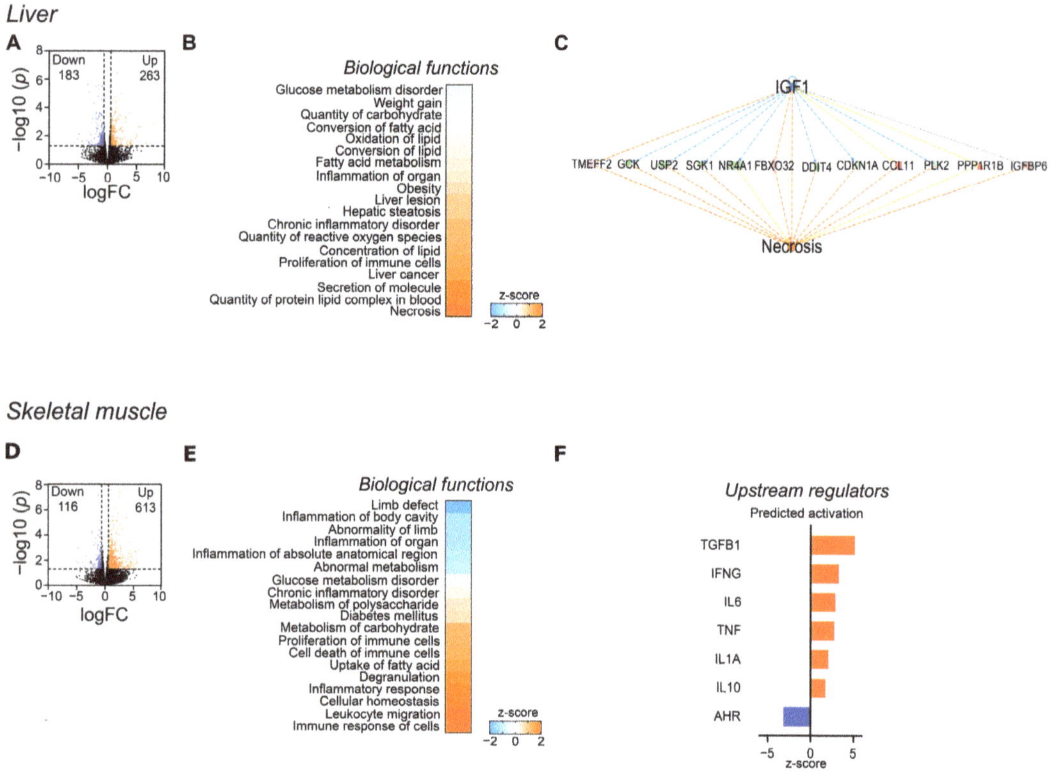

Figure 3. Effects of the WES diet on gene expression of liver and skeletal muscle at slaughter. Only CTR− and WES− animals were included in this analysis. (**A**) Volcano plot representing the RNAseq results in liver in WES− sows compared to CTR− sows. Orange dots indicate the differentially abundant transcripts (DATs) that are significantly up-regulated (cut-off of $p < 0.05$ and $-1.5 >$ fold change $> +1.5$), while the blue dots represent the down-regulated DATs (cut-off of $p < 0.05$ and $-1.5 >$ fold change $> +1.5$). (**B**) Heatmap representing the z-score of biological functions significantly affected by WES diet (calculation performed in IPA) in the liver. Blue indicates a negative z-score (inhibition) and orange a positive z-score (activation). (**C**) Regulatory effect of IGF-1 on the hepatic necrosis pathway. (**D**) Volcano plot representing the RNAseq results in skeletal muscle. Orange dots indicate the up-regulated DATs (cut-off of $p < 0.05$ and $-1.5 >$ fold change $> +1.5$), while the blue dots represent the down-regulated DATs (cut-off of $p < 0.05$ and $-1.5 >$ fold change $> +1.5$). (**E**) Heatmap representing the z-score of biological functions significantly affected by WES diet (calculation performed in IPA) in muscle. (**F**) Predicted activation of upstream regulators in skeletal muscle of WES− sows.

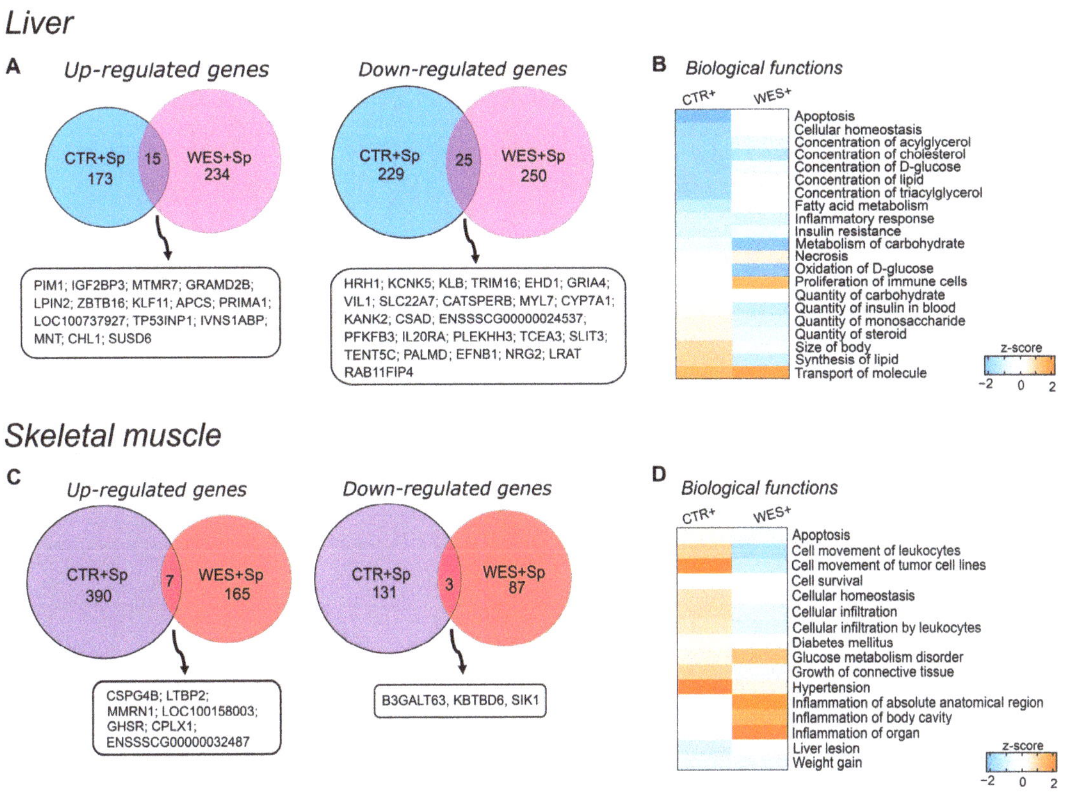

Figure 4. Effects of Sp supplementation on gene expression of liver and muscle at slaughter. (**A**) Venn diagram showing the intersection of differentially abundant transcripts in CTR+ and WES+ livers. Venn diagrams were designed using http://bioinformatics.psb.ugent.be/webtools/Venn/ (accessed on 1 September 2021). (**B**) Heatmap representing the z-score of biological functions significantly affected in both CTR+ and WES+ groups (calculation performed in IPA) in the liver. Blue indicates a negative z-score (inhibition) and orange a positive z-score (activation). (**C**) Venn diagram showing the intersection of differentially abundant transcripts in CTR+ and WES+ muscle. (**D**) Heatmap representing the z-score of biological functions significantly regulated in CTR+ and WES+ groups (calculation performed in IPA) in muscle.

3.4. Validation of RNAseq by qPCR

Selected DATs related to lipid and glucose metabolism, inflammation, and oxidative stress identified by RNAseq analysis were validated using qPCR. Expression of hepatic cytochrome P450 family 1 subfamily A member 2 (CYP1A2) and prostaglandin D2 synthase (PTDGS) was confirmed to be significantly up-regulated in WES groups compared to CTR groups (Figure 5A,B), while expression of glutathione S-transferase alpha 4 (GSTA4) was confirmed to be significantly down-regulated in WES vs. CTR sows (Figure 5C). In livers of Sp-supplemented groups, a significantly lower expression of GSTA4 (Figure 5C) and a significantly higher expression of the C-reactive protein (CRP) were observed compared to non-supplemented groups (Figure 5F). According to the RNAseq data, the hepatic expression of IGF1 did not differ due to diet or Sp supplementation, which was confirmed by qPCR (Figure 5H).

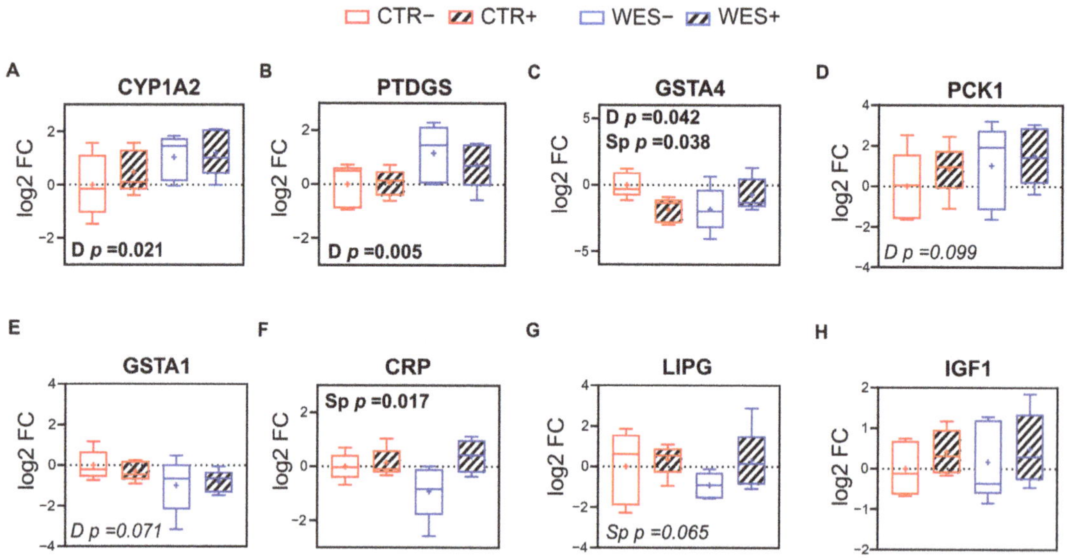

Figure 5. Validation of RNAseq results of selected differentially expressed genes in liver via qRT-PCR. Hepatic levels of expression of (**A**) cytochrome P450 family 1 subfamily A member 2 (*CYP1A2*), (**B**) prostaglandin D2 synthase (*PTDGS*), (**C**) glutathione S-transferase alpha 4 (*GSTA4*), (**D**) phosphoenolpyruvate carboxykinase 1 (*PCK1*), (**E**) glutathione S-transferase alpha 1 (*GSTA1*), (**F**) C-reactive protein (*CRP*), (**G**) lipase G (*LIPG*), and (**H**) insulin-like growth factor 1 (*IGF1*). CTR: control diet; WES: Western diet; D: diet; Sp: spirulina. Data are presented in boxplots with Spear style whiskers (min to max).

3.5. Organ Proportions, Composition, and Histology

The proportions and composition of liver, skeletal muscle, and visceral adipose tissue are presented in Table 3. The Sp-supplemented compared to non-supplemented sows had a significantly lower relative liver weight (10.0 vs. 11.6 ± 0.41 g/kg BW). The WES sows had a significantly higher cholesterol (55.8 vs. 53.6 ± 0.83 mg/100 g) and a significantly lower glycogen (7.0 vs. 13.3 ± 1.29 μM/g) content in muscle tissue. The Sp supplementation significantly increased the intramuscular cholesterol concentration compared to non-supplemented animals (55.9 vs. 53.5 ± 0.84 mg/100 g). No difference was observed concerning the relative VAT weight. A significant interaction of experimental diet and Sp supplementation was found for adipocytes with an area of 15–25 × 103 μm^2. Histological examination did not reveal any visible structural differences in liver, skeletal muscle, and VAT (Supplementary Figure S1).

3.6. Blood Biochemical Biomarkers

Spirulina supplementation resulted in a significantly lower serum glucose concentration (4.82 vs. 5.25 ± 0.179 mmol/L) during late gestation compared to non-supplemented animals (Figure 6A). Plasma IGF1 concentrations were significantly lower in WES compared to CTR sows (Figure 6D) at all three time points. Plasma total cholesterol concentrations were significantly higher in WES compared to CTR sows at slaughter (123 vs. 116 ± 2.3 mg/dL) but at none of the other time points (Figure 6B).

Table 3. Liver, skeletal muscle, and visceral adipose tissue composition of sows following 34 weeks of control (CTR) or Western (WES) diet feeding. Spirulina supplementation was started from week nine for half of the CTR/WES groups.

Diet (D)	Control Diet		Western Diet		SEM	Significance		
Spirulina (Sp)	−	+	−	+		D	Sp	D × Sp
Liver								
Proportion (g/kg BW)	11.1	10.1	11.8	10.8	1.10	#	*	n.s.
Glycogen (µM/g)	2.17	1.14	2.10	1.41	0.826	n.s.	n.s.	n.s.
Triglycerides (mg/g)	4.20	4.17	4.13	4.23	1.057	n.s.	n.s.	n.s.
Cholesterol (mg/g)	1.33	1.47	1.85	1.95	0.321	#	n.s.	n.s.
Skeletal muscle								
Intramuscular fat (g/kg DM)	55.7	82.7	56.0	52.3	17.30	n.s.	n.s.	n.s.
Protein (g/kg DM)	18.5	19.0	19.3	17.9	1.19	n.s.	n.s.	n.s.
Cholesterol (mg/100g)	51.1	54.9	55.1	56.7	2.53	*	*	n.s.
Glycogen (µM/g)	13.7	12.8	9.81	4.17	2.650	**	n.s.	n.s.
Visceral adipose tissue								
Proportion (g/kg BW)	12.6	11.6	10.1	12.2	1.86	n.s.	n.s.	n.s.
Adipocyte area ($\mu m^2 \times 10^3$)	16.3	12.2	14.2	16.5	3.23	n.s	n.s.	#
Adipocyte proportion (%)								
<5 ($\mu m^2 \times 10^3$)	5.10	5.86	4.05	6.00	6.300	n.s.	n.s.	n.s.
5–15	41.0	66.8	55.5	39.6	24.68	n.s.	n.s.	n.s.
15–25	52.0	23.4	36.1	49.0	15.76	n.s.	n.s.	*
>25	5.68	2.37	4.37	8.85	8.58	n.s.	n.s.	n.s.

Data are presented as least square mean ± standard error of the mean (SEM). Statistical significances were set at * $p < 0.05$ and ** $p < 0.01$. Trends were defined as # $0.05 < p < 0.10$. n.s.: not significant.

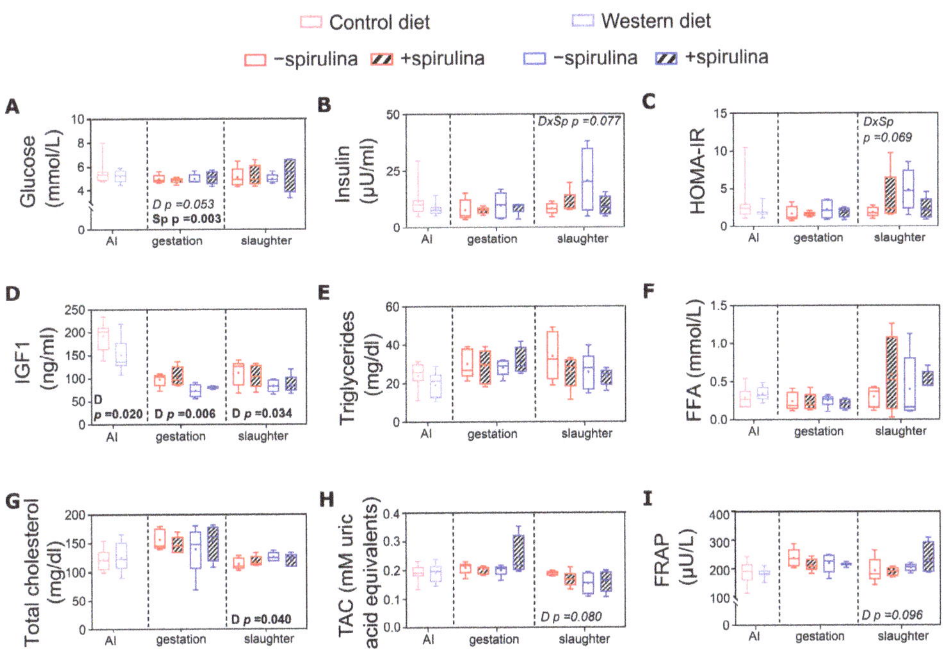

Figure 6. Systemic metabolic and oxidative stress markers of sows at artificial insemination (AI), 30 days before calculated farrowing (gestation), and 5 weeks post-lactation (slaughter). Concentrations of (**A**) serum glucose, (**B**) plasma insulin, (**C**) calculated index for insulin resistance (HOMA-IR), concentrations of (**D**) plasma IGF1, (**E**) plasma triglycerides, (**F**) serum free fatty acids (FFA); (**G**) plasma total cholesterol, and (**H**) serum total antioxidant capacity (TAC) as well as (**I**) serum ferric reducing antioxidant power (FRAP). CTR: control diet; WES: Western diet; D: diet; Sp: spirulina. Data are presented in boxplots with Spear style whiskers (min to max).

At the end of the pre-feeding period, serum bilirubin and ALT were significantly higher in WES compared to CTR sows (Figure 7A,B). Also during late gestation, significantly higher values were observed for serum bilirubin and ALT (Figure 7A,B) in WES compared to CTR animals (82 vs. 44 ± 5.2 U/L and 5.0 vs. 2.4 ± 0.43 µmol/L, respectively). For ALT and GLDH, a significant diet × Sp supplementation interaction was observed during late gestation (Figure 7B,D), with lower values detected in CTR+ compared to CTR− and higher values detected in WES+ compared to WES− animals.

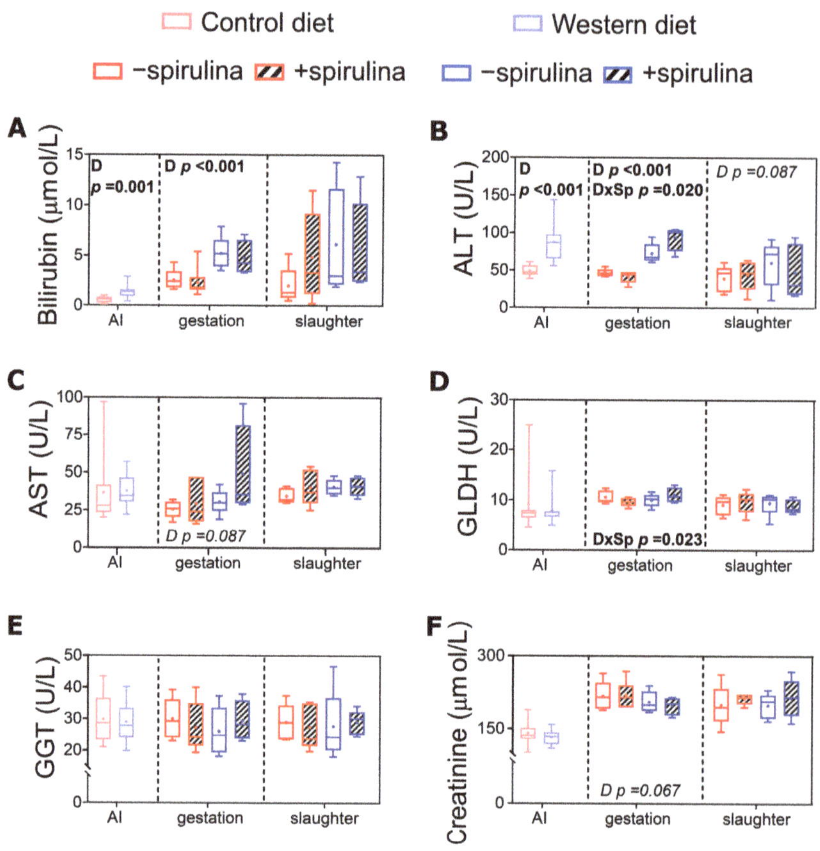

Figure 7. Biomarkers of organ functionality in serum of sows at artificial insemination (AI), 30 days before calculated farrowing (gestation), and 5 weeks post-lactation (slaughter). (**A**) Concentration of serum bilirubin, activities of (**B**) serum alanine transaminase (ALT), (**C**) serum aspartate transaminase (AST), (**D**) serum glutamate dehydrogenase (GLDH)[1], (**E**) serum gamma-glutamyl transferase (GGT), and (**F**) concentration of serum creatinine. CTR: control diet; WES: Western diet; D: diet; Sp: spirulina. Data are presented in boxplots with Spear style whiskers (min to max). Pairwise comparison showed no significant difference.

3.7. Fatty Acid Composition of Liver and Skeletal Muscle

Diet and Sp supplementation influenced the FA profile of liver and skeletal muscle at slaughter to different extents (Table 4). In liver tissue, the proportions of total SFA were significantly higher in WES compared to CTR sows. Significantly lower total n–6 FA were observed in the livers of WES compared to CTR animals without significantly affecting the n–6/n–3 FA ratio. The proportions of LA, C20:0, C20:1 n–9, C20:2 n–6, and C22:4 n–6 were significantly lower, while the proportions of DGLA and EPA were significantly higher in

WES compared to CTR sows (Supplementary Table S1). Supplementation with Sp hardly modified the liver FA profile with significantly lower proportions of C17:0 and DGLA (Supplementary Table S1).

Table 4. Effects of Western diet and spirulina supplementation on the main FA groups present in liver and skeletal muscle of sows (g/100 g total FA).

Diet (D)		Control Diet		Western Diet			Significance		
Spirulina (Sp)		–	+	–	+	**SEM**	**D**	**Sp**	**D × Sp**
	Tissue								
∑SFA	Liver	41.6	41.6	43.3	42.6	0.78	*	n.s.	n.s.
	Muscle	41.0	39.9	42.6	40.9	1.10	n.s.	#	n.s.
∑MUFA	Liver	18.8	18.4	18.3	21.8	3.47	n.s.	n.s.	n.s.
	Muscle	46.8	47.3	44.9	47.5	2.14	n.s.	n.s.	n.s.
∑PUFA	Liver	39.7	40.0	38.3	35.6	2.87	#	n.s.	n.s.
	Muscle	12.5	12.7	13.1	10.6	2.42	n.s.	n.s.	n.s.
∑n–3	Liver	3.60	3.72	3.90	3.58	4.500	n.s.	n.s.	n.s.
	Muscle	0.58	0.62	0.78	0.63	0.119	n.s.	n.s.	n.s.
∑n–6	Liver	36.0	36.1	34.3	31.9	2.49	*	n.s.	n.s.
	Muscle	11.7	12.0	12.3	9.8	2.31	n.s.	n.s.	n.s.
n–6/n–3	Liver	10.0	9.7	8.8	8.9	0.01	n.s.	n.s.	n.s.
	Muscle	20.2	19.4	15.8	15.6	0.003	***	n.s.	n.s.

SFA: saturated fatty acids; MUFA: monounsaturated fatty acids; PUFA: polyunsaturated fatty acids. Data are presented as least square mean ± standard error of the mean (SEM). Statistical significances were set at * $p < 0.05$ and *** $p < 0.001$. Trends were defined as # $0.05 < p < 0.10$. n.s.: not significant.

In muscle tissue, diet resulted in higher C17:0 and C18:0 proportions in WES compared to CTR sows (Supplementary Table S1). The proportions of C20:1 n–9 and C20:2 n–6 (Supplementary Table 1) as well as the n–3/n–6 FA ratio (Table 4) were significantly lower in WES compared to CTR sows′ muscle. The Sp supplementation resulted in significantly lower intramuscular C17:0 and C18:0 proportions (Supplementary Table S1). A significant interaction of diet and Sp supplementation was observed for C17:0, with WES− sows having higher proportions than all other groups (Supplementary Table S1). In addition, for intramuscular ALA proportions, a significant interaction of diet and Sp supplementation was observed. The ALA proportion was higher in CTR+ than in CTR− animals, while it was lower in WES+ than in WES− animals.

4. Discussion

Feeding a WD to domestic pigs before and during gestation as well as during and after lactation was not sufficient to induce an obese phenotype. Still, some metabolic dysfunctions occurred, as indicated by the significantly lower IGF1 concentrations as well as an activation of diabetes mellitus–related pathways in muscle tissue of WES compared to CTR animals. Moreover, serum hepatic biomarkers before and during gestation as well as hepatic cholesterol accumulation and gene expression at slaughter might indicate the onset of non-alcoholic steatohepatitis (NASH) in WES sows. The supplementation of Sp reduced plasma insulin levels in WES+ compared to WES− sows at slaughter to similarly low levels as observed in CTR− and CTR+ sows. However, Sp was not able to counteract the onset of NASH observed in WES sows.

4.1. Effects of a Western Diet on the Metabolic Phenotype of Female Domestic Pigs

4.1.1. Western Diet Intake Induced Regulatory Processes to Maintain Body Weight and Visceral Adipose Tissue Proportion

Despite the higher intake of gross energy, fat, and sugars in WES compared to CTR sows throughout the experiment, the BW, relative VAT weight, and SCAT thickness did not differ between WES and CTR animals at any time point. This contrasts with previous literature in other porcine models [37–43], where high energy and/or high-fat, high-sugar diets increased the BW by at least 60% and the VAT mass by twofold compared to control groups.

However, most of these studies used minipigs [37–41,43], which are more susceptible to BW gain and visceral adiposity in response to an obesogenic diet than domestic pigs [18].

The observed lack of BW differences in the present study might be explained by the IPA-predicted activation of lipid oxidation in the liver and FA uptake in the muscle of WES− compared to CTR− sows. Switching between glucose and FA as an energy source indicates metabolic flexibility, which allows an organism to successfully respond to changes in metabolic and energy demand [44]. The lack of both tissue triglyceride accumulation as well as increased BW of WES compared to CTR animals points towards an increased FA utilization in liver and skeletal muscle, which may be explained by a higher metabolic flexibility of the WES animals.

Even though the pigs in the present study did not develop an obese phenotype, it must be considered that the "lean metabolic syndrome" (LMS) has recently gained increasing attention in human medicine [45]. The LMS patients have a normal BW and VAT proportion but still show two or more characteristics of the metabolic syndrome [46,47]. Therefore, the WES pigs used in the present study might be a particularly suitable model for the human LMS as indicated by metabolic characteristics that are discussed in the following sections.

4.1.2. Liver Functionality Was Adversely Affected by the Western Diet

Elevated systemic transaminase (ALT, AST) and bilirubin levels as indicators for hepatic damage have been reported in rodents and human patients with obesity and fatty liver [48–51]. Serum bilirubin and ALT were about twofold higher in WES compared to CTR animals after the pre-feeding period at AI as well as one month before farrowing, indicating an impaired liver functionality in the WES compared to the CTR sows during gestation. Although transcriptome analysis of liver tissue predicted activation of pathways related to liver stress and damage in WES− compared to CTR− sows, these systemic markers did not significantly differ in our sows following the lactation period at slaughter. We suggest that catabolic processes occurring during lactation might have normalized systemic hepatic biomarkers at slaughter. Similarly, in women with a history of gestational diabetes, breastfeeding reset the maternal metabolism in the early post-partum period [52].

Enhanced inflammation and oxidative stress, as observed in obesity and LMS [53,54], were predicted by pathway analysis in livers of WES− compared to CTR− animals. Since the serum TAC and FRAP in the present study did not significantly differ between diet groups, we suggest that increased ROS levels and associated gene expression, as predicted by IPA of liver samples and confirmed by qPCR, might have been limited to liver tissue without affecting the systemic antioxidant status. The family of glutathione S-transferase (GST) enzymes is able to protect against ROS-induced damage by regulating tissue regeneration pathways in the liver [55–57]. A reduced hepatic expression of the GSTA1 isoform is considered a marker of hepatic injury [56]. Expression of both GSTA1 and GSTA4 was lower in the livers of WES compared to CTR sows, thus further confirming a certain extent of liver damage and impaired hepatic regenerative capacity in WES sows.

Oxidative stress and inflammation in liver tissue are caused, e.g., by a hepatic lipid accumulation [58]. Transcriptome activation of pathways related to hepatic steatosis was observed in WES− compared to CTR− sows. The higher proportions of hepatic SFA in WES compared to CTR sows reflected the higher SFA proportion of the WES feed. Continuous exposure to high proportions of dietary SFA has been shown to promote the accumulation of SFA in liver tissue and thus the development of fatty liver disorders by inducing pro-inflammatory pathways and stimulating apoptosis due to endoplasmic reticulum and oxidative stress [59]. The significantly higher hepatic SFA proportions in WES compared to CTR animals at slaughter thus further support a prevalent impaired liver functionality in WES sows. Similar to obese humans, a frequent occurrence of NASH or nonalcoholic fatty liver diseases has been observed in human LMS patients [60,61].

In line with this, the most strongly activated biological function identified by pathway analysis indicated the occurrence of necrotic processes in the liver of WES− compared to CTR− sows. This was also predicted to be induced by reduced IGF1 signaling. The IGF1 is a growth hormone that has been shown to exert hepatoprotective effects by promoting tissue regeneration in the case of injury [62]. In line with the predicted reduction of IGF1 signaling, lower plasma IGF1 concentrations were indeed observed in WES compared to CTR animals throughout the entire experimental period including slaughter. Reduced IGF1 abundance and thus signaling suggest an impaired regenerative capacity of the liver in WES sows, also matching the observed (temporary) increase in serum bilirubin and transaminases.

4.1.3. The Impaired Glucose Metabolism Induced by the Western Diet Was Likely Mediated by Metabolic Changes Predominantly in the Skeletal Muscle

Diabetes mellitus related pathways were activated in the muscle of WES− compared to CTR− sows. Interestingly, pathways for glucose metabolism disorder were instead inhibited in both liver and muscle. This may indicate that the increased intake of fat and sugars indeed activated diabetes mellitus–related pathways in the muscle but that in both muscle and liver, processes counteracting these negative WD-induced effects were activated. Previous studies in minipigs have similarly failed to induce diabetes mellitus through feeding [63]. The main reason why pigs do not easily develop diabetes lies in the evolutionary history of wild pigs, which had to store energy during the summer to be later used in the winter when food availability was lower, thus building a pancreatic resistance to diabetogenic diets [18]. Still, at all time points, WES compared to CTR sows had lower plasma IGF1 concentrations as observed in patients with a long history of diabetes [64]. A reduced glycogen content in muscle, as observed in WES compared to CTR sows, was also found in both type 1 and 2 diabetic patients [65]. Taken together, results from the present study point towards a mild diabetogenic-like phenotype that was induced by WD feeding in pigs.

4.2. *Effects of Spirulina Supplementation on the Metabolic Phenotype of Female Domestic Pigs*

4.2.1. Spirulina Supplementation Did Not Affect Body Weight but Potentially Aggravated WD-Induced Liver Damage

In humans, Sp supplementation has been shown to promote weight loss via regulation of food intake and adipokines [66]. In contrast, Sp supplementation did not affect either feed intake or BW of the adult but still growing sows in the present study. This also contrasts with the predicted inhibition of weight gain obtained from gene expression analysis in muscle tissue of Sp-supplemented compared to non-supplemented pigs.

Despite the similar BW and VAT proportion, Sp supplementation led to a diet type–independent significantly decreased relative liver weight compared to non-supplemented animals. Liver weight differences are usually caused by differences in fat or glycogen accumulation or by hepatic degeneration [67]. The similar hepatic lipid and glycogen concentrations in Sp-supplemented and non-supplemented groups contradict the IPA prediction of Sp-mediated higher hepatic lipid and lower hepatic carbohydrate concentration in WES compared to CTR animals. The significant interaction of diet and Sp supplementation for plasma ALT and GLDH might point towards a diet-dependent effect of Sp on liver functionality that even aggravated the adverse effects of the WD. In line with this, pathway analyses showed that Sp supplementation in the CTR+ group inhibited hepatic necrosis, while this pathway was activated in the WES+ group. Therefore, Sp in combination with the WES diet in pigs seems to not exert the hepatoprotective effects previously described in rodent models fed a WD [14,68].

4.2.2. Spirulina Supplementation Had Only Small Diet-Dependent Effects on Systemic Metabolic Biomarkers

Inhibition of insulin resistance and activation of glucose metabolic disorders were predicted by hepatic and muscle gene expression, respectively, in Sp-supplemented compared to non-supplemented animals, independent of diet type. However, a significant interaction of diet and Sp supplementation was observed for plasma insulin concentration at slaughter. In line with the gene expression prediction, insulin concentrations of WES+ sows were similarly low to those of the CTR− and CTR+ sows. These findings suggest that Sp supplementation in WD-fed pigs may maintain plasma insulin concentrations at a normal level via a yet unknown mechanism.

Plasma lipids and antioxidant capacity were not affected by Sp supplementation. In previous reports, Sp supplementation has been shown to exert hypolipemic effects in both rodent models of obesity and in obese patients [16,66]. These individuals were, however, neither pregnant nor lactating. Given the extensive metabolic changes during pregnancy and lactation, those study results can therefore not be directly compared to the results of the present study.

4.2.3. Spirulina Supplementation Had Only Minor Effects on the Tissue Fatty Acid Profiles

Although Sp is rich in GLA and LA as well as in other bioactive components potentially affecting FA metabolism, Sp supplementation had very little effect on the FA profiles of liver and muscle in the present study. Importantly, the proportions of GLA were higher in the liver of Sp-supplemented versus non-supplemented animals. Hepatic incorporation of GLA plays a pivotal role in reducing fat deposition by enhancing the activity of enzymes involved in FA oxidation [69]. However, no such effect was observed in the present experiment, as hepatic lipids were not affected in Sp-supplemented compared to non-supplemented animals.

4.3. Limitations of the Study

Feeding a WD to adult but still growing domestic sows did not induce a typical obesity phenotype as intended. Therefore, using adult sows as a translational model for human obesity does not seem feasible. However, pigs grow faster before puberty [70]. Starting the WD feeding at an earlier age and/or extending the experimental feeding period to potentially increase the magnitude of the adverse metabolic effects might thus have a considerable impact on body composition, as was shown by Fisher et al. [71], who fed the same proportions of fats and sugar to growing pigs.

The composition of the CTR diet might also have introduced a bias since it was rich in starch, and starch-rich diets were previously demonstrated to be also positively associated with metabolic syndrome [72]. Therefore, the lack of significance in some parameters may be due to the fact that the CTR group also consumed a diet that could potentially lead to similar disturbances and would not be considered particularly healthy in human nutrition.

Importantly, the metabolic changes naturally occurring during gestation and lactation likely interacted with the experimental diets. Despite the absence of an obesity phenotype, the sows presented with some metabolic abnormalities manifesting in the protein level during pre-gestation and gestation. The indications of metabolic abnormalities on the hepatic and muscle transcriptome level at the end of lactation could, however, not be confirmed on the protein level at the same time point. The experimental design limited tissue sample collection to the end of the 5-week lactation period, which is known to cause significant metabolic adaptations via catabolic processes that potentially ameliorated the detrimental effects that were indicated during gestation [52]. Analyses of liver and muscle biopsies during gestation would have provided more reliable insights since the current data do not allow us to exclude with certainty that significant differences in lipid accumulation and thus a fatty liver were present before or during gestation.

The few effects of Sp supplementation on the mitigation of detrimental WD effects might be related to the rather low dosage applied. Previous studies have shown that higher Sp dosages were used to achieve beneficial health effects in pigs [73]. Still, the amount supplemented in the present study corresponded to the amount recommended in human nutrition relative to BW. The lack of mitigating Sp effect might, however, also be due to the absence of an obesity phenotype in our WES sows. Therefore, future studies on the potential beneficial effects of Sp supplementation on obesity-induced metabolic disturbances should be performed in animals showing a pronounced obesity phenotype, as can be observed in humans.

Overall, it has to be mentioned that group sizes in the present study were rather small, and larger studies should be designed to investigate the observed effects in more detail.

5. Conclusions

Our hypothesis that (i) feeding a WD to domestic pigs induces an obesity-like metabolic phenotype similar to that observed in humans is only partially confirmed by the study results. The variable effects on metabolic biomarkers observed during the different physiological states confirm the hypothesis that (ii) the effects of the WD differ during pre-gestation, gestation, lactation, and post-lactation periods. The hypothesis that (iii) low-dose supplementation of Sp may attenuate or even prevent the unfavorable phenotype observed in pigs fed a WD was only partially confirmed by the normalized plasma insulin levels in WES+ animals, representing the only counter-regulatory effect observed. Further investigations are of interest to determine if Sp supplementation may prevent more pronounced adverse metabolic effects in animals with more severe diet-induced metabolic impairments and if higher Sp doses might be more effective. In addition, more in-depth analyses, particularly of liver and muscle tissue during gestation and lactation, are warranted to understand the nutritional impact of WD and Sp supplementation during such special metabolic events.

Supplementary Materials: The following supporting information can be downloaded at: https://www.mdpi.com/article/10.3390/nu14173574/s1, Figure S1: Morphology of liver, skeletal muscle and visceral adipose tissue collected at slaughter.; Table S1: Effects of Western diet and spirulina supplementation on the FA composition of liver and skeletal muscle of sows (g/100 g total FA).

Author Contributions: Conceptualization, R.L. and K.G.; methodology, R.L., S.R., A.L., R.B. and K.G.; formal analysis, R.L., S.R., A.L., R.B. and K.G.; investigation, R.L. and K.G.; resources, K.G.; data curation, R.L. and K.G.; writing—original draft preparation, R.L. and K.G.; writing—review and editing, R.L., S.R., E.W., A.L., R.B. and K.G.; supervision, K.G.; project administration, K.G.; funding acquisition, K.G. All authors have read and agreed to the published version of the manuscript.

Funding: The study was supported by the ETH Zurich Foundation and the World Food System Center of ETH Zurich.

Institutional Review Board Statement: The animal experiment was approved by the Cantonal Veterinary Office of Zurich, Switzerland (license no. ZH157/18).

Informed Consent Statement: Not applicable.

Data Availability Statement: The Fastq files from the RNAseq experiment were deposited in NCBI's Gene Expression Omnibus and are accessible through GEO Series accession number GSE197750. All other data are included within the article and supplementary data. Details can be made available upon reasonable request.

Acknowledgments: The authors are especially grateful to M. Kreuzer for his financial support and valuable comments on the manuscript. We are grateful to the staff of AgroVet-Strickhof and the Animal Nutrition group at ETH Zurich and UZH for their assistance and support during the animal experiment and in the lab. Acknowledgements go to C. Kunz, N. Perez, M. Mergani, N. Kachappilly, and C. Blechinger for assistance with laboratory analyses, to S. Amelchanka and R. Jendly in the animal facility, to W. Pendl for assistance during blood sampling and artificial insemination, to S. Greve, M. Keller, M. Heuel, and A. Birkinshaw for helping during sample collection, and to H. Renfer and his whole team for their assistance in the animal experiment. Special thanks go to S. Bauersachs (University of Zurich) and S.E. Ulbrich (ETH Zurich) for granting access to laboratory facilities and instruments and to F. Janett (University of Zurich) for providing liquid nitrogen at sampling. We are also grateful for the support of P. Müller and D. Heller of the University of Zurich abattoir. The authors are grateful to colleagues at the Functional Genomic Center Zurich, especially to Hubert Rehrauer and Timothy Sykes for assistance with transcriptomic and bioinformatic analyses and to Josep Monné Rodriguez at the Laboratory for Animal Model Pathology (University of Zurich) for assistance with histology analyses.

Conflicts of Interest: The authors declare no conflict of interest.

References

1. World Population Review. Most Obese Countries. Available online: https://worldpopulationreview.com/country-rankings/most-obese-countries (accessed on 1 October 2021).
2. Tovar, A.R.; Díaz-Villaseñor, A.; Cruz-Salazar, N.; Ordáz, G.; Granados, O.; Palacios-González, B.; Tovar-Palacio, C.; López, P.; Torres, N. Dietary Type and Amount of Fat Modulate Lipid Metabolism Gene Expression in Liver and in Adipose Tissue in High-fat Diet-fed Rats. *Arch. Med Res.* **2011**, *42*, 540–553. [CrossRef] [PubMed]
3. Parks, E.J.; Skokan, L.E.; Timlin, M.T.; Dingfelder, C.S. Dietary Sugars Stimulate Fatty Acid Synthesis in Adults. *J. Nutr.* **2008**, *138*, 1039–1046. [CrossRef]
4. Gabbia, D.; Roverso, M.; Guido, M.; Sacchi, D.; Scaffidi, M.; Carrara, M.; Orso, G.; Russo, F.P.; Floreani, A.; Bogialli, S.; et al. Western Diet-Induced Metabolic Alterations Affect Circulating Markers of Liver Function before the Development of Steatosis. *Nutrients* **2019**, *11*, 1602. [CrossRef]
5. Barnard, R.J.; Roberts, C.K.; Varon, S.M.; Berger, J.J. Diet-induced insulin resistance precedes other aspects of the metabolic syndrome. *J. Appl. Physiol.* **1998**, *84*, 1311–1315. [CrossRef] [PubMed]
6. Misra, B.B.; Upadhayay, R.P.; Mattern, V.; Parks, J.S.; Cox, L.A.; Comuzzie, A.G.; Olivier, M. Metabolic programming: Short term western diet exposure induces sustained changes in plasma metabolites. *bioRXiv* **2019**. [CrossRef]
7. Shi, Z.; Wu, X.; Yu, S.; Huynh, M.; Jena, P.K.; Nguyen, M.; Wan, Y.Y.; Hwang, S.T. Short-term exposure to a Western diet induces psoriasiform dermatitis by promoting accumulation of IL-17A-producing γδ T cells. *J. Investig. Dermatol.* **2020**, *140*, 1815–1823. [CrossRef] [PubMed]
8. Simon, A.; Pratt, M.; Hutton, B.; Skidmore, B.; Fakhraei, R.; Rybak, N.; Corsi, D.J.; Walker, M.; Velez, M.P.; Smith, G.N.; et al. Guidelines for the management of pregnant women with obesity: A systematic review. *Obes. Rev.* **2019**, *21*, e12972. [CrossRef]
9. Butte, N.F. Carbohydrate and lipid metabolism in pregnancy: Normal compared with gestational diabetes mellitus. *Am. J. Clin. Nutr.* **2000**, *71*, 1256S–1261S. [CrossRef]
10. Hyatt, H.; Zhang, Y.; Hood, W.R.; Kavazis, A.N. Lactation has persistent effects on a mother's metabolism and mitochondrial function. *Sci. Rep.* **2017**, *7*, 17118. [CrossRef]
11. Musial, B.; Vaughan, O.; Twinn, D.; Voshol, P.; Ozanne, S.; Fowden, A.; Sferruzzi-Perri, A. A Western-style obesogenic diet alters maternal metabolic physiology with consequences for fetal nutrient acquisition in mice. *J. Physiol.* **2017**, *14*, 4875–4892. [CrossRef]
12. Holman, B.W.B.; Malau-Aduli, A.E.O. Spirulina as a livestock supplement and animal feed. *J. Anim. Physiol. Anim. Nutr.* **2013**, *97*, 615–623. [CrossRef] [PubMed]
13. Wu, Q.; Liu, L.; Miron, A.; Klímová, B.; Wan, D.; Kuča, K. The antioxidant, immunomodulatory, and anti-inflammatory activities of Spirulina: An overview. *Arch. Toxicol.* **2016**, *90*, 1817–1840. [CrossRef] [PubMed]
14. Coué, M.; Tesse, A.; Falewée, J.; Aguesse, A.; Croyal, M.; Fizanne, L.; Chaigneau, J.; Boursier, J.; Ouguerram, K. Spirulina Liquid Extract Protects against Fibrosis Related to Non-Alcoholic Steatohepatitis and Increases Ursodeoxycholic Acid. *Nutrients* **2019**, *11*, 194. [CrossRef] [PubMed]
15. Heo, M.-G.; Choung, S.-Y. Anti-obesity effects of *Spirulina maxima* in high fat diet induced obese rats via the activation of AMPK pathway and SIRT1. *Food Funct.* **2018**, *9*, 4906–4915. [CrossRef]
16. Yang, Y.; Du, L.; Hosokawa, M.; Miyashita, K. Spirulina Lipids Alleviate Oxidative Stress and Inflammation in Mice Fed a High-Fat and High-Sucrose Diet. *Mar. Drugs* **2020**, *18*, 148. [CrossRef]
17. Deng, R.; Chow, T.J. Hypolipidemic, Antioxidant, and Anti-inflammatory Activities of Microalgae Spirulina. *Cardiovasc. Ther.* **2010**, *28*, e33–e45. [CrossRef]
18. Koopmans, S.J.; Schuurman, T. Considerations on pig models for appetite, metabolic syndrome and obese type 2 diabetes: From food intake to metabolic disease. *Eur. J. Pharmacol.* **2015**, *759*, 231–239. [CrossRef]
19. Mathias, P.C.F.; Elmhiri, G.; de Oliveira, J.; Delayre-Orthez, C.; Barella, L.F.; Tófolo, L.P.; Fabricio, G.S.; Chango, A.; Abdennebi-Najar, L. Maternal diet, bioactive molecules, and exercising as reprogramming tools of metabolic programming. *Eur. J. Nutr.* **2014**, *53*, 711–722. [CrossRef]
20. Pruis, M.G.M.; Lendvai, A.; Bloks, V.W.; Zwier, M.V.; Baller, J.F.W.; de Bruin, A.; Groen, A.K.; Plösch, T. Maternal western diet primes non-alcoholic fatty liver disease in adult mouse offspring. *Acta Physiol.* **2014**, *210*, 215–227. [CrossRef]
21. Kleinert, M.; Clemmensen, C.; Hofmann, S.; Moore, M.C.; Renner, S.; Woods, S.C.; Huypens, P.; Beckers, J.; de Angelis, M.H.; Schürmann, A.; et al. Animal models of obesity and diabetes mellitus. *Nat. Rev. Endocrinol.* **2018**, *14*, 140–162. [CrossRef]
22. Merrifiel, C.A.; Lewis, M.; Claus, S.P.; Beckonert, O.P.; Dumas, M.E.; Duncker, S.; Kochhar, S.; Rezzi, S.; Lindon, J.C.; Bailey, M.; et al. A metabolic system-wide characterization of the pig: A model for human physiology. *Mol. Bio. Syst.* **2011**, *7*, 2577–2588.
23. Bergen, W.G.; Mersmann, H.J. Comparative Aspects of Lipid Metabolism: Impact on Contemporary Research and Use of Animal Models. *J. Nutr.* **2005**, *135*, 2499–2502. [CrossRef] [PubMed]
24. Litten-Brown, J.C.; Corson, A.M.; Clarke, L. Porcine models for the metabolic syndrome, digestive and bone disorders: A general overview. *Animals* **2010**, *4*, 899–920. [CrossRef] [PubMed]
25. Lugarà, R.; Grześkowiak, Ł.; Zentek, J.; Meese, S.; Kreuzer, M.; Giller, K. A High-Energy Diet and Spirulina Supplementation during Pre-Gestation, Gestation, and Lactation do Not Affect the Reproductive and Lactational Performance of Primiparous Sows. *Animals* **2022**, *12*, 1171. [CrossRef] [PubMed]

26. Affentranger, P.; Gerwig, C.; Seewer, G.; Schwörer, D.; Künzi, N. Growth and carcass characteristics as well as meat and fat quality of three types of pigs under different feeding regimens. *Livest. Prod. Sci.* **1996**, *45*, 187–196. [CrossRef]
27. Dobin, A.; Davis, C.A.; Schlesinger, F.; Drenkow, J.; Zaleski, C.; Jha, S.; Batut, P.; Chaisson, M.; Gingeras, T.R. STAR: Ultrafast universal RNA-seq aligner. *Bioinformatics* **2013**, *29*, 15–21. [CrossRef]
28. Liao, Y.; Smyth, G.K.; Shi, W. The Subread aligner: Fast, accurate and scalable read mapping by seed-and-vote. *Nucleic Acids Res.* **2013**, *41*, e108. [CrossRef]
29. Xie, F.; Xiao, P.; Chen, D.; Xu, L.; Zhang, B. miRDeepFinder: A miRNA analysis tool for deep sequencing of plant small RNAs. *Plant Mol. Biol.* **2012**, *80*, 75–84. [CrossRef]
30. Madzlan, K. Determination of cholesterol in several types of eggs by gas chromatography. *J. Trop. Agric. Food Sci.* **2008**, *36*, 205–210.
31. Lan, F.; Misu, H.; Chikamoto, K.; Takayama, H.; Kikuchi, A.; Mohri, K.; Takata, N.; Hayashi, H.; Matsuzawa-Nagata, N.; Takeshita, Y.; et al. LECT2 Functions as a Hepatokine That Links Obesity to Skeletal Muscle Insulin Resistance. *Diabetes* **2014**, *63*, 1649–1664. [CrossRef]
32. Vicari, T.; Borne, J.V.D.; Gerrits, W.; Zbinden, Y.; Blum, J. Postprandial blood hormone and metabolite concentrations influenced by feeding frequency and feeding level in veal calves. *Domest. Anim. Endocrinol.* **2008**, *34*, 74–88. [CrossRef] [PubMed]
33. Matthews, D.R.; Hosker, J.P. Homeostasis model assessment: Insulin resistance and β-cell function from fasting plasma glucose and insulin concentrations in man. *Diabetologia* **1985**, *28*, 412–419. [CrossRef] [PubMed]
34. Wolf, C.; Ulbrich, S.; Kreuzer, M.; Berard, J.; Giller, K. Differential partitioning of rumen-protected n–3 and n–6 fatty acids into muscles with different metabolism. *Meat Sci.* **2018**, *137*, 106–113. [CrossRef]
35. Collomb, M.; Bühler, T. Analyse de la composition en acides gras de la graisse de lait. *Mitt. Lebensm. Hyg.* **2000**, *91*, 306–332.
36. R Core Team. *R: A Language and Environment for Statistical Computing;* R Foundation for Statistical Computing: Vienna, Austria, 2021. Available online: https://www.R-project.org/ (accessed on 19 April 2022).
37. Xi, S.; Yin, W.; Wang, Z.; Kusunoki, M.; Lian, X.; Koike, T.; Fan, J.; Zhang, Q. A minipig model of high-fat/high-sucrose diet-induced diabetes and atherosclerosis. *Int. J. Exp. Pathol.* **2004**, *85*, 223–231. [CrossRef]
38. Sébert, S.P.; Lecannu, G.; Sené, S.; Hucteau, S.; Chetiveaux, M.; Ouguerram, K.; Champ, M.M.-J. Obesity induced during sexual maturation is linked to LDL-triacylglycerols in Yucatan miniature swine. *Br. J. Nutr.* **2005**, *94*, 282–289. [CrossRef]
39. Dyson, M.C.; Alloosh, M.; Vuchetich, J.P.; A Mokelke, E.; Sturek, M. Components of metabolic syndrome and coronary artery disease in female Ossabaw swine fed excess atherogenic diet. *Comp. Med.* **2006**, *56*, 35–45.
40. Christoffersen, B.O.; Golozoubova, V.; Pacini, G.; Svendsen, O.; Raun, K. The young göttingen minipig as a model of childhood and adolescent obesity: Influence of diet and gender. *Obesity* **2007**, *21*, 149–158. [CrossRef] [PubMed]
41. Newell-Fugate, A.E.; Taibl, J.N.; Alloosh, M.; Sturek, M.; Bahr, J.M.; Nowak, R.A.; Krisher, R.L. Effects of Obesity and Metabolic Syndrome on Steroidogenesis and Folliculogenesis in the Female Ossabaw Mini-Pig. *PLoS ONE* **2015**, *10*, e0128749. [CrossRef]
42. Yang, X.; Ma, X.; Wang, L.; Gao, K.; Jiang, Z. 454 A high-fat diet expands body fat mass and up-regulates expression of genes involved in adipogenesis and inflammation in a genetically lean pig. *J. Anim. Sci.* **2018**, *95*, 223. [CrossRef]
43. Renner, S.; Blutke, A.; Dobenecker, B.; Dhom, G.; Müller, T.D.; Finan, B.; Clemmensen, C.; Bernau, M.; Novak, I.; Rathkolb, B.; et al. Metabolic syndrome and extensive adipose tissue inflammation in morbidly obese Göttingen minipigs. *Mol. Metab.* **2018**, *16*, 180–190. [CrossRef]
44. Goodpaster, B.H.; Sparks, L.M. Metabolic Flexibility in Health and Disease. *Cell Metab.* **2017**, *25*, 1027–1036. [CrossRef] [PubMed]
45. Buscemi, S.; Chiarello, P.; Buscemi, C.; Corleo, D.; Massenti, M.F.; Barile, A.M.; Rosafio, G.; Maniaci, V.; Settipani, V.; Cosentino, L.; et al. Characterization of Metabolically Healthy Obese People and Metabolically Unhealthy Normal-Weight People in a General Population Cohort of the ABCD Study. *J. Diabetes Res.* **2017**, *2017*, 9294038. [CrossRef] [PubMed]
46. Hamer, M.; Stamatakis, E. Metabolically Healthy Obesity and Risk of All-Cause and Cardiovascular Disease Mortality. *J. Clin. Endocrinol. Metab.* **2012**, *97*, 2482–2488. [CrossRef] [PubMed]
47. Stefan, N.; Schick, F.; Häring, H.-U. Causes, Characteristics, and Consequences of Metabolically Unhealthy Normal Weight in Humans. *Cell Metab.* **2017**, *26*, 292–300. [CrossRef] [PubMed]
48. Salvaggio, A.; Periti, M.; Miano, L.; Tavanelli, M.; Marzorati, D. Body mass index and liver enzyme activity in serum. *Clin. Chem.* **1991**, *37*, 720–723. [CrossRef]
49. Duval, C.; Thissen, U.; Keshtkar, S.; Accart, B.; Stienstra, R.; Boekschoten, M.V.; Roskams, T.; Kersten, S.; Müller, M. Adipose tissue dysfunction signals progression of hepatic steatosis towards non-alcoholic steatohepatitis in C57Bl/6 mice. *Diabetes* **2010**, *59*, 3181–31919. [CrossRef]
50. Hjelkrem, M.; Morales, A.; Williams, C.D.; Harrison, S.A. Unconjugated hyperbilirubinemia is inversely associated with non-alcoholic steatohepatitis (NASH). *Aliment. Pharmacol. Ther.* **2012**, *35*, 1416–1423. [CrossRef]
51. Bril, F.; McPhaul, M.J.; Caufield, M.P.; Clark, V.C.; Soldevilla-Pico, C.; Firpi-Morell, R.J.; Lai, J.; Shiffman, D.; Rowland, C.M.; Cusi, K. Performance of plasma biomarkers and diagnostic panels for non-alcoholic steatohepatitis and advanced fibrosis in patients with type 2 diabetes. *Diabetes Care* **2020**, *43*, 290–297. [CrossRef]
52. Stuebe, A.S.; Rich-Edwards, J.W. The reset hypothesis: Lactation and maternal metabolism. *Am. J. Perinatol.* **2010**, *26*, 81–88. [CrossRef]

53. Paczkowska-Abdulsalam, M.; Niemira, M.; Bielska, A.; Szałkowska, A.; Raczkowska, B.A.; Junttila, S.; Gyenesei, A.; Adamska-Patruno, E.; Maliszewska, K.; Citko, A.; et al. Evaluation of Transcriptomic Regulations behind Metabolic Syndrome in Obese and Lean Subjects. *Int. J. Mol. Sci.* **2020**, *21*, 1455. [CrossRef]
54. Biobaku, F.; Ghanim, H.; Batra, M.; Dandona, P. Macronutrient-Mediated Inflammation and Oxidative Stress: Relevance to Insulin Resistance, Obesity, and Atherogenesis. *J. Clin. Endocrinol. Metab.* **2019**, *104*, 6118–6128. [CrossRef]
55. Frova, C. Glutathione transferases in the genomics era: New insights and perspectives. *Biomol. Eng.* **2006**, *23*, 149–169. [CrossRef]
56. Giannini, E.; Risso, D.; Ceppa, P.; Botta, F.; Chiarbonello, B.; Fasoli, A.; Malfatti, F.; Romagnoli, P.; Lantieri, P.B.; Testa, R. Utility of α-Glutathione S-transferase assessment in chronic hepatitis C patients with near normal alanine aminotransferase levels. *Clin. Biochem.* **2000**, *33*, 297–301. [CrossRef]
57. Mavis, C.K.; Kinney, S.R.M.; Foster, B.A.; Karpf, A.R. Expression level and DNA methylation status of glutathione-*S*-transferase genes in normal murine prostate and TRAMP tumors. *Prostate* **2009**, *69*, 1312–1324. [CrossRef]
58. Bessone, F.; Razori, M.V.; Roma, M.G. Molecular pathways of nonalcoholic fatty liver disease development and progression. *Cell. Mol. Life Sci.* **2019**, *76*, 99–128. [CrossRef]
59. Gentile, C.L.; Pagliassotti, M.J. The role of fatty acids in the development and progression of nonalcoholic fatty liver disease. *J. Nutr. Biochem.* **2008**, *19*, 567–576. [CrossRef]
60. Younossi, Z.M.; Stepanova, M.; Negro, F.; Hallaji, S.; Younossi, Y.; Lam, B.; Srishord, M. Nonalcoholic Fatty Liver Disease in Lean Individuals in the United States. *Medicine* **2012**, *91*, 319–327. [CrossRef]
61. Tobari, M.; Hashimoto, E.; Taniai, M.; Ikarashi, Y.; Kodama, K.; Kogiso, T.; Tokushige, K.; Takayoshi, N.; Hashimoto, N. Characteristics of non-alcoholic steatohepatitis among lean patients in Japan: Not uncommon and not always benign. *J. Gastroenterol. Hepatol.* **2018**, *34*, 1404–1410. [CrossRef]
62. Adamek, A.; Kasprzak, A. Insulin-Like Growth Factor (IGF) System in Liver Diseases. *Int. J. Mol. Sci.* **2018**, *19*, 1308. [CrossRef]
63. Renner, S.; Dobenecker, B.; Blutke, A.; Zöls, S.; Wanke, R.; Ritzmann, M.; Wolf, E. Comparative aspects of rodent and nonrodent animal models for mechanistic and translational diabetes research. *Theriogenology* **2016**, *86*, 406–421. [CrossRef]
64. Teppala, S.; Shankar, A. Association Between Serum IGF-1 and Diabetes Among U.S. Adults. *Diabetes Care* **2010**, *33*, 2257–2259. [CrossRef]
65. He, J.; Kelley, D.E. Muscle glycogen content in type 2 diabetes mellitus. *Am. J. Physiol. Metab.* **2004**, *287*, E1002–E1007. [CrossRef] [PubMed]
66. Moradi, S.; Ziaei, R.; Foshati, S.; Mohammadi, H.; Nachvak, S.M.; Rouhani, M.H. Effects of Spirulina supplementation on obesity: A systematic review and meta-analysis of randomized clinical trials. *Complement. Ther. Med.* **2019**, *47*, 102211. [CrossRef]
67. Cattley, R.C.; Cullen, J.M. Liver and Gall bladder. In *Haschek and Rousseaux's Handbook of Toxicologic Pathology*, 3rd ed.; Academic Press: London, UK, 2013; pp. 1509–1566.
68. Yang, Y.; Du, L.; Hosokawa, M.; Miyashita, K. Effect of Spirulina lipids on high-fat and high-sucrose diet induced obesity and hepatic lipid accumulation in C57BL/6J mice. *J. Funct. Foods* **2020**, *65*, 103741. [CrossRef]
69. Ide, T.; Iwase, H.; Amano, S.; Sunahara, S.; Tachihara, A.; Yagi, M.; Watanabe, T. Physiological effects of γ-linolenic acid and sesamin on hepatic fatty acid synthesis and oxidation. *J. Nutr. Biochem.* **2017**, *41*, 42–55. [CrossRef]
70. Schinckel, A.; Ferrel, J.; Einstein, M.; Pearce, S.; Boyd, R. Analysis of Pig Growth from Birth to Sixty Days of Age1. *Prof. Anim. Sci.* **2004**, *20*, 79–86. [CrossRef]
71. Fisher, K.D.; Scheffler, T.L.; Kasten, S.C.; Reinholt, B.M.; van Eyk, G.R.; Escobar, J.; Scheffler, J.M.; Gerrard, D.E. Energy Dense, Protein Restricted Diet Increases Adiposity and Perturbs Metabolism in Young, Genetically Lean Pigs. *PLoS ONE* **2013**, *8*, e72320. [CrossRef] [PubMed]
72. Feng, R.; Du, S.; Chen, Y.; Zheng, S.; Zhang, W.; Na, G.; Li, Y.; Sun, C. High carbohydrate intake from starchy foods is positively associated with metabolic disorders: A Cohort Study from a Chinese population. *Sci. Rep.* **2015**, *5*, 16919. [CrossRef] [PubMed]
73. Liu, X.; Han, Y.S.; Kim, I.H. Growth performance, nutrient digestibility, antioxidant enzyme activity, and fecal microbiota flora in growing pigs fed diets containing Spirulina. *Can. J. Anim. Sci.* **2019**, *99*, 840–847. [CrossRef]

nutrients

MDPI

Article

Combination of Caloric Restriction and a Mixed Training Protocol as an Effective Strategy to Counteract the Deleterious Effects in Trabecular Bone Microarchitecture Caused by a Diet-Induced Obesity in Sprague Dawley Rats

Elena Nebot [1,†], Rosario Martínez [2,†], Garyfallia Kapravelou [2], Cristina Sánchez [2], Juan Llopis [2], Pilar Aranda [2], Jesús M. Porres [2,*], María López-Jurado [2,‡] and Peter Pietschmann [3,‡]

1 Department of Physiology, School of Medicine, Complutense University of Madrid, 28040 Madrid, Spain
2 Department of Physiology, Institute of Nutrition and Food Technology (INyTA), Biomedical Research Center (CIBM), University of Granada, Avda del Conocimiento s/n, 18100 Armilla, Spain
3 Institute of Pathophysiology and Allergy Research, Center for Pathophysiology, Infectiology and Immunology, Medical University of Vienna, 1090 Vienna, Austria
* Correspondence: jmporres@ugr.es
† These authors contributed equally to this work.
‡ These authors contributed equally to this work.

Citation: Nebot, E.; Martínez, R.; Kapravelou, G.; Sánchez, C.; Llopis, J.; Aranda, P.; Porres, J.M.; López-Jurado, M.; Pietschmann, P. Combination of Caloric Restriction and a Mixed Training Protocol as an Effective Strategy to Counteract the Deleterious Effects in Trabecular Bone Microarchitecture Caused by a Diet-Induced Obesity in Sprague Dawley Rats. *Nutrients* **2022**, *14*, 3672. https://doi.org/10.3390/nu14183672

Academic Editor: Connie Weaver

Received: 18 July 2022
Accepted: 19 August 2022
Published: 6 September 2022

Abstract: The association of obesity with changes in bone mass is not clear. Obese individuals tend to have an increased bone mineral density, but other studies have shown that obesity is a major risk factor for fractures. The mechanisms of bone response during a weight loss therapy as well as the possible osteoprotective effect of exercise should be analyzed. The aim of this study was to test the effects of a weight-loss program based on the combination of caloric restriction and/or a mixed training protocol on different parameters of bone morphology and functionality in a DIO rat model. Three stages were established over a 21-week period (obesity induction 0–12 w, weight loss intervention 12–15 w, weight maintenance intervention 15–21 w) in 88 male Sprague Dawley rats. Bone microarchitecture, total mineral and elemental composition, and bone metabolism parameters were assessed. Weight loss interventions were associated to healthy changes in body composition, decreasing body fat and increasing lean body mass. On the other hand, obesity was related to a higher content of bone resorption and inflammatory markers, which was decreased by the weight control interventions. Caloric restriction led to marked changes in trabecular microarchitecture, with a significant decrease in total volume but no changes in bone volume (BV). In addition, the intervention diet caused an increase in trabeculae number and a decrease in trabecular spacing. The training protocol increased the pore diameter and reversed the changes in cortical porosity and density of BV induced by the high protein diet at diaphysis level. Regarding the weight-maintenance stage, diminished SMI values indicate the presence of more plate-like spongiosa in sedentary and exercise groups. In conclusion, the lifestyle interventions of caloric restriction and mixed training protocol implemented as weight loss strategies have been effective to counteract some of the deleterious effects caused by a dietary induction of obesity, specifically in trabecular bone morphometric parameters as well as bone mineral content.

Keywords: bone microarchitecture; micro-CT; bone turnover markers; weight loss strategies; exercise; caloric restriction diet

1. Introduction

Overweight and obesity are defined as excessive and abnormal accumulation of fat that means a health risk. When combined with metabolic alterations such as hypertension, central obesity, insulin resistance, and/or dyslipidemia a cluster of pathologies shows up, denominated as metabolic syndrome (MetS). Due to their high prevalence in the world population, obesity and MetS are considered pandemics. Their incidence increases alarmingly every year mainly due to environmental factors, although genetic factors are also involved [1]. Regarding environmental factors, the regular intake of high fat and high fructose diets is directly related to the development of obesity, which is an important risk factor for the development of other associated chronic pathologies, such as insulin resistance, non-alcoholic fatty liver disease (NAFLD), as well as the alteration of bone functionality.

The development of obesity and MetS is favored by the consumption of unbalanced and hypercaloric diets. Therefore, the consumption of a balanced diet that provides adequate amounts of nutrients to treat or prevent these pathologies is highly recommended. Nevertheless, to implement a negative energy balance for weight-loss treatment, hypocaloric/hyper protein diets are usually prescribed. Moreover, increasing evidence on the beneficial effect of certain bioactive dietary components on both obesity and its comorbidities has been accumulating in recent years [2,3].

Related to lifestyle interventions, there are two core aspects to correct an altered energy balance: diet and physical exercise. The first step in the treatment of obesity is focused on losing extra weight and ameliorating the related metabolic alterations. Another important issue for patients who complete a weight loss program is to avoid the post-intervention rebound effect. The bodyweight regain usually takes place right after the end of weight loss intervention as weight loss programs are just transient [4]. A multidisciplinary approach is required, including lifestyle modifications [5] and, in some cases, the reinforcement with pharmacological treatment. On the other hand, physical activity plays an essential role in the prevention and treatment of obesity. It contributes to generating a negative energy balance, thus facilitating weight loss and avoiding the rebound effect and subsequent body weight regain [6]. It is well established that different training protocols induce changes in a variety of molecular mechanisms involved in numerous intracellular pathways related to glucose and lipid metabolism, inflammation, or antioxidant status [7]. The World Health Organization (WHO) recommends during adulthood at least 150–300 min of moderate-intensity aerobic physical activity; or at least 75–150 min of vigorous-intensity aerobic physical activity; or an equivalent combination of moderate- and vigorous-intensity activity throughout the week, for substantial health benefits [8]. Such exercise practice confers benefits for the following health outcomes: improved all-cause mortality, cardiovascular disease mortality, incident hypertension, site-specific cancers, and type-2 diabetes, as well as mental health (reduced symptoms of anxiety and depression); cognitive health, and sleep. In addition, adequate physical activity seems to be crucial for the proper development and maintenance of the skeleton, and it is necessary to clarify the effects of physical exercise combined with dietary interventions on structural parameters, histomorphometry, and bone metabolism.

The association of obesity with bone mass is contradictory; on the one hand, obese individuals tend to have an increased bone mineral density (BMD) mainly due to weight-induced loading of the bone [9]. On the other, many studies have shown that obesity is a major risk factor for fractures, and, especially, visceral adiposity is negatively associated with BMD and total mineral content in humans [10,11]. In rodent models, a high-fat diet (HFD) could massively affect bone health by reducing trabecular and/or cortical bone mass [12].

It has been reported that a weight-loss program induced by a caloric restrictive diet is linked to a concomitant accelerated bone loss. Studies conducted in obese women have found that such diets are associated with significant decreases in bone mass and total BMD, as well as an increased risk of fracture [13,14]. Moreover, it has been shown [15] that a moderate weight loss induced by a caloric restriction can increase bone resorption.

The association of a well-balanced diet with exercise is a key strategy to treat obesity. Regular exercise, known to induce beneficial effects on bone, could attenuate weight loss-induced bone loss. Nevertheless, the mechanisms of bone response during a weight loss therapy as well as the possible osteoprotective effect of exercise remain unclear [16].

Given the aforementioned, we hypothesized that our specific combined strategy of caloric restriction and physical exercise interventions could provide interesting benefits in the treatment of obesity and its related bone alterations. Thus, this study aimed to test the effects of a weight-loss program based on the combination of caloric restriction and/or a mixed training protocol on different parameters of bone morphology and functionality in a diet-induced obesity (DIO) model of Sprague Dawley rats.

2. Material and Methods

2.1. Animals, Diets, and Experimental Design

The experiment used 88 male Sprague Dawley rats with an average body weight of 184 ± 10 g (6-weeks old, Charles Rives, Barcelona, Spain) that were allocated into eleven different experimental groups (*n* = 8). We only used male rats to avoid sex differences. The experiments lasted for 21 weeks and were divided into three stages (obesity induction and development of related alterations 0–12, weight loss intervention 12–15, weight maintenance intervention 15–21 weeks) (Figure 1).

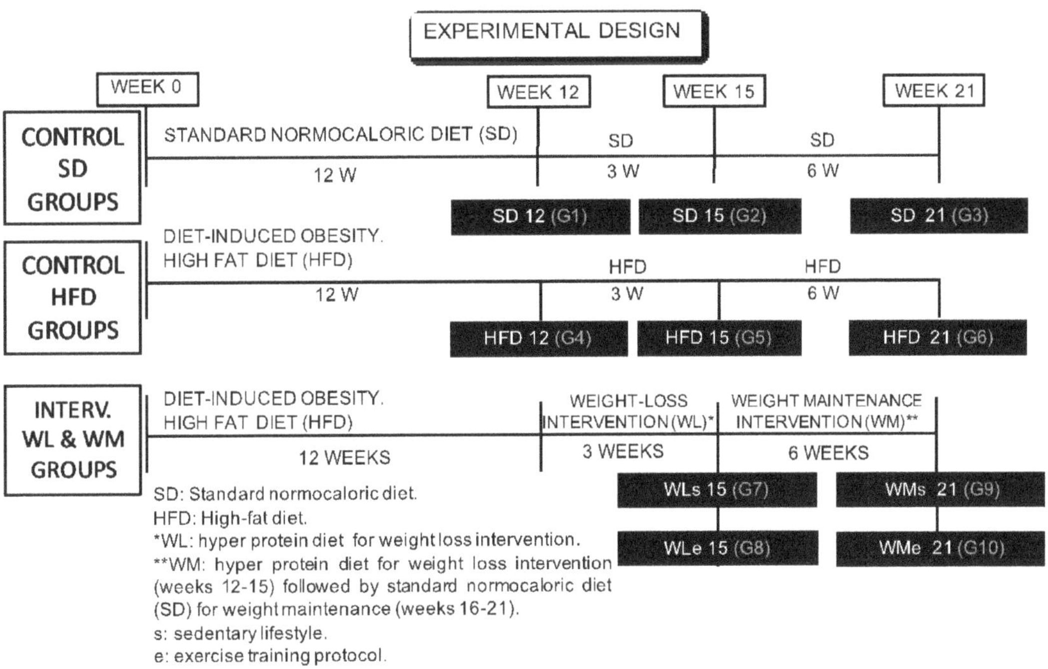

Figure 1. Experimental design: six control experiments were carried out for 21 weeks using a standard rat chow diet (Control SD groups: SD12, SD15 and SD21) or a high-fat diet to induce obesity (Control HFD groups: HFD12, HFD15 and HFD21). For intervention trials (Intervention WL & WM groups: WLs15, WLe15, WMs21, WMe21), rats were divided into 4 groups that were fed the hypercaloric diet to induce obesity for 12 weeks, continued by three weeks of intervention with a high-protein diet for weight loss (WL15), either following a sedentary lifestyle or combined with a training protocol (s or e, respectively). The intervention period was continued by an additional 6-week weight-maintenance stage of dietary treatment with a standard rat chow diet (WM21), either following a sedentary lifestyle or combined with a training protocol (s or e, respectively), in order to maintain the weight lost during the previous intervention period of three weeks.

Two control experiments that involved three groups of animals in each of them were organized with the following design:

1. Standard normocaloric groups (SD) fed a normocaloric standard rodent diet (Teklad Global Diet 2014; 2.4 Kcal/g) along the whole experiment:

G0. SD 0 weeks. Baseline control group.

G1. SD 12 weeks

G2. SD 15 weeks

G3. SD 21 weeks

2. Diet-induced obesity groups (HFD) fed a hypercaloric obesogenic diet containing 60% of Kcal as fat (Research diets D12492; 5.2 Kcal/g) along the whole experiment:

G4. HFD 12 weeks

G5. HFD 15 weeks

G6. HFD 21 weeks

In addition, four experimental groups were arranged with the following design:

3. Weight loss intervention groups (WL) based on caloric restriction and/or physical exercise during 3 weeks (weeks 13–15):

G7. A first stage for a period of 12 weeks consisting of dietary induction of obesity after ingestion of a hypercaloric diet (HFD), followed by a second 3-week stage (up to week 15) in which a caloric restriction dietary intervention to lose weight was implemented using an experimental diet designed to induce greater satiety combining the effects of high protein and soluble dietary fiber content (WLs 15) (2.9 Kcal/g). In addition to its satiating action, soluble dietary fiber is an effective therapeutic agent to treat many of the MetS components associated to obesity [17].

G8. Similar dietary intervention as in G7 complemented with a mixed training protocol implemented 5 days per week during weeks 13–15 of weight loss intervention (WLe 15).

For the following experimental groups, an additional third stage of the experiment ran for a period between weeks 16 to 21 and was designed for maintenance of lost body weight without any rebound effect:

G9. Similar dietary treatment as in G7 up to week 15 followed by ingestion of a SD diet for 6 weeks up to week 21 (WMs 21). The amount of food ingested during that 6-week period was *pair fed* to 23 g/d in order to achieve a 12–15% caloric reduction compared to the same period in control SD group as part of the strategy to maintain the lost weight during post-intervention stage avoiding the rebound effect.

G10. Similar dietary treatment as in G9 complemented during the intervention weight loss period (weeks 13–15) and weight-loss maintenance period (weeks 16–21) with a mixed training protocol implemented 5 days per week (Monday–Friday) (WMe 21).

The weight-loss intervention and body weight maintenance periods have been designed based on the information provided by Sengupta [18] who reported that laboratory rats live about 2–3.5 years (average 3 years), while the worldwide life expectancy of humans is 80 years. Thus, one human year almost equals two rat weeks (13.8 rat days) while correlating their entire life span. Under our experimental conditions, 3 weeks is equal to 1.5 years of human life, which is a long enough period to achieve an efficient weight loss. The maintenance period of 6 weeks is equal to 3 years, which is enough to demonstrate the success of our combined strategy against body weight regain.

The animals were housed in a well-ventilated, thermostatically controlled room (21 ± 2 °C) (Unidad de Experimentación Animal, CIC, Universidad de Granada). A reversed 12:12 light/dark cycle was implemented so the animals would perform the training protocol in darkness. Throughout the trial, animals had free access to type 2 water (resistivity 15 $M\Omega^{-cm}$) and consumed the diet *ad libitum*, with the exception of the intervention groups in the last stage of the experiment that were adapted to slightly lower food intake (23 g/d) compared to that of the normocaloric control during the same period (28 g/d) in order to keep a certain degree of caloric adaptation to avoid body weight rebound, as recommended by weight control programs [19]. The diet was provided for all four animals in each cage but the body weight control was registered individually. Caloric intake was recorded daily whereas body weight was measured once a week. At the end of each experimental period (obesity induction and development of related alterations 12th week, weight loss intervention 15th week, weight maintenance intervention 21th week), the animals were fasted for 8 h. Then, body composition was assessed with a whole-body composition analyzer based on magnetic resonance imaging (EchoMRI™; EchoMedical Systems, Houston 145 TX) prior to being anesthetized with ketamine (75 mg/kg body weight) and xylazine (10 mg/kg body weight) and euthanized by cannulation of the abdominal aorta. Blood was collected (with heparin as anticoagulant) and centrifuged at 3000 rpm for 15 min to separate the plasma, which was subsequently removed and frozen in liquid nitrogen and stored at −80 °C. Epydidimal and abdominal fat was extracted and weighted. The femur was extracted, weighted, measured, and immediately frozen in liquid nitrogen and stored at −80 °C until bone mass and microarchitectural analysis were conducted. Tibiae was also extracted, weighted, and measured. Bone marrow was extracted, frozen in liquid nitrogen, and stored at −80 °C for the determination of RANKL, interleukin 10, and leptin. All experiments were undertaken according to Directional Guides Related to Animal Housing and Care [20] and all procedures were approved by the Animal Experimentation Ethics Committee of the University of Granada, Spain (Project Reference DEP2014-58296R).

2.2. Training Protocol

Rats trained following a protocol based on interval aerobic training combined with strength exercise in the same session [21,22]. The animals ran on a specially designed treadmill (Panlab, LE 8710R) and all sessions were performed 5 days/week and during the dark cycle of the animals (active period). The training protocol was designed due to the beneficial effects of high-intensity interval aerobic training on obesity and parameters of lipid metabolism, and combined with an aerobic strength training protocol with effective action on insulin sensitivity and lipid profile. To establish the velocity that would correspond to the VO_2 max of each rat, a maximal incremental test was performed at the start of the study. A final incremental test was performed 96 h prior the end of the study to test the maximal aerobic capacity and physical performance achieved by the animals as a result of the intervention. All sessions of the mixed training protocol consisted of 60 min of effective work. The sessions started with a 10-min warm-up at 35–50% maximal oxygen consumption (Supplementary Table S1), followed by the strength training consisting on eight 2-min running bouts separated by 1 min of rest during which animals ran with an inclination, progressively increased every three weeks from 10° up to 20° at a constant slow speed (20–25 cm/s, equivalent to 30–40% maximal oxygen consumption). The strength exercise was followed by 30 min of aerobic interval exercise, alternating 4 min bouts at 50–65% maximal oxygen consumption with 3 min bouts at submaximal intensity at 65–85% maximal oxygen consumption.

2.3. Bone Marrow Analyses

Bone marrow was extracted from tibiae by cutting lower portion of the bone and centrifuging at 6000 rpm. RANKL was measured with the rat kit Milliplex Rat RANKL (MAP kit, Millipore, Burlington, MA, USA), the cytokine interleukin (IL-1β) was measured with the rat kit Milliplex Rat Cytokine (MAP kit, Millipore), and leptin was measured with the rat kit Milliplex Rat Leptin (MAP kit, Millipore) and calibrated with Luminex 100/200 Calibration kit.

2.4. Assessment of Bone Mass and Bone Microarchitecture

Bone microarchitecture parameters of the femora were analyzed by µCT using a µCT-50 device (ScancoMedical, CH, Wangen-Brüttisellen, Switzerland). The long axis of the biopsies was oriented along the rotation axis of the scanner. The X-ray tube was operated at 70 kV with an intensity of 200 µA, and an exposure time of 500 ms, resulting in a resolution of 10 µm/pixel. Femora were scanned in a cortical region (mid-shaft and extending a 10%

of the whole femur length) and in a trabecular region (proximal of the knee joint extending to a 10% at distal length of 75% of the whole femur length).

2.5. Ash Measurement and Elemental Composition of Femur

Femur samples were cleaned of flesh and debris before being weighed and their length measured. The samples were then freeze-dried, weighed, and processed for measurement of total mineral content after calcination in an oven at 450 °C for 5 days to a constant weight or processed for elemental analysis by wet digestion. The concentration of Mg, Ca, V, Mn, Fe, Co, Zn, As, and Se in femur was determined using inductively coupled plasma mass spectrometry (ICP-MS) following the protocol previously described by Sánchez-González et al. [23].

2.6. Statistical Analyses

For a more accurate description and interpretation of the data, the experimental period has been divided in three different stages: (i) dietary induction of obesity during weeks 0–12, (ii) individual or combined weight loss interventions: dietary (caloric restriction) and/or lifestyle (training protocol of mixed exercise during weeks 13–15, and (iii) post-intervention maintenance stage with normocaloric, diet combined or not with a training protocol weeks 16–21.

Significant differences in final body weight, body weight changes, caloric intake, body weight/tibial length ratio, body composition parameters, bone microarchitecture parameters and bone elemental composition were analyzed by *t*-test at 12 weeks of the experimental period, and by one-way ANOVA at 15 and 21 weeks of experimental period. Duncan's test was used to detect differences between treatment means. Statistical analysis was performed with the Statistical Package for Social Sciences (IBM SPSS for Windows®, version 22.0, Armonk, NY, USA), and the level of significance was set at $p < 0.05$.

3. Results and Discussion

3.1. Caloric Intake, Body Weight Changes and Body Composition

The effects of a high-fat diet intake (12 weeks) followed by an intervention high-protein diet (3 weeks) and weight maintenance normocaloric diet (6 weeks) combined or not with the training program on bodyweight change and body composition are shown in Figure 2a,b and in Table 1. As expected, caloric intake was significantly higher in the group fed the high fat diet compared to the group fed a standard normocaloric diet on weeks 12th, 15th and 21st, leading to higher body weight gain (expressed as g/week) only on week 12th. The treatment with high-protein diet during weeks 13–15 caused a significant weight loss, which was stronger when dietary treatment was combined with exercise.

During the weight maintenance stage (16–21 weeks), the intake of a normocaloric diet combined or not with exercise led to stabilization of body weight without any further gain or loss. Therefore, net body weight remained significantly lower in the intervention groups compared to both standard diet or high fat diet fed animals.

Regarding body composition, the significantly higher bodyweight found in obese groups (HFD 12, 15, and 21) was linked to similar lean body mass and total water but significantly higher total fat mass, abdominal fat and epididymal fat, than the groups fed the standard diet. The weight loss intervention (weeks 13–15) led to a significant reduction in weight (similar values were found in sedentary and exercised groups), whereas LBM was maintained, and fat mass was significantly reduced. Along the weeks 16–21 an effective maintenance of lost weight was achieved, preserving LBM levels with similar values to the group fed a standard diet (either in sedentary or exercised groups) and maintaining the low levels of total, abdominal and epididymal fat achieved in the previous stage. Exercise induced a further weight loss associated with an additional and significant loss of total, abdominal and epididymal fat, and increased the amount of LBM and the LBM to total body weight ratio on week 21. All these changes were reflected in the bodyweight to tibial length ratio that was significantly higher in obese vs. control normocaloric groups and returned to values similar to the normocaloric controls after the weight-loss and lost-weight maintenance interventions (Figure 2c).

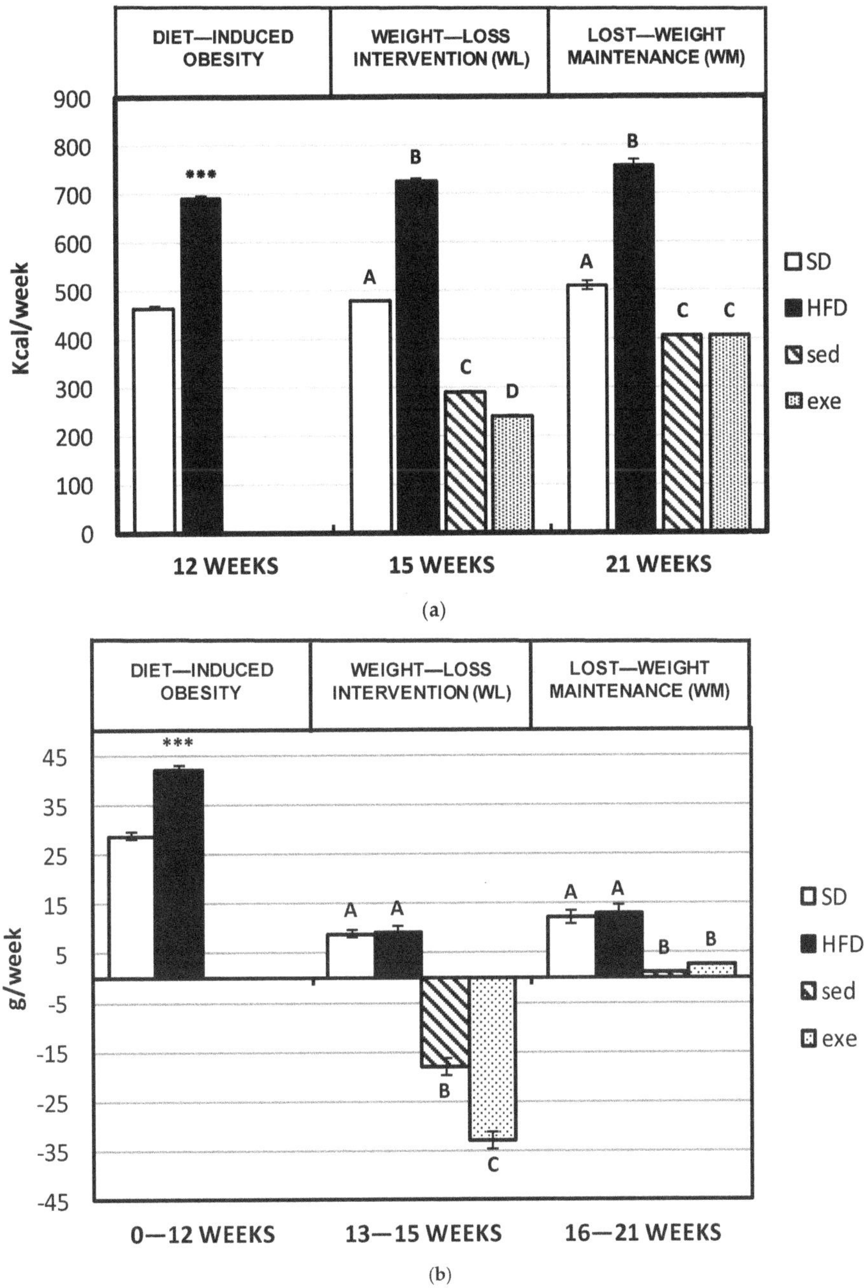

Figure 2. *Cont.*

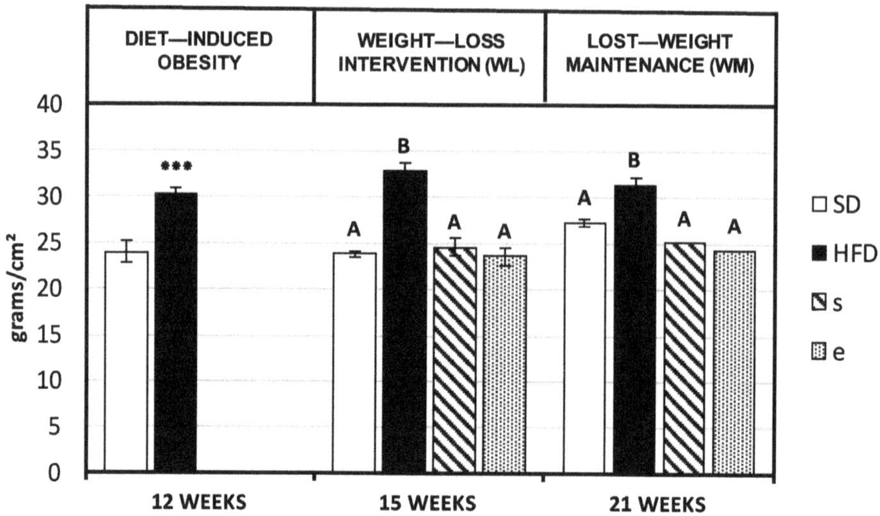

(c)

Figure 2. (**a**)Average weekly caloric intake (kcal/week) along the different experimental stages (DIO, weight-loss intervention, and weight-maintenance). Results are means of eight rats ± SEM depicted by vertical bars. *** $p < 0.001$ in *t*-test (12 weeks); A, B, C, D means with different letters are significantly different (ANOVA treatment, $p < 0.05$; 15 and 21 weeks). SD, standard normocaloric diet; HFD, high-fat diet; sed, sedentary lifestyle; exe, exercise training protocol. (**b**) Average weekly body weight changes (g/week) along the different experimental stages (DIO, weight-loss intervention, and weight-maintenance). Results are means of eight rats ± SEM depicted by vertical bars. *** $p < 0.001$ in *t*-test (12 weeks); A, B, C means with different letters are significantly different (ANOVA treatment, $p < 0.05$; 15 and 21 weeks). SD, standard normocaloric diet; HFD, high-fat diet; sed, sedentary lifestyle; exe, exercise training protocol. (**c**) Body weight/ tibial length ratio (g/cm^2) at the end of the different experimental stages (DIO, weight-loss intervention, and weight-maintenance). Results are means of eight rats ± SEM depicted by vertical bars. *** $p < 0.001$ in *t*-test (12 weeks); A, B means with different letters are significantly different (ANOVA treatment, $p < 0.05$; 15 and 21 weeks). SD, standard normocaloric diet; HFD, high-fat diet; sed, sedentary lifestyle; exe, exercise training protocol.

Discussion

At the beginning of the weight loss intervention period obese rats nearly doubled the percentage body fat (18.4% to 10.6%) and were significantly heavier compared with SD rats of the same age. In contrast, the contribution of LBM to total weight was lower in those animals. Obese rats responded to caloric restriction (CR) with an efficient weight loss, mainly due to fat loss, while lean body mass increased in percentual terms. Bertrand et al. [24] found that this reduction in fat content is due to a lifelong decrease in both the size of individual adipocytes and the number of adipocytes in the fat depots. Barzilai and Gupta [25] have reported that CR is particularly effective in decreasing visceral fat in rats, in agreement to what has been reported in this experiment at the end of week 15, in which the percentage of abdominal fat in treated animals is half of that in obese rats not subjected to caloric restriction (3.5% to 6.1%). Besides, the exercise protocol followed enhanced the positive changes induced by CR intervention. Several investigations have demonstrated numerous adaptations to WAT in response to exercise that result in improved whole-body metabolic health [26]. These adaptations include increased mitochondrial biogenesis and gene expression [27–30] as well as changes in adipokine secretion. Certain beneficial effects of exercise may be mediated by an altered adipokine profile [31]. Among all the adipokines, leptin is significantly affected by exercise and acts as a satiety hormone to regulate energy balance through inhibition of hunger. The amount of circulating leptin correlates with adipose tissue mass, and a loss of adipose tissue mass in rodents and humans results in decreased serum concentrations of leptin [32,33], supporting the idea that adipose tissue plays an important role as a major endocrine organ that can be stimulated by exercise.

Table 1. Effects of obesity induction and weight control interventions on bodyweight and body composition.

	0 WEEKS (Baseline)	Diet-Induced Obesity 12 WEEKS		Weight-Loss Intervention (WL) 15 WEEKS				Lost-Weight Maintenance Intervention (WM) 21 WEEKS			
		SD	HFD	SD	HFD	WLs	WLe	SD	HFD	WMs	WMe
Body weight (g)	172.7 (2.21)	516.5 (18.9)	664.3 *** (14.1)	503 a (8.44)	705 b (15.1)	566 c (11.2)	552 abc (23.4)	631 a (19.8)	742 b (24.6)	639 a (9.33)	574 a (19.9)
Lean body mass (g)	155.9 (2.13)	422.2 (12.9)	453.4 (14.8)	414.3 a (8.21)	479 b (5.80)	448.2 ab (11.4)	442 ab (16.9)	481.8 a (12.9)	500.4 a (14.6)	490.3 a (9.67)	483.8 a (11.2)
Total water (g)	136.3 (1.92)	351.8 (10.7)	377.7 (9.64)	339.3 a (7.69)	397.6 b (5.78)	375.8 ab (11.8)	372.6 ab (13.1)	405.0 a (12.0)	416.1 a (24.4)	408.6 a (5.77)	409.1 a (11.4)
ΔLBM/ΔBW		0.78 (0.02)	0.61 *** (0.02)	0.76 a (0.02)	0.61 b (0.02)	0.74 a (0.02)	0.76 a (0.03)	0.71 a (0.03)	0.60 b (0.02)	0.74 ac (0.03)	0.82 c (0.03)
LB/TW	1.14 (0.004)	1.20 (0.02)	1.20 (0.02)	1.221 (0.02)	1.205 (0.02)	1.193 (0.02)	1.186 (0.02)	1.190 (0.02)	1.203 (0.02)	1.200 (0.02)	1.183 (0.02)
Fat mass (g)	8.06 (0.75)	54.8 (5.07)	122.8 *** (9.34)	49.2 a (6.68)	178.7 b (12.6)	73.3 a (10.2)	63.5 a (10.8)	90.6 a (7.54)	161.7 b (15.1)	110.7 a (8.70)	39.7 c (7.25)
Abdominal fat (g)	0.69 (0.07)	12.9 (1.14)	28.8 *** (2.16)	11.2 a (1.46)	43.0 b (3.0)	18.8 a (2.65)	16.2 a (2.32)	19.7 a (2.02)	38.5 b (2.49)	27.2 a (2.44)	9.88 c (1.57)
Epydidimal fat (g)	1.05 (0.09)	8.73 (0.79)	16.7 *** (0.88)	8.19 a (0.74)	20.7 b (1.10)	12.8 a (1.57)	12.2 a (1.72)	11.6 a (0.75)	21.0 b (1.61)	12.5 a (0.68)	7.34 c (0.61)

SD, standard rat chow diet; HFD, hypercaloric diet for dietary induction of obesity; WLs, high-protein weight-loss intervention diet with sedentary lifestyle (weeks 12–15); WLe, high-protein weight-loss intervention diet with training protocol; WMs, high-protein weight-loss intervention diet with sedentary lifestyle (weeks 12–15) followed by weight-maintenance stage (weeks 15–21) with SD dietary treatment and sedentary lifestyle; WMe, high-protein weight-loss intervention diet with training protocol followed by weight-maintenance stage with SD dietary treatment and training protocol. ΔLBM/ΔBW, changes in lean body mass vs. changes in body weight with respect to control baseline group; LB/TW, lean body mass vs. total water ratio at each experimental stage. Results are expressed as means of 8 rats and standard error of the mean (in parenthesis). *** $p < 0.001$ in *t*-test (12 weeks); a,b,c, means within the same line of each experimental stage (15 and 21 weeks) with different letters are significantly different (ANOVA treatment, $p < 0.05$).

3.2. Bone Weight and Length, Metabolism Markers and Microarchitecture

The effects of obesity and weight-loss interventions on bone parameters are described in Tables 2–4. At the end of dietary induction of obesity stage (week 12), femur weight and length were higher in HFD vs. SD group. In contrast, no significant differences in tibial length or weight were observed between the former experimental groups. The intervention period with a high protein diet (13–15 w) lead to a decrease in femur weight compared to the HFD control, while a stabilization in femur length was observed in all experimental groups. The combination of exercise with a high protein diet did not induce any further changes in femur weight. During the weight maintenance period (16–21 w) femur weight results followed the same trend as in the previous stage, whereas exercise intervention tended to increase this parameter (Table 2).

Although due to high variability no statistically significant effects were established, a clear biological trend can be inferred to relate the development of obesity and increased content in medulla of the bone resorption marker RANKL, and the inflammatory markers IL-1β and leptin. In general, the weight control interventions exhibited a positive effect on such markers, returning them to levels similar to those of the SD controls.

The effects of DIO and weight control interventions on bone microarchitecture are presented in Tables 3 and 4. Obesity produced a significant decrease in BV/TV index, connectivity density (Conn.D), trabecular number (Tb.N), and mean density TV compared to the SD fed group. In contrast, it increased Structure Model Index (SMI) and trabecular spacing (Tb. Sp). The main changes induced by the intake of HFD along 12 weeks on cortical bone microarchitecture were higher TV and BV values as well as a significant decrease in mean pore diameter and Ct.Po associated to higher density of bone volume.

These effects were still observed at the end of 15 w in the group fed HFD compared to the group fed SD, although results were not significant.

Table 2. Effects of obesity and weight control interventions on bone anthropometry and bone markers of structure and functionality in bone marrow.

	0 WEEKS (Baseline)	Diet-Induced Obesity 12 WEEKS		Weight-Loss Intervention (WL) 15 WEEKS				Lost-Weight Maintenance Intervention (WM) 21 WEEKS			
		SD	HFD	SD	HFD	WLs	WLe	SD	HFD	WMs	WMe
Femur weight (g)	0.58 (0.02)	1.37 (0.04)	1.94 *** (0.09)	1.47 a (0.02)	1.82 b (0.03)	1.41 a (0.06)	1.40 a (0.06)	1.74 ab (0.04)	1.77 b (0.09)	1.57 a (0.03)	1.73 ab (0.06)
Femur length (cm)	2.89 (0.02)	4.15 (0.03)	4.35 ** (0.05)	4.21 a (0.04)	4.32 a (0.03)	4.28 a (0.05)	4.33 a (0.04)	4.36 a (0.04)	4.34 a (0.07)	4.36 a (0.04)	4.36 a (0.04)
Tibial weight (g)	0.74 (0.03)	1.44 (0.03)	1.54 (0.02)	1.31 (0.05)	1.56 (0.06)	1.58 (0.04)	1.52 (0.06)	1.57 (0.06)	1.72 (0.05)	1.53 (0.02)	1.68 (0.11)
Tibial length (cm)	3.36 (0.04)	4.66 (0.03)	4.64 (0.05)	4.59 (0.03)	4.65 (0.03)	4.79 (0.05)	4.77 (0.04)	4.82 (0.06)	4.92 (0.08)	4.73 (0.02)	4.84 (0.05)
					Bone marrow						
RANKL (pg/mL)	1339.5 (216.9)	990.3 (87.6)	1624.5 (310.6)	1154.9 a (260.0)	1621.5 a (233.5)	916.7 a (209.4)	1044.5 a (179.5)	1021.6 a (280.3)	848.9 a (137.5)	1024 a (96.2)	1032.3 a (260.4)
IL-1β (pg/mL)	69.1 (7.12)	133.6 (11.6)	117.8 (7.38)	128.1 a (13.6)	139.9 a (25.7)	153.2 a (15.3)	126.0 a (6.70)	134.0 a (13.1)	180.6 a (46.9)	151.0 a (13.2)	159.7 a (13.7)
Leptin (pg/mL)	5.29 (1.17)	32.4 (7.41)	27.4 (9.20)	26.5 a (11.8)	46.2 a (15.3)	23.5 a (8.45)	35.2 a (10.2)	39.9 a (9.1)	59.9 a (19.2)	44.5 a (7.84)	34.4 a (13.1)

SD, standard rat chow diet; HFD, hypercaloric diet for dietary induction of obesity; WLs, high-protein weight-loss intervention diet with sedentary lifestyle (weeks 12–15); WLe, high-protein weight-loss intervention diet with training protocol; WMs, high-protein weight-loss intervention diet with sedentary lifestyle (weeks 12–15) followed by weight-maintenance stage (weeks 15–21) with SD dietary treatment and sedentary lifestyle; WMe, high-protein weight-loss intervention diet with training protocol followed by weight-maintenance stage with SD dietary treatment and training protocol. Results are expressed as means of 8 rats and standard error of the mean (in parenthesis). ** $p < 0.01$, *** $p < 0.001$ in t-test (12 weeks); a,b, means within the same line of each experimental stage (15 and 21 weeks) with different letters are significantly different (ANOVA treatment, $p < 0.05$). RANKL, Receptor Activator for Nuclear Factor κB Ligand.

Table 3. Effects of obesity and weight control interventions on 3D outcomes for trabecular (metaphysis) bone microarchitecture in femur.

	0 WEEKS (Baseline)	Diet-Induced Obesity 12 WEEKS		Weight-Loss Intervention (WL) 15 WEEKS				Lost-Weight Maintenance Intervention (WM) 21 WEEKS			
		SD	HFD	SD	HFD	WLs	WLe	SD	HFD	WMs	WMe
TV (mm^3)	29.9 (0.9)	68.4 (3.3)	78.2 (4.0)	70.4 (2.5) ab	76.5 (1.9) b	65.1 (3.6) a	69.9 (2.9) ab	81.7 (3.2) b	76.6 (3.5) ab	70.9 (1.5) a	71.3 (2.7) a
BV (mm^3)	2.78 (0.3)	20.8 (1.6)	18.6 (1.9)	22.4 (1.4) b	13.6 (0.9) a	12.7 (1.2) a	14.3 (1.1) a	19.4 (1.3) b	14.5 (0.6) a	17.8 (1.7) ab	19.1 (2.2) ab
BV/TV	0.09 (0.01)	0.30 (0.02)	0.24 * (0.02)	0.32 (0.02) b	0.18 (0.01) a	0.20 (0.02) a	0.21 (0.01) a	0.24 (0.02) ab	0.19 (0.01) a	0.25 (0.02) a	0.27 (0.03) b
Conn. D ($1/mm^3$)	44.3 (10.4)	92.9 (4.1)	67.0 *** (3.5)	92.7 (4.9) b	52.9 (5.5) a	47.9 (2.3) a	48.2 (2.7) a	60.9 (3.1) b	45.5 (3.2) a	44.7 (4.2) a	44.8 (5.4) a
SMI	3.13 (0.1)	1.18 (0.2)	1.57 (0.1)	1.04 (0.1) a	1.91 (0.1) b	1.77 (0.1) b	1.69 (0.1) b	1.45 (0.1) bc	1.77 (0.1) c	1.19 (0.1) ab	0.98 (0.2) a
Tb. N (1/mm)	3.31 (0.23)	4.06 (0.13)	2.54 *** (0.28)	3.91 (0.13) c	2.08 (0.18) a	2.62 (0.12) b	2.64 (0.11) b	2.72 (0.23) b	2.06 (0.17) a	2.85 (0.18) b	2.84 (0.21) b
Tb. Th (mm)	0.051 (0.002)	0.096 (0.004)	0.099 (0.004)	0.101 (0.005) c	0.091 (0.002) b	0.074 (0.003) a	0.077 (0.002) a	0.098 (0.005) b	0.093 (0.003) ab	0.087 (0.002) a	0.093 (0.003) ab
Tb. Sp (mm)	0.32 (0.02)	0.24 (0.01)	0.45 ** (0.06)	0.25 (0.01) a	0.53 (0.05) b	0.31 (0.02) a	0.31 (0.02) a	0.40 (0.04) b	0.53 (0.04) c	0.27 (0.03) a	0.27 (0.03) a
Mean density TV (mg HA/cm^3)	143.9 (11.4)	295.4 (16.9)	232.2 * (16.8)	302.7 (12.8) b	177.8 (11.3) a	202.1 (11.9) a	208.4 (11.7) a	238.5 (18.5) ab	196.7 (10.4) a	237.8 (16.4) ab	253.6 (16.9)b
Mean density BV (mg HA/cm^3)	645.9 (2.9)	787.9 (6.0)	802.1 (6.9)	795.1 (7.6) a	791.8 (7.2) a	794.3 (5.0) a	792.1 (7.0) a	807.0 (8.6) a	811.4 (6.6) a	810.5 (5.1) a	817.8 (4.6) a

SD, standard rat chow diet; HFD, hypercaloric diet for dietary induction of obesity; WLs, high-protein weight-loss intervention diet with sedentary lifestyle (weeks 12–15); WLe, high-protein weight-loss intervention diet with training protocol; WMs, high-protein weight-loss intervention diet with sedentary lifestyle (weeks 12–15) followed by weight-maintenance stage (weeks 15–21) with SD dietary treatment and sedentary lifestyle; WMe, high-protein weight-loss intervention diet with training protocol followed by weight-maintenance stage with SD dietary treatment and training protocol. Results are expressed as means of 8 rats and standard error of the mean (in parenthesis). * $p < 0.05$, ** $p < 0.01$, *** $p < 0.001$ in t-test (12 weeks); a,b,c, means within the same line of each experimental stage (15 and 21 weeks) with different letters are significantly different (ANOVA treatment, $p < 0.05$). TV, total volume, BV, bone volume, BV/TV, bone volume density, Conn. D, connectivity density, SMI, Structure Model Index, Tb. N, trabecular number, Tb. Th, trabecular thickness, Tb. Sp, trabecular spacing, HA, hydroxyapatite.

Table 4. Effects of obesity and weight control interventions on 3D outcomes for cortical (diaphysis) bone microarchitecture in femur.

	0 WEEKS (Baseline)	Diet-Induced Obesity 12 WEEKS		Weight-Loss Intervention (WL) 15 WEEKS				Lost-Weight Maintenance Intervention (WM) 21 WEEKS			
		SD	HFD	SD	HFD	WLs	WLe	SD	HFD	WMs	WMe
TV (mm^3)	22.0 (0.4)	56.4 (2.8)	64.2 (3.5)	56.7 (1.36) a	64.1 (1.43) b	60.1 (2.80) b	62.9 (2.38) b	69.8 (2.1) a	69.7 (3.2) a	70.1 (2.5) a	71.5 (3.0) a
BV (mm^3)	9.26 (0.15)	33.8 (1.3)	39.3 (0.04) *	35.1 (0.82) a	39.9 (1.18) b	38.0 (1.58) b	39.4 (1.82) b	43.6 (0.9) a	44.2 (2.1) a	44.8 (1.3) a	45.6 (1.8) a
BV/TV	0.42 (0.09)	0.60 (0.01)	0.61 (0.003)	0.62 (0.01) a	0.62 (0.01) a	0.63 (0.01) a	0.63 (0.01) a	0.63 (0.01) a	0.63 (0.01) a	0.64 (0.01) a	0.64 (0.01) a
Ct. Th (mm)	0.35 (0.01)	0.77 (0.02)	0.78 (0.06)	0.80 (0.02) a	0.84 (0.02) a	0.85 (0.02) a	0.85 (0.02) a	0.87 (0.02) a	0.89 (0.02) a	0.90 (0.02) a	0.91 (0.03) a
Mean pore diameter (mm)	0.063 (0.002)	0.040 (0.005)	0.028 (0.003) *	0.047 (0.007) ab	0.035 (0.007) a	0.050 (0.009) ab	0.085 (0.02) b	0.038 (0.006) ab	0.031 (0.004) a	0.042 (0.008) ab	0.087 (0.02) b
Ct.Po (%)	0.148 (0.004)	0.049 (0.001)	0.041 (0.001) ***	0.044 (0.001) a	0.041 (0.001) a	0.050 (0.001) b	0.044 (0.002) a	0.038 (0.001) a	0.037 (0.001) a	0.043 (0.004) a	0.039 (0.002) a
Mean density of BV (mg HA/cm^3)	897.6 (4.8)	1050.6 (4.6)	1081.4 (4.9) ***	1074.3 (2.8) b	1083.9 (3.9) b	1022.8 (5.32) a	1068.2 (11.8) b	1107.2 (4.7) a	1098.5 (5.0) a	1088.9 (14.5) a	1101.5 (7.9) a

SD, standard rat chow diet; HFD, hypercaloric diet for dietary induction of obesity; WLs, high-protein weight-loss intervention diet with sedentary lifestyle (weeks 12–15); WLe, high-protein weight-loss intervention diet with training protocol; WMs, high-protein weight-loss intervention diet with sedentary lifestyle (weeks 12–15) followed by weight-maintenance stage (weeks 15–21) with SD dietary treatment and sedentary lifestyle; WMe, high-protein weight-loss intervention diet with training protocol followed by weight-maintenance stage with SD dietary treatment and training protocol. Results are expressed as means of 8 rats and standard error of the mean (in parenthesis). * $p < 0.05$, *** $p < 0.001$ in *t*-test (12 weeks); a,b, means within the same line of each experimental stage (15 and 21 weeks) with different letters are significantly different (ANOVA treatment, $p < 0.05$). TV, total volume, BV, bone volume, BV/TV, bone volume density, Ct. Th, cortical thickness, Ct.Po, cortical porosity, HA, hydroxyapatite.

The intake of a high protein diet during the weight loss intervention period (w13–15) led to marked changes in trabecular microarchitecture, with a significant decrease in total volume (TV) but no changes in bone volume (BV). Therefore, the BV/TV index remained similar to the group fed HFD and significantly lower than the control fed SD. The intervention diet caused a significant increase in trabecular number (Tr.N) and a decrease in trabecular spacing (Tr. Sp), while the trabecular thickness remained lower than the values of either the HFD or the SD groups. A trend to recover the mean density of TV was apparent. The main changes observed during the same period in cortical microarchitecture were a marked increase in Ct.Po that runs in parallel to lower mean density of BV. The training protocol significantly increased the pore diameter and reversed the changes in Ct. Po and density of BV induced by the high protein diet. However, no significant alterations were observed in both TV and BV, leading to no changes in the BV/TV index.

The effectiveness of the intervention period for maintaining the weight loss achieved in the previous stage was very high. Concerning trabecular microarchitecture, the increase in trabecular number compared to the HFD group was also clear, reaching similar values to that of the SD group. Moreover, the decrease in trabecular spacing achieved in the previous period remained, and a higher value for the density of BV was also observed that reached significantly higher values than the HFD group as a result of the training protocol. Related to cortical microarchitecture, no major changes in any of the measured parameters and indices were apparent among the different experimental groups at this stage except for a higher pore diameter in the femur of trained rats.

Discussion

To our knowledge, this is the first study that has investigated the effects of a high fat diet intake (12 weeks) followed by a weight-loss intervention high-protein diet (three weeks) and weight maintenance normocaloric diet (six weeks) combined or not with a training program on bone metabolism parameters and trabecular and cortical microarchitecture in the femur of male adult rats.

Obesity has been traditionally linked to greater bone mineral content that might be expected to protect the skeleton [34]. Nevertheless, animal studies have shown that a relationship exists between obesity and poor bone quality in diet-induced obese animals [35,36]. Because BMD only partially explains bone strength, we investigated the bone quality by means of its micro-architecture. It is well known that trabecular bone structure parameters could be affected in DIO [37,38]. In accordance with these findings, our results also demonstrated notable trabecular bone loss and microarchitecture deterioration in the HFD group, as evidenced by decreased BV/TV, Conn.D, Tb.N, and density of TV, as well as increased Tb.Sp. In this regard, a study observed similar aggravated results in the cancellous bone of rats fed HFD (58% fat of total kcal) after 16 weeks [39]. Gautam et al. [40] also showed marked deterioration at the trabecular region in mice after 10 weeks of HFD treatment (60% fat). However, in their study, HFD did not alter cortical bone mass. On the other hand, Li et al. [41] found reduced bone density, Tb.Th, and Tb.N in Wistar rats fed with HFD (40% fat) compared to the standard diet group at 10th week, while the BV/TV was not significantly affected. Nevertheless, Cao et al. [35] reported decreased cancellous but not cortical bone mass in tibia of mice fed a 45% HFD for 14 weeks. In our study, the main changes on cortical bone microarchitecture induced by HFD along 12 weeks were higher TV and BV, as well as a significant decrease in mean pore diameter and Ct.Po associated to higher density of bone volume. Our results are consistent with the study by Cao et al. [35] showing no significant effects of HFD on tibial cortical thickness. Other study [37] showed unaffected both trabecular and cortical thickness in the fourth lumbar vertebra (L4) in mice aged 31 weeks and fed high-fat chow (60% fat) for 24 weeks being coincident with our results. Differences in the diet composition, age or strain of the rodent, length of the study, and the site (femur, tibia, vertebra) could account for these discrepancies [35]. In this regard, the effects of obesity on bone quality are complex and appear to vary with several factors, including age, sex, and site [42]. Obesity has been linked to a site-specific increase in fracture risk [43]. This risk has been partially attributed to a decrease in mineral content, but the relationship between obesity and bone mineral density is incompletely understood. Animal models of obesity show varying bone responses to obesity, with some studies showing an increase in bone formation and others a decrease [44,45].

In the present study, weight-loss intervention led to marked changes in trabecular microarchitecture. We observed trabecular changes as a decreased TV, but not the BV, meaning that this therapy did not significantly modify the BV/TV index in comparison with HFD. In this sense, we also observed a trend to recover the density of TV. These data suggest that, at this time point, the weight-loss intervention (with or without exercise) does not vary the cancellous bone mass and density in the femur of male adult rats. Nevertheless, the intervention was effective at improving other trabecular parameters such an elevated Tb.N together with a diminished Tb.Sp, but we did not observe any particular effect of the training protocol. Interestingly, the thickness of trabeculae was significantly diminished, although a trend of some positive effect of the exercise is clear; nevertheless, it was not fully reversed to that of the SD group. Similarly, Gerbaix et al. [16] observed that a well-balanced diet alone failed to alter total and tibia bone mass and BMD in obese rats after 2 months. However, Tb.Th and bone volume density of metaphysis were decreased by the diet. The moderate intensity exercise performed significantly improved BMD possibly by inhibiting the bone resorption without any trabecular and cortical adaptation. It is known that exercise training added to a diet-induced weight loss can attenuate the weight loss-induced reduction in BMD and lean body mass in obese human [46,47]. Just as reduced loading of bone such as that experienced during weight loss induce dramatic decreases in BMD, high forces that are rapidly developed may increase bone density by increased loading. In this regard, jump exercise during hindlimb unloading protects against adverse changes in trabecular bone microarchitecture in young rats [48].

In our study, at diaphysis level, the weight-loss intervention during the same period led to no changes in both BV and TV, independently of the sedentary lifestyle or the training protocol, suggesting a non-altered cortical bone mass. This finding is associated with a higher density of BV in the femora of the WLe group. The training protocol seems to improve the density of BV reaching the values of the control SD and HFD groups, meaning a beneficial effect on bone cortical microarchitecture despite the non-beneficial effect of the weight-loss intervention. This effect is consistent with a marked increase in the mean pore diameter and Ct.Po. However, the training protocol reversed the detrimental changes in Ct. Po and density of BV induced by the high protein diet.

Regarding the weight-maintenance stage with SD dietary treatment (weeks 15–21), we clearly observed an effect of both, the diet and the training protocol on trabecular microarchitecture. Combination of the SD dietary therapy with exercise was effective on the BV/TV and on the density of TV, which were evidently increased in this group (WMe). Moreover, changes in the SMI were detectable in both groups (WMs and WMe). SMI describes if an examined volume of trabecular bone has either plate- or rod-like properties and is thus a suitable tool to describe subtle ongoing changes in bone microarchitecture. The values for mammalian spongiosa range from 0 to 3, with 0 being the ideal plate and 3 the ideal rod [49]. In our study, significantly decreased SMI values indicate the presence of more plate-like spongiosa in both groups. Interestingly, and in a different way to what happened in the weight-loss intervention, trabecular thickness remained relatively unaffected in all groups. This finding was already explained by Patsch et al. [37] considering the computation of the parameter itself: simplified, trabecular thickness is calculated as the most frequently occurring diameter of a virtual ball fitting into trabecular structures [50]. Furthermore, the increase in Tb.N compared to the HFD group, reaching similar values to that of the SD group, confirms the effectiveness of the weight-maintenance stage with SD dietary treatment. Also, the decrease in trabecular spacing achieved in the previous period remained. Related to cortical microarchitecture, the WM groups showed no significant alterations in any of the measured parameters and indices among the different experimental groups, but an exaggerated increase in the mean pore diameter in the femur of trained rats. Similar to our findings, Scheller et al. [12] investigated the impact of HFD (60%) and subsequent weight loss on skeletal parameters in six-week old male mice. They observed decreased trabecular bone volume fraction, mineral content, and number after 12, 16, or 20 weeks of HFD compared to normal chow diet controls, with only partial recovery after weight loss (HFD for 12 weeks and then normal chow for eight weeks to mimic weight loss).

In general terms, our study demonstrates that there are some reversible and some permanent changes in bone quality with HFD, followed by WL. Diet-induced obesity causes greater damage in growing bones [51]. Indeed, our animals started HFD at an age in which skeletal development is still highly active, likely contributing to impaired bone accrual during growth. For healthy aging, different regimens may be required to maintain bone health after WL, possibly with a focus on activity and diet. Training strategies that include heavy resistance training and high impact loading may be especially productive in maintaining, or even increasing bone density with weight loss [52].

3.3. Ash and Bone Mineral Content

The effects of DIO and bodyweight control interventions on the elemental composition of the femur are presented in Table 5. Although total mineral content represented by ash percentage did not differ significantly among all the groups at the different stages of the experimental period, some differences were apparent in specific elements. Generally, the dietary induction of obesity caused a decrease in femur content of macrominerals P, Ca, and Mg that was significant at week 12 of the experimental period but did not reach statistical significance at week 15. By week 21 such differences were not apparent. Obesity induction also caused a significant decrease in the content of certain microminerals and trace elements such as Cu, Mn, Co, or As that remained on weeks 15 and 21 except for Cu in the later stage. Besides, the femur content of K and Fe was significantly decreased by obesity from week 15 of the experimental period. The effects of weight control interventions were not consistent for all the minerals studied, but tended to reverse the previously described action of obesity in P, Ca, Mg, Fe and Mn.

Table 5. Influence of obesity and weight control interventions on mineral content of femur.

	0 WEEKS (Base-line)	**Diet-Induced Obesity 12 WEEKS**		**Weight-Loss Intervention (WL) 15 WEEKS**				**Lost-Weight Maintenance Intervention (WM) 21 WEEKS**			
		SD	**HFD**	**SD**	**HFD**	**WLs**	**WLe**	**SD**	**HFD**	**WMs**	**WMe**
Ash (%)	53.3	58.3	57.4	56.2 a	56.7 a	60.7 a	58.7 ab	58.4 a	58.5 a	59.5 a	58.2 a
	(0.9)	(1.1)	(1.5)	(0.4)	(0.9)	(1.4)	(0.8)	(0.7)	(0.9)	(0.7)	(1.4)
P (g/kg)	94.2	129.6	107.3 ***	116.6 a	107.8 a	114.3 a	132.0 a	119.6 a	131.3 b	130.6 b	117.9 a
	(2.3)	(3.8)	(3.1)	(5.1)	(3.0)	(5.3)	(3.4)	(3.1)	(4.1)	(1.7)	(2.8)
Ca (g/kg)	177.9	255.8	225.7 *	231.6 a	224.3 a	228.8 a	254.2 a	247.3 ab	237.5 a	259.5 b	237.1 a
	(4.7)	(7.3)	(6.7)	(9.4)	(6.3)	(1.2)	(3.8)	(6.6)	(7.2)	(3.2)	(5.5)
Mg (g/kg)	3.42	4.39	3.13 ***	3.54 ab	3.06 a	3.35 ab	4.17 b	3.72 b	4.17 c	3.78 b	3.36 a
	(0.08)	(0.15)	(0.12)	(0.20)	(0.10)	(0.16)	(0.17)	(0.13)	(0.15)	(0.06)	(0.11)
K (g/kg)	5.04	1.71	1.79	2.36 c	1.82 ab	1.95 b	1.51 a	1.62 a	1.74 a	1.58 a	1.76 a
	(0.14)	(0.07)	(0.16)	(0.20)	(0.12)	(0.10)	(0.07)	(0.10)	(0.08)	(0.08)	(0.12)
Fe (mg/kg)	88.5	45.2	44.0	68.6 ab	46.8 a	72.1 ab	84.5 b	51.0 a	69.4 a	70.8 a	66.5 a
	(6.4)	(7.1)	(7.8)	(6.7)	(4.3)	(9.1)	(17.4)	(3.5)	(8.4)	(11.4)	(7.7)
Zn (mg/kg)	221.3	219.1	216.8	200.4 a	215.8 a	244.9 b	241.5 b	186.2 a	235.8 bc	257.5 c	230.8 b
	(11.5)	(6.6)	(10.8)	(6.9)	(3.5)	(10.4)	(8.7)	(7.2)	(7.3)	(6.8)	(4.5)
Cu (mg/kg)	2.11	1.00	0.78 *	1.57 b	0.91 a	1.93 c	0.82 a	0.73 a	1.01 ab	1.24 bc	1.63 c
	(0.22)	(0.17)	(0.05)	(0.08)	(0.05)	(0.06)	(0.05)	(0.04)	(0.05)	(0.06)	(0.08)
Mn (µg/kg)	428.1	358.1	205.0 ***	326.2 ab	189.8 a	374.9 b	288.2 ab	219.3 a	186.4 a	401.9 c	309.8 b
	(16.3)	(86.7)	(21.5)	(43.4)	(15.2)	(22.8)	(28.3)	(17.4)	(9.0)	(32.5)	(11.7)
Se (µg/kg)	395.2	238.8	245.8	272.9 a	261.0 a	229.8 a	274.6 a	244.5 a	266.6 a	218.6 a	239.8 a
	(18.8)	(18.2)	(14.3)	(14.3)	(14.8)	(26.4)	(22.1)	(20.3)	(20.5)	(17.5)	(14.2)
V (µg/kg)	10.7	31.0	19.5	15.4 a	60.4 c	34.1 b	26.4 ab	48.9 ab	61.7 b	26.5 a	25.1 a
	(1.0)	(6.3)	(6.7)	(4.4)	(6.7)	(3.8)	(4.2)	(6.4)	(4.9)	(2.1)	(2.4)
Co (µg/kg)	56.9	76.5	62.8 ***	79.1 bc	67.4 ab	89.0 c	62.3 a	88.5 d	51.8 a	79.0 c	69.4 b
	(2.2)	(2.3)	(2.0)	(4.1)	(2.6)	(5.6)	(1.4)	(2.9)	(3.8)	(2.9)	(1.8)
Sc (µg/kg)	435.8	483.8	551.9	509.3 ab	565.5 bc	612.7 c	442.4 a	631.3 b	666.4 b	502.0 a	493.3 a
	(13.5)	(37.6)	(15.0)	(29.0)	(16.4)	(27.0)	(13.5)	(17.6)	(22.2)	(8.0)	(13.6)
As (µg/kg)	67.8	91.2	23.9 ***	84.5 c	24.6 a	40.5 b	52.0 b	75.3 c	40.9 a	58.0 b	58.2 b
	(4.5)	(11.4)	(2.7)	(5.3)	(1.8)	(5.1)	(4.9)	(4.8)	(3.1)	(5.2)	(7.0)

SD, standard rat chow diet; HFD, hypercaloric diet for dietary induction of obesity; WLs, high-protein weight-loss intervention diet with sedentary lifestyle (weeks 12–15); WLe, high-protein weight-loss intervention diet with training protocol; WMs, high-protein weight-loss intervention diet with sedentary lifestyle (weeks 12–15) followed by weight-maintenance stage (weeks 15–21) with SD dietary treatment and sedentary lifestyle; WMe, high-protein weight-loss intervention diet with training protocol followed by weight-maintenance stage with SD dietary treatment and training protocol. Results are expressed as means of 8 rats and standard error of the mean (in parenthesis). * $p < 0.05$, *** $p < 0.001$ in *t*-test (12 weeks); a,b,c, means within the same line of each experimental stage (15 and 21 weeks) with different letters are significantly different (ANOVA treatment, $p < 0.05$).

Discussion

Metabolic syndrome and osteoporosis have been described to share some common underlying pathways, such as regulation of calcium homeostasis, receptor activator of NF-κB ligand (RANKL)/receptor activator of the NF-κB (RANK)/osteoprotegerin (OPG), and Wnt-β-catenin signaling pathways [53]. Thus, metabolic syndrome may have a potential role in the development of osteoporosis. In this regard, mineral homeostasis is significantly affected in murine DIO models. The content of several elements in the liver, kidney, heart, and pancreas has been shown to decrease in response to high-fat diet feeding [54]. Such altered trace elements status is supposed to be a primary modification and precedes other metabolic obesity-related disturbances. Nevertheless, treatment of obesity-related symptoms alleviated the altered trace elements metabolism induced by HFD by modulating hyperglycemic and insulin resistance status [55]. Likewise, altered renal functionality derived from T2DM, one of the associated metabolic disturbances of obesity in metabolic syndrome, has been described to affect mineral metabolism and modify the elemental composition of plasma and bone [56]. Here, the development of obesity appeared to exert an inhibitory action on the bone mineral content of both macro and micro or trace elements, although no differences in total mineral content of bone were observed despite morphometric changes associated with obesity. In this regard, Ip et al. [57] reported no significant differences in femur ash content of obese Zucker rats vs. their lean counterparts although a smaller femur size and weight was apparent in the obese animals. On the other hand, Song and Sergeev [58] found a significant decrease in femur Ca and P content of HFD-treated mice. Such deleterious effects were reversed by high intakes of vitamin D-3 and calcium.

Modifications of femur mineral content in obese animals was correlated to the changes in microarchitecture parameters observed in trabecular rather than cortical bone. In the former, a significant decrease in bone volume, connectivity density, trabecular number, and BMD of total volume were caused by obesity. In addition to its possible relationship to microarchitecture changes, the lower concentration of minerals in the femur could also be due to a dilution effect caused by the accumulation of fat in the bone marrow.

Mathey et al. [59] eported that a continuous aerobic training protocol (35–40% VO_{2max}, 20–50 min/day, 6 days/week, 89 days) induced an exercise-induced increment in BMD, bone calcium content, diameter, and femoral failure load. The mixed training protocol implemented in our experiment showed a trend to increase P, Ca, and Mg content when compared to their sedentary controls. Nevertheless, such an effect was observable on week 15 but not on week 21 of the experimental period. Other authors [60] have pointed out the beneficial effects on fructose-induced obese rats of exercise (1-h running protocol a day, six days per week, ten weeks) that reduced visceral fat and ameliorated glucose intolerance, lowered blood lactic acid levels, improved lactic acid usage efficiency, and increased oxidative stress and hepatic levels of Mn, Fe, Cu, and Zn in the normal and obese animals.

4. Conclusions

The lifestyle interventions of caloric restriction and mixed training protocol implemented as weight loss strategies have been effective to counteract some of the deleterious effects caused by dietary induction of obesity in a Sprague Dawley rat model, specifically in trabecular bone morphometric parameters and indices as well as on bone mineral content. Thus, the interventions can be used as efficient strategies in the treatment of obesity, although some modifications in the training protocol could be of interest to maintain or even increase bone density with weight loss.

Supplementary Materials: The following supporting information can be downloaded at: https://www.mdpi.com/article/10.3390/nu14183672/s1, Table S1: Details of the mixed training protocol*.

Author Contributions: Conceptualization, J.M.P., J.L., P.A., M.L.-J. and P.P.; methodology, E.N., R.M., C.S., J.L. and J.M.P.; validation, P.A. and M.L.-J.; formal analysis, E.N., R.M., C.S. and G.K.; investigation, E.N., R.M. and G.K.; resources, P.A., J.M.P., M.L.-J. and P.P.; data curation, J.M.P.; writing—original draft preparation, E.N., R.M., J.M.P., M.L.-J. and P.P.; writing—review and editing, E.N., R.M., G.K., C.S., J.L., P.A., J.M.P., M.L.-J. and P.P.; visualization, E.N., R.M., G.K., P.A., J.M.P., M.L.-J. and P.P.; supervision, J.M.P.; project administration, P.A. and M.L.-J.; funding acquisition, P.A. and J.M.P. All authors have read and agreed to the published version of the manuscript.

Funding: This research was funded by the Spanish Ministry of Science, Innovation and Universities and the European Union through projects DEP2014-58296-R, RTC-2017-6540-1, RTI2018-100934-B-I00 and FEDER program, respectively.

Institutional Review Board Statement: All experiments were undertaken according to Directional Guides Related to Animal Housing and Care [20] and all procedures were approved by the Animal Experimentation Ethics Committee of the University of Granada, Spain (Project Reference DEP2014-58296R).

Informed Consent Statement: Not applicable.

Data Availability Statement: Not applicable.

Acknowledgments: The authors want to acknowledge the special contribution to the development of the experiments by the Experimental Animal Unit of CIC (Universidad de Granada). Authors want also to thank Carina Quack and Matthias Schmidt for their special contribution to this work.

Conflicts of Interest: The authors declare no conflict of interest. The funders had no role in the design of the study; in the collection, analyses, or interpretation of data; in the writing of the manuscript; or in the decision to publish the results.

References

1. Hruby, A.; Manson, J.E.; Qi, L.; Malik, V.S.; Rimm, E.B.; Sun, Q.; Willett, W.C.; Hu, F.B. Determinants and Consequences of Obesity. *Am. J. Public Health* **2016**, *106*, 1656–1662. [CrossRef]
2. Jakubczyk, A.; Kiersnowska, K.; Ömeroğlu, B.; Gawlik-Dziki, U.; Tutaj, K.; Rybczyńska-Tkaczyk, K.; Szydłowska-Tutaj, M.; Złotek, U.; Baraniak, B. The Influence of *Hypericum perforatum* L. Addition to Wheat Cookies on Their Antioxidant, Anti-Metabolic Syndrome, and Antimicrobial Properties. *Foods* **2021**, *10*, 1379. [CrossRef]
3. Koshovyi, O.; Granica, S.; Piwowarski, J.P.; Stremoukhov, O.; Kostenko, Y.; Kravchenko, G.; Krasilnikova, O.; Zagayko, A. Highbush Blueberry (*Vaccinium corymbosum* L.) Leaves Extract and Its Modified Arginine Preparation for the Management of Metabolic Syndrome-Chemical Analysis and Bioactivity in Rat Model. *Nutrients* **2021**, *13*, 2870. [CrossRef] [PubMed]
4. Maclean, P.S.; Bergouignan, A.; Cornier, M.A.; Jackman, M.R. Biology's response to dieting: The impetus for weight regain. *Am. J. Physiol. Regul. Integr. Comp. Physiol.* **2011**, *301*, R581–R600. [CrossRef] [PubMed]
5. Yumuk, V.; Tsigos, C.; Fried, M.; Schindler, K.; Busetto, L.; Micic, D.; Toplak, H. European Guidelines for Obesity Management in Adults. *Obes. Facts* **2015**, *8*, 402–424. [CrossRef]
6. Camacho, S.; Ruppel, A. Is the calorie concept a real solution to the obesity epidemic? *Glob. Health Action* **2017**, *10*, 1289650. [CrossRef]
7. Golbidi, S.; Badran, M.; Laher, I. Antioxidant and anti-inflammatory effects of exercise in diabetic patients. *Exp. Diabetes Res.* **2012**, *2012*, 941868. [CrossRef]
8. Bull, F.C.; Al-Ansari, S.S.; Biddle, S.; Borodulin, K.; Buman, M.P.; Cardon, G.; Carty, C.; Chaput, J.P.; Chastin, S.; Chou, R.; et al. World Health Organization 2020 guidelines on physical activity and sedentary behaviour. *Br. J. Sports Med.* **2020**, *54*, 1451–1462. [CrossRef] [PubMed]
9. Picke, A.K.; Sylow, L.; Møller, L.L.V.; Kjøbsted, R.; Schmidt, F.N.; Steejn, M.W.; Salbach-Hirsch, J.; Hofbauer, C.; Blüher, M.; Saalbach, A.; et al. Differential effects of high-fat diet and exercise training on bone and energy metabolism. *Bone* **2018**, *116*, 120–134. [CrossRef]
10. Nielson, C.M.; Srikanth, P.; Orwoll, E.S. Obesity and fracture in men and women: An epidemiologic perspective. *J. Bone Miner. Res.* **2012**, *27*, 1–10. [CrossRef]
11. Palermo, A.; Tuccinardi, D.; Defeudis, G.; Watanabe, M.; D'Onofrio, L.; Lauria Pantano, A.; Napoli, N.; Pozzilli, P.; Manfrini, S. BMI and BMD: The Potential Interplay between Obesity and Bone Fragility. *Int. J. Environ. Res. Public Health* **2016**, *13*, 544. [CrossRef]
12. Scheller, E.L.; Khoury, B.; Moller, K.L.; Wee, N.K.Y.; Khandaker, S.; Kozloff, K.M.; Abrishami, S.H.; Zamarron, B.F.; Singer, K. Changes in Skeletal Integrity and Marrow Adiposity during High-Fat Diet and after Weight Loss. *Front. Endocrinol.* **2016**, *7*, 102. [CrossRef] [PubMed]
13. Chao, D.; Espeland, M.A.; Farmer, D.; Register, T.C.; Lenchik, L.; Applegate, W.B.; Ettinger, W.H., Jr. Effect of voluntary weight loss on bone mineral density in older overweight women. *J. Am. Geriatr Soc.* **2000**, *48*, 753–759. [CrossRef] [PubMed]
14. Jensen, L.B.; Quaade, F.; Sørensen, O.H. Bone loss accompanying voluntary weight loss in obese humans. *J. Bone Miner. Res.* **1994**, *9*, 459–463. [CrossRef] [PubMed]
15. Ricci, T.A.; Heymsfield, S.B.; Pierson, R.N., Jr.; Stahl, T.; Chowdhury, H.A.; Shapses, S.A. Moderate energy restriction increases bone resorption in obese postmenopausal women. *Am. J. Clin. Nutr.* **2001**, *73*, 347–352. [CrossRef]
16. Gerbaix, M.; Metz, L.; Mac-Way, F.; Lavet, C.; Guillet, C.; Walrand, S.; Masgrau, A.; Vico, L.; Courteix, D. A well-balanced diet combined or not with exercise induces fat mass loss without any decrease of bone mass despite bone micro-architecture alterations in obese rat. *Bone* **2013**, *53*, 382–390. [CrossRef]
17. Galisteo, M.; Sánchez, M.; Vera, R.; González, M.; Anguera, A.; Duarte, J.; Zarzuelo, A. A diet supplemented with husks of Plantago ovata reduces the development of endothelial dysfunction, hypertension, and obesity by affecting adiponectin and TNF-alpha in obese Zucker rats. *J. Nutr.* **2005**, *135*, 2399–2404. [CrossRef]
18. Sengupta, P. A Scientific Review of Age Determination for a Laboratory Rat: How Old is it in Comparison with Human Age? *Biomed Int.* **2011**, *2*, 81–89.
19. Ramage, S.; Farmer, A.; Eccles, K.A.; McCargar, L. Healthy strategies for successful weight loss and weight maintenance: A systematic review. *Appl. Physiol. Nutr. Metab.* **2014**, *39*, 1. [CrossRef]
20. Directive 2010/63/EU of the European Parliament and of the Council of 22 September 2010 on the Protection of Animals used for Scientific Purposes. 2010, L276/233-279. Available online: https://eur-lex.europa.eu/LexUriServ/LexUriServ.do?uri=OJ%3AL%3A2010%3A276%3A0033%3A0079%3Aen%3APDF (accessed on 19 August 2022).
21. Coll-Risco, I.; Aparicio, V.A.; Nebot, E.; Camiletti-Moirón, D.; Martínez, R.; Kapravelou, G.; López-Jurado, M.; Porres, J.M.; Aranda, P. Effects of interval aerobic training combined with strength exercise on body composition, glycaemic and lipid profile and aerobic capacity of obese rats. *J. Sports Sci.* **2016**, *34*, 1452–1460. [CrossRef]

22. Lopez Trinidad, L.M.; Martinez, R.; Kapravelou, G.; Galisteo, M.; Aranda, P.; Porres, J.M.; Lopez-Jurado, M. Caloric restriction, physical exercise, and CB1 receptor blockade as an efficient combined strategy for bodyweight control and cardiometabolic status improvement in male rats. *Sci. Rep.* **2021**, *11*, 4286. [CrossRef] [PubMed]
23. Sánchez-González, C.; López-Chaves, C.; Gómez-Aracena, J.; Galindo, P.; Aranda, P.; Llopis, J. Association of plasma manganese levels with chronic renal failure. *J. Trace Elem. Med. Biol.* **2015**, *31*, 78–84. [CrossRef] [PubMed]
24. Bertrand, H.A.; Lynd, F.T.; Masoro, E.J.; Yu, B.P. Changes in adipose mass and cellularity through the adult life of rats fed ad libitum or a life-prolonging restricted diet. *J. Gerontol.* **1980**, *35*, 827–835. [CrossRef] [PubMed]
25. Barzilai, N.; Gupta, G. Interaction between aging and syndrome X: New insights on the pathophysiology of fat distribution. *Ann. N. Y. Acad Sci.* **1999**, *892*, 58–72. [CrossRef] [PubMed]
26. Lehnig, A.C.; Stanford, K.I. Exercise-induced adaptations to white and brown adipose tissue. *J. Exp. Biol.* **2018**, *221*, jeb161570. [CrossRef] [PubMed]
27. Stallknecht, B.; Vinten, J.; Ploug, T.; Galbo, H. Increased activities of mitochondrial enzymes in white adipose tissue in trained rats. *Am. J. Physiol.* **1991**, *261*, E410–E414. [CrossRef]
28. Stanford, K.I.; Middelbeek, R.J.; Goodyear, L.J. Exercise Effects on White Adipose Tissue: Beiging and Metabolic Adaptations. *Diabetes* **2015**, *64*, 2361–2368. [CrossRef]
29. Vernochet, C.; Mourier, A.; Bezy, O.; Macotela, Y.; Boucher, J.; Rardin, M.J.; An, D.; Lee, K.Y.; Ilkayeva, O.R.; Zingaretti, C.M.; et al. Adipose-specific deletion of TFAM increases mitochondrial oxidation and protects mice against obesity and insulin resistance. *Cell Metab.* **2012**, *16*, 765–776. [CrossRef]
30. Wu, M.V.; Bikopoulos, G.; Hung, S.; Ceddia, R.B. Thermogenic capacity is antagonistically regulated in classical brown and white subcutaneous fat depots by high fat diet and endurance training in rats: Impact on whole-body energy expenditure. *J. Biol. Chem.* **2014**, *289*, 34129–34140. [CrossRef]
31. Yaspelkis, B.B., 3rd; Singh, M.K.; Krisan, A.D.; Collins, D.E.; Kwong, C.C.; Bernard, J.R.; Crain, A.M. Chronic leptin treatment enhances insulin-stimulated glucose disposal in skeletal muscle of high-fat fed rodents. *Life Sci.* **2004**, *74*, 1801–1816. [CrossRef]
32. Bradley, R.L.; Jeon, J.Y.; Liu, F.F.; Maratos-Flier, E. Voluntary exercise improves insulin sensitivity and adipose tissue inflammation in diet-induced obese mice. *Am. J. Physiol. Endocrinol. Metab.* **2008**, *295*, E586–E594. [CrossRef] [PubMed]
33. Zachwieja, J.J.; Hendry, S.L.; Smith, S.R.; Harris, R.B. Voluntary wheel running decreases adipose tissue mass and expression of leptin mRNA in Osborne-Mendel rats. *Diabetes* **1997**, *46*, 1159–1166. [CrossRef] [PubMed]
34. Felson, D.T.; Zhang, Y.; Hannan, M.T.; Anderson, J.J. Effects of weight and body mass index on bone mineral density in men and women: The Framingham study. *J. Bone Miner. Res.* **1993**, *8*, 567–573. [CrossRef] [PubMed]
35. Cao, J.J.; Gregoire, B.R.; Gao, H. High-fat diet decreases cancellous bone mass but has no effect on cortical bone mass in the tibia in mice. *Bone* **2009**, *44*, 1097–1104. [CrossRef]
36. Cao, J.J.; Sun, L.; Gao, H. Diet-induced obesity alters bone remodeling leading to decreased femoral trabecular bone mass in mice. *Ann. N. Y. Acad. Sci.* **2010**, *1192*, 292–297. [CrossRef]
37. Patsch, J.M.; Kiefer, F.W.; Varga, P.; Pail, P.; Rauner, M.; Stupphann, D.; Resch, H.; Moser, D.; Zysset, P.K.; Stulnig, T.M.; et al. Increased bone resorption and impaired bone microarchitecture in short-term and extended high-fat diet-induced obesity. *Metabolism* **2011**, *60*, 243–249. [CrossRef]
38. Zhang, K.; Wang, C.; Chen, Y.; Ji, X.; Chen, X.; Tian, L.; Yu, X. Preservation of high-fat diet-induced femoral trabecular bone loss through genetic target of TNF-α. *Endocrine* **2015**, *50*, 239–249. [CrossRef]
39. Tang, L.; Gao, X.; Yang, X.; Liu, C.; Wang, X.; Han, Y.; Zhao, X.; Chi, A.; Sun, L. Ladder-Climbing Training Prevents Bone Loss and Microarchitecture Deterioration in Diet-Induced Obese Rats. *Calcif. Tissue Int.* **2016**, *98*, 85–93. [CrossRef]
40. Gautam, J.; Choudhary, D.; Khedgikar, V.; Kushwaha, P.; Singh, R.S.; Singh, D.; Tiwari, S.; Trivedi, R. Micro-architectural changes in cancellous bone differ in female and male C57BL/6 mice with high-fat diet-induced low bone mineral density. *Br. J. Nutr.* **2014**, *111*, 1811–1821. [CrossRef]
41. Li, W.; Xu, P.; Wang, C.; Ha, X.; Gu, Y.; Wang, Y.; Zhang, J.; Xie, J. The effects of fat-induced obesity on bone metabolism in rats. *Obes. Res. Clin. Pract.* **2017**, *11*, 454–463. [CrossRef]
42. Votava, L.; Schwartz, A.G.; Harasymowicz, N.S.; Wu, C.L.; Guilak, F. Effects of dietary fatty acid content on humeral cartilage and bone structure in a mouse model of diet-induced obesity. *J. Orthop. Res.* **2019**, *37*, 779–788. [CrossRef] [PubMed]
43. Walsh, J.S.; Vilaca, T. Obesity, Type 2 Diabetes and Bone in Adults. *Calcif. Tissue Int.* **2017**, *100*, 528–535. [CrossRef] [PubMed]
44. Lau, B.Y.; Fajardo, V.A.; McMeekin, L.; Sacco, S.M.; Ward, W.E.; Roy, B.D.; Peters, S.J.; Leblanc, P.J. Influence of high-fat diet from differential dietary sources on bone mineral density, bone strength, and bone fatty acid composition in rats. *Appl. Physiol. Nutr. Metab.* **2010**, *35*, 598–606. [CrossRef] [PubMed]
45. Wang, Y.; Dellatore, P.; Douard, V.; Qin, L.; Watford, M.; Ferraris, R.P.; Lin, T.; Shapses, S.A. High fat diet enriched with saturated, but not monounsaturated fatty acids adversely affects femur, and both diets increase calcium absorption in older female mice. *Nutr. Res.* **2016**, *36*, 742–750. [CrossRef] [PubMed]
46. Shah, K.; Armamento-Villareal, R.; Parimi, N.; Chode, S.; Sinacore, D.R.; Hilton, T.N.; Napoli, N.; Qualls, C.; Villareal, D.T. Exercise training in obese older adults prevents increase in bone turnover and attenuates decrease in hip bone mineral density induced by weight loss despite decline in bone-active hormones. *J. Bone Miner. Res.* **2011**, *26*, 2851–2859. [CrossRef]
47. Villareal, D.T.; Chode, S.; Parimi, N.; Sinacore, D.R.; Hilton, T.; Armamento-Villareal, R.; Napoli, N.; Qualls, C.; Shah, K. Weight loss, exercise, or both and physical function in obese older adults. *N. Engl. J. Med.* **2011**, *364*, 1218–1229. [CrossRef]

48. Ju, Y.I.; Sone, T.; Ohnaru, K.; Choi, H.J.; Choi, K.A.; Fukunaga, M. Jump exercise during hindlimb unloading protect against the deterioration of trabecular bone microarchitecture in growing young rats. *Springerplus* **2013**, *2*, 35. [CrossRef]
49. Hildebrand, T.; Rüegsegger, P. Quantification of Bone Microarchitecture with the Structure Model Index. *Comput. Methods Biomech. Biomed. Engin.* **1997**, *1*, 15–23. [CrossRef]
50. Hildebrand, T.; Rüegsegger, P. A new method for the model-independent assessment of thickness in three-dimensional images. *J. Microsc.* **1997**, *185*, 67–75. [CrossRef]
51. Wee, N.; Herzog, H.; Baldock, P. Diet-induced obesity alters skeletal microarchitecture and the endocrine activity of bone. In *Handbook of Nutrition and Diet in Therapy of Bone Diseases*; Wageningen Academic Publishers: Wageningen, The Netherlands, 2016; pp. 1602–1606.
52. Hunter, G.R.; Plaisance, E.P.; Fisher, G. Weight loss and bone mineral density. *Curr. Opin. Endocrinol. Diabetes Obes.* **2014**, *21*, 358–362. [CrossRef]
53. Wong, S.K.; Chin, K.Y.; Suhaimi, F.H.; Ahmad, F.; Ima-Nirwana, S. The Relationship between Metabolic Syndrome and Osteoporosis: A Review. *Nutrients* **2016**, *8*, 347. [CrossRef] [PubMed]
54. Tinkov, A.A.; Gatiatulina, E.R.; Popova, E.V.; Polyakova, V.S.; Skalvaya, A.A.; Agletdinov, E.F.; Nikonorov, A.A.; Radysh, I.V.; Kkarganov, M.Y.; Skalny, A.V. The impact of adipogenic diet on rats' tissue trace elements content. *Patol. Fiziol. Eksp. Ter.* **2016**, *60*, 79–85. [PubMed]
55. Antony Rathinasamy, J.I.R.; Uddandrao, V.V.S.; Raveendran, N.; Sasikumar, V. Antiobesity Effect of Biochanin-A: Effect on Trace Element Metabolism in High Fat Diet-Induced Obesity in Rats. *Cardiovasc. Hematol. Agents Med. Chem.* **2020**, *18*, 21–30. [CrossRef] [PubMed]
56. Malmgren, L.; McGuigan, F.; Christensson, A.; Akesson, K.E. Reduced kidney function is associated with BMD, bone loss and markers of mineral homeostasis in older women: A 10-year longitudinal study. *Osteoporos. Int.* **2017**, *28*, 3463–3473. [CrossRef]
57. Ip, T.Y.; Peterson, J.; Byrner, R.; Tou, J.C. Bone responses to body weight and moderate treadmill exercising in growing male obese (fa/fa) and lean Zucker rats. *J. Musculoskelet. Neuronal. Interact.* **2009**, *9*, 155–166.
58. Song, Q.; Sergeev, I.N. High vitamin D and calcium intakes increase bone mineral (Ca and P) content in high-fat diet-induced obese mice. *Nutr Res.* **2015**, *35*, 146–154. [CrossRef]
59. Mathey, J.; Horcajada-Molteni, M.N.; Chanteranne, B.; Picherit, C.; Puel, C.; Lebecque, P.; Cubizoles, C.; Davicco, M.J.; Coxam, V.; Barlet, J.P. Bone mass in obese diabetic Zucker rats: Influence of treadmill running. *Calcif. Tissue Int.* **2002**, *70*, 305–311. [CrossRef]
60. Miya, N.; Uratani, A.; Chikamoto, K.; Naito, Y.; Terao, K.; Yoshikawa, Y.; Yasui, H. Effects of exercise on biological trace element concentrations and selenoprotein P expression in rats with fructose-induced glucose intolerance. *J. Clin. Biochem. Nutr.* **2020**, *66*, 124–131. [CrossRef]

nutrients MDPI

Article

High-Calorie Food-Cues Impair Conflict Control: EEG Evidence from a Food-Related Stroop Task

Yong Liu [1,2,*], Jia Zhao [1,2], Yizhou Zhou [3], Ruiyu Yang [2], Beichen Han [2], Yufei Zhao [2], Yazhi Pang [2], Hong Yuan [1,2] and Hong Chen [1,2,*]

1 Key Laboratory of Cognition and Personality (Ministry of Education), School of Psychology, Southwest University, Chongqing 400715, China
2 School of Psychology, Southwest University, Chongqing 400715, China
3 School of Education, Chongqing Normal University, Chongqing 401331, China
* Correspondence: liuy0768@swu.edu.cn (Y.L.); chenhswu@163.com (H.C.)

Abstract: Long-term excessive intake of high-calorie foods might lead to cognitive impairments and overweight or obesity. The current study aimed to examine the effects of high-calorie foods on the behavioral and neurological correlates of food-related conflict control ability. A food-related Stroop task, which asked the participants to respond to the food images and ignore the calorie information, were employed. A total of 61 individuals were recruited and who completed the food-related Stroop task with event-related potentials (ERPs). Participants exhibited a slower reaction time and lower accuracy in high-calorie food stimuli than that in low-calorie food stimuli. The ERP results exhibited a reduction in N2 amplitudes when responding to high-calorie food stimuli compared to when responding to low-calorie food stimuli. In addition, time-frequency analysis revealed that theta power induced by low-calorie food stimuli was significantly greater than that of high-calorie food stimuli. The findings indicated that high-calorie foods impair food-related conflict control. The present study expands on the previous studies of the neural correlates of food cues and provides new insights into the processing and resolving of conflicting information for eating behavior and weight control.

Keywords: conflict control; high-calorie foods; food-related Stroop task; N2 and P3; theta power

Citation: Liu, Y.; Zhao, J.; Zhou, Y.; Yang, R.; Han, B.; Zhao, Y.; Pang, Y.; Yuan, H.; Chen, H. High-Calorie Food-Cues Impair Conflict Control: EEG Evidence from a Food-Related Stroop Task. *Nutrients* **2022**, *14*, 4593. https://doi.org/10.3390/nu14214593

Academic Editor: Fernando Fernandez-Aranda

Received: 30 June 2022
Accepted: 29 August 2022
Published: 1 November 2022

1. Introduction

We live in an environment filled with various food temptations at our fingertips and we are constantly struggling between maintaining a healthy eating style and giving in to immediate food temptations. The excessive intake of high-calorie foods is considered a contributing factor to obesity, while resisting high-calorie foods is considered to require successful executive function [1–4]. Being overweight and obese can, in turn, have negative effects on executive functions including response inhibition, conflict control, cognitive flexibility, and decision-making [5–11].

Conflict control refers to the information processing capability of the brain to monitor and respond to the emergence of conflicts and is considered a core component of the brain's executive function [12]. Previous behavioral data indicate that unhealthy food (high-calorie food) may have negative effects on individuals' conflict control. Nijs and colleagues (2010) aims to investigate the food-related conflict control in individuals with obesity. During a food-related Stroop task, participants were instructed to respond as fast as possible to the font color of the words [grouped as food-related words (such as chocolate, meatball) and neutral office-related words (such as scissor, desk)] shown on the screen and ignore the meaning of the words. They found greater reaction times to food-related words than to neutral words in participants of both obese and normal-weight groups, which indicated general attentional biases toward food [13]. Lyu et al. used the food-related flanker task to investigate the conflict control ability in Chinese females who report binge eating disorder characteristics and found that while the binge eating group displayed a slower reaction time to incongruent trials than that of congruent trials, the control group did not exhibit such a pattern, which indicated that the binge eating group had a relative conflict control ability deficit [14]. Smith et al. utilized a food-choice task to explore conflict control when making decisions between healthy and unhealthy food. All participants were asked to choose the healthier food during the task. They found that more self-reported cravings and desires for unhealthy food predicted a greater response conflict. The authors believed that greater conflict management would be required for individuals to complete the food-choice task efficiently [15]. In addition, during a food choice task, individuals who scored higher on restrained eating were more likely to choose healthier foods compared to high-calorie foods and exhibited less self-control conflict when choosing healthier foods compared to individuals who scored lower on restrained eating [1]. Zhang et al. (2021) investigated the effects of food-related thought suppression on conflict control ability in restrained-eating adults. They found that after inhibiting thoughts about eating, the restrained eaters chose more high-calorie foods. The findings also indicated that the suppression of food-related thoughts could lead to a reduction in the ability to monitor conflicts between current behaviors and goals, which might lead to unhealthy eating behaviors [16]. In addition, high-calorie food may have a negative effect on the other cores of executive function. For example, Meule et al. explored the effects of high-calorie food on working memory in high food cravers and low food cravers. All participants performed an n-back task (the stimuli consisted of high-calorie food and neutral items). They found slower reaction times and lower accuracy in response to high-calorie food than to neutral items regardless of reported cravings, which indicated that high-calorie food cues have immediate negative effects on working memory performance [17].

Adding to behavior studies, prior neurophysiology studies have also investigated the relationships between inhibition control and food-cues. Although most of the previous studies have shown that high-calorie foods have a wide range of effects on P2, N2, P3, N450, and late positive potential (LPP) amplitudes, they have revealed mixed findings. Asmaro and his colleagues compared the neural-difference induced by high-calorie foods (e.g., chocolate, milk.) and neutral objects (e.g., chair) during discrimination tasks using event-related potentials (ERPs, a neuroscience tool with high temporal resolution, which is often used in cognitive psychology research), and found that high-calorie foods elicited significant enhancement of P2 and LPP amplitudes. The findings indicated that participants exhibited attentional bias by allocating more attentional resources to high calorie food stimuli compared to general stimuli [18]. When asked to inhibit during a go/no-go or oddball task, the participants exhibited larger N2 amplitudes when inhibiting to high-calorie food compared to when inhibiting to low-calorie or non-food-related stimuli [19–22]. Liu et al. found greater N2b amplitudes in response to high-calorie food cues than those in response to low-calorie food cues during a food-related go/no-go task [7]. In addition, Carbine et al., found that during a food-related go/no-go task, the P3 amplitudes were greater when inhibiting to high-calorie food cues compared to when inhibiting to low-calorie food cues [23]. Franken et al. also found that sweet food elicited greater P3 amplitudes compared to neutral-flavored food [24]. However, in successful restrained eaters, Kong et al. found greater P3 amplitudes to neural image cues and low-calorie food cues than to high-calorie food cues [21]. However, using a go/no-go task, Lapenta et al. (2014) found no difference in N2 and P3 amplitudes when inhibiting toward food cues or furniture cues [25]. Nijs and colleagues (2010) set out to investigate food-related conflict control abilities in individuals with obesity. During a food-related words Stroop task, the participants were instructed to respond as fast as possible to the font color of the word shown on the screen and ignore the meaning of the word. They found enhanced P2 amplitudes in the obese group, which indicated that the obese participants' attentional bias towards food cues occurred at a heighted automatic, habitual, and preconscious level [13]. A study conducted by Liu et al. showed that there were decreases in N2 and N450 amplitudes and increases in P3 and LPP amplitudes in a food-related conflict task in individuals who were overweight [6]. Woltering and colleagues used an attention blink task to reveal decreased P3 amplitudes during the onset of food stimuli in obese adolescents compared to their normal-weight counterparts. Lower P3 amplitudes were related to a higher body mass index (BMI) assessment. The findings suggested that an automatic attentional bias to food should be considered as an key factor in the process of tackling the rising obesity crisis [26]. As the findings present, the existing mechanism of high-calorie food on conflict control ability remains unclear. It is beneficial to explore how high-calorie food cues influence executive function since it would provide insights into the challenges of developing healthier eating behaviors.

Theta power has also been a neural marker of focus for previous studies when discussing conflict control ability regarding food-related stimuli. Theta power of the medial frontal cortex is sensitive to perception conflicts and reaction conflicts. Theta band activities are usually involved in conflict processing, especially in the neural network that connects reaction conflicts [27,28]. In addition, frontal midline theta power is related to increased cognitive control [29–31]. A previous study aimed at examining the neural markers that are in support of the effectiveness of food-related go/no-go training on reducing food intake found that food cues requiring withholding responses induced larger increases in theta power compared to cues that do not relate to withholding responses [32].

Up to now, there have been few EEG studies that have utilized a food-related Stroop task to investigate the effects of high-calorie foods on conflict control. To add to previous studies, the present study examined the influence of high-calorie food cues on the neural correlates of conflict control. Based on previous studies, the neural correlates discussed in the present study are N2, P3, and theta power and we hypothesized that high-calorie foods would impair conflict control ability, which would be reflected by a slower reaction time, lower accuracy, a reduction in N2 amplitudes and theta power, as well as an increase in P3 amplitudes when reacting to high-calorie foods compared to when reacting to low-calorie foods during the food-related Stroop task.

2. Methods

2.1. Participants

Participants (n = 61; 29 females; M_{age} = 20.25 years, SD_{age} = 1.96; M_{BMI} = 21.75 kg/m^2, SD_{BMI} = 2.98) were recruited from Chongqing Normal University, Chongqing. Before participating in the study, all participants were required to abstain from substances (including tea and coffee) and medications that could potentially affect their concentration during the study, and were required to disclose any history of psychological disorders. They were also required to have normal or corrected-to-normal vision. The participants were given the chance to ask any questions about their participation before consenting to the study. The present study was approved by the Southwest University Ethics Committee.

2.2. Food-Related Stroop Task

The food-related Stroop task (Figure 1) employed in the current study was modified from the previous version used by Liu et al. [6]. The stimuli were food pictures (high- and low-calorie foods) and calorie information text (high-calorie or low-calorie) attached to them. Congruent and incongruent trials were generated by combining different stimuli. A congruent trial would be when the displayed food cue with text was consistent, such as a combination of a high-calorie food cue and the words "high-calorie", and vice versa for incongruent trials. Participants were instructed to respond to only the food pictures and ignore the words (calorie information). During the food-related Stroop task, the participants were instructed to press the "F" (or "J") key on the keyboard if the high-calorie foods were displayed and to press the "J" (or "F") key if the low-calorie foods were displayed. The responses were counterbalanced across the participants in the study. In the task, the trial proceeded as follows: after a fixation appeared on the screen for 500 ms, the stimuli would be presented on the monitor until the participants responded or would disappear automatically after 2000 ms of first appearing. The stimuli were then followed by an inter-stimuli interval of 1000 ms. The food-related Stroop task consisted of a practice block of 20 trials and an experimental block of 160 trials. Only data from the experimental trials were used in the analysis process. The participants were instructed to sit as still as possible and to minimize eye blinking during the task in order to reduce potential experimental artifacts of the EEG data collection.

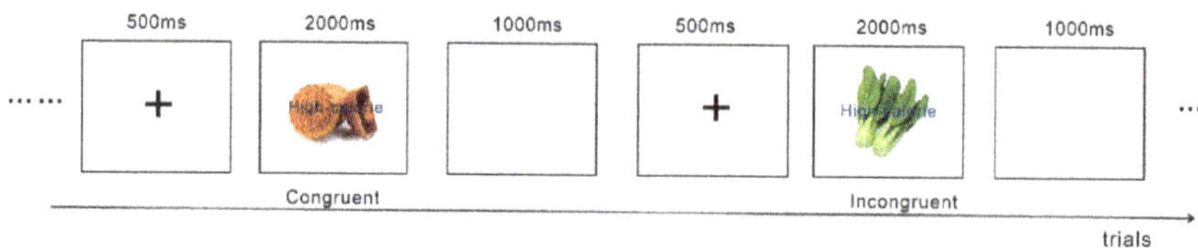

Figure 1. Two trials from the food-related Stroop task.

2.3. Behavioral Analysis

Two 2 (food: high- and low-calorie food) × 2 (congruence: congruent and incongruent) repeated measures ANOVA was conducted on the reaction time (RT) and accuracy (ACC). All analyses were conducted by SPSS 25.0. The *p*-values were computed for deviation in all analyses, based on the Greenhouse–Geisser method. Post-hoc t-tests were conducted with Bonferroni correction for multiple pairwise comparisons.

2.4. EEG Recording and Analyses

EEG data were recorded from 64 scalp sites using tin electrodes mounted in an elastic cap (brain products GmbH, Gilching, Germany), with the reference electrode placed on the fronto-central aspect and a ground electrode on the medial frontal aspect. All inter-electrode impedance was maintained below 5 KΩ.

EEG data preprocessing was performed via EEGLAB, an open-source toolbox available on the Matlab software. Individual and grand ERP averages were created for the food-related Stroop task. We first down sampled the data from 1000 Hz to 256 Hz and performed high-pass filtering at 0.1 Hz and low-pass filtering at 45 Hz. The mean values of the left and right mastoids were selected as the re-reference. Data were epoched from 200 ms prior to stimuli onset to 1000 ms after the onset, and were baseline corrected to the pre-stimuli interval. Trials with large fluctuations in amplitudes were removed before the independent component analysis (ICA). The components, including EOG artifacts (ocular movements and eye blinks) and head movement, were then removed from the results of the ICA. Based on the topographical distribution of the grand-average ERP activities, the ERP components and their time window were as follows: N2 (250–360 ms), and P3 (360–400 ms). Based on previous studies, the following electrode sites were selected, Fz, FCz, Cz, CPz, and Pz. Two 2 (food: high- and low-calorie food) × 2 (congruence: congruent and incongruent) × 5 (electrode site: Fz, FCz, Cz, CPz, and Pz) repeated measures ANOVAs were conducted on the mean amplitudes of N2 and P3.

The time-frequency analysis for the EEG data, employed a windowed Fourier transform (WFT) with a fixed 250 ms width Hanning window. The WFT yielded a complex time-frequency spectral estimate $F(t, f)$ at each point (t, f) of the time-frequency plane extending from −200 ms to 1000 ms in the time domain, and from 1 Hz to 30 Hz (in step of 1 Hz) in the frequency domain, for each single trial. A baseline correction was applied at the subject level using the pre-stimulus interval (pre-stimulus −200 to 0 ms) to calculate the change of power according to the formula:

$$\mathrm{TFD}(t,f) = P(t,f) - R(f)$$

where $P(t,f) = |F(t,f)|^2$ is the power spectral density at a given time-frequency point (t, f), and $R(f)$ is the averaged power spectral density of the signal enclosed within the pre-stimulus reference interval (−200 to 0 ms before the onset of the stimulation) for each estimated frequency *f*. Brain rhythmic activity of the theta (4–8 Hz, 80–250 ms) was selected in the current analysis. A 2 (food: high- and low-calorie food) × 2 (congruence: congruent and incongruent) × 5 (electrode site: Fz, FCz, Cz, CPz, and Pz) repeated measures ANOVA was conducted on the value of theta.

All analyses were conducted via SPSS 25.0. Based on the Greenhouse–Geisser method, *p*-values were computed for deviation in all analyses. Post-hoc t-tests were conducted with Bonferroni correction for multiple pairwise comparisons.

3. Results

3.1. Behavior Results

The results on RT (Figure 2A) showed a main effect of food, F (1, 60) = 39.30, $p < 0.001$, partial $\eta^2 = 0.40$, RT to high-calorie foods was significantly greater than that to low-calorie foods. The results on RT also showed a main effect of congruence, F (1, 60) = 39.30, $p = 0.001$, partial $\eta^2 = 0.16$, RT in incongruent trials was significantly greater than that in congruent trials.

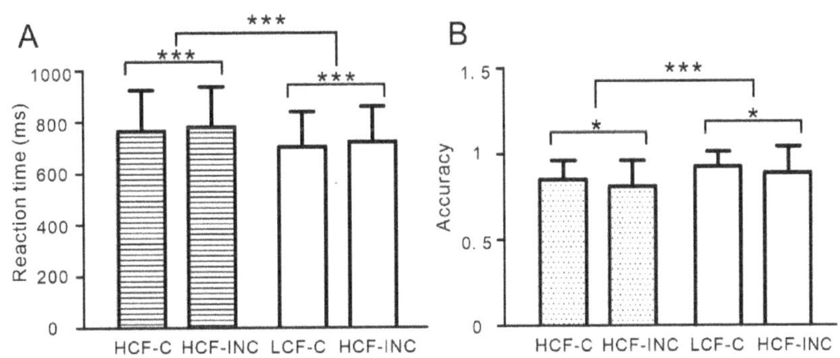

Figure 2. The behavioral results ((**A**). reaction time and (**B**). accuracy) from the food-related Stroop task. HCF-C, high-calorie foods in congruent trials; HCF-INC, high-calorie foods in incongruent trials; LCF-C, low-calorie foods in congruent trials; LCF-INC, low-calorie foods in incongruent trials. * $p < 0.05$; *** $p < 0.001$.

The results on ACC (Figure 2B) showed a main effect of food, F (1, 60) = 19.48, $p < 0.001$, partial $\eta^2 = 0.25$, ACC of high-calorie foods was significantly lower than that of low-calorie foods. The results on ACC also showed a main effect of congruence, F (1, 60) = 5.01, $p = 0.03$, partial $\eta^2 = 0.08$, ACC in congruent trials was significantly greater than that in incongruent trials.

3.2. EEG Results

Grand average ERPs for N2 and P3 at Fz and topography plots are shown in Figure 3A. The theta power at Fz is shown in Figure 3B.

N2

The results on N2 amplitudes showed an interaction of food and electrode site, F (4, 240) = 13.55, $p < 0.001$, partial $\eta^2 = 0.18$. The simple effect analysis showed that N2 amplitudes of low-calorie foods were significantly greater than those of high-calorie foods at Fz, F (1, 60) = 4.97, $p = 0.03$, partial $\eta^2 = 0.08$. The results also showed an interaction of congruence and electrode site, F (4, 240) = 3.18, $p = 0.048$, partial $\eta^2 = 0.05$. The simple effect analysis showed that N2 amplitudes in incongruent trials were significantly greater than those in congruent trials at Fz [F (1, 60) = 6.49, $p = 0.013$, partial $\eta^2 = 0.10$], FCz [F (1, 60) = 5.41, $p = 0.023$, partial $\eta^2 = 0.08$], and Cz [F (1, 60) = 7.09, $p = 0.01$, partial $\eta^2 = 0.11$].

P3

The results on P3 amplitudes showed a main effect of congruence, F (1, 60) = 24.55, $p < 0.001$, partial $\eta^2 = 0.29$, P3 amplitudes in congruent trials were significantly greater than those in incongruent trials.

Theta

The results on theta showed a main effect of food, F (1, 60) = 8.25, $p = 0.006$, partial $\eta^2 = 0.12$, theta power of low-calorie foods (0.77) was significantly greater than that of high-calorie foods (0.59). The results also showed an interaction of congruence and electrode site, F (4, 240) = 9.46, $p = 0.001$, partial $\eta^2 = 0.14$. The simple effect analysis showed that theta power in incongruent trials was significantly greater than that in congruent trials at Pz, F (1, 60) = 4.15, $p = 0.046$, partial $\eta^2 = 0.07$.

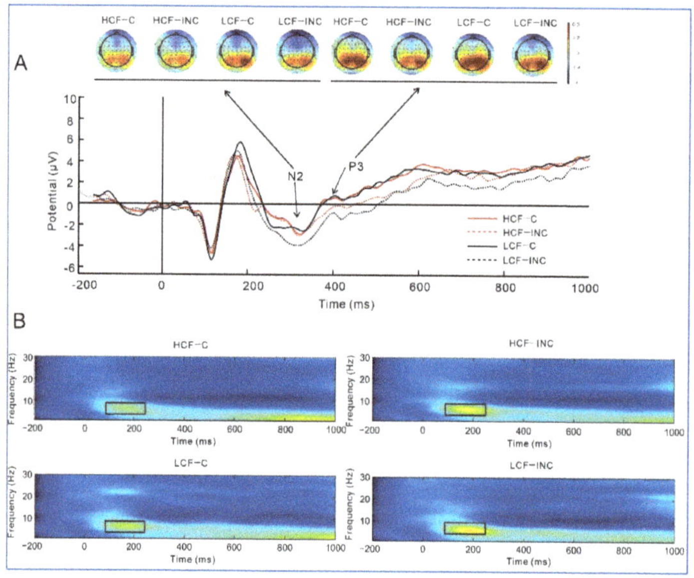

Figure 3. (**A**). Stimuli-locked, grand average waveforms of N2 and P3 at site Fz; (**B**). The theta power at site Fz. HCF−C, high-calorie foods in congruent trials; HCF−INC, high-calorie foods in incongruent trials; LCF−C, low-calorie foods in congruent trials; LCF−INC, low-calorie foods in incongruent trials.

4. Discussion

The present study used an originally modified food-related Stroop task to explore the effects of high-calorie foods on conflict control and the underlying neurophysiological responses. The hypothesis was partially confirmed by our original method of examination. The results showed that the RT to high-calorie food cues was greater than that to low-calorie food cues, whereas the ACC of high-calorie food cues was significantly lower than that of low-calorie food cues. We also found that high-calorie foods induced lower N2 amplitudes and theta power compared to low-calorie foods. No significant difference in P3 amplitudes were found between high-calorie and low-calorie food cues.

As our modified task elicited data from both the mental processing of food images combined with calorie information, as well as motor response performance, the current results support previous findings that high-calorie foods impair cognitive performance [17]. Adding to previous studies, the current study employed a novel paradigm in which the direct influences of food-related stimuli on reaction time and accuracy could be measured. A previous study showed that individuals with binge eating disorder exhibited a flanker effect to food-related stimuli while the controls did not [14], which might indicate that binge eating impairs food-related conflict control. It is also found that higher reported cravings for unhealthy foods (high-calorie foods) are related to greater conflict when making decisions between healthy and unhealthy foods [15], thus, indicating that high-calorie foods might have a negative effect on conflict control. Consistent with a previous study using an n-back task and food related flanker task [17,33], the current finding indicates that high-calorie foods effect individuals' conflict control, which was reflected by slower reaction time and lower accuracy. Longer reaction time in reaction to high-calorie foods might suggest that more attentional efforts are required when responding to high-calorie foods compared to low-calorie foods, thus, resulting in longer processing time [33,34]. According to the dual-process theories of self-control, a healthy lifestyle is depended on the balance between impulsive and reflective systems, while the excessive consumption of high-calorie foods indicate that this cognitive balance is not sufficiently maintained [3,35,36]. In return, deficiencies in the ability to monitor high-calorie food conflicts might also lead to overeating or obesity. Obese individuals displayed marked attentional bias to high-calorie foods and food-related stimuli compared to lean individuals [37]. Enhanced spatial memory for high-calorie foods also significantly predicted a higher BMI [38,39]. Therefore, existing evidence indicates that conflict control ability plays a beneficial role in adopting healthier eating behaviors.

The present study showed a reduction in N2 amplitudes in high-calorie foods compared to low-calorie foods. Depending on the stimuli and tasks displayed, N2 amplitudes reflect sensory processing, response inhibition, and conflict monitoring [20,40]. Folstein and Van concluded that the fronto-central N2 is relevant to cognitive control, which included response inhibition, response conflict, and error monitoring [40]. In addition, N2 reflects the detection and monitoring of conflicts when confronted with choosing between correct and incorrect options [40,41]. Previous studies showed that obese and overweight individuals exhibited decreased N2 amplitudes compared to normal-weight individuals [5–7]. Decreased N2 also indicates a deficit in the ability to recruit neural resources [42]. Thus, the reduced N2 amplitudes in reaction to high-calorie foods would indicate that the individuals are less able or less effective when allocating attentional and neural resources to monitor food-specific conflicts when encountering high-calorie foods.

The result of decreased theta power in high-calorie foods was consistent with the result of reduced N2 amplitudes in high-calorie foods. Frontal midline theta is located in the anterior cingulate cortex (ACC) and the auxiliary frontal motor area of the medial frontal cortex (MFC) [29] and is part of the physiological mechanism that operates cognitive control through a series of coupling mechanisms [43]. Greater frontal midline theta power is found to be related to increased cognitive control [29–31]. The significant increase in theta-band oscillation of the frontal midline is related to an increased capability for adaptation, behavioral regulation and overcoming conflicts [44,45]. Previous studies have found that theta initiate a greater increase when experiencing stimuli such as stimulus-response conflicts and response errors [27,46]. Therefore, the theta power is significantly lower when reacting to high-calorie foods compared to when reacting to low-calorie foods, which might indicate that high-calorie foods affect individuals' ability for conflict monitoring, regulation, and/or adaptation.

However, we did not find a difference in P3 amplitudes between high-calorie foods and low-calorie foods. P3 reflects various cognitive processes including working memory and attention allocation [47]. A previous study has shown that the onset of food-related cues elicited a significant increase in P3 amplitudes, which would suggest that the salience of food-related cues influence cognitive control [22]. Food-related stimuli elicited greater P3 amplitudes than neutral stimuli [22], and high-calorie foods elicited greater P3 amplitudes compared to low-calorie foods [23], which would indicate that more cognitive resources were required when responding to food-related stimuli, especially high-calorie foods. Therefore, we hypothesized that high-calorie foods would affect the individuals' cognitive processing, which would be reflected by enhanced P3 amplitudes. The slower reaction time and lower accuracy of the current study supported the hypothesis that more cognitive resources and efforts are needed when reacting to high-calorie foods compared to when reacting to low-calorie foods. However, no significant difference in P3 amplitudes between high-calorie and low-calorie foods was found in the present study. We believe a reason might account for the finding. P3 represents a late motor inhibition process, which occurs when a decision to withhold a response is made [48–50]. The food-related Stroop task did not ask participants to withhold their response. Therefore, high-calorie foods did not have effects on the individual's response inhibition ability during the food-related Stroop task.

The present study has several limitations. First, the participants' food cravings were not measured. The relationship between food craving and food-related conflict control was therefore not explored in the current study. Future research could measure the state of the food craving (e.g., food craving questionnaire state) before and after the food-related Stroop task. Second, the current study focused on the temporal measures of brain activity of the EEG data, hence the lack of spatial measures of brain activity. Future studies could explore both the temporal and spatial neural indices in a food-related Stroop task by collecting both EEG and fMRI data. Third, general stimuli (e.g., flowers or office supplies) were not adopted in the current study, and the influence of other characteristics such as the stimuli colors or nutrient density on the experimental findings was not considered in the present study, which might be worth exploring in future studies.

5. Conclusions

In conclusion, this study uses an originally modified food-specific Stroop task and EEG to complement previous literatures. The current original version of the food-related Stroop task may be a promising instrument to be implicated in future brain imaging studies to further demonstrate the nuances of various types of foods on cognitive processing. Furthermore, the task could be adopted to explore the underpinnings of the relationship between high-calorie foods and conflict control, including its behavioral consequences. The study has shown that high-calorie foods impair food-related conflict control ability, expanding the studies of the neural mechanism of food cues while providing new insights into the processing and solving of conflicting information for healthier eating behavior and weight control.

Author Contributions: Conceptualization, Y.L. and H.C.; Methodology, Y.L. and H.C.; Software, Y.L. and J.Z.; Validation, Y.L., J.Z., Y.Z. (Yizhou Zhou), R.Y., B.H., Y.Z. (Yufei Zhao), Y.P., H.Y. and H.C.; Formal Analysis, Y.L. and J.Z.; Investigation, Y.L.; Resources, Y.L. and H.C.; Data Curation, Y.L. and Y.Z. (Yizhou Zhou); Writing—Original Draft Preparation, Y.L., R.Y., B.H. and Y.Z. (Yufei Zhao); Writing—Review & Editing, Y.L., Y.P., H.Y. and H.C.; Visualization, Y.P. and H.Y.; Supervision, H.C.; Project Administration, H.C.; Funding Acquisition, Y.L. All authors have read and agreed to the published version of the manuscript.

Funding: The work was supported by the Chinese National Natural Science Foundation (No. 32200849); Natural Science Foundation of Chongqing (No. CSTB2022NSCQ-MSX0788); the Fundamental Research Funds for the Central Universities (SWU2209501); Southwest University Training Program of Innovation and Entrepreneurship for Undergraduates (X202210635235).

Institutional Review Board Statement: The study was conducted in accordance with the Declaraion of Helsinki, and approved by the Ethics Committee of the Southwest University (H22020, 28.02.2022).

Informed Consent Statement: Informed consent was obtained from all participants involved in the study.

Data Availability Statement: The data presented in this study are available on request from the corresponding author.

Acknowledgments: We would especially like to thank the participants of the Chongqing Normal University.

Conflicts of Interest: The authors declare that they have no known competing financial interest or personal relationships that could have appeared to influence the work in the paper.

References

1. Georgii, C.; Schulte-Mecklenbeck, M.; Richard, A.; Van Dyck, Z.; Blechert, J. The dynamics of self-control: Within-participant modeling of binary food choices and underlying decision processes as a function of restrained eating. *Psychol. Res.* **2020**, *84*, 1777–1788. [CrossRef] [PubMed]
2. Baumeister, R.F.; Vohs, K.D.; Tice, D.M. The strength model of self-control. *Curr. Dir. Psychol. Sci.* **2007**, *16*, 351–355. [CrossRef]
3. Hofmann, W.; Friese, M.; Strack, F. Impulse and self-control from a dual-systems perspective. *Perspect. Psychol. Sci.* **2009**, *4*, 162–176. [CrossRef] [PubMed]
4. Volkow, N.D.; Wise, R.A. How can drug addiction help us understand obesity? *Nat. Neurosci.* **2005**, *8*, 555–560. [CrossRef] [PubMed]
5. Liu, Y.; Zhao, J.; Zhang, X.; Gao, X.; Xu, W.; Chen, H. Overweight adults are more impulsive than normal weight adults: Evidence from ERPs during a chocolate-related delayed discounting task. *Neuropsychologia* **2019**, *133*, 107181. [CrossRef]
6. Liu, Y.; Quan, H.; Song, S.; Zhang, X.; Yang, C.; Chen, H. Decreased conflict control in overweight Chinese females: Behavioral and event-related potentials evidence. *Nutrients* **2019**, *11*, 1450. [CrossRef] [PubMed]
7. Liu, Y.; Gao, X.; Zhao, J.; Zhang, L.; Chen, H. Neurocognitive correlates of food-related response inhibition in overweight/obese adults. *Brain Topogr.* **2020**, *33*, 101–111. [CrossRef] [PubMed]
8. Song, S.; Li, Q.; Jiang, Y.; Liu, Y.; Xu, A.; Liu, X.; Chen, H. Do Overweight People Have Worse Cognitive Flexibility? Cues-Triggered Food Craving May Have a Greater Impact. *Nutrients* **2022**, *14*, 240. [CrossRef]
9. Syan, S.K.; Owens, M.M.; Goodman, B.; Epstein, L.H.; Meyre, D.; Sweet, L.H.; MacKillop, J. Deficits in executive function and suppression of default mode network in obesity. *NeuroImage Clin.* **2019**, *24*, 102015. [CrossRef]
10. Yang, Y.; Shields, G.S.; Guo, C.; Liu, Y. Executive function performance in obesity and overweight individuals: A meta-analysis and review. *Neurosci. Biobehav. Rev.* **2018**, *84*, 225–244. [CrossRef]
11. Chen, S.; Jia, Y.; Woltering, S. Neural differences of inhibitory control between adolescents with obesity and their peers. *Int. J. Obes.* **2018**, *42*, 1753–1761. [CrossRef] [PubMed]
12. Diamond, A. Executive functions. *Annu. Rev. Psychol.* **2013**, *64*, 135. [CrossRef] [PubMed]
13. Nijs, I.M.; Franken, I.H.; Muris, P. Food-related Stroop interference in obese and normal-weight individuals: Behavioral and electrophysiological indices. *Eat. Behav.* **2010**, *11*, 258–265. [CrossRef] [PubMed]
14. Lyu, Z.; Zheng, P.; Lu, S.; Qin, M. Impaired Conflict Monitoring to Food Cues in Women Who Binge Eat. *Front. Psychol.* **2018**, *9*, 2585. [CrossRef] [PubMed]
15. Smith, R.; Alkozei, A.; Killgore, W.D. Conflict-related dorsomedial frontal cortex activation during healthy food decisions is associated with increased cravings for high-fat foods. *Brain Imaging Behav.* **2018**, *12*, 685–696. [CrossRef]
16. Zhang, X.; Wang, S.; Liu, Y.; Chen, H. More restriction, more overeating: Conflict monitoring ability is impaired by food-thought suppression among restrained eaters. *Brain Imaging Behav.* **2021**, *15*, 2069–2080. [CrossRef]
17. Meule, A.; Skirde, A.K.; Freund, R.; Vögele, C.; Kübler, A. High-calorie food-cues impair working memory performance in high and low food cravers. *Appetite* **2012**, *59*, 264–269. [CrossRef]
18. Asmaro, D.; Jaspers-Fayer, F.; Sramko, V.; Taake, I.; Carolan, P.; Liotti, M. Spatiotemporal dynamics of the hedonic processing of chocolate images in individuals with and without trait chocolate craving. *Appetite* **2012**, *58*, 790–799. [CrossRef]
19. Carbine, K.A.; Christensen, E.; LeCheminant, J.D.; Bailey, B.W.; Tucker, L.A.; Larson, M.J. Testing food-related inhibitory control to high-and low-calorie food stimuli: Electrophysiological responses to high-calorie food stimuli predict calorie and carbohydrate intake. *Psychophysiology* **2017**, *54*, 982–997. [CrossRef]
20. Carbine, K.A.; Rodeback, R.; Modersitzki, E.; Miner, M.; LeCheminant, J.D.; Larson, M.J. The utility of event-related potentials (ERPs) in understanding food-related cognition: A systematic review and recommendations. *Appetite* **2018**, *128*, 58–78. [CrossRef]
21. Kong, F.; Zhang, Y.; Chen, H. Inhibition ability of food cues between successful and unsuccessful restrained eaters: A two-choice oddball task. *PLoS ONE* **2015**, *10*, e0120522. [CrossRef]
22. Watson, T.D.; Garvey, K.T. Neurocognitive correlates of processing food-related stimuli in a Go/No-go paradigm. *Appetite* **2013**, *71*, 40–47. [CrossRef]
23. Carbine, K.A.; Duraccio, K.M.; Kirwan, C.B.; Muncy, N.M.; LeCheminant, J.D.; Larson, M.J. A direct comparison between ERP and fMRI measurements of food-related inhibitory control: Implications for BMI status and dietary intake. *NeuroImage* **2018**, *166*, 335–348. [CrossRef]
24. Franken, I.H.; Huijding, J.; Nijs, I.M.; van Strien, J.W. Electrophysiology of appetitive taste and appetitive taste conditioning in humans. *Biol. Psychol.* **2011**, *86*, 273–278. [CrossRef]
25. Lapenta, O.M.; Di Sierve, K.; de Macedo, E.C.; Fregni, F.; Boggio, P.S. Transcranial direct current stimulation modulates ERP-indexed inhibitory control and reduces food consumption. *Appetite* **2014**, *83*, 42–48. [CrossRef]
26. Woltering, S.; Chen, S.; Jia, Y. Neural Correlates of Attentional Bias to Food Stimuli in Obese Adolescents. *Brain Topogr.* **2021**, *34*, 182–191. [CrossRef]
27. Nigbur, R.; Cohen, M.X.; Ridderinkhof, K.R.; Stürmer, B. Theta dynamics reveal domain-specific control over stimulus and response conflict. *J. Cogn. Neurosci.* **2012**, *24*, 1264–1274. [CrossRef]
28. Womelsdorf, T.; Johnston, K.; Vinck, M.; Everling, S. Theta-activity in anterior cingulate cortex predicts task rules and their adjustments following errors. *Proc. Natl. Acad. Sci. USA* **2010**, *107*, 5248–5253. [CrossRef]
29. Cavanagh, J.F.; Frank, M.J. Frontal theta as a mechanism for cognitive control. *Trends Cogn. Sci.* **2014**, *18*, 414–421. [CrossRef]

30. Cavanagh, J.F.; Shackman, A.J. Frontal midline theta reflects anxiety and cognitive control: Meta-analytic evidence. *J. Physiol.* **2015**, *109*, 3–15. [CrossRef]
31. Schmidt, B.; Kanis, H.; Holroyd, C.B.; Miltner, W.H.; Hewig, J. Anxious gambling: Anxiety is associated with higher frontal midline theta predicting less risky decisions. *Psychophysiology* **2018**, *55*, e13210. [CrossRef] [PubMed]
32. van de Vijver, I.; van Schie, H.T.; Veling, H.; van Dooren, R.; Holland, R.W. Go/no-go training affects frontal midline theta and mu oscillations to passively observed food stimuli. *Neuropsychologia* **2018**, *119*, 280–291. [CrossRef] [PubMed]
33. Forestell, C.A.; Lau, P.; Gyurovski, I.I.; Dickter, C.L.; Haque, S.S. Attentional biases to foods: The effects of caloric content and cognitive restraint. *Appetite* **2012**, *59*, 748–754. [CrossRef] [PubMed]
34. Meule, A.; Kübler, A. Double trouble. Trait food craving and impulsivity interactively predict food-cue affected behavioral inhibition. *Appetite* **2014**, *79*, 174–182. [CrossRef] [PubMed]
35. Hofmann, W.; Friese, M.; Roefs, A. Three ways to resist temptation: The independent contributions of executive attention, inhibitory control, and affect regulation to the impulse control of eating behavior. *J. Exp. Soc. Psychol.* **2009**, *45*, 431–435. [CrossRef]
36. Strack, F.; Deutsch, R. Reflective and impulsive determinants of social behavior. *Personal. Soc. Psychol. Rev.* **2004**, *8*, 220–247. [CrossRef]
37. Castellanos, E.H.; Charboneau, E.; Dietrich, M.S.; Park, S.; Bradley, B.P.; Mogg, K.; Cowan, R.L. Obese adults have visual attention bias for food cue images: Evidence for altered reward system function. *Int. J. Obes.* **2009**, *33*, 1063–1073. [CrossRef]
38. Allan, K.; Allan, J.L. An obesogenic bias in women's spatial memory for high calorie snack food. *Appetite* **2013**, *67*, 99–104. [CrossRef]
39. de Vries, R.; Morquecho-Campos, P.; de Vet, E.; de Rijk, M.; Postma, E.; de Graaf, K.; Boesveldt, S. Human spatial memory implicitly prioritizes high-calorie foods. *Sci. Rep.* **2020**, *10*, 1–6. [CrossRef]
40. Folstein, J.R.; van Petten, C. Influence of cognitive control and mismatch on the N2 component of the ERP: A review. *Psychophysiology* **2008**, *45*, 152–170. [CrossRef]
41. Yeung, N.; Botvinick, M.M.; Cohen, J.D. The neural basis of error detection: Conflict monitoring and the error-related negativity. *Psychol. Rev.* **2004**, *111*, 931. [CrossRef]
42. Senderecka, M.; Grabowska, A.; Szewczyk, J.; Gerc, K.; Chmylak, R. Response inhibition of children with ADHD in the stop-signal task: An event-related potential study. *Int. J. Psychophysiol.* **2012**, *85*, 93–105. [CrossRef]
43. Narayanan, N.S.; Cavanagh, J.F.; Frank, M.J.; Laubach, M. Common medial frontal mechanisms of adaptive control in humans and rodents. *Nat. Neurosci.* **2013**, *16*, 1888–1895. [CrossRef]
44. Cavanagh, J.F.; Eisenberg, I.; Guitart-Masip, M.; Huys, Q.; Frank, M.J. Frontal theta overrides pavlovian learning biases. *J. Neurosci.* **2013**, *33*, 8541–8548. [CrossRef]
45. Cohen, M.X.; Wilmes, K.A.; van de Vijver, I. Cortical electrophysiological network dynamics of feedback learning. *Trends Cogn. Sci.* **2011**, *15*, 558–566. [CrossRef]
46. Cohen, M.X.; van Gaal, S. Dynamic interactions between large-scale brain networks predict behavioral adaptation after perceptual errors. *Cereb. Cortex* **2013**, *23*, 1061–1072. [CrossRef]
47. Polich, J. Updating P300: An integrative theory of P3a and P3b. *Clin. Neurophysiol.* **2007**, *118*, 2128–2148. [CrossRef]
48. Gajewski, P.D.; Falkenstein, M. Effects of task complexity on ERP components in Go/Nogo tasks. *Int. J. Psychophysiol.* **2013**, *87*, 273–278. [CrossRef]
49. Schmajuk, M.; Liotti, M.; Busse, L.; Woldorff, M.G. Electrophysiological activity underlying inhibitory control processes in normal adults. *Neuropsychologia* **2006**, *44*, 384–395. [CrossRef]
50. Albert, J.; López-Martín, S.; Hinojosa, J.A.; Carretié, L. Spatiotemporal characterization of response inhibition. *Neuroimage* **2013**, *76*, 272–281. [CrossRef]

nutrients

Article

Effects of Fermented Milk Containing *Bifidobacterium animalis* Subsp. *lactis* MN-Gup (MN-Gup) and MN-Gup-Based Synbiotics on Obesity Induced by High Fat Diet in Rats

Chenyuan Wang [1,2], Shusen Li [2], Erna Sun [2], Ran Xiao [1,2], Ran Wang [3], Yimei Ren [1], Jingjing He [1], Qi Zhang [1] and Jing Zhan [1,*]

1 Key Laboratory of Precision Nutrition and Food Quality, Department of Nutrition and Health, China Agricultural University, Beijing 100193, China; drwangchenyuan@163.com (C.W.); xiaoran881230@163.com (R.X.); renyimei123@cau.edu.cn (Y.R.); hejingjing_89@163.com (J.H.); zhangqi@cau.edu.cn (Q.Z.)
2 Mengniu Hi-tech Dairy Product Beijing Co., Ltd., Beijing 101100, China; lishusen@mengniu.cn (S.L.); sunerna@mengniu.cn (E.S.)
3 Key Laboratory of Functional Dairy, Department of Nutrition and Health, China Agricultural University, Beijing 100193, China; wangran@cau.edu.cn
* Correspondence: jingzhan@cau.edu.cn; Tel./Fax: +86-010-6273-8589

Abstract: Given the probiotic effects previously found in *Bifidobacterium animalis* subsp. *lactis* MN-Gup (MN-Gup) and its great application potential in dairy products, this study aimed to investigate the effects of fermented milk containing MN-Gup or MN-Gup-based synbiotics on high fat diet (HFD)-induced obesity in rats. Galacto-oligosaccharides (GOS) and xylo-oligosaccharides (XOS) were selected as the tested prebiotics in MN-Gup-based synbiotics due to their promotion of MN-Gup growth in vitro. After nine weeks of HFD feeding, the obese rats were intervened with fermented milk containing MN-Gup (MN-Gup FM) or its synbiotics (MN-Gup + GOS FM, MN-Gup + XOS FM) for eight weeks. The results showed that the interventions could alleviate HFD-induced body weight gain, epididymal fat deposition, adipocyte hypertrophy, dyslipidemia and inflammation, but GOS and XOS did not exhibit significant synergies with MN-Gup on those alleviations. Furthermore, the interventions could regulate the HFD-affected gut microbiota and microbial metabolites, as shown by the increases in short chain fatty acids (SCFAs) and alterations in obesity-related bile acids (BAs), which may play important roles in the mechanism underlying the alleviation of obesity. This study revealed the probiotic effects of MN-Gup on alleviating obesity and provided the basis for MN-Gup applications in the future.

Keywords: *Bifidobacterium animalis* subsp. *lactis* MN-Gup; obesity; gut microbiota; galacto-oligosaccharides; xylo-oligosaccharides

Citation: Wang, C.; Li, S.; Sun, E.; Xiao, R.; Wang, R.; Ren, Y.; He, J.; Zhang, Q.; Zhan, J. Effects of Fermented Milk Containing *Bifidobacterium animalis* Subsp. *lactis* MN-Gup (MN-Gup) and MN-Gup-Based Synbiotics on Obesity Induced by High Fat Diet in Rats. *Nutrients* **2022**, *14*, 2631. https://doi.org/10.3390/nu14132631

Academic Editor: Dennis Savaiano

Received: 21 May 2022
Accepted: 22 June 2022
Published: 24 June 2022

1. Introduction

The prevalence of overweight and obesity has become a serious health problem in the world. Obesity is usually associated with a high risk of chronic diseases, such as type 2 diabetes, gallbladder disease, cardiovascular disease and cancer [1,2]. People continue to seek various methods to alleviate obesity, such as diet regulation, physical exercise and bariatric surgery [3]. Recently, obesity has been found to be correlated with an imbalance in gut microbiota, and gut microbiota has contributed to a new perspective for regulating host energy metabolism [4]. A number of mechanisms have been proposed for gut microbiota-influenced obesity [5]. For instance, some microbiota can promote the production of short chain fatty acids (SCFAs), which can activate G protein-coupled receptors and induce subsequent intestinal hormone secretion to regulate energy metabolism [6].

Probiotics, defined as "live microorganisms which give a health benefit to the host when administered in adequate amounts" [7], have exerted excellent effects on modulating gut homeostasis via balancing beneficial and harmful bacteria, improving intestinal barrier function, and regulating immune responses [8]. It is estimated that the probiotics market grew 37% globally from 2016 to 2020 [9]. Accumulating animal and human data have shown that probiotics exhibit great potential in anti-obesity applications [8]. *Bifidobacterium lactis* and *Lactobacillus paracasei* are common probiotics that have been reported to regulate gut microbiota and attenuate obesity in high fat diet (HFD)-fed mice [10–12]. Meanwhile, the effects of probiotics or their applications in food on obesity were investigated in randomized controlled trials (RCTs). For example, daily consumption of fermented milk containing 2×10^9 CFU of *Lactobacillus gasseri* SBT2055 for 12 weeks was found to significantly reduce body mass index (BMI), waist, and hip circumferences in Japanese adults with large visceral fat areas [13]. A meta-analysis study based on twenty-six RCTs showed that there were significant effects of probiotics on reducing body weight, waist circumference, fat mass, total cholesterol (TC), tumor necrosis factor-α (TNF-α) level of overweight/obese subjects [14]. Therefore, the dietary intervention using probiotics may provide a potential strategy to alleviate obesity.

Furthermore, prebiotics, some non-digestible food ingredients that can selectively stimulate the growth and/or activity of beneficial bacteria in the intestinal tract, have been suggested to regulate gut microbiota and exhibit a synergy with probiotics [15]. As a result, the combinations of probiotics and prebiotics, namely synbiotics, have attracted extensive attention as tools to help humans maintain optimal health [16]. To date, prebiotics such as inulin, fructo-oligosaccharides (FOS), galacto-oligosaccharides (GOS), xylo-oligosaccharides (XOS) and polydextrose have been commonly used with probiotics as synbiotics in obesity treatment [17,18]. For example, a synbiotic containing *Lactobacillus fermentum* CECT5716 and FOS could reverse HFD-induced gut microbial dysbiosis and metabolic syndrome in rats [19]. A clinical study showed that *Bifidobacterium animalis* subsp. *lactis* 420 with polydextrose could alter gut microbiota and microbial metabolism, which may support the alleviation of obesity [20]. Synbiotic supplementation could not only reduce body weight, but also improve obesity-related phenotypes like blood lipids, cytokines, and oxidative stress [21]. Thus, it is of great significance to explore synbiotics for alleviating obesity.

Bifidobacterium animalis subsp. *lactis* MN-Gup (MN-Gup), an aerospace mutant and screened according to high oxygen tolerance coefficient from the Shenzhou-11 re-entry spacecraft, has been demonstrated to benefit health associated with gut microbiota [22]. Given the wide applications of probiotics in fermented milk, the current study aimed to investigate the effects of fermented milk containing MN-Gup or MN-Gup-based synbiotics on alleviation of obesity in HFD-fed rats and elucidate the underlying mechanism from the perspectives of gut microbiota and its metabolites. This study will help to explain the anti-obesity function of MN-Gup and its synbiotics, and provide the theoretical bases for the applications of MN-Gup in foods.

2. Materials and Methods

2.1. Bacterial Strain and Its Prebiotics Screening In Vitro

Bifidobacterium animalis subsp. *lactis* MN-Gup (MN-Gup, CGMCC No. 15578) was provided by Mengniu Hi-tech Dairy Product Beijing Co., Ltd. (Beijing, China). The strain was activated in De Man, Rogosa, Sharpe (MRS) broth (Difco Laboratories, Detroit, MI, USA) and then incubated at 37 °C for 24 h and repeated three times. Individual prebiotics including fructo-oligosaccharides (FOS), resistance dextrin (RD), inulin, xylo-oligosaccharides (XOS), and galacto-oligosaccharides (GOS) were purchased from Baoling Bao Biology Company (Dezhou, China) and sterilized before use. The prebiotic index (PI) was used to evaluate the relative growth-promoting capability of a prebiotic to that of glucose [23]. The activated MN-Gup was placed in the MRS containing glucose or 1.5% prebiotics as a carbon source, or in a glucose-free MRS medium, and statically cultivated at

37 °C. The absorbance of the fermentation broth was measured at 660 nm at 0 h, 4 h, 8 h, 12 h, and 24 h, respectively, and the PI was calculated according to the following formula.

$$PI = \frac{(A_{PP24} - A_{PP0}) - (A_{PN24} - A_{PN0})}{(A_{PG24} - A_{PG0}) - (A_{PN24} - A_{PN0})}$$

A_{PP24} and A_{PP0} represent the absorbance of fermentation broth of MN-Gup cultured for 24 h and 0 h when the carbon source is a prebiotic in medium; A_{PG24} and A_{PG0} represent the absorbance of fermentation broth of MN-Gup cultured for 24 h and 0 h when the carbon source is glucose in medium; A_{PN24} and A_{PN0} represent the absorbance of fermentation broth of probiotics strains cultured for 24 h and 0 h in a glucose-free MRS medium.

2.2. Preparation of the Test Fermented Milk

Fermented milk was prepared using a commonly used procedure for conventional production of fermented milk with a modification [13], and the main ingredients of fermented milk were provided by Mengniu Hi-tech Dairy Product Beijing Co., Ltd. (Beijing, China). The test fermented milk was added with 1×10^8 CFU/g of MN-Gup or synbiotics consisting of 1×10^8 CFU/g of MN-Gup with either 1.5% GOS or 1.5% XOS. Erythritol was used as calorie-free sweetener in the fermented milk, and per 100 g of fermented milk contained 3.0 g protein, 3.5 g fat, and 5.3 g carbohydrate and its total energy was 78 kcal. The test fermented milk was kept in cold storage and delivered weekly.

2.3. Animals and Experimental Design

Male SD rats (six weeks old) were purchased from Charles River (Beijing, China) and housed in an SPF-grade laboratory animal facility under a standard light/dark (12 h/12 h) cycle with free access to water and food in Pony Testing International Group Co., LTD (Beijing, China). The study was approved by the institutional animal ethics committee (Approval Number: PONY-2021-FL-16). After a one-week acclimatization period, all rats were divided into five groups, and each group consisted of five rats. One group was continuously fed with a normal-chow diet (ND group) and the other four groups of rats were changed to a refined HFD diet (45% fat, TROPHIC, Nantong, China). After nine weeks, HFD induced obesity in rats based on 15% greater body weight gain than ND group, and three groups of obese rats were intragastrically administered with 2 mL the test fermented milk containing MN-Gup (MN-Gup FM), MN-Gup FM with GOS (MN-Gup + GOS FM), and MN-Gup FM with XOS (MN-Gup + XOS FM), and the other group of HFD-induced obese rats was intragastrically administered 2 mL of 0.9% sterile saline (HFD group). The body weight of rats was recorded once a week, and the food intake was replaced and recorded twice a week. After eight weeks of intervention, the rats were euthanised and their blood samples were collected to prepare serum. The epididymal fat pads were weighed and fixed in 4% paraformaldehyde, and colonic feces were also collected and stored at −80 °C until analysis.

2.4. Biochemical Parameters Analysis in Serum

The levels of triglycerides (TG), total cholesterol (TC), low-density lipoprotein (LDL) and high-density lipoprotein (HDL) cholesterol in serum were determined by commercial kits according to the manufacturer's recommendations. Cytokines concentration of tumor necrosis factor-α (TNF-α), interleukin (IL)-6, IL-4 and IL-10 were measured by commercial enzyme-linked immunosorbent assay (ELISA) kits.

2.5. Histological Analysis

The fixed epididymal fat tissues were embedded in paraffin and stained with haematoxylin and eosin (H&E). Finally, the histopathology of epididymal adipocyte was visualized using an optical microscope (DMi8, Leica, Weztlar, Germany), and adipocyte size was analyzed by Image-Pro Plus 6.0 software (Media Cybernetics, Silver Spring, MD, USA) according a previous study.

2.6. Gut Microbiota Analysis

Feces bacterial analysis was performed on the V3-V4 region of the 16S rRNA gene using Illumina MiSeq sequencing as described previously [22]. Briefly, DNA was extracted from the fecal samples using soil DNA extraction kits (Omega Bio-tek, Norcross, GA, USA) according to the manufacturer's protocols. The hypervariable region V3-V4 of 16S rRNA genes were amplified by PCR using the primers 338F (5′-GTACTCCTACGGGAGGCAGCA-3′) and 806R (5′-GTGGACTACHVGGGTWTCTAAT-3′). Sequencing of the PCR amplification products was performed on an Illumina Miseq platform (PE300). Operational taxonomic units (OTUs) were clustered using UPARSE (version 7.1) based on 97% similarity. The taxonomy of each 16S rRNA gene sequence was analyzed by an RDP Classifier algorithm.

2.7. Measurement of Short Chain Fatty Acids (SCFAs) in Colonic Feces

Acetate, propionate and butyrate were measured as previously described [24]. Briefly, 20 mg of fecal sample was acidified with 50 μL of 50% sulfuric acid and homogenised with 500 μL of diethyl ether (containing 19.8 mM 2-ethylbutyric acid as the internal standard) for 1 min using a Mini Bead beater-16 (BioSpec Products, Inc., Bartlesville, OK, USA). Then, the samples were centrifugated at 14,000 rpm for 5 min, and then the supernatant was collected. Extraction was repeated, and the supernatant was combined. After filtering through a 0.22 μm membrane, SCFA content was detected by a gas chromatograph (GC, Shimadzu, GC-2010 Plus, Kyoto, Japan) equipped with the DB-FFAP column (30 m × 0.25 mm × 0.25 μm, Agilent Technologies, Inc., Santa Clara, CA, USA) and FID detector. The splitless injector temperature was 250 °C and the injection volume was 1 μL. The carrier gas was nitrogen. The GC oven temperature program was set as below: maintained at 60 °C for 4 min, increased by 20 °C/min to 180 °C for 1 min, then increased by 50 °C/min to 220 °C for 1 min. The temperature of the FID detector was maintained at 250 °C.

2.8. Measurement of Bile Acids (BAs) in Colonic Feces

An aliquot of 15 mg fecal sample was homogenised with 1 mL of ethanol containing 5% ammonia water and isotope internal standards for five cycles of 15 s. After centrifugation at 14,000 rpm for 10 min at 4 °C, the supernatant was collected. Extraction was repeated three times, and the supernatant was combined and dried under a gentle nitrogen flow. The residue was accurately re-dissolved with 1.8 mL of solution consisting of acetonitrile and water (75:25, *v:v*), and finally passed through a 0.22 μm of filter for UPLC-Q-TOF (G6540, Agilent Technologies, Inc., Santa Clara, CA, USA) analysis. Bile acids were separated with a BEH C_{18} (2.1 mm × 100 mm, 1.7 μm) UPLC column (Waters Inc., Milford, CT, USA), which was kept at 30 °C with a 0.3 mL/min flow rate. The injection volume was 5 μL and the separation was carried out using a gradient elution with water (containing 0.01% formic acid and 2 mM acetic acid) and acetonitrile. The parameters of mass spectrum were set at as follow: electrospray ionization (ESI) model, the detector operated in a low mass range (1700 m/z) and a 2 GHz extended dynamic range, the MS scan range was 100–1200 m/z, and the acquisition rate was two spectrum per second. Twenty-four BAs presented in Table S1 were identified and quantified, but the concentrations of some BAs were too low to reach the detection limits. The detectable BAs included taurocholic acid (TCA), alpha-muricholic acid (α-MCA), tauro-alpha-muricholic acid (T-α-MCA), beta-muricholic acid (β-MCA), tauro-beta-muricholic acid (T-β-MCA), omega-muricholic acid (ω-MCA), deoxycholic acid (DCA), taurodeoxycholic acid (TDCA), hyodeoxycholic acid (HDCA) and tauroursodeoxycholic acid (TUDCA).

2.9. Statistical Analysis

Data were shown as means ± standard deviations (SD). Statistical analysis was carried out using SPSS software (version 23.0, Chicago, IL, USA). One-way ANOVA with Duncan's post hoc test was used to evaluate differences among groups, and significant difference was set at $p < 0.05$.

3. Results

3.1. Selection of Prebiotics with the Capability to Promote the Growth of MN-Gup In Vitro

FOS, RD, inulin, XOS and GOS are five well-defined prebiotics, and their capability to promote the growth of MN-Gup was investigated. As shown in Figure S1A, the growth of MN-Gup was very slow in the glucose-free medium. While in the medium containing glucose, MN-Gup exponentially grew over time. FOS, RD, inulin, XOS and GOS all promoted MN-Gup growth to varying degrees, indicating they could be used as carbon sources for MN-Gup growth. During the 24-h incubation period, the trend of promotion on MN-Gup growth of XOS was the most similar to that of glucose; and GOS had a weak effect on the growth of MN-Gup before 12 h, but its growth-promoting effect was suddenly accelerated after 12 h (Figure S1A). The prebiotic index (PI) of these five prebiotics was GOS > XOS > inulin > RD > FOS (Figure S1B). Therefore, GOS and XOS were the potential prebiotics that could be applied with MN-Gup in symbiotics, and selected for the following study.

3.2. Effects of Fermented Milk Containing MN-Gup or MN-Gup-Based Synbiotics on Body Weight and Body Fat in HFD-Induced Obese Rats

As shown in Figure S2A, body weight of all groups of rats had no significant difference before HFD feeding, and the obese rat model was successfully established through a 9-week HFD feeding base on a gain of at least 15% of body weight relative to ND-fed rats. During the intervention period, HFD induced durative body weight gain, and interventions with MN-Gup FM, MN-Gup + GOS FM and MN-Gup + XOS FM curbed the weight gain of rats (Figure 1A), which may be related to the fewer energy intake in intervention groups (Figure S2B). After eight weeks of intervention, HFD-fed rats showed a significant body weight gain compared to ND-fed rats ($p < 0.05$), and MN-Gup FM, MN-Gup + GOS FM and MN-Gup + XOS FM resulted in less body weight gain in HFD-fed rats (Figure 1B). The body weight gains in MN-Gup + GOS FM and MN-Gup + XOS FM group were significantly decreased than that in HFD group ($p < 0.05$), whereas the body weight gain in MN-Gup FM group was not significantly different from that in HFD group (Figure 1B). These results showed that fermented milk containing MN-Gup and MN-Gup-based synbiotics could reverse HFD-induced weight gain, and MN-Gup with GOS and XOS have a better performance that MN-Gup alone on suppressing body weight gain.

Obesity is highly connected to body fat accumulation and adipocyte hypertrophy [25]. HFD significantly increased the percentage of epididymal fat mass, which was reversed by interventions with MN-Gup FM, MN-Gup + GOS FM and MN-Gup + XOS FM ($p < 0.05$, Figure 1C). As shown in Figure 1D,E, HFD caused epididymal adipocyte hypertrophy, and MN-Gup FM, MN-Gup + GOS FM and MN-Gup + XOS FM all significantly reversed the increased adipocyte sizes ($p < 0.05$). These results suggested that fermented milk containing MN-Gup and its synbiotics with GOS or XOS could reduce the body fat induced by HFD, but GOS and XOS did not exhibit a significant synergy with MN-Gup.

3.3. Effects of Fermented Milk Containing MN-Gup or MN-Gup-Based Synbiotics on Lipid Profile Levels in HFD-Induced Obese Rats

Dyslipidemia is a common complication of obesity. The levels of TG, TC and LDL were significantly elevated in serum of HFD-fed rats, whereas they were significantly decreased by MN-Gup FM and MN-Gup + GOS FM interventions ($p < 0.05$, Figure 2A–C). MN-Gup + XOS FM did not obviously influence the concentrations of TG and TC compared to HFD group, but resulted in significant down-regulation of LDL concentration ($p < 0.05$). No significant changes were found in HDL levels in all groups (Figure 2D). These results indicated that fermented milk containing MN-Gup and its synbiotics with GOS exerted better effects than its synbiotics with XOS in alleviating dyslipidemia.

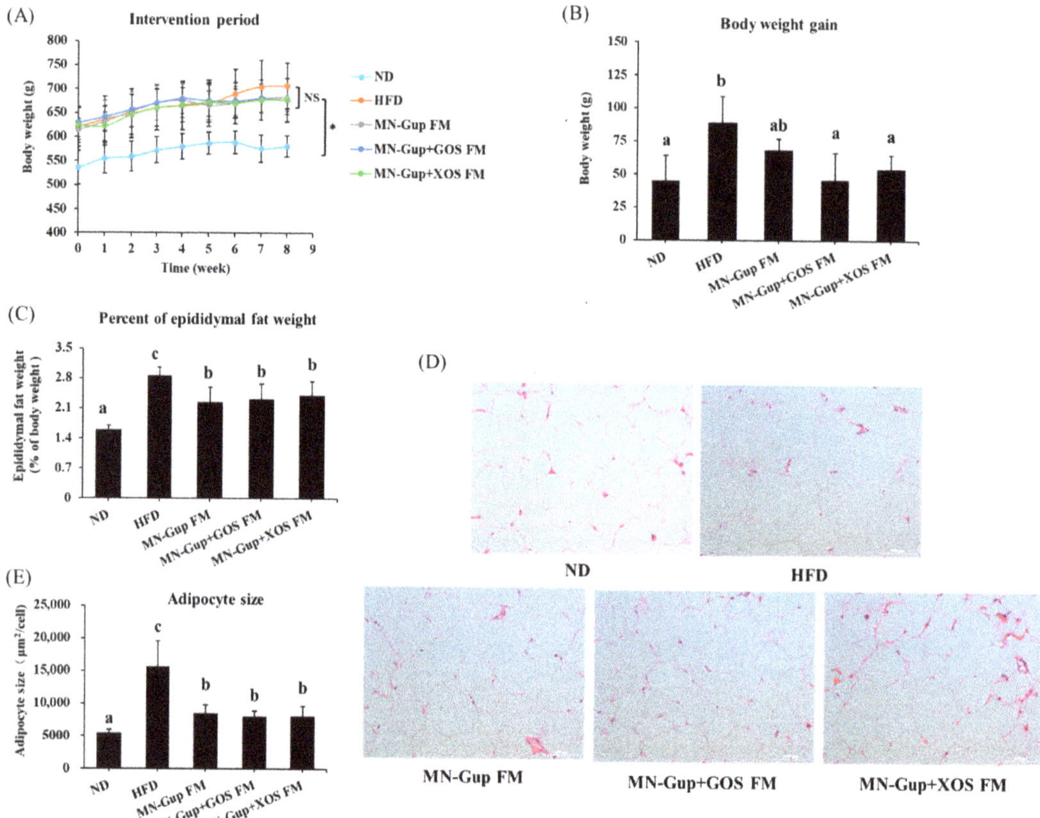

Figure 1. Effects of fermented milk containing MN-Gup or MN-Gup-based synbiotics on body weight and body fat in HFD-fed rats. (**A**) changes of body weight during intervention period (* indicates that ND group is significantly different from other groups); (**B**) body weight gain, (**C**) percent of epididymal fat weight; (**D**) representative images of hematoxylin and eosin (H&E) staining in epididymal adipose tissues; and (**E**) epididymal adipocyte size. Different lowercase letters (e.g., a, b, c) indicate significant differences, $p < 0.05$, ($n = 5$).

3.4. Effects of Fermented Milk Containing MN-Gup or MN-Gup-Based Synbiotics on Pro-Inflammatory and Anti-Inflammatory Cytokines in HFD-Induced Obese Rats

Obesity is associated with a chronic low-grade inflammation state of the host [26]. As shown in Figure 3A,B, HFD induced a significant increase in the level of pro-inflammatory cytokines IL-6 and TNF-α and a significant decrease in the level of anti-inflammatory cytokines IL-10 and IL-4 in rat serum, which were significantly reversed by interventions with MN-Gup FM and MN-Gup + GOS FM ($p < 0.05$). MN-Gup + XOS FM had no significant effects on IL-6 and TNF-α levels, but significantly increased IL-4 and IL-10 levels compared to the HFD group ($p < 0.05$). These results indicated that MN-Gup FM and MN-Gup + GOS FM could reverse HFD-induced inflammation by down regulating pro-inflammatory cytokines and up regulating anti-inflammatory cytokines, and MN-Gup + XOS FM may alleviate HFD-induced inflammation by elevating anti-inflammatory cytokines.

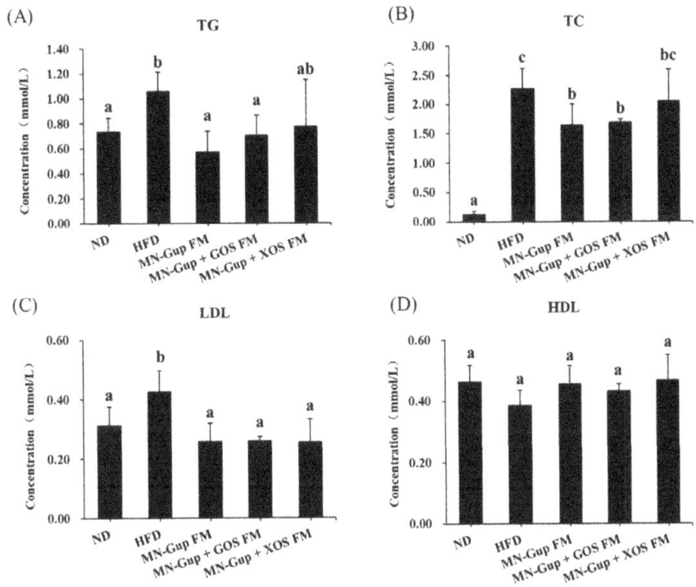

Figure 2. Effects of fermented milk containing MN-Gup or MN-Gup-based synbiotics on lipid profile levels in HFD-induced obese rats. The concentrations of (**A**) triglyceride (TG), (**B**) total cholesterol (TC), (**C**) low-density lipoprotein (LDL) cholesterol, and (**D**) high-density lipoprotein (HDL) cholesterol were measured in rat serum. Different lowercase letters (e.g., a, b, c) indicate significant differences, $p < 0.05$, ($n = 5$).

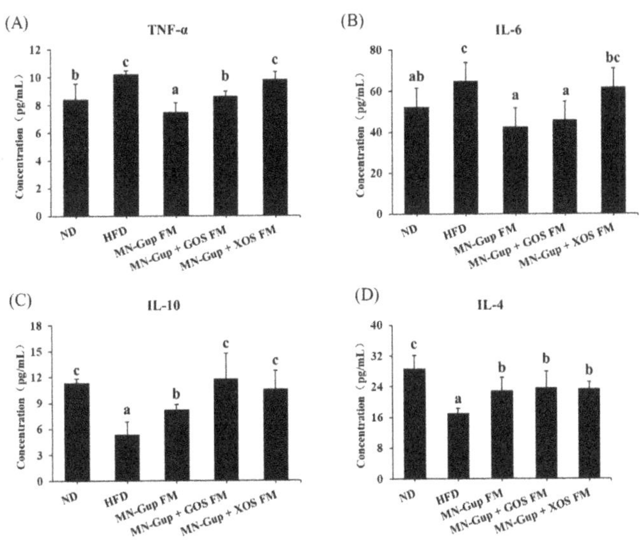

Figure 3. Effects of fermented milk containing MN-Gup or MN-Gup-based synbiotics on pro-inflammatory and anti-inflammatory cytokines in HFD-induced obese rats. The levels of (**A**) tumor necrosis factor α (TNF-α), (**B**) interleukin 6 (IL-6), (**C**) IL-10, and (**D**) IL-4 in serum were measured. Different lowercase letters (e.g., a, b, c) indicate significant differences, $p < 0.05$, ($n = 5$).

3.5. Effects of Fermented Milk Containing MN-Gup or MN-Gup-Based Synbiotics on Gut Microbiota in HFD-Induced Obese Rats

There has been a lot of evidence for the important role of gut microbiota in obesity. The changes in gut microbiota were investigated via 16S rDNA sequencing of colon fecal genes. As shown in Figure 4A, principal coordinates analysis (PCoA) based on operational taxonomic units (OTUs) presented that HFD feeding caused a distinct clustering from ND feeding, indicating that the gut microbiota composition was apparently changed by HFD feeding. The clustering of MN-Gup FM, MN-Gup + GOS FM and MN-Gup + XOS FM groups was different from that of HFD group but has not been completely separated (Figure 4A). A Venn diagram showed that there were 215 OTUs shared in all groups, and 31 OTUs were specific in the ND group and 1 OTU was respectively specific in MN-Gup FM and the MN-Gup + GOS FM group (Figure 4B). The correlations between gut microbiota (based on OTUs) and obesity related indexes (including body weight gain, percent of epididymal fat weight, the levels of TG, TC, HDL, LDL, TNF-α, IL-6, IL-10, and IL-4) were investigated using the Pearson correlation analysis. The heatmap in Figure 4C showed the top 50 OTUs for total abundances associated with obesity and these OTU assignments were shown in Table S2. Most of OTUs, such as OUT684, OUT790, OUT777, OUT469, OUT810, OUT555, OUT625, OUT657, were positively correlated with body weight gain, blood lipids and negatively correlated with anti-inflammatory cytokines (Figure 4C). In particular, OTU479 and OUT545 were significantly positively correlated with the percent of epididymal fat weight, TG, TC, LDL and TNF-α, and negatively correlated with IL-10, and IL-4 levels, revealing that their changes were closely connected with changed obesity in rats.

There were notable differences between gut microbiota of HFD and ND groups from the taxonomic profiling at phylun and genus levels, and interventions with MN-Gup FM, MN-Gup + GOS FM and MN-Gup + XOS FM had influences on the gut microbiota composition induced by HFD (Figure S3). Most obviously, the relative abundance of *Bacteroidota* was significantly reduced by HFD, which was not reversed by interventions (Figure 4D). *Firmicutes*, the dominant phylum, were significantly elevated in the HFD group compared to the ND group, whereas its relative abundance was reduced by interventions, especially by MN-Gup + XOS FM with significance ($p < 0.05$, Figure 4E). At the genus level, the relative abundance of *Lachnoclostridium* and *Allobaculum* was found to be significantly increased in the HFD group, but significantly reduced by MN-Gup FM ($p < 0.05$, Figure 4F,G). These results indicate that fermented milk containing MN-Gup and its synbiotics with GOS and XOS could regulate HFD-affected gut microbiota and may alleviate obesity through the regulation of gut microbiota.

3.6. Effects of Fermented Milk Containing MN-Gup or MN-Gup-Based Synbiotics on the Profiles of SCFAs and BAs in HFD-Induced Obese Rats

Gut microbiota-derived metabolites, such as SCFAs and BAs, have been considered as key molecular mediators between the microbiota and host. Gut microbiota can produce SCFAs via fermentation with dietary fiber, and modulate bile acid pool by metabolizing primary BAs to their secondary forms [6,27]. To give an insight into the potential mechanism associated with gut microbiota by which the interventions alleviated HFD-caused obesity, the levels of SCFAs and BAs were detected in colonic feces. As shown in Figure 5A–C, SCFAs, including acetate, propionate and butyrate, were significantly reduced by HFD feeding, which was significantly elevated by all interventions ($p < 0.05$), suggesting that fermented milk containing MN-Gup or MN-Gup-based synbiotics could promote the production of SCFAs.

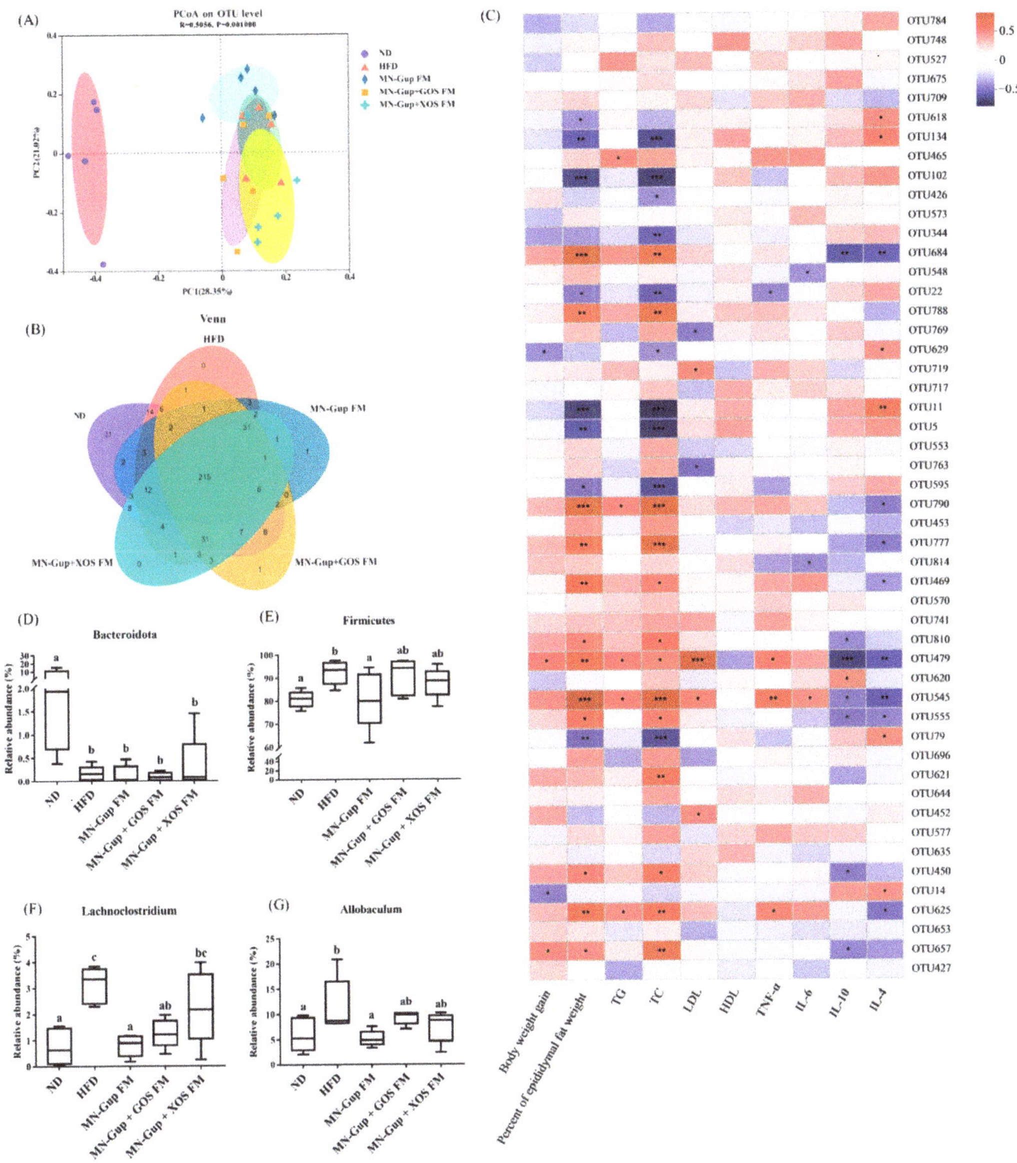

Figure 4. Effects of fermented milk containing MN-Gup or MN-Gup-based synbiotics on gut microbiota in HFD-induced obese rats. (**A**) Principal coordinate analysis (PCoA) based on bacterial operational taxonomic units (OTUs) using Bray-Curtis calculation; (**B**) Venn diagram; (**C**) the heatmap of correlation analysis between gut microbiota (based on OTUs) and obesity related indexes (significant correlations were noted by * $p < 0.05$, ** $p < 0.01$, and *** $p < 0.001$); (**D**,**E**) the significantly altered bacteria at phylum level; (**F**,**G**) the significantly altered bacteria at genus level (different lowercase letters (e.g., a, b, c) indicate significant differences, $p < 0.05$, $n = 5$).

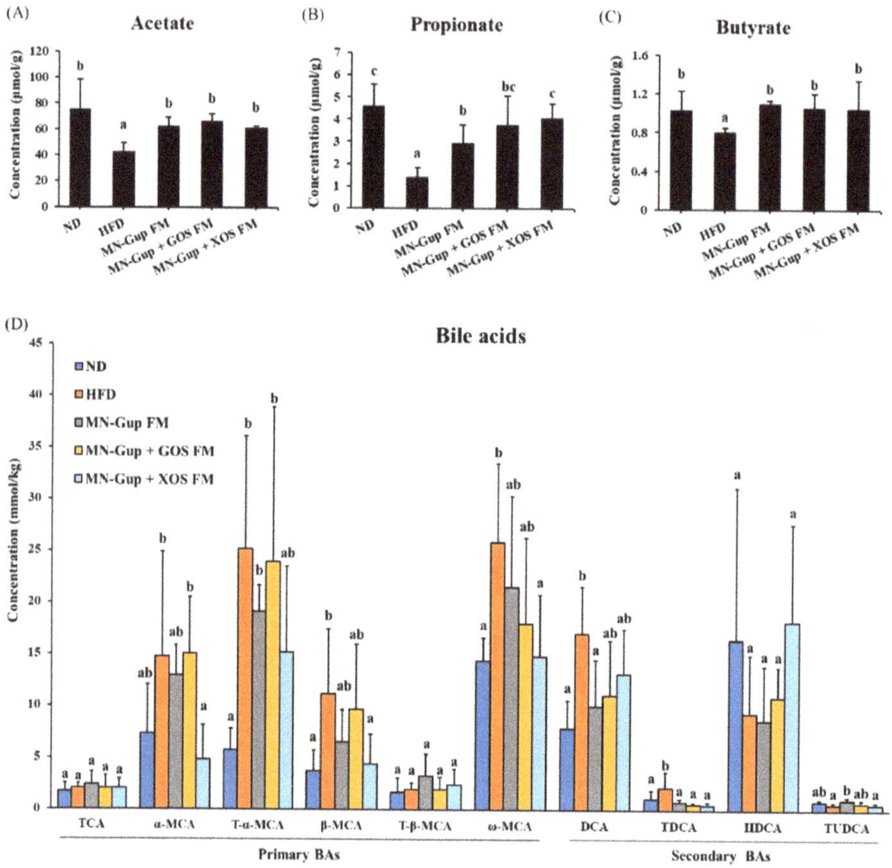

Figure 5. Effects of fermented milk containing MN-Gup or MN-Gup-based synbiotics on the profiles of short chain fatty acids (SCFAs) and bile acids (BAs) in HFD-induced obese rats. The concentration of (**A**) acetate, (**B**) propionate, and (**C**) butyrate was detected in colonic feces (n = 5). (**D**) The detectable bile acids in colonic feces, taurocholic acid (TCA), alpha-muricholic acid (α-MCA), tauro-alpha-muricholic acid (T-α-MCA), beta-muricholic acid (β-MCA), tauro-beta-muricholic acid (T-β-MCA), omega-muricholic acid (ω-MCA), deoxycholic acid (DCA), taurodeoxycholic acid (TDCA), hyodeoxycholic acid (HDCA) and tauroursodeoxycholic acid (TUDCA). Different lowercase letters (e.g., a, b, c) indicate significant differences, $p < 0.05$, (n = 4 in ND group, and n = 5 in the rest groups).

Meanwhile, HFD feeding decreased the ratio of secondary BAs compared to ND feeding, revealing that HFD may suppress the biotransformation of primary BAs (Figure S4). MN-Gup FM and MN-Gup + GOS FM did not apparently change the ratio of secondary BAs, while an increased ratio of secondary BAs was observed in the MN-Gup + XOS FM group (Figure S4). Some primary BAs including α-MCA, T-α-MCA, β-MCA, and ω-MCA were significantly elevated in the HFD group, most of which were significantly reduced in the MN-Gup + XOS FM group ($p < 0.05$, Figure 5D). The levels of DCA and TDCA (secondary BAs) were significantly higher in the HFD group compared to the ND group ($p < 0.05$), but were lowered by all interventions. In contrast, HDCA and TUDCA (secondary BAs) levels were decreased in HFD, and interventions had inapparent reversions (Figure 5D). These results suggested that fermented milk containing MN-Gup or

MN-Gup-based synbiotics could affect BAs levels and composition, and MN-Gup + XOS FM may show the best performance on regulating the HFD-changed bile acid pool.

4. Discussion

Given the high prevalence of obesity, finding an effective intervention to alleviate obesity is important for public health. Accumulating evidence demonstrates that the gut microbiota plays a crucial role in the development of obesity, and modifying gut microbiota through a diet rich in probiotics has become one of the most potential interventions for obesity [8]. Additionally, the synergism in the combination of prebiotics and probiotics has been attracting more and more attention due to the beneficial effects of prebiotic on gut microbiota and intestinal function [28]. *Bifidobacterium animalis* subsp. *lactis* MN-Gup (MN-Gup) has been found to regulate gut microbiota and relieve constipation [22], but its effects on obesity are still unknown. In this study, we evaluated the effects of MN-Gup-containing fermented milk on alleviating obesity using a therapeutic obesity model based on HFD-fed rats, and investigated whether prebiotics including GOS and XOS had the synergistic effects with MN-Gup on obesity.

The results showed that treatment with MN-Gup-containing fermented milk significantly reduced body weight gain and fat mass in HFD-induced obese rats (Figure 1). Although XOS and GOS could apparently promote the proliferation of MN-Gup in vitro (Figure S1), their supplement did not promote a more significant weight/fat-loss than MN-Gup alone in fermented milk, suggesting that they had no significant synergies with MN-Gup on reducing obesity (Figure 1). In fact, current opinions support that it is more important to make obese people healthier, as with improvements in blood lipids and chronic inflammation, rather than just focusing on body weight loss [29]. Similarly, MN-Gup FM, MN-Gup + GOS FM and MN-Gup + XOS FM could reduce the levels of TG, TC, LDL, TNF-α and IL-6, but GOS and XOS did not cause a significant synergistic effect with MN-Gup (Figures 2 and 3). Pro-inflammatory cytokines, such as TNF-α and IL-6, could contribute to systemic insulin resistance or metabolic disorder in obesity [30], while anti-inflammatory cytokines, such as IL-10 and IL-4, could prevent diet-induced obesity and insulin resistance [31,32]. In particular, MN-Gup + XOS FM did not significantly reverse the blood lipids and pro-inflammatory cytokines, which was not as effective as the MN-Gup alone and MN-Gup + GOS FM. However, MN-Gup + XOS FM significantly enhanced the level of IL-10 and IL-4 (Figure 3), revealing that XOS may have had a promotion on health perhaps in a unique way. These results suggested that prebiotics did not always have synergistic effects with probiotics, and it is necessary to confirm the effectiveness of prebiotics for probiotic strains through more experimental data including animal and clinical data before commercial use.

In previous studies, gut microbiota showed differences in obese individuals compared to normal individuals [33]. The results from PCoA (Figure 4A) and taxonomic composition (Figure S3) analysis showed that HFD feeding apparently altered the composition of gut microbiota in rats, which was consistent with previous reports [34,35]. Numerous OTUs were found to correlate with the obesity-related indexes (Figure 4C). Notably, OTU479 and OTU545, which were respectively annotated as *Enterococcus* and *Lachnoclostridium* genus (Table S2), were significantly positively correlated with most phenotypes pro-obesity and were negatively correlated with the anti-obesity phenotypes (Figure 4C). *Enterococcus* and *Lachnoclostridium* have been previously reported to be elevated in HFD-fed mice and their increases were associated with obesity and related inflammation [36,37]. At the phylum level, the *Bacteroidota* proportion was largely reduced by HFD and was hardly changed by interventions (Figure 4D), suggesting that a long-term high fat diet could cause an irreversible effect on *Bacteroidota*. A significant increase of *Firmicutes* proportion was observed in the HFD group, and only MN-Gup FM could significantly reverse it (Figure 4E), revealing that a GOS and XOS supplement may benefit the growth of *Firmicutes*. At the genus level, the relative abundance of *Lachnoclostridium* was significantly elevated by HFD and reversed by the interventions (Figure 4F), which was consistent with the correlation analysis (Figure 4C). Additionally, similar results were also observed in the *Allobaculum* proportion (Figure 4G). *Allobaculum* has been demonstrated to be increased in HFD and involved in lipid metabolism and fat deposition [38]. These results indicated that fermented milk containing MN-Gup and MN-Gup based synbiotics may attenuate HFD-induced obesity through the regulation of gut microbiota.

Previous research has considered that one potential mechanism underlying the regulation of obesity by gut microbiota is through microbiome-derived bioactive metabolites like SCFAs and BAs and their regulatory effects on the host metabolism [39,40]. The major SCFAs, including acetate, propionate, and butyrate, have been demonstrated to reduce lipogenesis and inflammation [6], and their significant increases in MN-Gup FM, MN-Gup + GOS FM and MN-Gup + XOS FM (Figure 5A–C) may be the explanations for the alleviation of obesity. Interestingly, MN-Gup + XOS FM led to significantly higher propionate concentration than MN-Gup FM and MN-Gup + GOS FM (Figure 5B). Propionate was found to promote the secretion of a gut hormone related to energy metabolism and to regulate appetite via the gut-brain axis [6]. It was speculated that XOS may benefit the growth of propionate-producing bacteria, which may help to alleviate obesity.

Recent studies had also shown that gut microbiota could metabolize primary BAs to secondary Bas, and that BAs levels and composition are associated with metabolic diseases such as obesity, dyslipidemia, and diabetes [41]. Our results indicated that HFD decreased the proportion of secondary BAs compared to ND (Figure S4), which may result from the HFD-altered gut microbiota. Notably, significant increases of muricholic acids (including α-MCA, T-α-MCA, β-MCA, and ω-MCA) were observed in the HFD group, which were consistent with previous studies [42], and revealed the suppression of primary BAs metabolism. MN-Gup + XOS FM showed significantly lower concentrations of α-MCA, T-α-MCA, β-MCA, and ω-MCA than the HFD group (Figure 5D), suggesting its potential role in regulation lipid and glucose metabolism. Moreover, DCA and its conjugated form TDCA have been demonstrated to trigger the production of proinflammatory cytokines like TNF-α and IL-6 [42,43], so their significant increase in HFD and decrease in the intervention groups (especially MN-Gup + XOS FM) were consistent with the obesity-related inflammation and the alleviation by interventions. Taken together, fermented milk containing MN-Gup or MN-Gup-based synbiotics may alleviate HFD-induced obesity through microbiome-associated SCFAs and BAs. In the MN-Gup-based synbiotics, MN-Gup + GOS FM and MN-Gup + XOS FM had their own advantages and characteristics in the regulation of gut microbiota and SCFAs and BAs.

5. Conclusions

In conclusion, our study demonstrated that fermented milk containing MN-Gup or its synbiotics (MN-Gup + GOS, MN-Gup + XOS) had the potential effects on alleviating obesity in HFD-induced obese rats, as shown by the decreases in body weight gain, epididymal fat mass, adipocyte sizes and the improvement in dyslipidemia and obesity-related inflammation. Furthermore, they could regulate the HFD-affected gut microbiota through lowing obesity-related bacteria like *Firmicutes*, *Lachnoclostridium* and *Allobaculum*, and modify the microbiota-driven metabolites including SCFAs and BAs, which may be the potential mechanism for obesity alleviation. Although GOS and XOS could promote the growth of MN-Gup in vitro, in vivo results did not show that GOS and XOS had significant synergistic effects with MN-Gup on alleviating obesity, revealing that the symbiotic applications based on MN-Gup will need more research and evidence. This study will provide the underpinning for the future clinical trial of MN-Gup and scientific supports for applications of MN-Gup in probiotic products.

Supplementary Materials: The following supporting information can be downloaded at: https://www.mdpi.com/article/10.3390/nu14132631/s1, Figure S1: Screening prebiotics with the capability to promote the growth of MN-Gup in vitro. (A)The growth curve of MN-Gup in 24 h; and (B) Prebiotic index (PI) of prebiotics. FOS, fructo-oligosaccharides; RD, resistance dextrin; XOS, xylo-oligosaccharides; GOS, galacto-oligosaccharides; Figure S2: (A) The body weight of rats before high-fat diet (HFD) feeding, and the beginning of end of interventions; (B) The intake of feed calories during intervention period (n = 5). ND, normal diet; HFD, fed high-fat diet; MN-Gup FM, fermented milk containing MN-Gup; MN-Gup FM + GOS, fermented milk containing MN-Gup and galacto-oligosaccharides (GOS); MN-Gup FM + XOS, fermented milk containing MN-Gup and xylo-oligosaccharides (XOS); Figure S3: The average relative abundance of bacteria at (A) phylum and (B) genus level (n = 5). ND, normal diet; HFD, fed high-fat diet; MN-Gup FM, fermented milk containing MN-Gup; MN-Gup FM + GOS, fermented milk containing MN-Gup and galacto-oligosaccharides (GOS); MN-Gup FM + XOS, fermented milk containing MN-Gup and xylo-oligosaccharides (XOS); Figure S4: Effects of fermented milk containing MN-Gup or MN-Gup-based synbiotics on the proportion of primary and secondary bile acids (BAs) in colonic feces. ND, normal diet; HFD, fed high-fat diet; MN-Gup FM, fermented milk containing MN-Gup; MN-Gup FM + GOS, fermented milk containing MN-Gup and galacto-oligosaccharides (GOS); MN-Gup FM + XOS, fermented milk containing MN-Gup and xylo-oligosaccharides (XOS) (n = 4 in ND group, and n = 5 in the rest groups); Table S1: The tested bile acids and their abbreviation; Table S2: OTU assignments in the correlation analysis.

Author Contributions: C.W., S.L., R.W. and J.Z. designed the experiments. C.W., Y.R., Q.Z. and R.X. performed the experiments. S.L., R.W., J.H. and E.S. provided the materials. C.W. and J.Z. were primarily responsible for writing the manuscript. All authors have read and agreed to the published version of the manuscript.

Funding: The research was supported in part by a grant from the Inner Mengolia Science & Technology Project Plan (2020GG0111) and Beijing Postdoctoral Research Foundation (2021ZZ141).

Institutional Review Board Statement: The study was conducted according to the Guide for the Care and Use of Laboratory Animals of Pony Testing International Group Co., LTD. (Beijing, China) and approved by the institutional animal ethics committee (Approval Number: PONY-2021-FL-16).

Informed Consent Statement: Not applicable.

Data Availability Statement: Data are contained within the article or Supplementary Materials.

Conflicts of Interest: The authors declare that they have no conflict of interest.

References

1. Must, A.; Spadano, J.; Coakley, E.H.; Field, A.E.; Colditz, G.; Dietz, W.H. The disease burden associated with overweight and obesity. *JAMA J. Am. Med. Assoc.* **1999**, *282*, 1523–1529. [CrossRef] [PubMed]
2. O'Neill, S.; O'Driscoll, L. Metabolic syndrome: A closer look at the growing epidemic and its associated pathologies. *Obes. Rev.* **2015**, *16*, 1–12. [CrossRef] [PubMed]
3. Franz, M.J.; VanWormer, J.J.; Crain, A.L.; Boucher, J.L.; Histon, T.; Caplan, W.; Bowman, J.D.; Pronk, N.P. Weight-loss outcomes: A systematic review and meta-analysis of weight-loss clinical trials with a minimum 1-year follow-up. *J. Am. Diet. Assoc.* **2007**, *107*, 1755–1767. [CrossRef]
4. Cuevas-Sierra, A.; Ramos-Lopez, O.; Riezu-Boj, J.I.; Milagro, F.I.; Martinez, J.A. Diet, gut microbiota, and obesity: Links with host genetics and epigenetics and potential applications. *Adv. Nutr.* **2019**, *10*, S17–S30. [CrossRef]
5. Riedl, R.A.; Atkinson, S.N.; Burnett, C.M.L.; Grobe, J.L.; Kirby, J.R. The gut microbiome, energy homeostasis, and implications for hypertension. *Curr. Hypertens. Rep.* **2017**, *19*, 27. [CrossRef]
6. Koh, A.; De Vadder, F.; Kovatcheva-Datchary, P.; Backhed, F. From dietary fiber to host physiology: Short-chain fatty acids as key bacterial metabolites. *Cell* **2016**, *165*, 1332–1345. [CrossRef]
7. Kwon, M.; Lee, J.; Park, S.; Kwon, O.H.; Seo, J.; Roh, S. Exopolysaccharide isolated from lactobacillus plantarum L-14 has anti-inflammatory effects via the toll-like receptor 4 pathway in LPS-induced RAW 264.7 cells. *Int. J. Mol. Sci.* **2020**, *21*, 9283. [CrossRef]
8. Abenavoli, L.; Scarpellini, E.; Colica, C.; Boccuto, L.; Salehi, B.; Sharifi-Rad, J.; Aiello, V.; Romano, B.; De Lorenzo, A.; Izzo, A.A.; et al. Gut microbiota and obesity: A role for probiotics. *Nutrients* **2019**, *11*, 2690. [CrossRef]
9. Mikolaj, K.; Igor, L.; Wojciech, M. Global internet data on the interest in antibiotics and probiotics generated by Google trends. *Antibiotics* **2019**, *8*, 147.
10. Wang, J.J.; Tang, H.; Zhang, C.H.; Zhao, Y.F.; Derrien, M.; Rocher, E.; Vlieg, J.; Strissel, K.; Zhao, L.P.; Obin, M.; et al. Modulation of gut microbiota during probiotic-mediated attenuation of metabolic syndrome in high fat diet-fed mice. *ISME J.* **2015**, *9*, 1–15. [CrossRef]
11. Huo, Y.X.; Zhao, G.P.; Li, J.W.; Wang, R.; Ren, F.Z.; Li, Y.X.; Wang, X.Y. *Bifidobacterium animalis* subsp. *lactis* A6 enhances fatty acid beta-oxidation of adipose tissue to ameliorate the development of obesity in mice. *Nutrients* **2022**, *14*, 598. [CrossRef]

12. Song, W.; Song, C.; Li, L.; Wang, T.Y.; Hu, J.H.; Zhu, L.N.; Yue, T.L. Lactobacillus alleviated obesity induced by high-fat diet in mice. *J. Food Sci.* **2021**, *86*, 5439–5451. [CrossRef]
13. Kadooka, Y.; Sato, M.; Ogawa, A.; Miyoshi, M.; Uenishi, H.; Ogawa, H.; Ikuyama, K.; Kagoshima, M.; Tsuchida, T. Effect of Lactobacillus gasseri SBT2055 in fermented milk on abdominal adiposity in adults in a randomised controlled trial. *Brit. J. Nutr.* **2013**, *110*, 1696–1703. [CrossRef]
14. Pontes, K.S.D.; Guedes, M.R.; da Cunha, M.R.; Mattos, S.D.; Silva, M.I.B.; Neves, M.F.; Marques, B.; Klein, M. Effects of probiotics on body adiposity and cardiovascular risk markers in individuals with overweight and obesity: A systematic review and meta-analysis of randomized controlled trials. *Clin. Nutr.* **2021**, *40*, 4915–4931. [CrossRef]
15. Bindels, L.B.; Delzenne, N.M.; Cani, P.D.; Walter, J. Towards a more comprehensive concept for prebiotics. *Nat. Rev. Gastro. Hepat.* **2015**, *12*, 303–310. [CrossRef]
16. Schrezenmeir, J.; de Vrese, M. Probiotics, prebiotics, and synbiotics-approaching a definition. *Am. J. Clin. Nutr.* **2001**, *73*, 361S–364S. [CrossRef]
17. Cerdo, T.; Garcia-Santos, J.A.; Bermudez, M.G.; Campoy, C. The role of probiotics and prebiotics in the prevention and treatment of obesity. *Nutrients* **2019**, *11*, 635. [CrossRef]
18. Álvarez-Arraño, V.; Martín-Peláez, S. Effects of probiotics and synbiotics on weight Loss in subjects with overweight or obesity: A systematic review. *Nutrients* **2021**, *13*, 3627. [CrossRef]
19. Rivero-Gutierrez, B.; Gamez-Belmonte, R.; Dolores Suarez, M.; Luis Lavin, J.; Maria Aransay, A.; Olivares, M.; Martinez-Augustin, O.; Sanchez de Medina, F.; Zarzuelo, A. A synbiotic composed of Lactobacillus fermentum CECT5716 and FOS prevents the development of fatty acid liver and glycemic alterations in rats fed a high fructose diet associated with changes in the microbiota. *Mol. Nutr. Food Res.* **2017**, *61*, 1600622. [CrossRef]
20. Hibberd, A.A.; Yde, C.C.; Ziegler, M.L.; Honore, A.H.; Saarinen, M.T.; Lahtinen, S.; Stahl, B.; Jensen, H.M.; Stenman, L.K. Probiotic or synbiotic alters the gut microbiota and metabolism in a randomised controlled trial of weight management in overweight adults. *Benef. Microbes* **2019**, *10*, 121–135. [CrossRef]
21. Chaiyasut, C.; Sivamaruthi, B.S.; Kesika, P.; Khongtan, S.; Khampithum, N.; Thangaleela, S.; Peerajan, S.; Bumrungpert, A.; Chaiyasut, K.; Sirilun, S.; et al. Synbiotic supplementation improves obesity index and metabolic biomarkers in Thai obese adults: A randomized clinical trial. *Foods* **2021**, *10*, 1580. [CrossRef] [PubMed]
22. Wang, R.; Sun, J.; Li, G.; Zhang, M.; Niu, T.; Kang, X.; Zhao, H.; Chen, J.; Sun, E.; Li, Y. Effect of *Bifidobacterium animalis* subsp. *lactis* MN-Gup on constipation and the composition of gut microbiota. *Benef. Microbes* **2021**, *12*, 31–42. [PubMed]
23. Figueroa-Gonzalez, I.; Rodriguez-Serrano, G.; Gomez-Ruiz, L.; Garcia-Garibay, M.; Cruz-Guerrero, A. Prebiotic effect of commercial saccharides on probiotic bacteria isolated from commercial products. *Food Sci. Tech.* **2019**, *39*, 747–753. [CrossRef]
24. Scortichini, S.; Boarelli, M.C.; Silvi, S.; Fiorini, D. Development and validation of a GC-FID method for the analysis of short chain fatty acids in rat and human faeces and in fermentation fluids. *J. Chromatogr. B* **2020**, *1143*, 121972. [CrossRef] [PubMed]
25. Verboven, K.; Wouters, K.; Gaens, K.; Hansen, D.; Bijnen, M.; Wetzels, S.; Stehouwer, C.D.; Goossens, G.H.; Schalkwijk, C.G.; Blaak, E.E.; et al. Abdominal subcutaneous and visceral adipocyte size, lipolysis and inflammation relate to insulin resistance in male obese humans. *Sci. Rep.* **2018**, *8*, 4677. [CrossRef]
26. Gregor, M.F.; Hotamisligil, G.S. Inflammatory mechanisms in obesity. *Annu. Rev. Immunol.* **2011**, *29*, 415–445. [CrossRef]
27. Wei, Y.X.; Zheng, K.Y.; Wang, Y.G. Gut microbiota-derived metabolites as key mucosal barrier modulators in obesity. *World J. Gastroenterol.* **2021**, *27*, 5555–5565. [CrossRef]
28. Suzumura, E.A.; Bersch-Ferreira, A.C.; Torreglosa, C.R.; da Silva, J.T.; Coqueiro, A.Y.; Kuntz, M.G.F.; Chrispim, P.P.; Weber, B.; Cavalcanti, A.B. Effects of oral supplementation with probiotics or synbiotics in overweight and obese adults: A systematic review and meta-analyses of randomized trials. *Nutr. Rev.* **2019**, *77*, 430–450. [CrossRef]
29. Linda Bacon, L.A. Weight Science: Evaluating the Evidence for a Paradigm Shift. *Nutr. J.* **2011**, *10*, 13.
30. Wellen, K.E. Inflammation, stress, and diabetes. *J. Clin. Investig.* **2005**, *115*, 1111–1119. [CrossRef]
31. Hong, E.G.; Ko, H.J.; Cho, Y.R.; Kim, H.J.; Ma, Z.; Yu, T.Y.; Friedline, R.H.; Kurt-Jones, E.; Finberg, R.; Fischer, M.A.; et al. Interleukin-10 prevents diet-induced insulin resistance by attenuating macrophage and cytokine response in skeletal muscle. *Diabetes* **2009**, *58*, 2525–2535. [CrossRef] [PubMed]
32. Choi, E.W.; Lee, M.; Song, J.W.; Kim, K.; Lee, J.; Yang, J.; Lee, S.H.; Kim, I.Y.; Choi, J.-H.; Seong, J.K. Fas mutation reduces obesity by increasing IL-4 and IL-10 expression and promoting white adipose tissue browning. *Sci. Rep.* **2020**, *10*, 12001. [CrossRef] [PubMed]
33. Turnbaugh, P.J.; Ley, R.E.; Mahowald, M.A.; Magrini, V.; Mardis, E.R.; Gordon, J.I. An obesity-associated gut microbiome with increased capacity for energy harvest. *Nature* **2006**, *444*, 1027–1031. [CrossRef] [PubMed]
34. Chiou, W.-C.; Chang, B.-H.; Tien, H.-H.; Cai, Y.-L.; Fan, Y.-C.; Chen, W.-J.; Chu, H.-F.; Chen, Y.-H.; Huang, C. Synbiotic intervention with an adlay-based prebiotic and probiotics improved diet-induced metabolic disturbance in mice by modulation of the gut microbiota. *Nutrients* **2021**, *13*, 3161. [CrossRef] [PubMed]
35. Zhao, L.; Zhang, Q.; Ma, W.; Tian, F.; Shen, H.; Zhou, M. A combination of quercetin and resveratrol reduces obesity in high-fat diet-fed rats by modulation of gut microbiota. *Food Funct.* **2017**, *8*, 4644–4656. [CrossRef]
36. Zhang, J.; Yi, C.; Han, J.; Ming, T.; Zhou, J.; Lu, C.; Li, Y.; Su, X. Novel high-docosahexaenoic-acid tuna oil supplementation modulates gut microbiota and alleviates obesity in high-fat diet mice. *Food Sci. Nutr.* **2020**, *8*, 6513–6527. [CrossRef]

37. Wu, G.-D.; Pan, A.; Zhang, X.; Cai, Y.-Y.; Wang, Q.; Huang, F.-Q.; Alolga, R.N.; Li, J.; Qi, L.-W.; Liu, Q. Cordyceps improves obesity and its related inflammation via modulation of Enterococcus cecorum abundance and bile acid metabolism. *Am. J. Chin. Med.* **2022**, *50*, 817–838. [CrossRef]
38. Zheng, Z.; Lyu, W.; Ren, Y.; Li, X.; Zhao, S.; Yang, H.; Xiao, Y. Allobaculum involves in the modulation of intestinal ANGPTLT4 expression in mice treated by high-fat diet. *Front. Nutr.* **2021**, *8*, 690138. [CrossRef]
39. Den Besten, G.; van Eunen, K.; Groen, A.K.; Venema, K.; Reijngoud, D.J.; Bakker, B.M. The role of short-chain fatty acids in the interplay between diet, gut microbiota, and host energy metabolism. *J. Lipid Res.* **2013**, *54*, 2325–2340. [CrossRef]
40. Staley, C.; Weingarden, A.R.; Khoruts, A.; Sadowsky, M.J. Interaction of gut microbiota with bile acid metabolism and its influence on disease states. *Appl. Microbiol. Biotechnol.* **2017**, *101*, 47–64. [CrossRef]
41. Chen, L.; van den Munckhof, I.C.L.; Schraa, K.; Ter Horst, R.; Koehorst, M.; van Faassen, M.; van der Ley, C.; Doestzada, M.; Zhernakova, D.V.; Kurilshikov, A.; et al. Genetic and microbial associations to plasma and fecal bile acids in obesity relate to plasma lipids and liver fat content. *Cell Rep.* **2020**, *33*, 108212. [CrossRef] [PubMed]
42. Zeng, H.; Larson, K.J.; Cheng, W.-H.; Bukowski, M.R.; Safratowich, B.D.; Liu, Z.; Hakkak, R. Advanced liver steatosis accompanies an increase in hepatic inflammation, colonic, secondary bile acids and Lactobacillaceae/Lachnospiraceae bacteria in C57BL/6 mice fed a high-fat diet. *J. Nutr. Biochem.* **2020**, *78*, 108336. [CrossRef] [PubMed]
43. Jia, W.; Xie, G.; Jia, W. Bile acid-microbiota crosstalk in gastrointestinal inflammation and carcinogenesis. *Nat. Rev. Gastroenterol. Hepatol.* **2018**, *15*, 111–128. [CrossRef] [PubMed]

nutrients

MDPI

Article

Swimming Suppresses Cognitive Decline of HFD-Induced Obese Mice through Reversing Hippocampal Inflammation, Insulin Resistance, and BDNF Level

Hu Zhang [1,†], Ji-Ling Liang [1,†], Qiu-Yue Wu [1], Jin-Xiu Li [2], Ya Liu [1], Liang-Wen Wu [2], Jie-Lun Huang [1], Xiao-Wen Wu [1], Ming-Hui Wang [1] and Ning Chen [2,*]

1 Graduate School, Wuhan Sports University, Wuhan 430079, China; 17858503393@163.com (H.Z.); ljl19930210@163.com (J.-L.L.); wuqiuyue1252@163.com (Q.-Y.W.); ly13515234752@163.com (Y.L.); hjl19960713@163.com (J.-L.H.); wxw1123wxw@163.com (X.-W.W.); wangminghui290335@163.com (M.-H.W.)
2 Tianjiu Research and Development Center for Exercise Nutrition and Foods, Hubei Key Laboratory of Exercise Training and Monitoring, College of Health Science, Wuhan Sports University, Wuhan 430079, China; ljx973960148@163.com (J.-X.L.); hancock120@163.com (L.-W.W.)
* Correspondence: nchen510@gmail.com; Tel.: +86-27-67846140
† These authors contributed equally to this work.

Abstract: Obesity is an important public health problem nowadays. Long-term obesity can trigger a series of chronic diseases and impair the learning and memory function of the brain. Current studies show that scientific exercise can effectively improve learning and memory capacity, which also can provide benefits for obese people. However, the underlying mechanisms for the improvement of cognitive capacity under the status of obesity still need to be further explored. In the present study, the obesity-induced cognition-declined model was established using 4-week-old mice continuously fed with a high-fat diet (HFD) for 12 weeks, and then the model mice were subjected to an 8-week swimming intervention and corresponding evaluation of relevant indicators, including cognitive capacity, inflammation, insulin signal pathway, brain-derived neurotrophic factor (BNDF), and apoptosis, for exploring potential regulatory mechanisms. Compared with the mice fed with regular diets, the obese mice revealed the impairment of cognitive capacity; in contrast, swimming intervention ameliorated the decline in cognitive capacity of obese mice by reducing inflammatory factors, inhibiting the JNK/IRS-1/PI3K/Akt signal pathway, and activating the PGC-1α/BDNF signal pathway, thereby suppressing the apoptosis of neurons. Therefore, swimming may be an important interventional strategy to compensate for obesity-induced cognitive impairment.

Keywords: obesity; swimming; cognitive capacity; inflammation; insulin resistance; BDNF

Citation: Zhang, H.; Liang, J.-L.; Wu, Q.-Y.; Li, J.-X.; Liu, Y.; Wu, L.-W.; Huang, J.-L.; Wu, X.-W.; Wang, M.-H.; Chen, N. Swimming Suppresses Cognitive Decline of HFD-Induced Obese Mice through Reversing Hippocampal Inflammation, Insulin Resistance, and BDNF Level. *Nutrients* **2022**, *14*, 2432. https://doi.org/10.3390/nu14122432

Academic Editor: Éric Goulet

Received: 13 May 2022
Accepted: 6 June 2022
Published: 11 June 2022

1. Introduction

Obesity, as a sub-health stage with the excessive accumulation of body fat, is mainly affected by genetics and lifestyles. Due to the long-term chronic inflammatory state and abnormal metabolism as the hotbed for metabolic, endocrine, cardiovascular, and nervous system diseases, obesity has become a public health problem that is a current research focus all over the world. Currently, nearly one-third of the world's population is overweight or obese [1]. According to the Epidemiological Survey in China from 2015–2019, the obesity rate in China reveals a significant increase, and children and adolescents have also become the hardest-hit areas for obesity [2]. Unhealthy diets such as high-fat and high-sugar diets may cause damage of brain tissues, thus resulting in the decline in cognitive and learning capacity [3]. Therefore, reducing high-fat diet (HFD)-induced obesity and alleviating the impaired brain function have become an urgent problem to be solved.

There are many ways to intervene obesity, including exercise, diets, drugs, and surgery. Previous studies have demonstrated that obesity-induced impairment of cognitive and memory function can be suppressed and rescued after exercise [4], ketogenic diet [5], and surgical interventions due to body weight loss [6]. Therefore, the gain of body weight may be the culprit. However, adolescents are in the rapid growth and development stage, and surgical, drug, and dietary interventions may adversely affect the body. Exercise is an economical and green intervention strategy for health promotion, physical fitness, and obesity management. Considering the impact of different exercise methods on obesity, swimming intervention that is more suitable for obesity was selected in this study. In addition, previous studies on molecular mechanisms have shown that the decline in cognitive function may be correlated with insulin resistance, neurotrophic factor deficiency, apoptosis, and reduced synaptic plasticity in obesity models, as well as the long-term chronic inflammation at the status of obesity [7]. Insulin resistance in the hippocampus is an important factor that leads to the decline of cognitive function. Insulin resistance may also indirectly result in the adverse impact on synaptic plasticity by reducing the brain-derived neurotrophic factor (BDNF), eventually leading to the decline of cognitive and memory function [8]. Many studies using HFD-induced obesity mouse models have also documented the increased inflammation, insulin resistance, and reduced BDNF level in the hippocampus [9–12]. Therefore, reduced inflammation, increased insulin sensitivity, up-regulated BDNF, and suppressed apoptosis of neurons may be the critical molecular mechanisms for ameliorating obesity-induced cognitive decline during the process of body weight loss.

It is well known that scientific and reasonable exercise can significantly promote the health of the body and reduce the occurrence of sports injuries. Relevant studies have shown that swimming can effectively improve learning and memory functions and increase hippocampal BDNF level [13]. Therefore, in this study, the relevant indicators associated with inflammation, insulin resistance, BDNF level, and apoptosis in hippocampal tissues were detected to uncover the molecular mechanisms for alleviating obesity-induced cognitive impairment upon a swimming intervention in HFD-induced obese mice.

2. Materials and Methods

2.1. Animals, Experimental Design, and Ethics

Thirty 4-week-old specific pathogen-free (SPF) grade C57BL/6 male mice (body weight: 19.9 ± 1.4 g) were purchased from the Experimental Animal Center of Hubei Provincial Center for Disease Control and Prevention (Certificate No. 42010200005338). The mice were randomly divided into the regular diet group (10 mice) as the normal control (NC) group and the HFD group (20 mice) to establish the obesity model. The obese mice were then divided into two subgroups, which were the obesity-control (OC) group without swimming and the obesity-exercise (OE) group with loading-free swimming for 8 weeks (Figure 1A). The mice involved in this study were kept under the conditions with a standard light and dark cycle (12:00–12:00) and temperature (22 ± 2 °C). The regular diet had the energy ratio of 63.4% from carbohydrates, 22.8% from protein, and 13.8% from fat, and the high-fat diet had the energy ratio of 60% from fat, 20% from protein, and 20% from carbohydrates.

2.2. Swimming Training Protocol

The obese mice were subjected to non-intermittent swimming training in a swimming tank (30 ± 1 °C) without weight bearing for 8 weeks, five times a week and 60 min each time. In order to relieve the stress from swimming, the mice were provided with incremental exercise time for adaptation from 15, 20, 30, 45, and 60 min in the first week. The swimming time was set up in the early morning in accordance with the mouse biological clock (18:00–20:00). During the exercise intervention in this study, all mice were abstained from other additional physical stimulations.

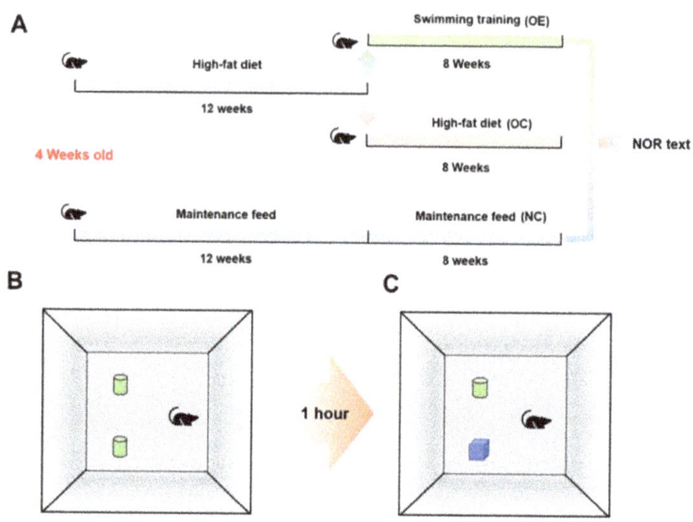

Figure 1. The diagram for describing animal grouping with corresponding feeding and exercise intervention, as well as new object recognition test. (**A**) Flow chart of mouse grouping with corresponding feeding and exercise intervention; (**B**) A 10-min test environment adaptation; (**C**) learning and memory capacity test.

2.3. New Object Recognition Test

After the swimming intervention for 8 weeks, all mice were subjected to a new object recognition test for evaluating the learning and memory function. The mice to be tested were placed in the test environment for 24 h in advance for environmental adaptation, and the training began to place two identical objects on the same side. The mouse was placed in the field with its back facing the two same objects, and the heads of the mice were at the same distance from the two objects. After the mice were put into the test environment for 10 min to adapt to the test environment, the mice were immediately put back into the rearing box, and after 1 h of rest, one of the objects was replaced with another object with a different color and shape to conduct new object recognition tests within 5 min, as shown in the general process (Figure 1B,C).

2.4. Histological Examination of Hippocampal Tissues

After the novel object recognition test, 3 mice in each group were sequentially perfused with 0.9% saline and 4% paraformaldehyde (pH 7.4) under anesthesia. After perfusion, the mice were sacrificed by the dislocation of cervical vertebrae, and the brain was removed and placed in 4% paraformaldehyde at 4 °C overnight for paraffin embedding and staining. The paraffin block of mouse brain tissue in each group was cut coronally into sections with the thickness of 5 μm using a microtome, and the brain tissue sections were then subjected to de-paraffinization and rehydration, Nissl staining, dehydration sealing, and microscopic examination. HE staining and antigen retrieval were performed after de-paraffinization and rehydration, endogenous peroxidase blocking, primary antibody incubation at 4 °C overnight, secondary antibody incubation at room temperature, color development, hematoxylin staining, dehydration, clearing, mounting, and microscopic evaluation. In immunofluorescence, the primary antibody and corresponding secondary antibody, CY3-TSA, were added for the probing. Nuclei were counterstained with DAPI. The images were acquired by an imaging system (Eclipse-E100, Nikon, Japan) and fluorescence was analyzed using ImageJ software (NIH, Bethesda, MD, USA).

2.5. Western Blot

The hippocampal tissues of the mice were collected on a low-temperature plate and placed in liquid nitrogen immediately and then transferred to a −80 °C refrigerator for storage and future use. Hippocampal tissue samples were subjected to pre-cooled cell lysis in the presence of protease inhibitors and to homogenization in ice for 30 min. The supernatant was harvested by centrifuging at 10,000× *g* for 5 min at 4 °C. After protein concentration of the supernatant was detected by the BCA method, the aliquots of the supernatant were mixed with 2× loading buffer and then subjected to a metal-plate bath at 95 °C for 5 min to denature proteins. Approximately 25 µg of total protein in the prepared samples were separated using 10% or 12% sodium dodecyl sulfate-polyacrylamide gel electrophoresis (SDS-PAGE) and then transferred to a polyvinylidene fluoride (PVDF) membrane. The target protein in the membrane was probed by a specific primary antibody against IL-6, TNF-α and BDNF (GeneTex, Irvine, CA, USA), p-IRS-1^{ser307}, IRS-1, p-Akt^{ser473}, Akt, p-PI3K, PI3K, JNK, p-JNK, PSD95, and β-actin (Cell Signaling Technology, Danvers, MA, USA), as well as the corresponding secondary antibody (Cell Signaling Technology, Danvers, MA, USA). Protein bands were visualized by using an enhanced chemiluminescence (ECL) reagent and imaged using an ultra-sensitive fluorescence/chemiluminescence imaging system ChemiScope6300 (CLiNX Science Instruments, Shanghai, China).

2.6. Statistical Analysis

All data were expressed as mean ± standard deviation (M ± SD). Statistical analysis was conducted by GraphPad Prism software (La Jolla, CA, USA). A one-way analysis of variance (ANOVA) was used to analyze statistically significant differences between multiple groups for parametric data. Otherwise, a nonparametric Kruskal-Wallis analysis was performed. The significant difference was considered at $p < 0.05$.

3. Results

3.1. Swimming Reduced Lee's Index and Body Weight of HFD-Induced Obese Mice

In order to examine the effect of HFD on physical development and body weight, we regularly measured the body weights and the trunk length from the nose to the anus of the mice in each group once every two weeks to calculate Lee's index (Figure 2A). Compared with the normal control (NC) group, the mice from the HFD group revealed extremely significant differences in Lee's index ($p < 0.001$) from 6 to 12 weeks and extremely significant differences in body weight from 4 to 20 weeks ($p < 0.001$), indicating that HFD can rapidly promote the body development of the mice. After obesity modeling for 12 weeks, the obese mice were divided into the obesity-exercise (OE) group and the obesity-control (OC) group. Compared with the OC group, the body weights of the mice from the OE group showed a significant decrease after a swimming intervention for 8 weeks ($p < 0.05$) (Figure 2B). Therefore, 12 weeks of HFD administration significantly induced the development of obesity in young mice, while the 8-week swimming intervention reduced the body weights of the obese mice.

3.2. Swimming Intervention Enhanced Learning and Memory Capacity of Obese Mice

Usually, learning and memory functions reveal the rapid development during the adolescent period, and the long-term consumption of high-fat diets has adverse effects on the brain and nervous system [14]. Therefore, we used a novel object recognition test to evaluate the changes in learning and memory capacity of obese mice (Figure 2A,B). Compared with the NC group, the recognition capability to novel objects of the mice from the OC and OE groups was significantly lower, indicating that a high-fat diet could suppress the learning and memory capacity of the mice ($p < 0.05$). In contrast, after a swimming intervention for 8 weeks, the learning and memory capacity of the mice from the OE group was significantly higher than that in the OC group ($p < 0.05$) (Figure 2C). These results suggest that the long-term consumption of a high-fat diet during the adolescent period can lead to the impaired learning and memory capacity of the mice, while a swimming intervention can be beneficial to the improvement or rescuing of the declined learning and memory capacity of the obese mice.

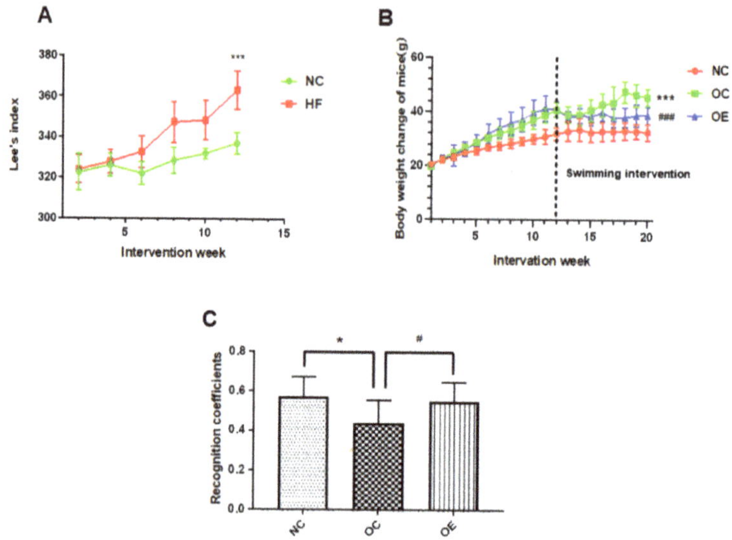

Figure 2. Effect of a swimming intervention on body weights of the mice after 12 weeks of normal chow and high-fat feeding for 8 weeks, as well as the assessment of the learning and memory capacity for the mice through a novel object recognition test. All data were presented as mean ± standard deviation (M ± SD) ($n = 7$ mice per group). (**A**) statistical analysis for the changes in Lee's index of the mice within the 12-week feeding for establishment of the obesity model; (**B**) statistical analysis of the changes in body weights of the mice in NC, OC, and OE groups upon swimming intervention; (**C**) statistical analysis of recognition coefficients of the mice from the NC, OC, and OE groups, and the recognition coefficient is equal to the times of recognizing new objects to the total times of recognizing old and new objects. Compared with the NC group, * $p < 0.05$, *** $p < 0.001$; compared with the OE group, # $p < 0.05$, ### $p < 0.001$.

3.3. *Swimming Suppressed Hippocampal Neuronal Degeneration of Obese Mice*

Long-term consumption of a high-fat diet may lead to the damage and functional decline of hippocampal neurons [15]. In order to further understand the effect of obesity on hippocampal neurons, we conducted HE, Nissl, and NeuN staining to evaluate the morphological and pathological changes of the hippocampal neurons of the obese mice upon the swimming intervention (Figure 3A–C). Compared with the NC group, the neuronal damage, disordered and sparse neuron arrangement, and decreased number of mature neurons in the mice from the OC group were observed. In contrast, the swimming intervention rescued these obesity-induced impairments or abnormal changes of hippocampal neurons, or it suppressed the hippocampal neuronal degeneration of the obese mice, as shown in the deep and compact neuronal borders and nuclei with Nissl staining (Figure 3B) and more mature neurons under the same background intensity (Figure 3D). These results suggest that obesity can impair the maturation process and the corresponding functions of neurons, while a swimming intervention can ameliorate hippocampal neuronal damage and stimulate the maturation of neurons.

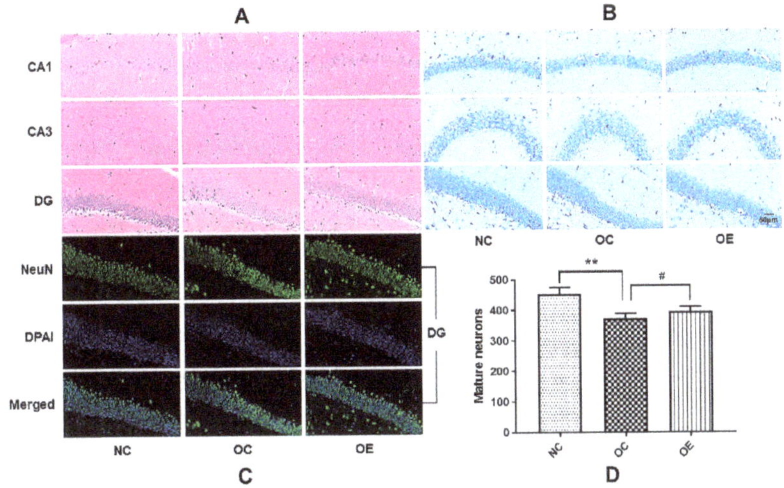

Figure 3. Representative images for the HE, Nissl, and NeuN staining of hippocampal tissues. All data were presented as mean ± standard deviation (M ± SD) (n = 3 mice per group). (**A**) HE staining of the CA1, CA3, and DG hippocampal regions of the mice from the NC, OC, and OE groups; (**B**) Nissl staining of the CA1, CA3, and DG hippocampal regions of the mice from the NC, OC, and OE groups; (**C**) NeuN staining of the DG hippocampal region of the mice from the NC, OC, and OE groups. All images were acquired under a 400× optical microscope. (**D**) Statistical analysis of mature neurons in the DG regions of hippocampal tissues from the NC, OC, and OE groups. Compared with the NC group, ** $p < 0.01$; compared with the OE group, # $p < 0.05$.

3.4. Swimming Suppressed Hippocampal Neuroinflammation of Obese Mice

Obesity can trigger chronic inflammation, and same changes can be observed in the nervous system [16]. In order to determine the level of inflammation in hippocampal tissue under the status of obesity, we evaluated inflammation-related proteins by Western blot (Figure 4A). Compared with the NC group, the expression of IL-6 and TNF-α in hippocampal tissues of the mice from the OC group showed a significant increase ($p < 0.05$); on the contrary, swimming intervention reversed the obesity-induced increase of IL-6 and TNF-α ($p < 0.05$; $p < 0.01$) (Figure 4B,C). The immunofluorescence results showed that the obviously higher expression of NF-κB p65 in the hippocampal tissues of mice from the OC and OE groups was observed when compared with the NC group; in contrast, swimming intervention down-regulated the expression of NF-κB p65 in hippocampal tissues in comparison with the OC group (Figure 4D), indicating that the swimming intervention has an obviously inhibitory effect on neuroinflammation, thereby executing the suppression of the neuroinflammation-induced reduction of learning and memory capacity.

3.5. Swimming Activated Hippocampal Insulin Signaling in Obese Mice

Long-term obesity may lead to inflammation in the body, thus inducing insulin resistance and impairing learning and memory functions [17]. To further explore the changes in inflammation and insulin-related signaling in hippocampal tissues of obese mice, Western blotting was used to evaluate the expression of the proteins associated with inflammation and the insulin signal pathway (Figure 5A). Compared with the NC group, p-JNK/JNK and p-IRS-1^{ser307}/IRS-1 ratios in hippocampal tissues of the mice from the OC group revealed a significant increase ($p < 0.05$); in contrast, p-PI3K/PI3K and p-Akt^{ser473}/Akt ratios exhibited an obvious decrease ($p < 0.05$), suggesting that a high-fat diet contributes to inflammation-induced insulin resistance. On the other hand, the swimming intervention suppressed the increase of p-JNK/JNK and p-IRS-1^{ser307}/IRS-1 ratios ($p < 0.05$) and the reduction of p-PI3K/PI3K and p-Akt^{ser473}/Akt ratios ($p < 0.05$, $p < 0.01$) (Figure 5B–D), thereby stimulating the activation of the insulin signal pathway. These results suggest that elevated levels of inflammation induced by long-term obesity can lead to insulin resistance in the hippocampal tissue, while a swimming intervention can attenuate inflammation-induced insulin resistance.

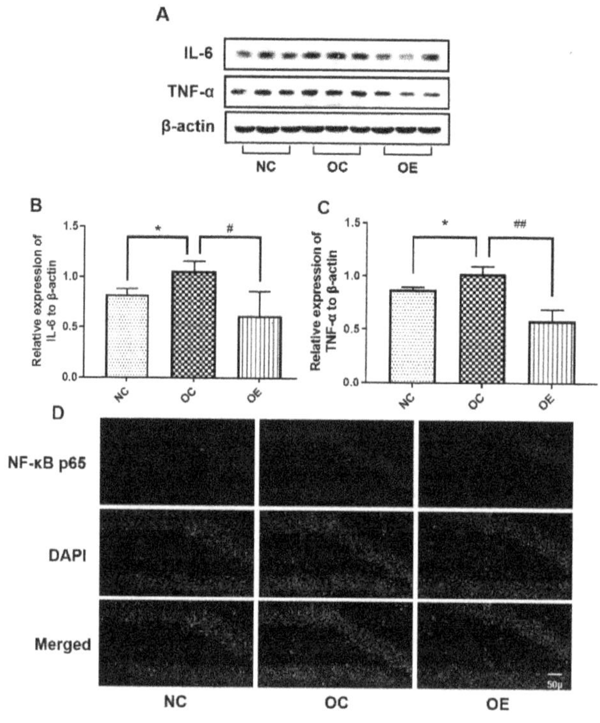

Figure 4. The Western blot of proteins associated with neuroinflammation (**A**) and corresponding statistical analysis of IL-6 (**B**) and TNF-α (**C**) expression levels, as well as the immunofluorescence of NF-κB p65 (**D**) in hippocampal tissues. All data were presented as mean ± standard deviation (M ± SD) ($n = 3$ mice per group). Immunofluorescence images were acquired under a 400× optical microscope. Compared with the NC group, * $p < 0.05$; compared with the OE group, # $p < 0.05$, ## $p < 0.01$.

3.6. Swimming Up-Regulated The Proteins Associated with Neurotrophic Factors and Synaptic Plasticity in the Hippocampal Tissue of Obese Mice

The reduction of neurotrophic factors and synaptic plasticity in hippocampal tissues of the mice with obesity may be the inducers for poor learning and memory capacity [18]. To understand the effects of neurotrophic factors on the regeneration of hippocampal neurons in an obese state, we conducted the evaluation of the corresponding protein expression associated with neurotrophic factors and synaptic plasticity in the hippocampal tissue of the mice through Western blot. The expression of PGC-1α, BDNF, and PSD95 in the hippocampal tissues of the mice from the OC group revealed a significant decrease when compared with the NC group (Figure 6A–D) ($p < 0.05$, $p < 0.01$), while a swimming intervention could rescue the down-regulation of these proteins ($p < 0.05$, $p < 0.01$). Therefore, a swimming intervention is beneficial to the rescuing of down-regulated PGC-1α, BDNF, and PSD95 in hippocampal tissues of the mice caused by the long-term consumption of a high-fat diet, thereby enhancing neurotrophic factors and synaptic plasticity.

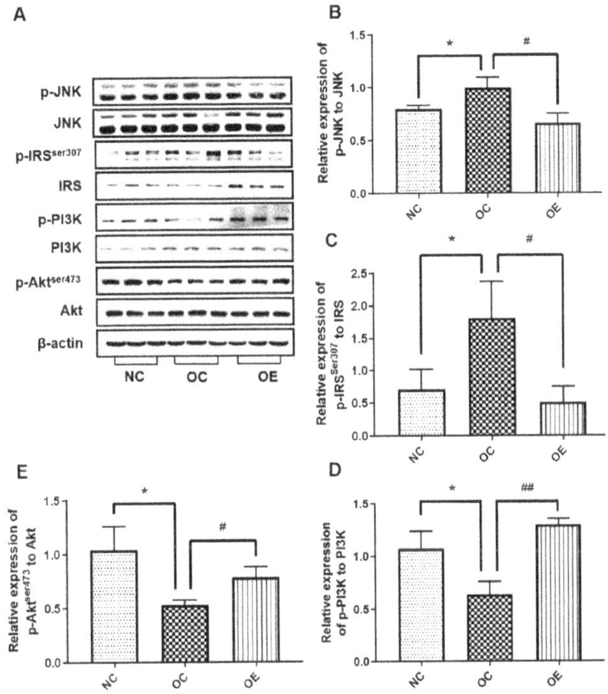

Figure 5. The Western blot of proteins associated with insulin signal pathway (**A**) and corresponding statistical analysis of p-JNK/JNK (**B**), p-IRS-1^{ser307}/IRS-1 (**C**), p-PI3K p85/PI3K (**D**), and p-Akt^{ser473}/Akt (**E**) ratios in the hippocampal tissues. All data were presented as mean ± standard deviation (M ± SD) (n = 3 mice per group). Compared with the NC group, * $p < 0.05$; compared with the OE group, # $p < 0.05$, ## $p < 0.01$.

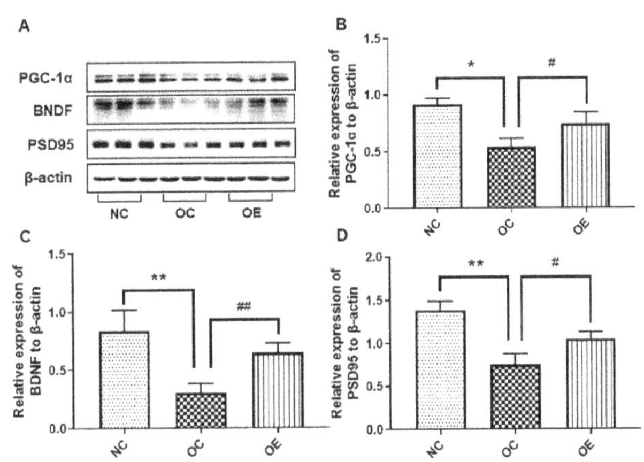

Figure 6. The Western blot of proteins associated with neurotrophic factors and synaptic plasticity (**A**) and corresponding statistical analysis of PGC-1α (**B**), BDNF (**C**), and PSD95 (**D**) expression levels in hippocampal tissues of the mice. All data were presented as mean ± standard deviation (M ± SD) (n = 3 mice per group). Compared with the NC group, * $p < 0.05$, ** $p < 0.01$; compared with the OE group, # $p < 0.05$, ## $p < 0.01$.

3.7. Swimming Inhibited Hippocampal Neuronal Apoptosis in Obese Mice

A high-fat diet can induce the oxidative stress for stimulating apoptosis of the hippocampal neurons in mice, thereby triggering a reduced learning and memory capacity [19]. To further explore the effect of a long-term high-fat diet on apoptosis of the hippocampal neurons in obese mice, we examined the expression of anti-apoptotic and apoptotic proteins. Compared with the NC group, a significant decrease in Bcl-2 and an increase in Bax in hippocampal tissues of the mice from the OC group were observed ($p < 0.05$); on the contrary, a swimming intervention could up-regulate Bcl-2 and down-regulate Bax to execute the suppression of hippocampal neuronal apoptosis (Figure 7A–C). Although Bax did not appear to be significantly different between the OC and OE groups, the expression of Bax in the OE group showed a downward trend when compared to the OC group, and it could significantly reverse the Bcl-2/Bax ratio to suppress apoptosis of the hippocampal neurons in obese mice (Figure 7D) ($p < 0.05$). These results suggest that the long-term consumption of a high-fat diet may trigger apoptosis of the hippocampal neurons, and swimming can rescue this phenomenon to some extent.

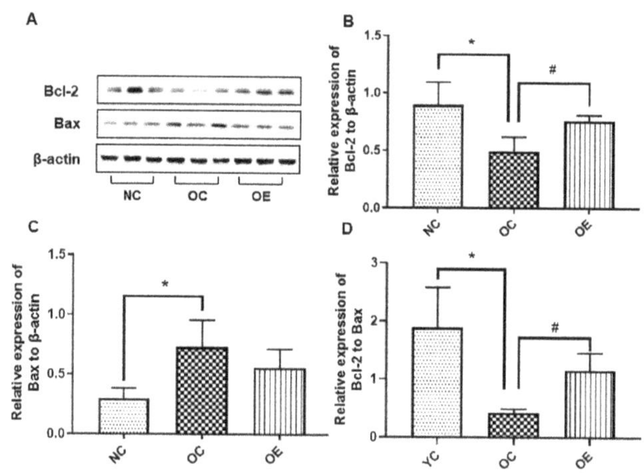

Figure 7. The Western blot of proteins associated with apoptosis (**A**) and corresponding statistical analysis of Bcl-2 (**B**) and Bax (**C**) expression levels, as well as Bcl-2/Bax (**D**) in hippocampal tissues of the mice. All data were presented as mean ± standard deviation (M ± SD) ($n = 3$ mice per group). Compared with the NC group, * $p < 0.05$; compared with the OE group, # $p < 0.05$.

4. Discussion

Obesity can lead to the impairment of learning and memory capacity, which is confirmed by increasing studies. However, the molecular mechanisms for the impairment of learning and memory capacity caused by obesity are unclear. The 1–6-months-old C5BL/6 mice are equivalent to 12–30-year-old humans during the rapid development period to adult period [20]. Therefore, this study highly mimicked the development of the adolescent population with obesity using 4-week-old young male mouse models fed with a high-fat diet to obesity for 3 months, which can avoid the uncertain factors such as the physiological cycle of female mice as soon as possible. Exercise is an important external means to promote the development of brain function accompanied by the suppression of obesity [21]. However, considering the risk of exercise injury caused by larger body weight, this study adopted a swimming intervention for 2 months. The experimental results were consistent with the expectation. Swimming intervention significantly reduced body weight, alleviated neuroinflammation and insulin resistance, and up-regulated neurotrophic factors to achieve the reversal of learning and memory function in obese mice.

The occurrence of obesity may lead to a significant decline in memory, but some studies have found that the decline in cognitive function caused by a high-fat diet may precede the occurrence of obesity [22]. Relevant studies have shown that short-term [23] and long-term exposure [15] to high-fat diets can result in the obesity-induced impairment of learning and memory capacity. Only 4 days of high-fat and high-sugar diets in humans can impair hippocampal-dependent learning and memory function to some extents [24]. The 7–9-year-old adolescents with the long-term intake of saturated fatty acids present a negative correlation with learning and memory function [25], as confirmed by the same performance in Zebrafish [26], which stimulates the interest in exploring the beneficial effect of regular exercise intervention on the mitigation of the HFD-induced impairment of learning and memory function and precision mechanisms through the mouse model, thereby providing a reference or guidance for health promotion. However, in 12-month-term HFD feeding, rats showed better learning and memory capacity and larger hippocampal volume [27], and the consumption of HFD for 6 months showed no spatial memory impairment [28]. This phenomenon, that a long-term high-fat diet does not trigger the changes in learning and memory capacity, may be related to stress, self-resistance, and the adaptation of neurons to high-fat diets, but the specific mechanisms need to be further explored. Similarly, the level of inflammation from high-fat diets can be altered with a change in the duration of diet interventions [29]. Although there is disagreement about the relationship between the duration of high-fat diet intervention and the impairment of learning and memory capacity, it has been widely reported and recognized that obesity can impair learning and memory capacity.

From the point of view of molecular mechanisms, the impaired learning and memory capacity may be related to the increased level of hippocampal inflammation caused by high-fat diets [8]. It has been widely proven that obesity can increase inflammation in the body, with a similar effect on brain tissue [30]. A cross-sectional study with 10,000 persons in the USA has demonstrated that the level of inflammation is negatively correlated with memory, and it also has the reference value in complex environments [31]. The exposure of juvenile rats to HFD-induced inflammation and reduced learning and cognitive function [32,33]. In a clinical trial with a large volume of samples involving more than 8000 adolescents and children, the results showed that the level of inflammation in obese adolescents is usually accompanied by poor learning and memory capacity [34]. The mice subjected to the consumption of short-term high-fat diets can result in the elevation of pro-inflammatory cytokines such as IL-6 and TNF-α in hippocampal tissues [35]. Once JNK is activated, it can act on NF-κB to enter the nucleus to promote the transcription of inflammatory genes, thus resulting in the release of more inflammatory factors, eventually forming a vicious circle and aggravating the degree of inflammation and impaired learning and cognitive capacity [36]. After reducing inflammation, the dietary intervention can fully or partially rescue the impaired cognition of obese mice. Therefore, reducing inflammation may be a solid inducer to delay or suppress the reduction of learning and memory function [37]. In the present study, the impaired learning and memory capacity of the obese mice was highly modulated by the expression levels of IL-6, TNF-α, NF-κB p65, and JNK in hippocampal tissues, and a regular swimming intervention can significantly down-regulate these inflammatory proteins. This is consistent with the significant reduction of inflammation levels in the hippocampal tissues of animal models subjected to treadmill running, swimming, or voluntary wheeling running interventions [38–40].

Inflammation caused by obesity may be an important factor associated with insulin resistance [41]. Previous studies have also documented that adolescent mice fed with high-fat diets present an increased level of inflammation in hippocampal tissues, and insulin signaling is also significantly blocked [42,43]. As a bridge between inflammation and insulin, inflammatory factors promote the phosphorylation of IRS-1 at the Ser307 site by activating JNK phosphorylation, thereby hindering insulin signaling and exacerbating insulin resistance. Long-term chronic inflammation can activate JNK and eventually lead to the occurrence of insulin resistance [44]. Insulin resistance in the hippocampal tissue is considered one of the important triggers for the decline in learning and memory function; therefore, activating the insulin signaling pathway can rescue impaired learning and memory capacity [45]. In our study, the similar results with increased expression levels of the proteins associated with the signal pathways of insulin resistance such as JNK/IRS-1/PI3K/Akt in the hippocampal tissues of the obese mice were observed, and the swimming intervention rescued the abnormal expression of these proteins (Figure 5), further suggesting impaired learning and memory capacity due to the insulin resistance from high-fat diets, and alleviated the obesity state and recovered insulin resistance for enhancing learning and memory capacity upon regular exercise intervention.

In addition, neurotrophic factors and synaptic plasticity are closely related to improving learning and memory functions. In the present study, the expression levels of PGC-1α, BDNF, and PSD95 showed a downward trend in the hippocampal tissues of obese mice, as consistency with the literature reports describing the alleviation of learning and memory impairment caused by soybean oil-induced BDNF reduction [39]. As an important neurotrophic factor, BDNF can play a critical role in the release and reception of neurotransmitters in presynaptic and postsynaptic membranes, thus promoting the connection between synapses and even the regeneration of nerves, thereby improving learning and memory functions. As an intracellular protein, PSD95 plays a vital role in neuronal synaptic plasticity and learning and memory functions [46]. A previous study has also found that exercise can inhibit the reduction of BNDF in the hippocampal tissue of an obesity model [47]. The feeding of high-fat diets for 3-month-old mice can trigger the reduction of the PSD95 level [48], and both aerobic exercise and resistance exercise training can up-regulate hippocampal PSD95 expression [49,50]. Therefore, a swimming intervention for enhancing the learning and memory capacity of obese adolescents may partly depend on the activation of the PGC-1α/BDNF signal pathway.

The long-term consumption of high-fat diets may be detrimental to the survival of hippocampal neurons, and corresponding studies have shown that exercise interventions can promote the improvement of hippocampal function by suppressing the apoptosis of hippocampal neurons, thereby enhancing learning and memory capacity [19]. Exercise can inhibit the apoptosis of hippocampal neurons in obese offspring [51], which is similar to our experimental results with the up-regulated anti-apoptotic protein Bcl-2, down-regulated pro-apoptotic protein Bax, and increased Bcl-2/Bax ratio (Figure 6), as well as the reduced expression of the proteins in hippocampal tissues associated with inflammation and insulin resistance upon the swimming intervention (Figures 4 and 5). However, endoplasmic reticulum stress, mitochondrial dysfunction, elevated ROS level caused by high-fat diets, gut microbiome change, and combinatorial stress responses under different dietary conditions may also be important triggers for neuronal apoptosis [33,51–55]. Consistently, aerobic exercise plays a positive role in reducing hippocampal neuronal apoptosis.

As is well known, dietary modification is another important way to mitigate obesity and enhance learning and cognitive capacity. In order to better understand the effect of exercise intervention on regulating the progression of obesity, in our study, the benefits of exercise intervention were significantly highlighted. Swimming intervention can alleviate inflammatory responses, promote insulin sensitivity, up-regulate neurotrophic factors, increase synaptic plasticity, and suppress apoptosis (Figures 4B,C and 5A), as well as rescue insulin resistance signaling (Figure 5B–D), suggesting that better nutritional supplementation combined with regular exercise training during the rapid development stage of the body may have more benefits, including the rescuing of the high-fat diets-induced impairment of learning and memory capacity.

5. Conclusions

Adolescents require more nutrients and energy, but the consumption of long-term high-fat diets can also lead to body weight gain and impaired learning and memory capacity, which is highly correlated with neuroinflammation, insulin resistance, reduced neurotrophic factors, and increased neuronal apoptosis. Swimming intervention can reverse these abnormal changes to rescue the impaired learning and memory capacity in obese mice by reducing obesity, alleviating hippocampal neuroinflammation, activating insulin signaling, promoting the generation and secretion of neurotrophic factors, and suppressing hippocampal neuronal apoptosis in adolescent mice, with the involvement of the JNK/IRS-1/PI3K/Akt and PGC-1α/BDNF signal pathways (Figure 8).

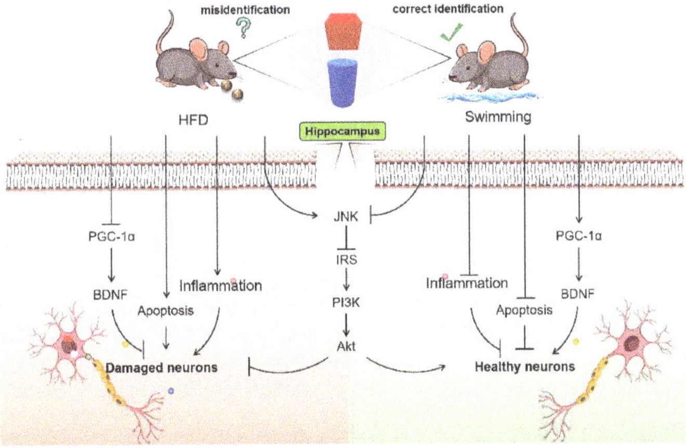

Figure 8. Swimming rescued the impaired cognitive capacity in adolescent mice caused by high-fat diets by suppressing inflammation, alleviating insulin resistance, up-regulating BDNF, and inhibiting apoptosis.

Author Contributions: H.Z. and N.C. designed the experiments; H.Z. and J.-L.L. conducted experiments, data analysis, and manuscript writing; Q.-Y.W., J.-X.L., J.-L.L., Y.L. and L.-W.W. conducted animal management, feeding, and interventions; J.-L.H., X.-W.W. and M.-H.W. participated in experimental data collection and analysis; N.C. completed final editing and revision of the manuscript. All authors have read and agreed to the published version of the manuscript.

Funding: This work was supported by the National Natural Science Foundation of China (No. 31771318); the Key Special Project of Disciplinary Development, Hubei Superior Discipline Group of Physical Education and Health Promotion; and the 14th Five-Year-Plan Advantageous and Characteristic Disciplines (Groups) of Colleges and Universities in Hubei Province for Exercise and Brain Science; as well as the Chutian Scholar Program and Innovative Start-up Foundation from Wuhan Sports University to Ning Chen.

Institutional Review Board Statement: This experimental followed China's animal welfare regulations strictly, and the animal experiment operators were certified by China's laboratory animal breeding and technical professional qualifications. All procedures performed in this study involving animals were in accordance with the ethical standards of the Animal Care and Use Committee at Wuhan Sports University and approved by this committee (Code No. S087-202105-05).

Informed Consent Statement: Not applicable.

Data Availability Statement: The data are available from the corresponding author.

Conflicts of Interest: The authors declare no conflict of interest.

References

1. Chooi, Y.C.; Ding, C.; Magkos, F. The epidemiology of obesity. *Metabolism* **2019**, *92*, 6–10. [CrossRef]
2. Pan, X.F.; Wang, L.; Pan, A. Epidemiology and determinants of obesity in China. *Lancet Diabetes Endocrinol.* **2021**, *9*, 373–392. [CrossRef]
3. Reichelt, A.C.; Rank, M.M. The impact of junk foods on the adolescent brain. *Birth Defects Res.* **2017**, *109*, 1649–1658. [CrossRef]
4. Tran, D.M.D.; Westbrook, R.F. A high-fat high-sugar diet-induced impairment in place-recognition memory is reversible and training-dependent. *Appetite* **2017**, *110*, 61–71. [CrossRef]
5. Mohorko, N.; Černelič-Bizjak, M.; Poklar-Vatovec, T.; Grom, G.; Kenig, S.; Petelin, A.; Jenko-Pražnikar, Z. Weight loss, improved physical performance, cognitive function, eating behavior, and metabolic profile in a 12-week ketogenic diet in obese adults. *Nutr. Res.* **2019**, *62*, 64–77. [CrossRef]
6. Gunstad, J.; Strain, G.; Devlin, M.J.; Wing, R.; Cohen, R.A.; Paul, R.H.; Crosby, R.D.; Mitchell, J.E. Improved memory function 12 weeks after bariatric surgery. *Surg. Obes. Relat. Dis.* **2011**, *7*, 465–472. [CrossRef]

7. Crispino, M.; Trinchese, G.; Penna, E.; Cimmino, F.; Catapano, A.; Villano, I.; Perrone-Capano, C.; Mollica, M.P. Interplay between peripheral and central inflammation in obesity-promoted disorders: The impact on synaptic mitochondrial functions. *Int. J. Mol. Sci.* **2020**, *21*, 5964. [CrossRef]
8. Mi, Y.; Qi, G.; Fan, R.; Qiao, Q.; Sun, Y.; Gao, Y.; Liu, X. EGCG ameliorates high-fat- and high-fructose-induced cognitive defects by regulating the IRS/AKT and ERK/CREB/BDNF signaling pathways in the CNS. *FASEB J.* **2017**, *31*, 4998–5011. [CrossRef]
9. Rivera, P.; Martos-Moreno, G.; Barrios, V.; Suárez, J.; Pavón, F.J.; Chowen, J.A.; de Fonseca, F.R.; Argente, J. A combination of circulating chemokines as biomarkers of obesity-induced insulin resistance at puberty. *Pediatr. Obes.* **2021**, *16*, e12711. [CrossRef]
10. Mulati, A.; Ma, S.; Zhang, H.; Ren, B.; Zhao, B.; Wang, L.; Liu, X.; Zhao, T.; Kamanova, S.; Sair, A.T.; et al. Sea-buckthorn flavonoids alleviate high-fat and high-fructose diet-induced cognitive impairment by inhibiting insulin resistance and neuroinflammation. *J. Agric. Food Chem.* **2020**, *68*, 5835–5846. [CrossRef]
11. Zhuang, J.; Lu, J.; Wang, X.; Wang, X.; Hu, W.; Hong, F.; Zhao, X.X.; Zheng, Y.L. Purple sweet potato color protects against high-fat diet-induced cognitive deficits through AMPK-mediated autophagy in mouse hippocampus. *J. Nutr. Biochem.* **2019**, *65*, 35–45. [CrossRef] [PubMed]
12. Liu, Z.; Patil, I.Y.; Jiang, T.; Sancheti, H.; Walsh, J.P.; Stiles, B.L.; Yin, F.; Cadenas, E. High-fat diet induces hepatic insulin resistance and impairment of synaptic plasticity. *PLoS ONE* **2015**, *10*, e0128274.
13. Khabour, O.F.; Alzoubi, K.H.; Alomari, M.A.; Alzubi, M.A. Changes in spatial memory and BDNF expression to simultaneous dietary restriction and forced exercise. *Brain Res. Bull.* **2013**, *90*, 19–24. [CrossRef] [PubMed]
14. Tsan, L.; Décarie-Spain, L.; Noble, E.E.; Kanoski, S.E. Westerndiet consumption during development: Setting thestage for neurocognitive dysfunction. *Front. Neurosci.* **2021**, *15*, 632312. [CrossRef]
15. Zhuang, H.; Yao, X.; Li, H.; Li, Q.; Yang, C.; Wang, C.; Xu, D.; Xiao, Y.; Gao, Y.; Gao, J.; et al. Long-term high-fat diet consumption by mice throughout adulthood induces neurobehavioral alterations and hippocampal neuronal remodeling accompanied by augmented microglial lipid accumulation. *Brain Behav. Immun.* **2022**, *100*, 155–171. [CrossRef]
16. Rohm, T.V.; Meier, D.T.; Olefsky, J.M.; Donath, M.Y. Inflammation in obesity, diabetes, and related disorders. *Immunity* **2022**, *55*, 31–55. [CrossRef]
17. Jeon, B.T.; Jeong, E.A.; Shin, H.J.; Lee, Y.; Lee, D.H.; Kim, H.J.; Kang, S.S.; Cho, G.J.; Choi, W.S.; Roh, G.S. Resveratrol attenuates obesity-associated peripheral and central inflammation and improves memory deficit in mice fed a high-fat diet. *Diabetes* **2012**, *61*, 1444–1454. [CrossRef]
18. Fusco, S.; Spinelli, M.; Cocco, S.; Ripoli, C.; Mastrodonato, A.; Natale, F.; Rinaudo, M.; Livrizzi, G.; Grassi, C. Maternal insulin resistance multigenerationally impairs synaptic plasticity and memory via gametic mechanisms. *Nat. Commun.* **2019**, *10*, 4799. [CrossRef]
19. Park, H.S.; Cho, H.; Skim, T.W. Physical exercise promotes memory capability by enhancing hippocampal mitochondrial functions and inhibiting apoptosis in obesity-induced insulin resistance by high fat diet. *Metab. Brain Dis.* **2018**, *33*, 283–292. [CrossRef]
20. Kevin, F.; Joanne, M.C.; Harrison, D.E. The mouse in biomedical research. In *American College of Laboratory Animal Medicine, American*, 2nd ed.; Academic Press: Cambridge, MA, USA, 2006; pp. 637–672.
21. Mäkinen, E.; Lensu, S.; Honkanen, M.; Laitinen, P.; Wikgren, J.; Koch, L.G.; Britton, S.L.; Kainulainen, H.; Pekkala, S.; Nokia, M.S. Rats bred for low intrinsic aerobic exercise capacity link obesity with brain inflammation and reduced structural plasticity of the hippocampus. *Brain Behav. Immun.* **2021**, *97*, 250–259. [CrossRef]
22. Davidson, T.L.; Hargrave, S.L.; Swithers, S.E.; Sample, C.H.; Fu, X.; Kinzig, K.P.; Zheng, W. Inter-relationships among diet, obesity and hippocampal-dependent cognitive function. *Neuroscience* **2013**, *253*, 110–122. [CrossRef] [PubMed]
23. Wang, Z.; Ge, Q.; Wu, Y.; Zhang, J.; Gu, Q.; Han, J. Impairment of long-term memory by a short-term high-fat diet via hippocampal oxidative stress and alterations in synaptic plasticity. *Neuroscience* **2020**, *424*, 24–33. [CrossRef] [PubMed]
24. Attuquayefio, T.; Stevenson, R.J.; Oaten, M.J.; Francis, H.M. A four-day Western-style dietary intervention causes reductions in hippocampal-dependent learning and memory and interoceptive sensitivity. *PLoS ONE* **2017**, *12*, e0172645. [CrossRef] [PubMed]
25. Baym, C.L.; Khan, N.A.; Monti, J.M.; Raine, L.B.; Drollette, E.S.; Moore, R.D.; Scudder, M.R.; Kramer, A.F.; Hillman, C.H.; Cohen, N.J. Dietary lipids are differentially associated with hippocampal-dependent relational memory in prepubescent children. *Am. J. Clin. Nutr.* **2014**, *99*, 1026–1032. [CrossRef] [PubMed]
26. Picolo, V.L.; Quadros, V.A.; Canzian, J.; Grisolia, C.K.; Goulart, J.T.; Pantoja, C.; de Bem, A.F.; Rosemberg, D.B. Short-term high-fat diet induces cognitive decline, aggression, and anxiety-like behavior in adult zebrafish. *Prog. Neuropsychopharmacol. Biol. Psychiatry* **2021**, *110*, 110288. [CrossRef]
27. Setkowicz, Z.; Gaździńska, A.; Osoba, J.J.; Karwowska, K.; Majka, P.; Orzeł, J.; Kossowski, B.; Bogorodzki, P.; Janeczko, K.; Wyleżoł, M.; et al. Does long-term high fat diet always lead to smaller hippocampi volumes, metabolite concentrations, and worse learning and memory? A magnetic resonance and behavioral study in wistar rats. *PLoS ONE* **2015**, *10*, e0139987. [CrossRef]
28. Abbasnejad, Z.; Nasseri, B.; Zardooz, H.; Ghasemi, R. Time-course study of high fat diet induced alterations in spatial memory, hippocampal JNK, P38, ERK and Akt activity. *Metab. Brain Dis.* **2019**, *34*, 659–673. [CrossRef]
29. Thaler, J.P.; Yi, C.X.; Schur, E.A.; Guyenet, S.J.; Hwang, B.H.; Dietrich, M.O.; Zhao, X.; Sarruf, D.A.; Izgur, V.; Maravilla, K.R.; et al. Obesity is associated with hypothalamic injury in rodents and humans. *J. Clin. Investig.* **2012**, *122*, 153–162. [CrossRef]
30. Guillemot-Legris, O.; Muccioli, G.G. Obesity-induced neuroinflammation: Beyond the hypothalamus. *Trends Neurosci.* **2017**, *40*, 237–253. [CrossRef]

31. Yang, Y.; Shields, G.S.; Wu, Q.; Liu, Y.; Chen, H.; Guo, C. The association between obesity and lower working memory is mediated by inflammation: Findings from a nationally representative dataset of U.S. adults. *Brain Behav. Immun.* **2020**, *84*, 173–179. [CrossRef]
32. Boitard, C.; Cavaroc, A.; Sauvant, J.; Aubert, A.; Castanon, N.; Layé, S.; Ferreira, G. Impairment of hippocampal-dependent memory induced by juvenile high-fat diet intake is associated with enhanced hippocampal inflammation in rats. *Brain Behav. Immun.* **2014**, *40*, 9–17. [CrossRef] [PubMed]
33. Nakandakari, S.; Muñoz, V.R.; Kuga, G.K.; Gaspar, R.C.; Sant'Ana, M.R.; Pavan, I.C.B.; da Silva, L.G.S.; Morelli, A.P.; Simabuco, F.M.; da Silva, A.S.R.; et al. Short-term high-fat diet modulates several inflammatory, ER stress, and apoptosis markers in the hippocampus of young mice. *Brain Behav. Immun.* **2019**, *79*, 284–293. [CrossRef] [PubMed]
34. Shields, G.S.; Deer, L.K.; Hastings, P.D.; Hostinar, C. Adiposity, inflammation, and working memory: Evidence for a vicious cycle. *Brain Behav. Immun.* **2021**, *13*, 100202. [CrossRef] [PubMed]
35. de Paula, G.C.; Brunetta, H.S.; Engel, D.F.; Gaspar, J.M.; Velloso, L.A.; Engblom, D.; de Oliveira, J.; de Bem, A.F. Hippocampal function is impaired by a short-term high-fat diet in mice: Increased blood-brain barrier permeability and neuroinflammation as triggering events. *Front. Neurosci.* **2021**, *15*, 734158. [CrossRef]
36. Nie, J.; Chang, Y.; Li, Y.; Zhou, Y.; Qin, J.; Sun, Z.; Li, H. Caffeic acid phenethyl ester (propolis extract) ameliorates insulin resistance by inhibiting JNK and NF-κB inflammatory pathways in diabetic mice and HepG2 cell models. *J. Agric. Food Chem.* **2017**, *65*, 9041–9053. [CrossRef]
37. Fan, R.; Hua, Y.; Shen, J.; Xiao, R.; Ma, W. Dietary fatty acids affect learning and memory ability via regulating inflammatory factors in obese mice. *J. Nutr. Biochem.* **2022**, *103*, 108959. [CrossRef]
38. Mokhtari-Zaer, A.; Hosseini, M.; Roshan, N.M.; Boskabady, M.H. Treadmill exercise ameliorates memory deficits and hippocampal inflammation in ovalbumin-sensitized juvenile rats. *Brain Res. Bull.* **2020**, *165*, 40–47. [CrossRef]
39. Cheng, M.; Cong, J.; Wu, Y.; Xie, J.; Wang, S.; Zhao, Y.; Zang, X. Chronic Swimming exercise ameliorates low-soybean-oil diet-induced spatial memory impairment by enhancing BDNF-mediated synaptic potentiation in developing spontaneously Hypertensive Rats. *Neurochem. Res.* **2018**, *43*, 1047–1057. [CrossRef]
40. Ko, Y.J.; Ko, I.G. Voluntary wheel running improves spatial learning memory by suppressing inflammation and apoptosis via inactivation of nuclear factor kappa B in brain inflammation rats. *Int. Neurourol. J.* **2020**, *24*, 96–103. [CrossRef]
41. Wu, H.; Ballantyne, C.M. Metabolic inflammation and insulin resistance in obesity. *Circ. Res.* **2020**, *126*, 1549–1564. [CrossRef]
42. Denver, P.; Gault, V.A.; McClean, P.L. Sustained high-fat diet modulates inflammation, insulin signalling and cognition in mice and a modified xenin peptide ameliorates neuropathology in a chronic high-fat model. *Diabetes Obes. Metab.* **2018**, *20*, 1166–1175. [CrossRef] [PubMed]
43. Mucellini, A.B.; Fonseca, N.; Manfro, G.G.; Silveira, P.P. Hippocampal insulin resistance and altered food decision-making as players on obesity risk. *Neurosci. Biobehav. Rev.* **2017**, *77*, 165–176. [CrossRef] [PubMed]
44. Feng, J.; Lu, S.; Ou, B.; Liu, Q.; Dai, J.; Ji, C.; Zhou, H.; Huang, H.; Ma, Y. The role of JNK signaling pathway in obesity-driven insulin resistance. *Diabetes Metab. Syndr. Obes.* **2020**, *13*, 1399–1406. [CrossRef] [PubMed]
45. Jeong, J.H.; Kang, E.B. Effects of treadmill exercise on PI3K/AKT/GSK-3β pathway and tau protein in high-fat diet-fed rats. *J. Exerc. Nutr. Biochem.* **2018**, *22*, 9–14. [CrossRef] [PubMed]
46. Zhou, M.; Liu, Z.; Yu, J.; Li, S.; Tang, M.; Zeng, L.; Wang, H.; Xie, H.; Peng, L.; Huang, H. Quantitative proteomic analysis reveals synaptic dysfunction in the amygdala of rats susceptible to chronic mild stress. *Neuroscience* **2018**, *376*, 24–39. [CrossRef] [PubMed]
47. Kim, T.W.; Baek, K.W.; Yu, H.S.; Ko, I.G.; Hwang, L.; Park, J.J. High-intensity exercise improves cognitive function and hippocampal brain-derived neurotrophic factor expression in obese mice maintained on high-fat diet. *J. Exerc. Rehabil.* **2020**, *16*, 124–131. [CrossRef]
48. Lizarbe, B.; Soares, A.F.; Larsson, S.; Duarte, J.M.N. Neurochemical modifications in the hippocampus, cortex and hypothalamus of mice exposed to long-term high-fat diet. *Front. Neurosci.* **2018**, *12*, 985. [CrossRef]
49. Cheng, J.; Chen, L.; Han, S.; Qin, L.; Chen, N.; Wan, Z. Treadmill running and rutin reverse high fat diet induced cognitive impairment in diet induced obese mice. *J. Nutr. Health Aging* **2016**, *20*, 503–508. [CrossRef]
50. Fernandes, J.; Soares, J.C.; do Amaral Baliego, L.G.; Arida, R.M. A single bout of resistance exercise improves memory consolidation and increases the expression of synaptic proteins in the hippocampus. *Hippocampus* **2016**, *26*, 1096–1103. [CrossRef]
51. Sikalidis, A.K.; Maykish, A. The gut microbiome and type 2 diabetes mellitus: Discussing a complex relationship. *Biomedicines* **2020**, *8*, 8. [CrossRef]
52. Kim, S.H.; Ko, I.G.; Jin, J.J.; Hwang, L.; Baek, S.S. Treadmill exercise ameliorates impairment of spatial learning memory in pups born to old and obese mother rats. *J. Exerc. Rehabil.* **2021**, *17*, 234–240. [CrossRef] [PubMed]
53. Shi, Z.; Li, C.; Yin, Y.; Yang, Z.; Xue, H.; Mu, N.; Wang, Y.; Liu, M.; Ma, H. Aerobic interval training regulated SIRT3 attenuates high-fat-diet-associated cognitive dysfunction. *Biomed. Res. Int.* **2018**, *2018*, 2708491. [CrossRef] [PubMed]
54. Pipatpiboon, N.; Pratchayasakul, W.; Chattipakorn, N.; Chattipakorn, S.C. PPARγ agonist improves neuronal insulin receptor function in hippocampus and brain mitochondria function in rats with insulin resistance induced by long term high-fat diets. *Endocrinology* **2012**, *153*, 329–338. [CrossRef] [PubMed]
55. Lee, J.I.; Dominy, J.J.; Sikalidis, A.K.; Hirschberger, L.L.; Wang, W.; Stipanuk, M.H. HepG2/C3A cells respond to cysteine deprivation by induction of the amino acid deprivation/integrated stress response pathway. *Physiol. Genom.* **2008**, *33*, 218–229. [CrossRef] [PubMed]

nutrients

MDPI

Article

Effects of Oats, Tartary Buckwheat, and Foxtail Millet Supplementation on Lipid Metabolism, Oxido-Inflammatory Responses, Gut Microbiota, and Colonic SCFA Composition in High-Fat Diet Fed Rats

Yong Wang [1], Wentao Qi [1,*], Xiaoxuan Guo [2], Ge Song [1], Shaojie Pang [1], Wei Fang [1] and Zhenzhen Peng [1]

[1] Academy of National Food and Strategic Reserves Administration, Beijing 100037, China; wangy@ags.ac.cn (Y.W.); sg@ags.ac.cn (G.S.); psj@ags.ac.cn (S.P.); fw@ags.ac.cn (W.F.); 13974364725@163.com (Z.P.)

[2] Institute of Quality Standard and Testing Technology for Agro-Products, Chinese Academy of Agricultural Sciences, Beijing 100081, China; guoxiaoxuan@caas.cn

* Correspondence: qwt@ags.ac.cn; Tel.: +86-10-56452656

Abstract: Coarse cereals rich in polyphenols, dietary fiber, and other functional components exert multiple health benefits. We investigated the effects of cooked oats, tartary buckwheat, and foxtail millet on lipid profile, oxido-inflammatory responses, gut microbiota, and colonic short-chain fatty acids composition in high-fat diet (HFD) fed rats. Rats were fed with a basal diet, HFD, oats diet (22% oat in HFD), tartary buckwheat diet (22% tartary buckwheat in HFD), and foxtail millet diet (22% foxtail millet in HFD) for 12 weeks. Results demonstrated that oats and tartary buckwheat attenuated oxidative stress and inflammatory responses in serum, and significantly increased the relative abundance of *Lactobacillus* and *Romboutsia* in colonic digesta. Spearman's correlation analysis revealed that the changed bacteria were strongly correlated with oxidative stress and inflammation-related parameters. The concentration of the butyrate level was elevated by 2.16-fold after oats supplementation. In addition, oats and tartary buckwheat significantly downregulated the expression of sterol regulatory element-binding protein 2 and peroxisome proliferator-activated receptors γ in liver tissue. In summary, our results suggested that oats and tartary buckwheat could modulate gut microbiota composition, improve lipid metabolism, and decrease oxidative stress and inflammatory responses in HFD fed rats. The present work could provide scientific evidence for developing coarse cereals-based functional food for preventing hyperlipidemia.

Keywords: coarse cereals; lipid metabolism; oxido-inflammatory responses; gut microbiota; short-chain fatty acids

Citation: Wang, Y.; Qi, W.; Guo, X.; Song, G.; Pang, S.; Fang, W.; Peng, Z. Effects of Oats, Tartary Buckwheat, and Foxtail Millet Supplementation on Lipid Metabolism, Oxido-Inflammatory Responses, Gut Microbiota, and Colonic SCFA Composition in High-Fat Diet Fed Rats. *Nutrients* **2022**, *14*, 2760. https://doi.org/10.3390/nu14132760

Academic Editors: Maria Luz Fernandez and Antoni Pons

Received: 12 May 2022
Accepted: 29 June 2022
Published: 4 July 2022

1. Introduction

Excessive intake of a high-fat diet (HFD) leads to abnormal lipid metabolism, which further gives rise to various chronic diseases, such as obesity, hyperlipidemia, and type 2 diabetes. By 2030, the number of people who are obese or overweight might reach 867.06 million in China [1], and obesity has become a major public health concern throughout the world. Meanwhile, gut microbiota and their metabolites such as short-chain fatty acids (SCFAs) impact the lipids and carbohydrate metabolism in the liver and adipose tissue, which is becoming a new target for anti-obesity treatment [2]. Thus, food intervention aiming at modulating the gut microbiota is available to improve metabolic health [3].

The gut microbiome has emerged as a key factor affecting human health and disease [4,5]. Elucidating the changes in gut microbiota composition has contributed to understanding the development and progress of obesity, metabolic syndrome, and type 2 diabetes mellitus [6]. The majority of gut microbiota serve as vital vehicles in the host metabolism by increasing energy harvest from the diet, which contributes to becoming overweight or obese [7]. In addition, microbiota-derived SCFA acetate can promote metabolic syndrome through regulating the gut-brain axis [8]. Numerous studies have reported that gut microbiome dysbiosis is directly or indirectly related to insulin resistance in type 2 diabetes mellitus [9]. Nevertheless, effective manipulation of the gut microbiota through diet has the potential to reduce metabolic diseases such as obesity and diabetes [10].

Epidemiological studies suggest that higher whole-grain cereals consumption reduces the risk of chronic diseases, including cardiovascular disease, type 2 diabetes, and certain types of cancer [11,12]. For instance, a barley kernel-based bread improved glucose metabolism that was associated with increased *Prevotella* abundance [13]. Previous studies reported that different varieties of coarse cereals reduced the blood glucose levels in rats with streptozotocin-induced hyperglycemia [14,15]. Buckwheat and foxtail millet could also improve lipid metabolism and blood glucose tolerance in vivo [16,17]. Moreover, our previous study suggested that the consumption of whole grain rice and wheat could regulate gut microbiota and improve SCFAs composition in rats [18].

Coarse cereals exert health-promoting benefits, which are largely due to their functional components such as dietary fiber, polyphenols, polysaccharides, and peptides [15,19,20]. Dietary fiber and phytochemicals in coarse cereals have the potential to regulate the gut microbiota, the balance of which is crucial to preserve gut homeostasis [21,22]. Furthermore, dietary fibers promote SCFAs production through intestinal bacterial fermentation [23]. Specifically, our recent research revealed that administration of oats increased gut microbiota diversity and the concentrations of SCFAs in normal rats [24]. In addition, tartary buckwheat protein markedly increased the SCFAs and the abundances of *Lactobacillus* and *Bifidobacterium* in HFD fed mice [25].

Oats, tartary buckwheat, and foxtail millet are common cereal grains in China. The impact of these three kinds of coarse cereal on gut health has gained great attention, but the comparative effects of them on gut microbiota and lipid metabolism remain unclear. Thus, the aim of this study was to determine the effects of these whole-grain cereals with matrix on the gut microflora and lipid metabolism in HFD fed rats.

2. Materials and Methods

2.1. Materials and Reagents

Oats (Variety name: Bayou No.1), tartary buckwheat (Variety name: Chuanqiao No.1), and foxtail millet (Variety name: Jingu No.21) were purchased from Hebei, Sichuan, and Shanxi Province, respectively. Main nutritional contents of cooked oats, tartary buckwheat and foxtail millet were shown in Table S1. The level of β-glucan was 4.19 ± 0.42 g/100 g in oats. The contents of total flavonoids and resistant starch were 1.76 ± 0.06 and 0.39 ± 0.01 g/100 g in tartary buckwheat, respectively. The levels of vitamin B1 was 0.27 ± 0.04 mg/100 g, and total dietary fiber content was 2.16 ± 0.02 g/100 g in foxtail millet. The diets containing oats, tartary buckwheat, and foxtail millet were ordered from Trophic Animal Feed High-tech Co. Ltd. (Jiangsu, China). All of the coarse cereals were cooked using a commercially available cooker as previously described [26]. All the cooked coarse cereals were ground prior to use in the experimental diets. Composition and energy density of the diets are presented in Table 1. Acetic, propionic, butyric, isobutyric, isovaleric, and valeric acid standards were purchased from Sigma-Aldrich Chemical Co. (Purity ≥ 99.5%, St. Louis, MO, USA). The primary antibody, anti-SREBP-2, was purchased from Santa Cruz Biotechnology (Dallas, TX, USA). All other antibodies were procured from Abcam (Cambridge, MA, USA).

Table 1. Composition and energy density of the diets.

	Con	HFD	HFD + Oat	HFD + Buc	HFD + Mil
Oats	0	0	220	0	0
Tartary buckwheat	0	0	0	220	0
Foxtail millet	0	0	0	0	220
Casein	190	242	206	202	213
Corn starch	504	193	25	38	11
Dextrin	81	103	103	103	103
Sucrose	95	121	121	121	121
Soybean oil	24	30	30	30	30
Lard	16	195	183	187	186
Cellulose	48	61	51	41	56
Mineral mixture	27	36	35	34	35
Vitamin mixture	10	12	12	12	12
L-Cystine	3	4	4	4	4
Choline chloride	2	3	3	3	3
TBHQ	0.008	0.045	0.045	0.045	0.045
Total	1000	1000	993	995	994
Content of coarse cereals/%	0	0	22.2	22.1	22.1
Energy density (Kcal/g)	3.5	4.5	4.6	4.7	4.6
Protein (% energy)	19	19	19	19	19
Carbohydrate (% energy)	71	36	36	36	36
Fat (% energy)	10	45	45	45	45

2.2. *Animals, Diets and Experimental Design*

Sixty male Sprague Dawley rats (5 weeks old) were purchased from Vital River Laboratory Animal Technology Co., Ltd. (Beijing, China). After acclimating for 1 week, the rats were randomly divided into five groups (n = 12 per group, three per cage): a basal diet group (Con group), a high-fat diet group (HFD group), HFD containing 22% oat group (Oat group), HFD containing 22% tartary buckwheat group (Buc group), and HFD containing 22% foxtail millet group (Mil group). The initial body weight of all groups was almost the same, and the rats were treated with the aforementioned diets for 12 weeks. Food intake was monitored daily and body weight was measured every week throughout the experiment. The blood samples (5 mL) were collected by cardiac puncture, centrifuged at 4000 rpm for 10 min, and stored at −80 °C. Liver and epididymal fat were carefully collected, weighed, quickly frozen in liquid nitrogen, and stored at −80 °C for further study. Pancreas and colon were also collected and frozen in liquid nitrogen.

2.3. *Biochemical Analysis in Serum*

The levels of total triglyceride (TG), total cholesterol (TC), high-density lipoprotein cholesterol (HDL-C), low-density lipoprotein cholesterol (LDL-C), aspartate transaminase (AST), alanine aminotransferase (ALT), and free fatty acid (FFA) in serum were analyzed by the corresponding reagent kits (Zhongsheng Beikong Bioengineering Institute, Beijing, China). The levels of fasting glucose were determined using a glucometer (Johnson & Johnson, Milpitas, CA, USA).

2.4. *Oral Glucose Tolerance Test (OGTT)*

Rats were fasted 12-h overnight. Following that, OGTT was implemented 1 day prior to sacrifice. Then, gavage with glucose (2 g/kg body weight) was performed on rats. Tail blood was collected at 0, 30, 60, and 120 min using a glucometer. OGTT were calculated by the trapezoidal method and expressed as area under the curve (AUC) [27].

2.5. *Histological Analysis*

Liver, pancreas, epididymal adipose, and colon tissues were fixed in 10% formalin, embedded in paraffin, sliced into 4 μm thick sections, and then stained with hematoxylin-eosin (H&E). For the oil red O staining, frozen liver tissues were sliced and stained with oil red O. The stained sections were observed by using an optical microscope (Olympus, Tokyo, Japan). Pathology scoring was performed in a blinded manner by a pathologist in Peking University.

2.6. Western Blot Analysis

Protein samples of liver tissues were extracted, separated by sodium dodecyl sulfate-polyacrylamide gel electrophoresis, and loaded onto a nitrocellulose membrane according to our previous procedures [28]. The membranes were blocked with 5% defatted milk and then incubated with primary and secondary antibodies. Protein bands were visualized by fluorescence using an ODYSSEY FC Imaging System (LI-COR Biosciences, Lincoln, NE, USA).

2.7. Analysis of Cytokines and Antioxidant Status in Serum

The levels of tumor necrosis factor-alpha (TNF-α), interleukine-6 (IL-6), interleukin-1β (IL-1β), adiponectin, and insulin in serum and the concentration of zonula occludens-1 (ZO-1) in the colon were measured using ELISA kits (R&D Systems, Minneapolis, MN, USA). The levels of malondialdehyde (MDA) and total antioxidant capacity (T-AOC) and the activities of superoxide dismutase (SOD) in serum were determined with commercial kits (Nanjing Jiancheng Biological Technology Institute, Nanjing, China) following the manufacturer's instructions.

2.8. Determination of Colonic SCFAs

Colonic content samples (80 mg, n = 6 per group) were mixed with 500 µL of methanol/water (v/v = 1:1) mixture and shaken by vortex for 30 min. Then, 50 µL supernatant that was collected after centrifugation (14,000 rpm, 10 min) was mixed with 50 µL of internal standard (5 µg/mL propionic acid-d2). Samples were derivatized with 50 µL 3-nitrophenylhydrazine (200 µM) for 30 min, and the supernatant was collected after centrifugation (14,000 rpm, 10 min). The resulting supernatant (5 µL) was analyzed on an Acquity UPLC I-Class system equipped with a BEH C18 chromatographic column (100 mm × 2.1 mm, 1.7 µm; Waters, Milford, MA, USA) coupled to a Xevo TQ-S mass spectrometer [29].

2.9. Gut Microbiota Analysis

Microbial DNA in colonic content samples was extracted according to the protocol described previously [18]. The V3-V4 hypervariable regions of 16S rRNA genes were amplified with primers 515F and 806R by a T100™ thermal cycler (Bio-Rad, Hercules, CA, USA). TruSeq DNA PCR-Free Library Preparation Kit (Illumina, San Diego, CA, USA) was used to construct the sequencing library. The sequencing was obtained from the Illumina NovaSeq 6000 platform at Novogene Bioinformatics Technology Co., Ltd. (Beijing, China). The Quantitative Insight into Microbial Ecology (QIIME) was used for further analysis. The sequences were classified into operational taxonomic units (OTUs) with 97% similarity. Then, the OTUs were subjected to taxonomy-based analysis by RDP Classifier (v2.13). Alpha diversity metrics and beta diversity were calculated using QIIME and displayed with R software. The principal coordinates analysis (PCoA), Venn diagram, and heatmap analysis of the major genus were performed by R software.

2.10. Statistical Analysis

All statistical analyses were performed using R 4.0.2. Continuous variables with normal distribution were presented as mean ± standard deviation (SD) and were performed by one-way ANOVA followed by Tukey's post hoc multiple comparison test. Skewed distribution variables were indicated with P_{50} (P_{25}, P_{75}) and were examined by non-parametric statistical hypothesis test. Spearman's correlation analysis between gut microbiota and biochemical parameters was performed and a heat map was generated. $p < 0.05$ was considered as statistically significant.

3. Results

3.1. Food Intake, Energy Intake, Body Weight, and Organ Weights

There were no significant differences in food intake among the five groups of rats during 12 weeks (Figure 1A). Energy intake in HFD group was significantly increased compared to Con group (Figure 1B), while Oat, Buc, and Mil groups were not significantly different from HFD group. At the end of study, weight gain in the HFD group was significantly higher than in the Con group (Figure 1C), but weight gain in the Oat, Buc, and Mil groups was not significantly different from in the HFD group. The rats in the HFD group showed a 45.57% increase in epididymal fat index compared to Con group (Figure 1D). However, epididymal fat index was significantly lower in Buc group compared with the HFD group. The liver indexes among the five groups of rats were not significantly different (data not shown).

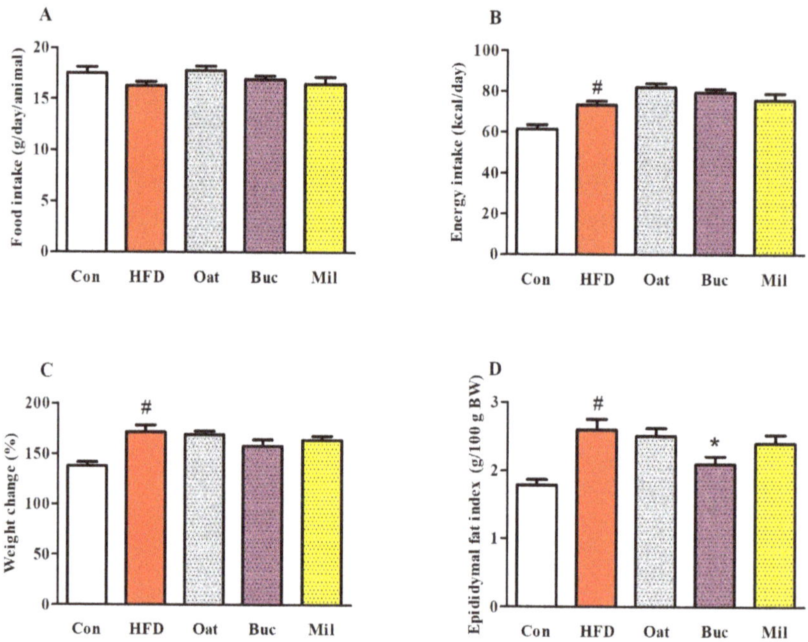

Figure 1. Effect of oats, tartary buckwheat, and foxtail millet supplementation on food intake (**A**), energy intake (**B**), percent change in body weight (**C**), and epididymal fat index (**D**) in high-fat diet fed rats. Con group, a basal diet group; HFD group, a high-fat diet group; Oat group, HFD containing 22% oat group; Buc group, HFD containing 22% tartary buckwheat group; Mil group, HFD containing 22% foxtail millet group. Data are presented as the mean ± SD (n = 12). # $p < 0.05$, compared to Con group, * $p < 0.05$, compared to HFD group.

3.2. Lipid Metabolism-Related Parameters in Serum, Fasting Blood Glucose, and OGTT

HFD feeding resulted in a significant increase in TC and LDL-C levels in serum (Figure 2B,D), but there were no significant differences in the TG and HDL-C levels in serum (Figure 2A,C) compared with the Con group. Serum TC and LDL-C in the Oat and Buc groups were remarkably decreased compared with that in the HFD group. There were no significant differences in the serum AST and ALT levels among the five groups (Figure S1). The HFD diet induced a higher level of FFA in the serum compared with the Con diet (Figure 2E), but the level of FFA concentration was significantly reduced in the Oat group compared with the HFD group. However, there were no significant differences in fasting blood glucose and AUC of OGTT among the five groups (Figure 3).

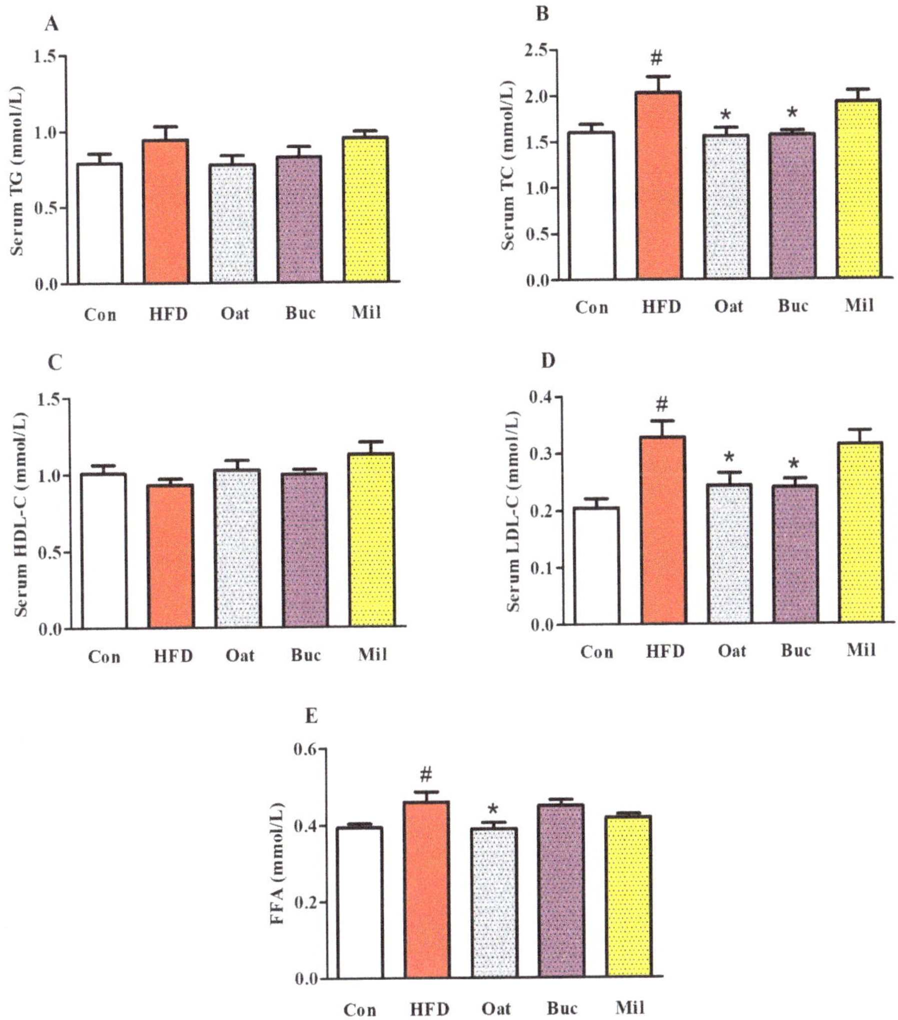

Figure 2. Effect of oats, tartary buckwheat, and foxtail millet supplementation on serum TG (**A**), TC (**B**), HDL-C (**C**), LDL-C (**D**), and FFA (**E**) in high-fat diet fed rats. Con group, a basal diet group; HFD group, a high-fat diet group; Oat group, HFD containing 22% oat group; Buc group, HFD containing 22% tartary buckwheat group; Mil group, HFD containing 22% foxtail millet group; TG, total triglyceride; TC, total cholesterol; HDL-C, high-density lipoprotein cholesterol; LDL-C, low-density lipoprotein cholesterol; FFA, free fatty acid. Data are presented as the mean ± SD (n = 12). # $p < 0.05$, compared to Con group, * $p < 0.05$, compared to HFD group.

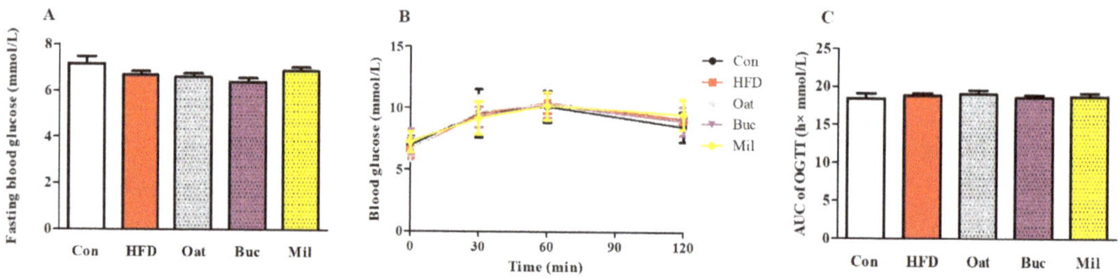

Figure 3. Effect of oats, tartary buckwheat, and foxtail millet supplementation on fasting blood glucose (**A**), OTGG (**B**), and AUC of OGTT (**C**) in high-fat diet fed rats. Con group, a basal diet group; HFD group, a high-fat diet group; Oat group, HFD containing 22% oat group; Buc group, HFD containing 22% tartary buckwheat group; Mil group, HFD containing 22% foxtail millet group; AUC, area under the curve; OGTT, oral glucose tolerance test. Data are presented as the mean ± SD (n = 12).

3.3. Pro-Inflammatory Cytokine Levels, Antioxidant Capability, Adiponectin, and Insulin in Serum

Oat and Buc groups exerted beneficial effects on lipid metabolism-related parameters in serum (TC, LDL-C, and FFA), so serum pro-inflammatory cytokine levels, antioxidant capability, adiponectin and insulin, and colonic ZO-1 protein were measured for these two kinds of coarse cereal. In HFD group, the levels of TNF-α, IL-6, and IL-1β were significantly increased in serum, while oats supplementation significantly decreased their levels (Figure 4A–C). Moreover, TNF-α and IL-6 were significantly down-regulated in Buc group. HFD significantly reduced the activities of SOD and the concentrations of T-AOC, and increased the level of MDA in serum (Figure 4D–F), indicating that HFD diet decreased antioxidant capacity of the serum in rats. However, oat and tartary buckwheat supplementation significantly lowered MDA levels and increased SOD activities, and oats supplementation also significantly increased T-AOC when compared with the HFD treatment. Although HFD did not result in a significant decrease in adiponectin level, oats administration markedly increased adiponectin level (Figure 4G). In addition, fasting blood insulin was significantly higher in the HFD group compared with the Con group, but oats supplementation significantly reduced its level relative to the HFD group (Figure 4H).

3.4. Histological Analysis of the Liver, Pancreas, Epididymal Adipose, and Colon Tissues

The histological analysis of the liver tissue showed that the liver steatosis induced by HFD was greatly attenuated following Buc supplementation (Figure 5A,B,F). An HFD diet induced inflammatory cell infiltration in the pancreas of rats, while oats supplementation significantly restored pancreatic islet morphology (Figure 5C,G). In line with the epididymal fat index shown in Figure 1D, the H&E staining of epididymal fat tissue revealed a drastically larger adipocyte size in HFD rats, while the oats and tartary buckwheat supplementation suppressed the increase of the adipocyte size (Figure 5D). Inflammatory cell infiltration in the colon can be observed in the HFD group, while oats supplementation lowered inflammatory cell infiltration (Figure 5E). Colonic ZO-1 protein expression in the HFD group was significantly lower than that in the Con group (Figure 5H), while ZO-1 protein was significantly higher in the Oat and Buc groups compared with the HFD group.

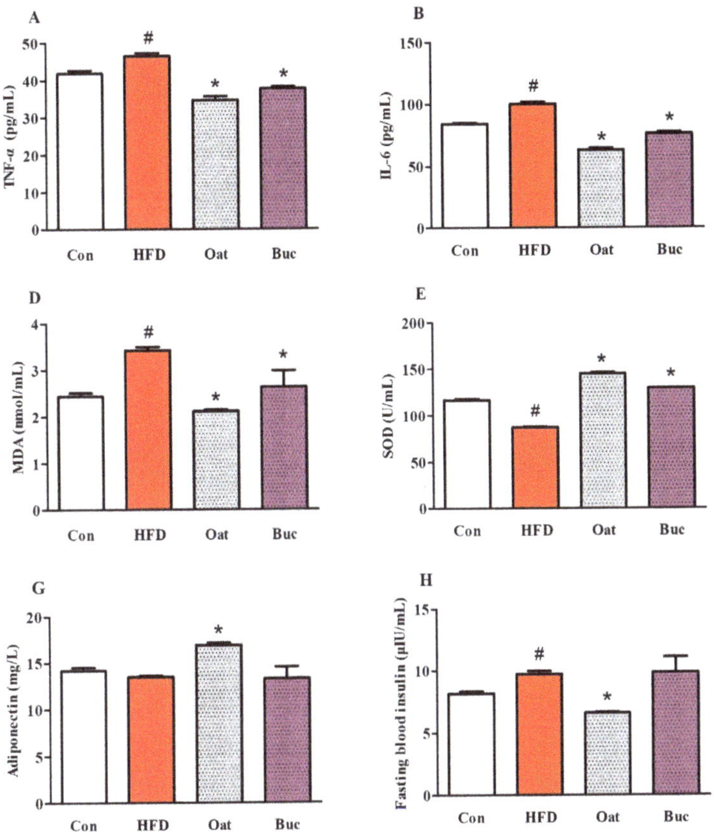

Figure 4. Effect of oats and tartary buckwheat supplementation on serum TNF-α (**A**), IL-6 (**B**), IL-1β (**C**), MDA (**D**), SOD (**E**), T-AOC (**F**), adiponectin (**G**), and fasting blood insulin (**H**) in high-fat diet fed rats. Con group, a basal diet group; HFD group, a high-fat diet group; Oat group, HFD containing 22% oat group; Buc group, HFD containing 22% tartary buckwheat group; TNF-α, tumor necrosis factor-alpha; IL-6, interleukine-6; IL-1β, interleukin-1β; MDA, malondialdehyde; SOD, superoxide dismutase; T-AOC, total antioxidant capacity. Data are presented as the mean ± SD ($n = 10$). # $p < 0.05$, compared to Con group, * $p < 0.05$, compared to HFD group.

3.5. Expression of Proteins Involved in Lipid Metabolism

The expression levels of sterol regulatory element-binding protein 2 (SREBP-2) and peroxisome proliferator-activated receptors γ (PPAR-γ) in liver, which were involved in lipid metabolism, were not significantly changed in HFD group compared to Con group (Figure 5I–K). However, oats and tartary buckwheat supplementation significantly decreased the expression of these proteins.

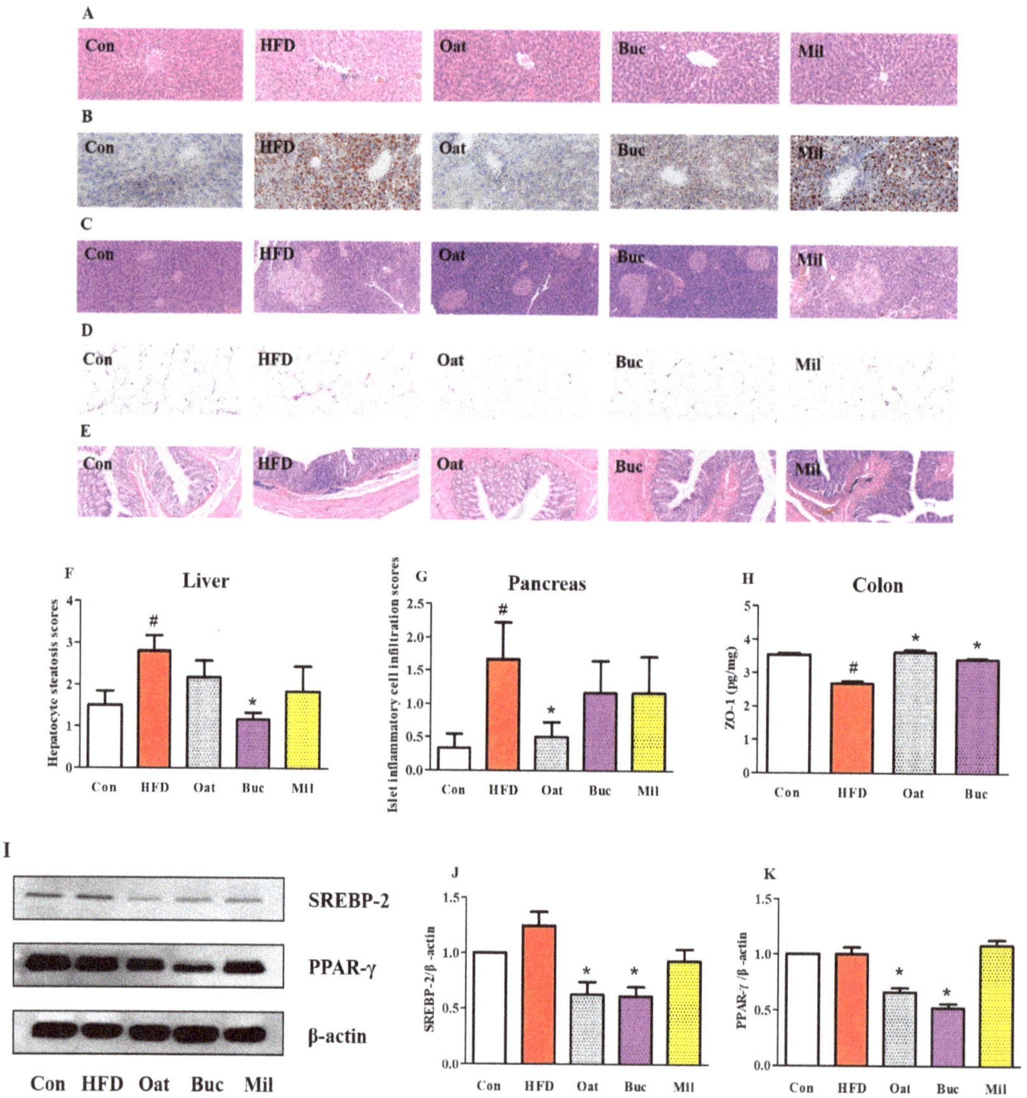

Figure 5. Effect of oats, tartary buckwheat, and foxtail millet supplementation on liver (**A**,**B**), pancreas (**C**), epididymal adipose (**D**), and colon (**E**) tissues in high-fat diet fed rats. Representative samples of liver tissue were stained with H&E (**A**) and oil red O (**B**). (**F**) Scores based on hepatocyte steatosis in the liver. (**G**) Scores based on inflammatory cell infiltration in the pancreas. (**H**) The protein expression levels of ZO-1 in the colon. (**I**) Western blot bands of SREBP-2 and PPAR-γ in the liver. (**J**,**K**) Densitometric analysis of SREBP-2 (**J**) and PPAR-γ (**K**) expressions relative to the loading control. Con group, a basal diet group; HFD group, a high-fat diet group; Oat group, HFD containing 22% oat group; Buc group, HFD containing 22% tartary buckwheat group; Mil group, HFD containing 22% foxtail millet group; ZO-1, zonula occludens-1; SREBP-2, sterol regulatory element-binding protein 2; PPAR-γ, peroxisome proliferator-activated receptors γ. Data are presented as the mean ± SD (n = 6). # $p < 0.05$, compared to Con group, * $p < 0.05$, compared to HFD group.

3.6. Coarse Cereals Supplementation Modulated the Gut Microbiota

HFD and coarse cereals supplementation has no significant effect on alpha diversity of colonic microbiota (data not shown). PCoA analysis demonstrated that it generated a distinct difference of microbial communities between HFD and Con groups (Figure 6A). A total of 388 OTUs (38.3%) were common among different groups (Figure 6B). The Mil group had more unique OTUs (67) than the Con (39), HFD (58), Oat (26), and Buc (32) groups. The influences of coarse cereals on the composition of gut microbiota at the phylum and genus levels was shown in Figure 6C,D. Furthermore, heatmap was used to show the abundance of 35 key genera of different groups in order to evaluate the impact of coarse cereals supplementation on the bacterial community of HFD fed rats (Figure 6E). At the phylum level, Firmicutes, Verrucomicrobia, and Bacteroidetes were the absolute dominated taxa in all groups, and the relative abundance of Firmicutes was much higher in the HFD group compared with the Con group. Tartary buckwheat and foxtail millet supplementation increased the Firmicutes (8.21% and 16.94%, respectively) and decreased Verrucomicrobia (46.64% and 79.54%, respectively) compared with the HFD group. At the genus level, HFD increased the relative abundance of unidentified *Ruminococcaceae*, *Blautia*, *Faecalitalea*, *Romboutsia*, and *Staphylococcus*, and decreased the relative abundance of *Akkermansia* and *Bacteroides* compared with the Con group. Oats and tartary buckwheat significantly increased the abundance of *Lactobacillus* and *Romboutsia*, and tartary buckwheat and foxtail millet significantly decreased the abundance of *Akkermansia* and *Blautia*.

3.7. Coarse Cereals Supplementation Regulated the Specific Bacteria

Figure 7 showed the significant differences of gut microbiota in the taxa among five groups, including 9 genera. The Con group was dominated by *Akkermansia*, while the *Faecalitalea* was the most prevalent bacteria in the HFD group. The Oat group exhibited a predominance of *Blautia*, *Romboutsia*, *Staphylococcus*, and *Lactobacillus* at the genus level. The Mil group featured the genera unidentified *Ruminococcaceae*, *Nitrosomonas*, and *Enterobacter* at the genus level.

3.8. Correlation between Gut Microbiota and Pro-Inflammatory Cytokine, Antioxidant Capability and Lipid Metabolism-Related Indices

Spearman's correlation analysis showed that *Lactobacillus* abundance was positively associated with SOD and adiponectin, and negatively associated with TNF-α, IL-6 and MDA (Figure 8). *Akkermansia* abundance was positively associated with ZO-1, but *Faecalitalea* and unidentified *Ruminococcaceae* was negatively associated with ZO-1. *Staphylococcus* was negatively associated with TNF-α and IL-6 and positively associated with SOD. Unidentified *Ruminococcaceae* abundance was positively associated with LDL-C and FFA. *Faecalitalea* and *Caproiciproducens* abundance was positively correlated with TNF-α, IL-6, and MDA, but negatively correlated with SOD and T-AOC.

3.9. Coarse Cereals Supplementation Regulated the SCFAs Production

There were no significant differences in SCFAs production between Con and HFD groups (Table 2). However, butyrate concentration was significantly increased by oat supplementation when compared with the HFD treatment. In Oat and Buc groups, the butyrate level was elevated by 2.16- and 1.77-fold when compared with the HFD group, respectively.

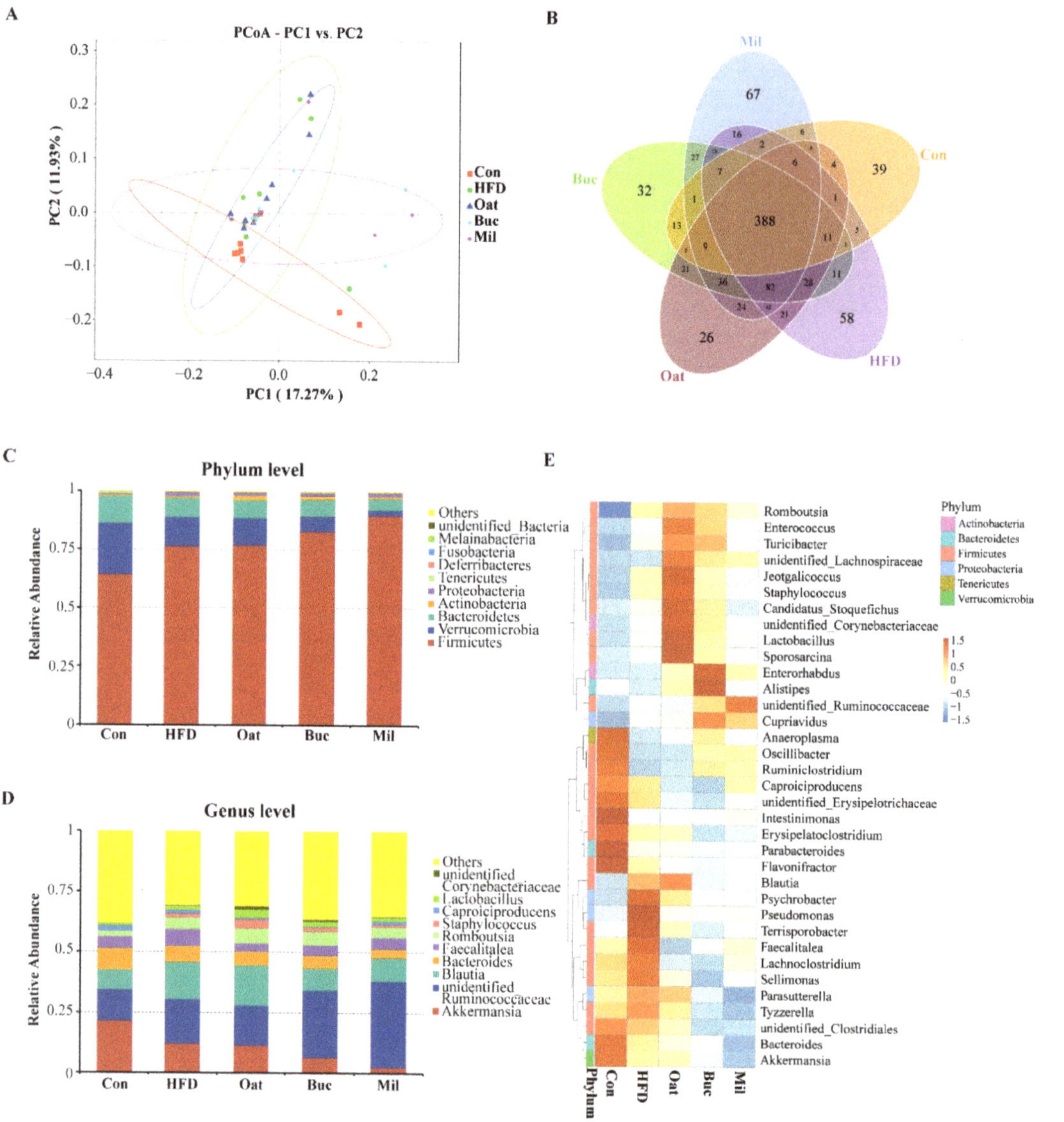

Figure 6. Effect of oats, tartary buckwheat, and foxtail millet supplementation on gut microbiota composition in high-fat diet fed rats. (**A**) Principal coordinate analysis (PCoA) of colonic microbiota ($n = 8$). (**B**) Venn diagram showing the shared and unique OTUs among the five groups. (**C**) Microbial community composition at the phylum level. (**D**) Microbial community composition at the genus level. (**E**) Heatmap of clustering analysis at the genus level. Con group, a basal diet group; HFD group, a high-fat diet group; Oat group, HFD containing 22% oat group; Buc group, HFD containing 22% tartary buckwheat group; Mil group, HFD containing 22% foxtail millet group.

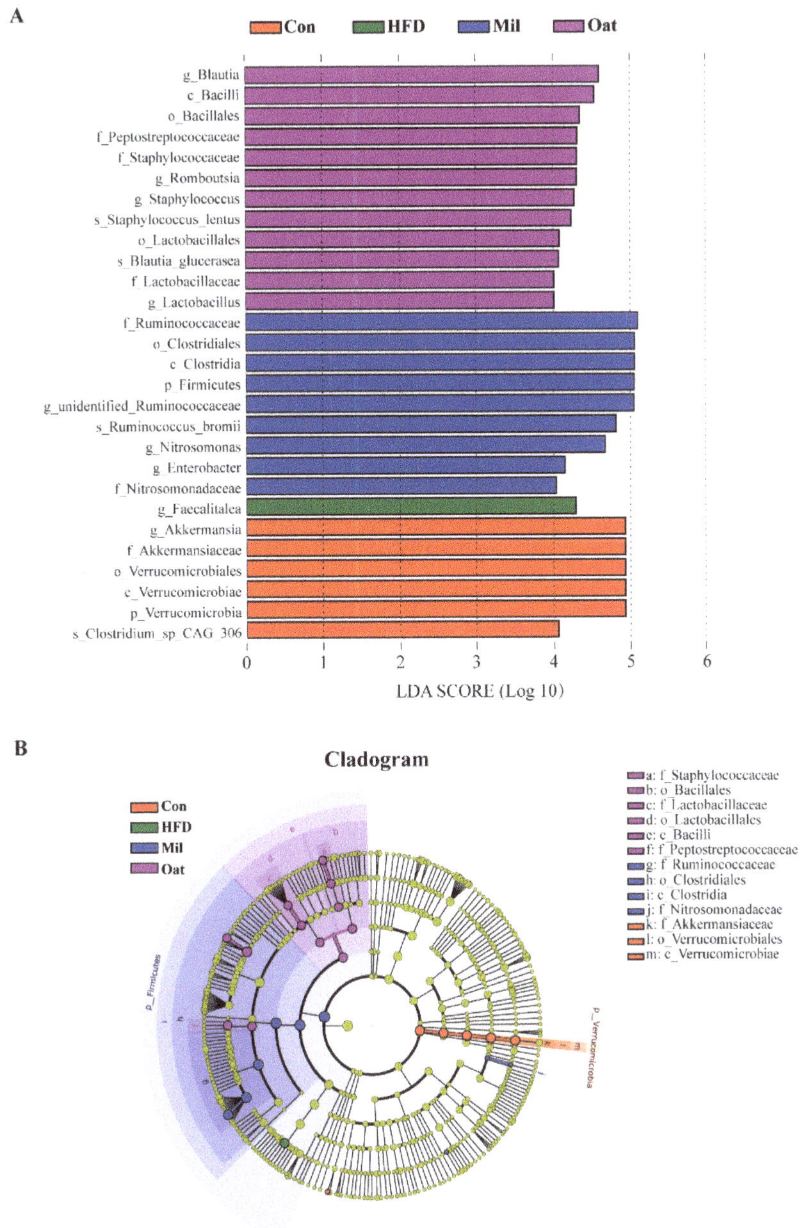

Figure 7. Linear discriminant analysis (LDA) effect size (LEfSe) analysis of the key genera of gut microbiota in high-fat diet fed rats. (**A**) Histogram of the LDA scores for differentially abundant bacterial taxa. (**B**) Cladogram visualizing the phylogenetic distribution of the gut microbiota community. Con group, a basal diet group; HFD group, a high-fat diet group; Oat group, HFD containing 22% oat group; Mil group, HFD containing 22% foxtail millet group; p, phylum; c, class; o, order; f, family; and g, genus.

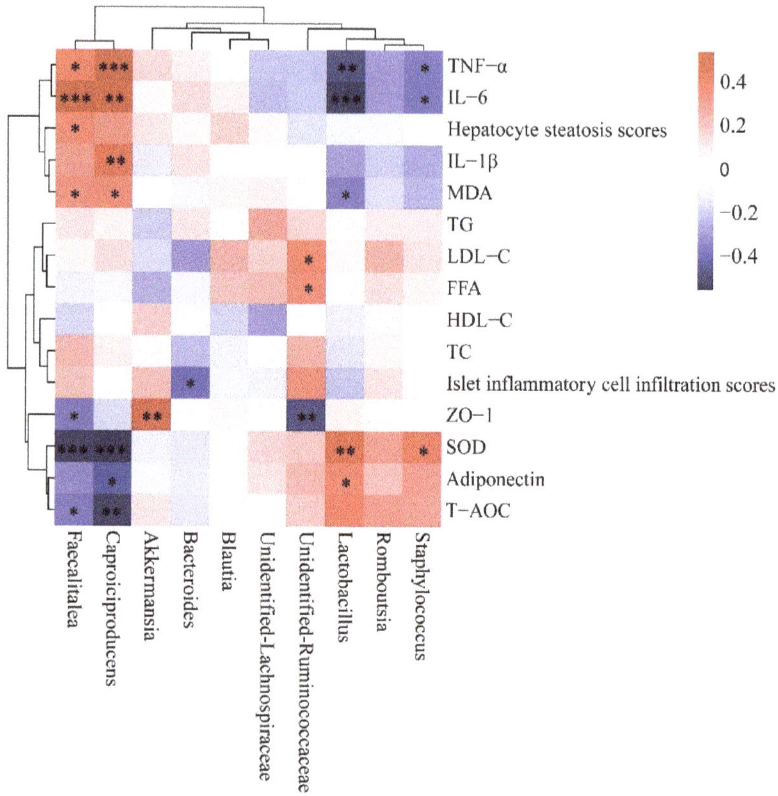

Figure 8. Heatmap of Spearman's correlation between gut microbiota (top 10 at genus level) and pro-inflammatory cytokine, antioxidant capability, and lipid metabolism-related indices. * $p < 0.05$, ** $p < 0.01$, *** $p < 0.001$. TG, total triglyceride; TC, total cholesterol; HDL-C, high-density lipoprotein cholesterol; LDL-C, low-density lipoprotein cholesterol; FFA, free fatty acid; TNF-α, tumor necrosis factor-alpha; IL-6, interleukine-6; IL-1β, interleukin-1β; MDA, malondialdehyde; SOD, superoxide dismutase; T-AOC, total antioxidant capacity; ZO-1, zonula occludens-1.

Table 2. Effects of oats, tartary buckwheat, and foxtail millet supplementation on SCFAs in colonic contents of rats at the end of 12 weeks.

	Con	HFD	Oat	Buc	Mil
Acetic acid (μg/g)	44.22 ± 7.11	69.71 ± 18.29	76.17 ± 7.685	74.89 ± 13.54	66.82 ± 10.55
Propionic acid (μg/g)	22.41 ± 4.69	34.56± 8.99	41.28 ± 4.72	34.50 ± 6.02	33.42 ± 4.90
Butyric acid (μg/g)	16.77 ± 3.01^{a}	15.69 ± 1.98^{a}	33.92 ± 6.97^{b}	$27.71 \pm 3.49^{a,b}$	15.96 ± 1.75^{a}
Isobutyric acid (μg/g)	5.20 ± 0.77	7.30 ± 1.25	8.56 ± 0.76	7.31 ± 0.84	6.90 ± 0.89
Isovaleric acid (μg/g)	4.19± 0.59	4.97 ± 0.60	5.85 ± 1.05	4.60 ± 0.63	4.43 ± 0.37
Valeric acid (μg/g)	5.08 ± 0.86	7.52 ± 1.43	9.31 ± 1.37	7.12 ± 1.03	6.74 ± 0.66
Total SCFAs (μg/g)	104.25 (71.95, 120.45)	126.82 (72.34, 223.53)	163.67 (141.09, 206.58)	159.76 (95.48, 212.05)	127.03 (99.15, 163.81)

The values are the means ± SD or P_{50} (P_{25}, P_{75}). Con group, a basal diet group; HFD group, a high-fat diet group; Oat group, HFD containing 22% oat group; Buc group, HFD containing 22% tartary buckwheat group; Mil group, HFD containing 22% foxtail millet group. Mean values not sharing the same superscript letter (a and b) within a row are significantly different at $p < 0.05$ in a post hoc test.

4. Discussion

In this study, intake of oat and tartary buckwheat attenuated oxidative stress, inflammatory responses, and hyperlipidemia in serum, downregulated the expression of lipid metabolism-related protein in liver, and increased the relative abundance of *Lactobacillus* and *Romboutsia* in colonic digesta of HFD-induced rats. Spearman's correlation analysis showed that the changed bacteria were strongly correlated with oxidative stress and inflammation-related parameters. Histological analysis also demonstrated that intake of tartary buckwheat reduced the excessive lipid droplet accumulation in the liver and inhibited lipid accumulation in epididymal adipose tissue. In previous studies, oat and tartary buckwheat consumption was associated with the cholesterol-lowering effect [30,31]. The increased concentrations of SCFAs and alterations in the gut microbiota probably have an important role in ameliorating dyslipidemia [18].

Coarse cereals have potential health benefits through modulating intestinal flora. *Lactobacillus*, *Bifidobacterium*, and *Akkermansia* have beneficial effects on metabolism, stimulating fatty acid oxidation and inhibiting lipoprotein lipase activity in HFD fed mice [32,33]. Intake of embryo-remaining oat rice enhanced the production of SCFAs and increased the abundance of *Bifidobacterium* and *Akkermansia* in HFD fed rats [34]. Tartary buckwheat consumption alleviated hyperlipidemia and gastritis by maintaining intestinal homeostasis [35,36] and markedly increased the abundances of *Lactobacillus*, *Blautia*, and *Akkermansia* in HFD fed mice [16]. Foxtail millet supplementation ameliorated colorectal cancer by the microbial metabolites (tryptophan metabolites and SCFAs) activating related gut receptors [37]. Whole barley decreased cholesterol accumulation and counteracted gut dysbiosis in obese mice [38]. In addition, dietary intake of mixture coarse cereals reduced fat accumulation, decreased serum lipids levels, and increased the abundances of *Lactobacillus* and *Bifidobacterium* in HFD fed mice [39]. To the best of our knowledge, few comparative studies on the effects of these three kinds of coarse cereal on gut microbiota and lipid metabolism in HFD fed rats.

In this study, *Lactobacillus* and *Staphylococcus* were thought to contribute to inhibiting inflammatory response, while the abundances of *Faecalitalea* and *Caproiciproducens* were positively correlated with levels of inflammatory factors. Furthermore, our work suggested that oat supplementation increased the production of butyrate, which was an important metabolite of gut microbiota because it promoted host intestinal barrier function and alleviated inflammation [40]. Gao et al., (2020) also showed that dietary oat fiber was conducive to produce SCFAs and increased the abundance of the gut bacteria that generated anti-inflammatory metabolites and improved gut barrier function in $LDLR^{-/-}$ mice [41]. Moreover, quinoa and buckwheat protein-rich flours decreased the plasma TC and LDL-C and elevated the production of SCFAs in rats [42].

Coarse cereals are rich in functional active substances such as dietary fiber and phytochemicals, which are considered to be natural prebiotics. Epidemiological study demonstrated that cardiometabolic risk was inversely associated with dietary fiber intake [43]. In a randomized controlled trial, dietary fiber (oat bran, 30 g/d) supplement decreased blood pressure and modulated the gut microbiota in patients with essential hypertension [44]. In another clinical trial, supplementation with 3 g/day of oat β-glucan reduced LDL-C, TC, and non-HDL-C in mildly hypercholesterolemic subjects [45]. In addition, increased consumption of coarse cereals containing dietary fiber contributes to healthy functioning of the intestine. Oat β-glucan mitigated the inflammatory status in colon, enhanced colonic barrier function, and increased gut microbiota-derived SCFAs (butyrate) in vivo [46,47]. Tartary buckwheat-resistant starch improved intestinal health by regulating the gut microbiota composition and increasing the yield of SCFAs in HFD fed mice [48]. In this study, oat and tartary buckwheat supplementation were significantly increased the ZO-1 protein expression, indicating they improved colonic barrier function in HFD-induced rats.

It is worth noting that purified fiber supplementation has no effect on gut microbiota diversity in previous study [49]. The health benefits of whole grains were probably because of the synergistic action of multiple components in grains [50]. For example, oat bran as a complex food matrix increased *Bifidobacteria* and SCFA production, rather than its main functional compounds such as β-glucan and polyphenols in an in vitro fermentation model [49]. Tartary buckwheat is rich in resistant starch and flavonoids, which may have synergistic effects on lipid metabolism and gut microbiota [19]. Sorghum polyphenols and fructooligosaccharides worked synergistically to increase the abundances of *Bifidobacterium* and *Lactobacillus* during in vitro fermentation [51].

In addition to dietary fiber, phytochemicals (phenolic acids, flavonoids, etc.) in coarse cereals also had the potential to combat common nutrition-related diseases including cardiovascular disease, diabetes, and obesity. Flavonoids from oat exhibited an anti-hyperlipidemic effect via regulating gut microbiota that increasing *Akkermansia* and decreasing *Blautia* in mice [52]. Polyphenols from foxtail millet bran and shell could remodel the gut microbiota to prevent tumor and atherosclerosis in mice, respectively [53,54]. Furthermore, vitexin as a millet-derived flavonoid suppressed HFD-induced brain oxidative stress and inflammation by regulating gut microbiota in HFD fed mice [32].

There are some limitations of our study. First, the rodent gut microbiota differs from the human gut microbiota [55]. Second, the analysis of lipid metabolism-related protein expression is not enough, and relevant analysis is conducive to further elucidate the mechanism of anti-hyperlipidemic effect of coarse cereals. Third, the synergistic health benefits of functional components in coarse cereals such as dietary fiber and polyphenols is unclear. Further studies are needed to investigate cereal dietary fiber and polyphenols synergistic alleviating obesity via regulating gut microbiota.

5. Conclusions

Consumption of coarse cereals such as oats and tartary buckwheat was able to improve lipid metabolism and modulate gut microbiota composition. Oats and tartary buckwheat increased the abundance of *Lactobacillus* and *Romboutsia*, which was strongly correlated with anti-oxidant and anti-inflammatory effects. The colonic level of SCFAs such as butyrate was significantly increased after oats supplementation. This study can provide scientific evidence for the development of cereal-based functional foods.

Supplementary Materials: The following supporting information can be downloaded at: https://www.mdpi.com/article/10.3390/nu14132760/s1. Figure S1: Effect of oats, tartary buckwheat and foxtail millet supplementation on serum AST (A) and ALT (B) in high-fat diet fed rats. Con group, a basal diet group; HFD group, a high-fat diet group; Oat group, HFD containing 22% oat group; Buc group, HFD containing 22% tartary buckwheat group; Mil group, HFD containing 22% foxtail millet group; AST, aspartate aminotransferase; ALT, alanine aminotransferase. Data are presented as the mean ± SD (n = 12); Table S1: Main nutritional contents of cooked oats, tartary buckwheat and foxtail millet.

Author Contributions: Conceptualization, Y.W.; data curation, Y.W., G.S. and S.P.; formal analysis, Y.W. and S.P.; investigation, Y.W., G.S., S.P. and Z.P.; funding acquisition, Y.W.; methodology, Y.W. and X.G.; project administration, W.Q.; software, S.P. and W.F.; writing—original draft preparation, Y.W.; writing—review and editing, Y.W., W.Q. and X.G. All authors have read and agreed to the published version of the manuscript.

Funding: This work was funded by the Special Funds of Basic Research of Central Public Welfare Institute (No. ZX1908), National Natural Science Foundation of China (No. 31901698), and Young Elite Scientists Sponsorship Program by the China Association for Science and Technology (No. 2019QNRC001).

Institutional Review Board Statement: All animal studies were approved by the Ethical Committee for Animal Experimentation of the Academy of National Food and Strategic Reserves Administration with utilization permission from Beijing Municipal Science & Technology Commission (No. SYXK (Jing) 2019-0015).

Informed Consent Statement: Not applicable.

Data Availability Statement: Data presented in this study are available on request from the corresponding author.

Conflicts of Interest: The authors declare no conflict of interest.

References

1. Wang, Y.; Zhao, L.; Gao, L.; Pan, A.; Xue, H. Obesity in China 3 Health policy and public health implications of obesity in China. *Lancet Diabetes Endocrinol.* **2021**, *9*, 446–461. [CrossRef]
2. Song, X.; Wang, L.; Liu, Y.; Zhang, X.; Weng, P.; Liu, L.; Zhang, R.; Wu, Z. The gut microbiota-brain axis: Role of the gut microbial metabolites of dietary food in obesity. *Food Res. Int.* **2022**, *153*, 110971. [CrossRef] [PubMed]
3. Aron-Wisnewsky, J.; Warmbrunn, M.V.; Nieuwdorp, M.; Clément, K. Metabolism and Metabolic Disorders and the Microbiome: The Intestinal Microbiota Associated with Obesity, Lipid Metabolism, and Metabolic Health-Pathophysiology and Therapeutic Strategies. *Gastroenterology* **2021**, *160*, 573–599. [CrossRef] [PubMed]
4. Gao, X.; Zhao, J.; Zhang, H.; Chen, W.; Zhai, Q. Modulation of gut health using probiotics: The role of probiotic effector molecules. *J. Future Foods* **2022**, *2*, 1–12. [CrossRef]
5. Wang, X.; Yang, S.; Li, S.; Zhao, L.; Hao, Y.; Qin, J.; Zhang, L.; Zhang, C.; Bian, W.; Zuo, L.; et al. Aberrant gut microbiota alters host metabolome and impacts renal failure in humans and rodents. *Gut* **2020**, *69*, 2131–2142. [CrossRef] [PubMed]
6. Ortega, M.A.; Fraile-Martínez, O.; Naya, I.; García-Honduvilla, N.; Álvarez-Mon, M.; Buján, J.; Asúnsolo, Á.; de la Torre, B. Type 2 Diabetes Mellitus Associated with Obesity (Diabesity). The Central Role of Gut Microbiota and Its Translational Applications. *Nutrients* **2020**, *12*, 2749. [CrossRef] [PubMed]
7. Zmora, N.; Suez, J.; Elinav, E. You are what you eat: Diet, health and the gut microbiota. *Nat. Rev. Gastroenterol. Hepatol.* **2019**, *16*, 35–56. [CrossRef]
8. Perry, R.J.; Peng, L.; Barry, N.A.; Cline, G.W.; Zhang, D.; Cardone, R.L.; Petersen, K.F.; Kibbey, R.G.; Goodman, A.L.; Shulman, G.I. Acetate mediates a microbiome-brain-beta-cell axis to promote metabolic syndrome. *Nature* **2016**, *534*, 213–217. [CrossRef]
9. Sharma, S.; Tripathi, P. Gut microbiome and type 2 diabetes: Where we are and where to go? *J. Nutr. Biochem.* **2019**, *63*, 101–108. [CrossRef] [PubMed]
10. Chen, J.; He, X.; Huang, J. Diet Effects in Gut Microbiome and Obesity. *J. Food Sci.* **2014**, *79*, R442–R451. [CrossRef] [PubMed]
11. Ye, E.Q.; Chacko, S.A.; Chou, E.L.; Kugizaki, M.; Liu, S. Greater Whole-Grain Intake Is Associated with Lower Risk of Type 2 Diabetes, Cardiovascular Disease, and Weight Gain. *J. Nutr.* **2012**, *142*, 1304–1313. [CrossRef]
12. Wu, W.; Qiu, J.; Wang, A.; Li, Z. Impact of whole cereals and processing on type 2 diabetes mellitus: A review. *Crit. Rev. Food Sci. Nutr.* **2020**, *60*, 1447–1474. [CrossRef]
13. Kovatcheva-Datchary, P.; Nilsson, A.; Akrami, R.; Lee, Y.S.; De Vadder, F.; Arora, T.; Hallen, A.; Martens, E.; Björck, I.; Bäckhed, F. Dietary Fiber-Induced Improvement in Glucose Metabolism Is Associated with Increased Abundance of Prevotella. *Cell Metab.* **2015**, *22*, 971–982. [CrossRef]
14. Qi, W.; Wang, Y.; Song, G.; Sun, H.; Pang, S.; Li, A. Effects of four coarse cereals on blood glucose levels in rats with STZ-induced hyperglycemia. *Food Agric. Immunol.* **2019**, *30*, 487–496. [CrossRef]
15. Ren, X.; Wang, L.; Chen, Z.; Hou, D.; Xue, Y.; Diao, X.; Shen, Q. Foxtail Millet Improves Blood Glucose Metabolism in Diabetic Rats through PI3K/AKT and NF-kappa B Signaling Pathways Mediated by Gut Microbiota. *Nutrients* **2021**, *13*, 1837. [CrossRef]
16. Huang, Z.R.; Deng, J.C.; Li, Q.Y.; Cao, Y.J.; Lin, Y.C.; Bai, W.D.; Liu, B.; Rao, P.F.; Ni, L.; Lv, X.C. Protective Mechanism of Common Buckwheat (*Fagopyrum esculentum* Moench.) against Nonalcoholic Fatty Liver Disease Associated with Dyslipidemia in Mice Fed a High-Fat and High-Cholesterol Diet. *J. Agric. Food Chem.* **2020**, *68*, 6530–6543. [CrossRef]
17. Ren, X.; Wang, L.; Chen, Z.; Zhang, M.; Hou, D.; Xue, Y.; Diao, X.; Liu, R.; Shen, Q. Foxtail millet supplementation improves glucose metabolism and gut microbiota in rats with high-fat diet/streptozotocin-induced diabetes. *Food Sci. Hum. Wellness* **2021**, *11*, 119–128. [CrossRef]
18. Han, F.; Wang, Y.; Han, Y.; Zhao, J.; Han, F.; Song, G.; Jiang, P.; Miao, H. Effects of Whole-Grain Rice and Wheat on Composition of Gut Microbiota and Short-Chain Fatty Acids in Rats. *J. Agric. Food Chem.* **2018**, *66*, 6326–6335. [CrossRef]
19. Ren, G.; Fan, X.; Teng, C.; Li, Y.; Everaert, N.; Blecker, C. The Beneficial Effect of Coarse Cereals on Chronic Diseases through Regulating Gut Microbiota. *Foods* **2021**, *10*, 2891. [CrossRef]
20. Tian, M.; Li, D.; Ma, C.; Feng, Y.; Hu, X.; Chen, F. Barley Leaf Insoluble Dietary Fiber Alleviated Dextran Sulfate Sodium-Induced Mice Colitis by Modulating Gut Microbiota. *Nutrients* **2021**, *13*, 846. [CrossRef]
21. Fu, J.; Zhang, Y.; Hu, Y.; Zhao, G.; Tang, Y.; Zou, L. Concise review: Coarse cereals exert multiple beneficial effects on human health. *Food Chem.* **2020**, *325*, 126761. [CrossRef]
22. Guo, H.; Wu, H.; Sajid, A.; Li, Z. Whole grain cereals: The potential roles of functional components in human health. *Crit. Rev. Food Sci. Nutr.* **2021**, 1928596. [CrossRef]
23. Shang, W.; Si, X.; Zhou, Z.; Li, Y.; Strappe, P.; Blanchard, C. Characterization of fecal fat composition and gut derived fecal microbiota in high-fat diet fed rats following intervention with chito-oligosaccharide and resistant starch complexes. *Food Funct.* **2017**, *8*, 4374–4383. [CrossRef]
24. Wang, Y.; Song, G.; Pang, S.; Qi, W. Effects of three kinds of coarse cereals on gut microbiota of rats explored by Illumina NovaSeq sequencing technology. *Shipin Kexue/Food Sci.* **2021**, *42*, 100–106.

25. Zhou, X.-L.; Yan, B.-B.; Xiao, Y.; Zhou, Y.-M.; Liu, T.-Y. Tartary buckwheat protein prevented dyslipidemia in high-fat diet-fed mice associated with gut microbiota changes. *Food Chem. Toxicol.* **2018**, *119*, 296–301. [CrossRef]
26. Han, F.; Han, F.; Wang, Y.; Fan, L.; Song, G.; Chen, X.; Jiang, P.; Miao, H.; Han, Y. Digestible indispensable amino acid scores of nine cooked cereal grains. *Br. J. Nutr.* **2019**, *121*, 30–41. [CrossRef]
27. Chen, H.H.; Nie, Q.X.; Hu, J.L.; Huang, X.J.; Zhang, K.; Nie, S.P. Glucomannans Alleviated the Progression of Diabetic Kidney Disease by Improving Kidney Metabolic Disturbance. *Mol. Nutr. Food Res.* **2019**, *63*, 1801008. [CrossRef]
28. Wang, Y.; Zhao, L.; Huo, Y.; Zhou, F.; Wu, W.; Lu, F.; Yang, X.; Guo, X.; Chen, P.; Deng, Q.; et al. Protective Effect of Proanthocyanidins from Sea Buckthorn (*Hippophae rhamnoides* L.) Seed against Visible Light-Induced Retinal Degeneration in vivo. *Nutrients* **2016**, *8*, 245. [CrossRef]
29. Wang, Y.; An, Y.; Ma, W.; Yu, H.; Lu, Y.; Zhang, X.; Wang, Y.; Liu, W.; Wang, T.; Xiao, R. 27-Hydroxycholesterol contributes to cognitive deficits in APP/PS1 transgenic mice through microbiota dysbiosis and intestinal barrier dysfunction. *J. Neuroinflamm.* **2020**, *17*, 199. [CrossRef]
30. Joyce, S.A.; Kamil, A.; Fleige, L.; Gahan, C.G.M. The Cholesterol-Lowering Effect of Oats and Oat Beta Glucan: Modes of Action and Potential Role of Bile Acids and the Microbiome. *Front. Nutr.* **2019**, *6*, 171. [CrossRef]
31. Huang, Z.R.; Chen, M.; Guo, W.L.; Li, T.T.; Liu, B.; Bai, W.D.; Ai, L.Z.; Rao, P.F.; Ni, L.; Lv, X.C. Monascus purpureus-fermented common buckwheat protects against dyslipidemia and non-alcoholic fatty liver disease through the regulation of liver metabolome and intestinal microbiome. *Food Res. Int.* **2020**, *136*, 109511. [CrossRef]
32. Li, S.; Liang, T.; Zhang, Y.; Huang, K.; Yang, S.; Lv, H.; Chen, Y.; Zhang, C.; Guan, X. Vitexin alleviates high-fat diet induced brain oxidative stress and inflammation via anti-oxidant, anti-inflammatory and gut microbiota modulating properties. *Free Radic. Biol. Med.* **2021**, *171*, 332–344. [CrossRef]
33. Geng, J.; Ni, Q.; Sun, W.; Li, L.; Feng, X. The links between gut microbiota and obesity and obesity related diseases. *Biomed. Pharmacother.* **2022**, *147*, 112678. [CrossRef]
34. Huang, K.; Yu, W.; Li, S.; Guan, X.; Liu, J.; Song, H.; Liu, D.; Duan, R. Effect of embryo-remaining oat rice on the lipid profile and intestinal microbiota in high-fat diet fed rats. *Food Res. Int.* **2020**, *129*, 108816. [CrossRef]
35. Li, Y.; Li, W.; Wang, X.; Ding, C.; Liu, J.; Li, Y.; Li, W.; Sun, Y. High-Salt Diet-Induced Gastritis in C57BL/6 Mice is Associated with Microbial Dysbiosis and Alleviated by a Buckwheat Diet. *Mol. Nutr. Food Res.* **2020**, *64*, e1900965. [CrossRef]
36. Ren, Y.; Wu, S.; Xia, Y.; Huang, J.; Ye, J.; Xuan, Z.; Li, P.; Du, B. Probiotic-fermented black tartary buckwheat alleviates hyperlipidemia and gut microbiota dysbiosis in rats fed with a high-fat diet. *Food Funct.* **2021**, *12*, 6045–6057. [CrossRef]
37. Zhang, B.; Xu, Y.; Liu, S.; Lv, H.; Hu, Y.; Wang, Y.; Li, Z.; Wang, J.; Ji, X.; Ma, H.; et al. Dietary Supplementation of Foxtail Millet Ameliorates Colitis-Associated Colorectal Cancer in Mice via Activation of Gut Receptors and Suppression of the STAT3 Pathway. *Nutrients* **2020**, *12*, 2367. [CrossRef]
38. Gong, L.; Wang, T.; Sun, C.; Wang, J.; Sun, B. Whole barley prevents obesity and dyslipidemia without the involvement of the gut microbiota in germ free C57BL/6J obese mice. *Food Funct.* **2019**, *10*, 7498–7508. [CrossRef]
39. Ji, Y.; Ma, N.; Zhang, J.; Wang, H.; Tao, T.; Pei, F.; Hu, Q. Dietary intake of mixture coarse cereals prevents obesity by altering the gut microbiota in high-fat diet fed mice. *Food Chem. Toxicol.* **2020**, *147*, 111901. [CrossRef]
40. Liu, W.; Luo, X.; Tang, J.; Mo, Q.; Zhong, H.; Zhang, H.; Feng, F. A bridge for short-chain fatty acids to affect inflammatory bowel disease, type 1 diabetes, and non-alcoholic fatty liver disease positively: By changing gut barrier. *Eur. J. Nutr.* **2021**, *60*, 2317–2330. [CrossRef]
41. Gao, H.; Song, R.; Li, Y.; Zhang, W.; Wan, Z.; Wang, Y.; Zhang, H.; Han, S. Effects of Oat Fiber Intervention on Cognitive Behavior in LDLR−/−Mice Modeling Atherosclerosis by Targeting the Microbiome-Gut-Brain Axis. *J. Agric. Food Chem.* **2020**, *68*, 14480–14491. [CrossRef] [PubMed]
42. Fotschki, B.; Juśkiewicz, J.; Jurgoński, A.; Amarowicz, R.; Opyd, P.; Bez, J.; Muranyi, I.; Petersen, I.L.; Llopis, M.L. Protein-Rich Flours from Quinoa and Buckwheat Favourably Affect the Growth Parameters, Intestinal Microbial Activity and Plasma Lipid Profile of Rats. *Nutrients* **2020**, *12*, 2781. [CrossRef] [PubMed]
43. Lie, L.; Brown, L.; Forrester, T.E.; Plange-Rhule, J.; Bovet, P.; Lambert, E.V.; Layden, B.T.; Luke, A.; Dugas, L.R. The Association of Dietary Fiber Intake with Cardiometabolic Risk in Four Countries across the Epidemiologic Transition. *Nutrients* **2018**, *10*, 628. [CrossRef]
44. Xue, Y.; Cui, L.; Qi, J.; Ojo, O.; Du, X.; Liu, Y.; Wang, X. The effect of dietary fiber (oat bran) supplement on blood pressure in patients with essential hypertension: A randomized controlled trial. *Nutr. Metab. Cardiovasc. Dis.* **2021**, *31*, 2458–2470. [CrossRef]
45. Cicero, A.F.; Fogacci, F.; Veronesi, M.; Strocchi, E.; Grandi, E.; Rizzoli, E.; Poli, A.; Marangoni, F.; Borghi, C. A Randomized Placebo-Controlled Clinical Trial to Evaluate the Medium-Term Effects of Oat Fibers on Human Health: The Beta-Glucan Effects on Lipid Profile, Glycemia and inTestinal Health (BELT) Study. *Nutrients* **2020**, *12*, 686. [CrossRef]
46. Bai, J.; Zhao, J.; Waleed, A.A.; Wang, J.; Xue, L.; Liu, J.; Wang, Y.; Fan, M.; Qian, H.; Li, Y.; et al. Oat beta-glucan alleviates DSS-induced colitis via regulating gut microbiota metabolism in mice. *Food Funct.* **2021**, *12*, 8976–8993. [CrossRef]
47. Cheng, W.Y.; Lam, K.L.; Li, X.; Kong, A.P.S.; Cheung, P.C.K. Circadian disruption-induced metabolic syndrome in mice is ameliorated by oat beta-glucan mediated by gut microbiota. *Carbohydr. Polym.* **2021**, *267*, 118216. [CrossRef]
48. Zhou, Y.; Zhao, S.; Jiang, Y.; Wei, Y.; Zhou, X. Regulatory Function of Buckwheat-Resistant Starch Supplementation on Lipid Profile and Gut Microbiota in Mice Fed with a High-Fat Diet. *J. Food Sci.* **2019**, *84*, 2674–2681. [CrossRef]

49. Kristek, A.; Wiese, M.; Heuer, P.; Kosik, O.; Schär, M.Y.; Soycan, G.; Alsharif, S.; Kuhnle, G.G.C.; Walton, G.; Spencer, J.P.E. Oat bran, but not its isolated bioactive beta-glucans or polyphenols, have a bifidogenic effect in an in vitro fermentation model of the gut microbiota. *Br. J. Nutr.* **2019**, *121*, 549–559. [CrossRef]
50. Fardet, A. New hypotheses for the health-protective mechanisms of whole-grain cereals: What is beyond fibre? *Nutr. Res. Rev.* **2010**, *23*, 65–134. [CrossRef]
51. Ashley, D.; Marasini, D.; Brownmiller, C.; Lee, J.A.; Carbonero, F.; Lee, S.O. Impact of Grain Sorghum Polyphenols on Microbiota of Normal Weight and Overweight/Obese Subjects during In Vitro Fecal Fermentation. *Nutrients* **2019**, *11*, 217. [CrossRef] [PubMed]
52. Duan, R.; Guan, X.; Huang, K.; Zhang, Y.; Li, S.; Xia, J.A.; Shen, M. Flavonoids from Whole-Grain Oat Alleviated High-Fat Diet-Induced Hyperlipidemia via Regulating Bile Acid Metabolism and Gut Microbiota in Mice. *J. Agric. Food Chem.* **2021**, *69*, 7629–7640. [CrossRef] [PubMed]
53. Yang, R.; Shan, S.; Zhang, C.; Shi, J.; Li, H.; Li, Z. Inhibitory Effects of Bound Polyphenol from Foxtail Millet Bran on Colitis-Associated Carcinogenesis by the Restoration of Gut Microbiota in a Mice Model. *J. Agric. Food Chem.* **2020**, *68*, 3506–3517. [CrossRef] [PubMed]
54. Liu, F.; Shan, S.; Li, H.; Shi, J.; Hao, R.; Yang, R.; Li, Z. Millet shell polyphenols prevent atherosclerosis by protecting the gut barrier and remodeling the gut microbiota in ApoE−/− mice. *Food Funct.* **2021**, *12*, 7298–7309. [CrossRef]
55. Heinritz, S.N.; Mosenthin, R.; Weiss, E. Use of pigs as a potential model for research into dietary modulation of the human gut microbiota. *Nutr. Res. Rev.* **2013**, *26*, 191–209. [CrossRef]

nutrients

MDPI

Article

Long-Term High-Fat High-Fructose Diet Induces Type 2 Diabetes in Rats through Oxidative Stress

Yue Zhao [1,2], Qing-Yu Wang [1], Lv-Tao Zeng [3], Jing-Jing Wang [4], Zhen Liu [1,2], Guo-Qing Fan [1,2], Jin Li [1] and Jian-Ping Cai [1,2,*]

1 The Key Laboratory of Geriatrics, Beijing Institute of Geriatrics, Institute of Geriatric Medicine, Chinese Academy of Medical Sciences, Beijing Hospital/National Center of Gerontology of National Health Commission, Beijing 100730, China; vita1023@163.com (Y.Z.); daisyqingyu@outlook.com (Q.-Y.W.); jane_computer@sina.com (Z.L.); missyueying@163.com (G.-Q.F.); niyani88@126.com (J.L.)
2 Graduate School of Peking Union Medical College, Beijing 100730, China
3 Peking University Fifth School of Clinical Medicine, Beijing Hospital/National Center of Gerontology of National Health Commission, Beijing 100730, China; zengtiancai@sina.com
4 Department of Clinical Laboratory, Henan Provincial People's Hospital, People's Hospital of Zhengzhou University, Zhengzhou 450066, China; 15133919987@163.com
* Correspondence: caijianping3200@bjhmoh.cn; Tel.: +86-010-58115080

Abstract: Long-term consumption of a Western diet is a major cause of type 2 diabetes mellitus (T2DM). However, the effects of diet on pancreatic structure and function remain unclear. Rats fed a high-fat, high-fructose (HFHF) diet were compared with rats fed a normal diet for 3 and 18 months. Plasma biochemical parameters and inflammatory factors were used to reflect metabolic profile and inflammatory status. The rats developed metabolic disorders, and the size of the islets in the pancreas increased after 3 months of HFHF treatment but decreased and became irregular after 18 months. Fasting insulin, C-peptide, proinsulin, and intact proinsulin levels were significantly higher in the HFHF group than those in the age-matched controls. Plasmatic oxidative parameters and nucleic acid oxidation markers (8-oxo-Gsn and 8-oxo-dGsn) became elevated before inflammatory factors, suggesting that the HFHF diet increased the degree of oxidative stress before affecting inflammation. Single-cell RNA sequencing also verified that the transcriptional level of oxidoreductase changed differently in islet subpopulations with aging and long-term HFHF diet. We demonstrated that long-term HFHF diet and aging-associated structural and transcriptomic changes that underlie pancreatic islet functional decay is a possible underlying mechanism of T2DM, and our study could provide new insights to prevent the development of diet-induced T2DM.

Keywords: Western dietary pattern; metabolic disorders; pancreas; islets; single-cell RNA sequencing analysis

Citation: Zhao, Y.; Wang, Q.-Y.; Zeng, L.-T.; Wang, J.-J.; Liu, Z.; Fan, G.-Q.; Li, J.; Cai, J.-P. Long-Term High-Fat High-Fructose Diet Induces Type 2 Diabetes in Rats through Oxidative Stress. *Nutrients* **2022**, *14*, 2181. https://doi.org/10.3390/nu14112181

Academic Editor: Lindsay Brown

Received: 1 April 2022
Accepted: 20 May 2022
Published: 24 May 2022

1. Introduction

Diabetes mellitus, also known as diabetes, is a serious metabolic disorder with multiple etiologies characterized by disturbances in carbohydrate, fat, and protein metabolism [1,2]. It is estimated that diabetes causes 11.3% of global deaths annually [3]. Overall, 4.2 million people in the 20–79-year age group died of diabetes in 2019 [3]. The increasing mortality associated with diabetes demonstrates the urgency to explore the pathogenesis of the disease.

The WHO proposed several dietary practices according to the confirmed correlation between obesity and type 2 diabetes mellitus (T2DM) risk [1,4]. The consumption of high amounts of fat and refined carbohydrates in the diet may be risk factors for dyslipidemia [5], obesity [6–8], insulin resistance [9], and heart disease [10,11]. In addition, many cross-sectional studies and meta-analyses have reported a positive correlation between strong adherence to a Western dietary pattern and an increased risk of T2DM [12–23]. However, the target organs and pathogenic mechanisms of diet in T2DM remain unclear.

In 2001, Brownlee first proposed that oxidative stress (OS) may play a major role in the pathophysiology of diabetes and its complications [24]. OS is caused by an imbalance between the production of reactive oxygen species (ROS) and the activity of the antioxidant defense system [25]. Epidemiological studies have also confirmed the elevation of OS biomarker levels in T2DM compared with those in healthy conditions, including antioxidant enzyme activity, lipid peroxidation, and nucleic acid oxidation markers. In a study that included 80 T2DM patients and 79 apparently healthy controls, the ferric reducing ability of plasma, γ-glutamyltransferase, and plasma glutathione reductase (GR) levels were significantly higher in patients with diabetes than those in nondiabetic people [26]. Bandeira et al. found that the total superoxide dismutase (SOD) activity and lipid peroxidation levels were higher in diabetes patients than those in non-diabetes patients through their case–control study (in the 40 to 86 year old group). In further stratified groups, the rate of lipid peroxidation was significantly higher in both groups of diabetes patients (hypertensive and normotensive) than in the prediabetic groups and controls, revealing that increased lipid peroxidation is closely related to T2DM [27]. Furthermore, the levels of the commonly used nucleic acid oxidation markers 8-dihydroguanosine (8-oxo-Gsn) and 8-hydroxy-2-deoxyguanosine (8-oxo-dGsn) were also higher in T2DM patients than those in healthy controls. Broedbaek et al. found that the level of the RNA oxidation marker 8-oxo-Gsn in urine independently predicted diabetes-related mortality [28,29]. Our previous study also verified that 8-oxo-Gsn level in urine was correlated with diabetic nephropathy, and that 8-oxo-Gsn might be a sensitive marker of diabetic nephropathy [30]. However, the mechanisms underlying OS targets in T2DM remain unclear.

Many T2DM animal models, including spontaneous and induced diabetes models, have been developed [31]. These models include characteristics such as insulin deficiency, insulin resistance, and/or beta cell failure [31]. Among induced rodent models, streptozotocin- or alloxan-induced diabetic models are the most common [31,32]. Both chemicals act as cytotoxic glucose analogs that accumulate in pancreatic beta cells through glucose transporter 2 (GLUT2), causing selective destruction of islet beta cells [31,32]. Although these models are helpful for studying the pathogenesis of diabetes, they are not in line with the internal environmental changes induced by long-term dietary treatments. Through the administration of a combination of fructose and fat, Lozano et al. first developed an ideal model for researching the effect of diet on the development of T2DM [33]. They noted marked hepatic vacuolar degeneration and increased ROS levels in the liver and vascular system of diabetic rats that had been fed a high-fat, high-fructose (HFHF) diet [33]. Dal et al. [34] and Werf et al. [35] found that diabetes symptoms and increased levels of ROS in the liver and vascular system induced by HFHF diet were partially corrected by supplementing the antioxidant-rich foods, such as cherries and red cabbages. Of note, existing studies have mainly focused on the HFHF diet-induced impairment of the liver, vascular system, cognition, and retina [36–40]. As an essential organ for sensing blood glucose concentrations and synthesizing and releasing glucose-regulating hormones, islets in the pancreas play a vital part in blood glucose homeostasis [41,42]. However, the effects of an HFHF diet on pancreatic structure and function, as well as their potential mechanisms, remain unclear.

Therefore, the present study explored the impact of a long-term HFHF diet on the rat pancreas. An 18 month HFHF diet was used to trigger the onset of T2DM in rats, which was confirmed by assessing metabolic, oxidative, and inflammatory parameters. After observing the changes in pancreatic structure and function, we used single-cell RNA-sequencing (scRNA-seq) to explore the distribution and transcriptomic patterns in pancreatic cells and islet cells. We decoded long-term HFHF diet and aging-associated structural and transcriptomic changes that underlie pancreatic islet functional decay at single-cell resolution and indicated that inflammation occurs after oxidative stress, suggesting a possible underlying mechanism of T2DM and providing new insights to prevent the development of diet-induced T2DM.

2. Materials and Methods

2.1. Ethics Statement

The animal use protocol was reviewed and approved by the Animal Ethics and Welfare Committee of Beijing Stomatological Hospital (Approval No. KQYY-202103-001). All efforts were made to minimize animal suffering and to reduce the number of animals used.

2.2. Animals and Induction of Diabetes

Forty male Wistar rats (male; 2 months old; 280.75 ± 8.19 g; SPF (Beijing Biotechnology Co., Ltd., Beijing, China), were fed in a temperature-controlled room with a 12 h light/dark cycle and ad libitum access to water and food. After 1 week, eight rats were randomly selected and sacrificed (baseline), and the other rats were then randomly divided into two groups of 16 rats each (control and HFHF groups). The control and HFHF groups were fed a normal diet and HFHF diet (SPF (Beijing) Biotechnology Co., Ltd.), respectively, as previously reported [33]. The primary ingredients of each diet are shown in Table S1.

Eight rats per group were sacrificed 3 months after the start of the administration of each diet. The rats were intraperitoneally anesthetized using Zoletil®50 (Boehringer Ingelheim, Ingelheim, Rhein, Germany). Blood was collected from the heart, and plasma was stored at −80 °C for subsequent analysis. All tissues were rinsed, weighed, and embedded in Tissue-Tek® Optimal Cutting Temperature compound (Sakura Finetek USA, Torrance, CA, USA) or frozen in liquid nitrogen. Spontaneous metabolism was assessed using standard rodent metabolic cages, and then 24 h urine was collected. After 18 months of feeding, six rats in each group were anesthetized and sacrificed according to the above procedures. The remaining two rats per group had their islets isolated for single-cell RNA sequencing analysis. The study outline is shown in Figure 1.

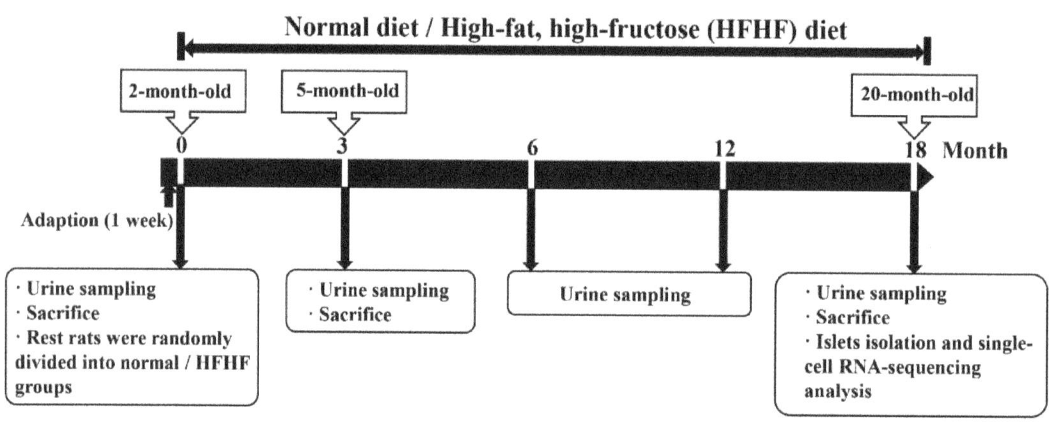

Figure 1. The study outline.

2.3. Metabolic, Inflammatory, and Oxidative Parameter Analyses

2.3.1. Glucose Measurement

Blood glucose levels were evaluated on the basis of the fasting glucose value (One-Touch Ultra Easy® glucometer; Johnson & Johnson, New Brunswick, NJ, USA) and intraperitoneal glucose tolerance test (IpGTT). A 20% glucose solution was injected intraperitoneally in rats at a 2 g/kg dose. A glucometer measured the caudal capillary glycemia of rats at 0, 15, 30, 60, and 120 min after injection.

2.3.2. Plasmatic Metabolic and Oxidative Parameters Assessments

All procedures for determining glycated serum protein (GSP; A037-2-1; Nanjing Jiancheng Bioengineering Institute, Nanjing, China), pyruvic acid (A081), glucose (A154-1-1), total triglycerides (TG; A110-1-1), total cholesterol (T-CHO; A111-1-1), low-density lipoprotein cholesterol (LDL-C; A113-1-1), SOD (A001-3), xanthine oxidase (XOD; A002-1-1), total antioxidant capacity (T-AOC; A05-3-1), lactate dehydrogenase (LDH; A020-2), T-GSH (A061-1), GR (A062-1-1), hydrogen peroxide (H_2O_2; A064-1-1), and lipid hydroperoxide (LPO; A106-1) were performed according to the manufacturer's instructions (Nanjing Jiancheng Bioengineering Institute, Nanjing, China).

2.3.3. Detection of 8-oxo-Gsn and 8-oxo-dGsn in Urine

First, 100 μL of urine was mixed with 400 μL of 30% methanol (Sigma-Aldrich, St. Louis, MO, USA) solution to acquire a 1:5 diluted urine sample for oxidized guanine nucleoside analysis. Then, 10 μL of 80 pg/μL [$_{15}N^5$] 8-hydroxy-2-deoxyguanosine (8-oxo-dGsn; Cambridge Isotope Laboratories, Andover, MA, USA) and 10 μL of 80 pg/μL [$_{15}N^2{}_{13}C^1$] 8-dihydroguanosine (8-oxo-Gsn; Toronto Research Chemicals, Toronto, ON, Canada) were added as internal controls and centrifuged at 12,000× *g* for 15 min at 4 °C.

Next, 100 μL of supernatant was loaded in the liquid chromatograph with tandem mass spectrometry (LC–MS/MS) (Agilent Technologies, Santa Clara, CA, USA) for the analysis of 8-oxo-dGsn and 8-oxo-Gsn [30]. The mobile phase consisted of 20 mM ammonium acetate (Thermo Fisher Scientific, Waltham, MA, USA) (A) and 100% methanol (B), which was added in a Waters Sunfire C18 column (5 μm, 4.6 × 250 mm). Gradient elution was performed from 0 to 6.0 min with 8% B at 0.8 mL/min, from 6.0 to 6.01 min changing from 8% B to 50% B at 1 mL/min (sixth curve), from 6.01 to 8 min in 50% B at 1 mL/min, from 8.0 to 8.01 min changing from 50% B to 8% B (sixth curve), and then from 8.01 to 10.0 min in 8% B at 0.8 mL/min (sixth curve). The injection volume was 20 μL, and the UV detection wavelength was 233 nm. The final concentrations of 8-oxo-dGsn and 8-oxo-Gsn in urine were calculated on the standard curve.

2.3.4. Plasmatic Inflammatory Parameters

Interleukin (IL)-1β, IL-6, tumor necrosis factor (TNF)-α, vascular endothelial growth factor (VEGF), intercellular cell adhesion molecule (ICAM)-1, and monocyte chemotactic protein (MCP)-1 were detected using enzyme-linked immunosorbent assay (ELISA) at the Nanjing Jiancheng Bioengineering Institute.

2.4. Histological Studies

To estimate histological changes, embedded pancreatic tissues were sectioned to a thickness of 8 μm for hematoxylin/eosin (HE) staining (G1121; Beijing Solarbio Science & Technology Co., Ltd., Beijing, China), Masson's trichrome staining (G1346; Beijing Solarbio Science s& Technology Co., Ltd., Beijing, China), and immunofluorescent staining. Mouse anti-insulin (rat, 1:500, ab181547; Abcam, Waltham, MA, USA), rabbit anti-glucagon (rabbit, 1:150, 15954-1-AP; Proteintech, Rosemont, IL, USA), and the corresponding secondary antibodies (donkey anti-rabbit IgG (H + L) Highly Cross-Adsorbed Secondary Antibody, Alexa Fluor 555; donkey anti-mouse IgG (H + L) Highly Cross-Adsorbed Secondary Antibody, Alexa Fluor 488(REF.A21202); Life Technologies, Carlsbad, CA, USA) were used to detect alpha and beta islet cells. As described in Chansela et al.'s study, the islets could be categorized as small (<10,000 μm^2), medium (10,000–50,000 μm^2), large (50,000–100,000 μm^2), and extra large (>100,000 μm^2) on the basis of the total area of islets [43]. The islet density is calculated as follows: the number of each islet size/total number of islets ×100 [43].

2.5. Pancreas-Specific Endocrine Protein Expression in Plasma

To evaluate the function of the pancreatic islets, fasting plasma insulin (Mercodia High Range Rat Insulin ELISA, Lot10-1145-01; Mercodia, Uppsala, Sweden) and C-peptide levels (Mercodia Rat C-peptide ELISA, Lot10-1172-01; Mercodia) were tested using a double-antibody sandwich enzyme-linked immunoassay. Proinsulin (Mercodia High Range Rat Proinsulin ELISA, Lot 10-1232-01; Mercodia) and intact proinsulin (Mouse/Rat Intact Proinsulin Assay Kit, No. 27706; Immuno-Biological Laboratories Co., Ltd., Naka, Fujioka-Shi, Japan) were examined according to the manufacturer's instructions.

2.6. 10X Genomics Single-Cell RNA Sequencing Analysis of Pancreas in 18 Month HFHF Rats

2.6.1. Preparation of Single-Cell Suspension

Pancreases of two rats per group (baseline, HFHF, and control groups after 18 months of feeding) were subjected to scRNA-seq. The islets were isolated as previously reported [44,45]. Briefly, precooled 0.5 mg/mL collagenase P solution was injected into the common bile duct to expand the pancreas, and then the pancreas was quickly extracted and transferred into a digestion bottle. After incubation at 38 °C for 10 min, the suspension was centrifuged at 1200× *g* for 2–3 min at 4 °C. The precipitate was washed twice with Hanks' solution and resuspended in Histopaque-1077. After gently adding RPMI 1640 (containing 10% fetal bovine serum) solution to the liquid surface, the tube was centrifuged at 3000× *g* for 20 min at 4 °C. Granules scattered between the interfaces were collected and resuspended in 1 mL TrypLE (12604013; Thermo Fisher Scientific) for further digestion. DAPI staining was used to evaluate cell viability (Count Star, TANON SCIENCE & TECHONLOGY CO, Shanghai, China). When cell viability reached 80% and the proportion of doublets was less than 20%, an estimated 10,000 single cells per sample were suspended in phosphate-buffered saline solution (containing 0.04% bovine serum albumin).

2.6.2. Cell Capture and cDNA Synthesis

Cell suspensions were loaded on a chromium single-cell controller (10× Genomics, San Francisco, CA, USA) to generate single-cell gel beads in emulsion (GEMs). A Chromium Next GEM Single Cell 3′ GEM, Library & Gel Bead Kit v3.1 (1000121; 10× Genomics, Pleasanton, CA, USA) and Single-Cell B Chip Kit (1000127; 10× Genomics) were used according to the manufacturer's protocol. As previously described, single-cell droplet collection, cDNA amplification, and sequencing library preparation were performed [46,47]. The cDNA libraries were sequenced on an Illumina Novaseq6000 sequencer with a sequencing depth of at least 30,000 reads per cell and a pair end of 150 bp (PE150; Capitalbio Technology Corporation, Beijing, China).

2.6.3. scRNA-Seq Data Preprocessing

Data were analyzed by Capitalbio Technology Corporation (Beijing, China). Cell Ranger software (ver 6.1.2 accessed on 25 October 2021; https://support.10xgenomics.com/single-cell-gene-expression/software/downloads/latest) was used to perform alignment, filtering, barcode counting, and unique molecular identifier (UMI) counting and then to generate a feature barcode matrix and determine clusters. Cells whose gene number was <200, whose gene number was ranked in the top 1%, or whose mitochondrial gene ratio was >25% were regarded as abnormal and filtered out. Data were then loaded into Seurat 3.0 (accessed on 6 June 2019; http://satijalab.org/seurat) without normalization, scaling, or centering. In addition to the expression data, metadata for each cell were collected, including clone identity, cell-cycle phase, and sample source. Highly variable genes were identified and used as inputs for dimensionality reduction using principal component analysis. The resulting principal components (PCs) and correlated genes were examined to determine the number of components included in the downstream analysis. T-distributed stochastic neighbor embedding and uniform manifold approximation and projection (UMAP) plots (based on the first 10 PCs) were used to visualize the cells in a two-dimensional space. Gene Ontology (GO) enrichment analysis was performed using the Rstudio (Version 4.1.2) software program with Benjamini–Hochberg multiple testing adjustment, which was based on all differentially expressed genes (p-adj < 0.05, avg_$\log_2$FC > 0.585).

2.7. Statistical Analyses

Results are expressed as the mean ± standard deviation (SD) or median (interquartile range; IQR). Statistical analyses were performed using Student's t-test, Mann–Whitney U test for unpaired data (SPSS®, 22.0, Chicago, IL, USA), or two-way analysis of variance and

Tukey's post hoc test (GraphPad Prism®, 8.3.1, San Diego, CA, USA). Statistical significance was set at $p < 0.05$.

3. Results

3.1. Long-Term HFHF Diet Induced Diabetes

Compared to those of the control group, after 3 months of HFHF treatment, the HFHF diet induced a sharp increase in body weight ($p < 0.0001$) and body mass index (BMI; $p < 0.0001$) in the rats of the HFHF group (Table S2, Figures 2A and S1). Fasting glycemia increased significantly between groups ($p < 0.05$) (Figure 2B). Lipid-related biochemical parameters increased in the HFHF group (Table S2). Obvious vacuolar degeneration in the livers of HFHF diet-fed rats was also observed (Figure S2).

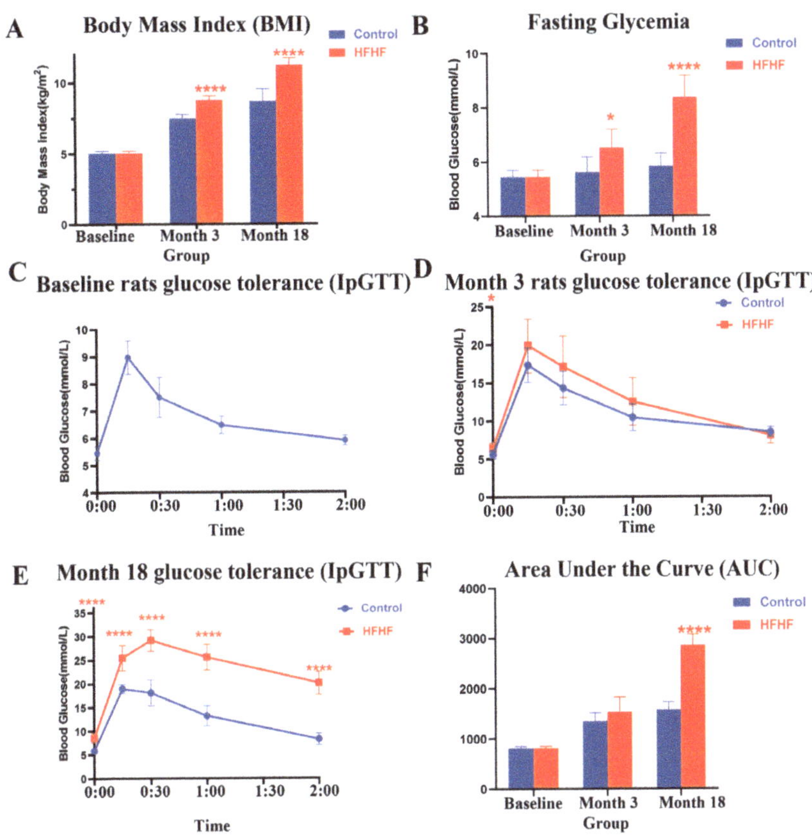

Figure 2. BMI, fasting glucose, and glucose tolerance measured in this study. (**A**) The body mass index (BMI) in this study for the normal diet (control, blue) and high-fat, high-fructose diet (HFHF, red) groups. (**B**) Evolution of fasting glycemia. (**C–E**) Evolution of glucose tolerance measured via the intraperitoneal glucose tolerance test (IpGTT) in all groups at the beginning (baseline) and after 3 (month 3) and 18 (month 18) months of feeding. (**F**) The area under the curve (AUC) during the IpGTT. Results are shown as the mean ± SD. * Significant results versus the age-matched control group: * $p < 0.05$, **** $p < 0.0001$.

After 18 months of treatment, the HFHF group exhibited more severe metabolic disorders, particularly dyslipidemia and hyperglycemia. Compared to the age-matched control group, the fasting glycemia level ($p < 0.0001$) and AUC of the IpGTT test ($p < 0.0001$) were significantly higher in the HFHF group (Figure 2). Lipid droplets showed further accumulation, and hepatic fibrosis was observed in the liver of HFHF rats (Figure S2). These findings suggest that long-term HFHF induces T2DM.

3.2. HFHF Diet Influenced Islet Cell Function and Structure

Insulin, secreted by islet cells, is the essential hormone responsible for maintaining glucose homeostasis [48]. Proinsulin, a precursor molecule of insulin, is intracellularly cleaved by prohormone convertases, which hydrolyze positions 32–33 and 65–66 in proinsulin, and the carboxypeptidase E, which removes the basic residues 64–65 and 31–32 to form insulin and C-peptide [41,49–52]. Therefore, the total proinsulin consists of several partially hydrolyzed forms of proinsulin intermediate (32–33 split proinsulin, 65–66 split proinsulin, des 64–65 proinsulin, and des 31–32 proinsulin) and intact proinsulin, which is the unhydrolyzed form of proinsulin [51–54]. Hence, the detection of total proinsulin and intact proinsulin could reflect the hormone-processing ability of islet cells.

To assess the functional consequences of an HFHF diet on the pancreas, levels of the above hormones were measured using ELISA (Figure 3A–D). After 3 months of receiving the appropriate diet, the fasting insulin level of HFHF rats was higher than that of the age-matched control group, whereas there were no differences in C-peptide, proinsulin, and intact proinsulin. After 18 months, the levels of these four hormones were significantly higher in the HFHF group than in the control group. Furthermore, during the IpGTT, the insulin, C-peptide, and proinsulin levels were also higher in the HFHF group than in the control group at each timepoint (Figure S3).

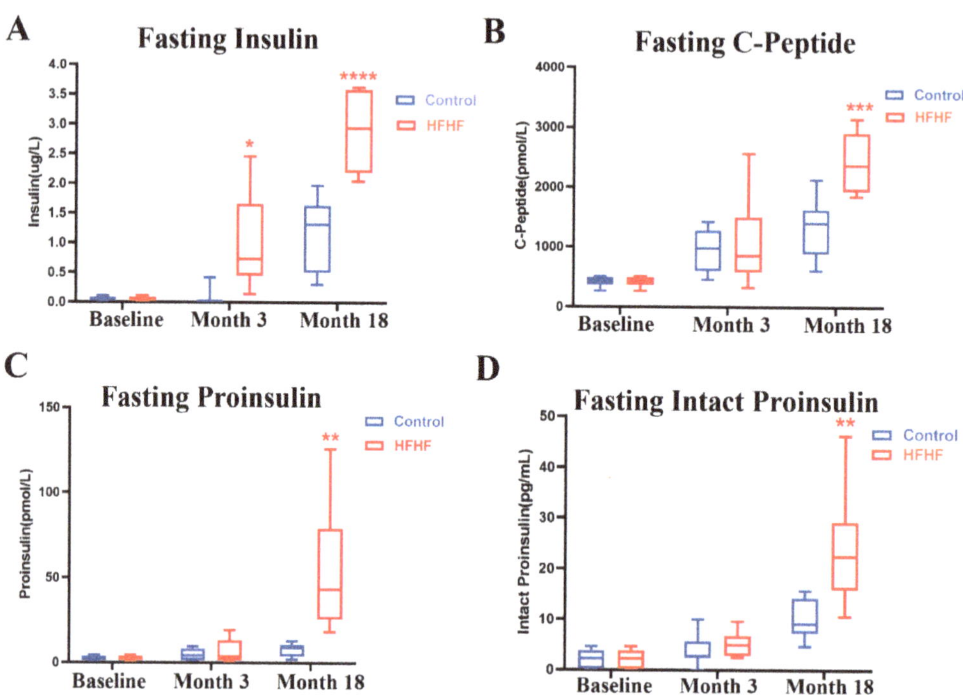

Figure 3. *Cont.*

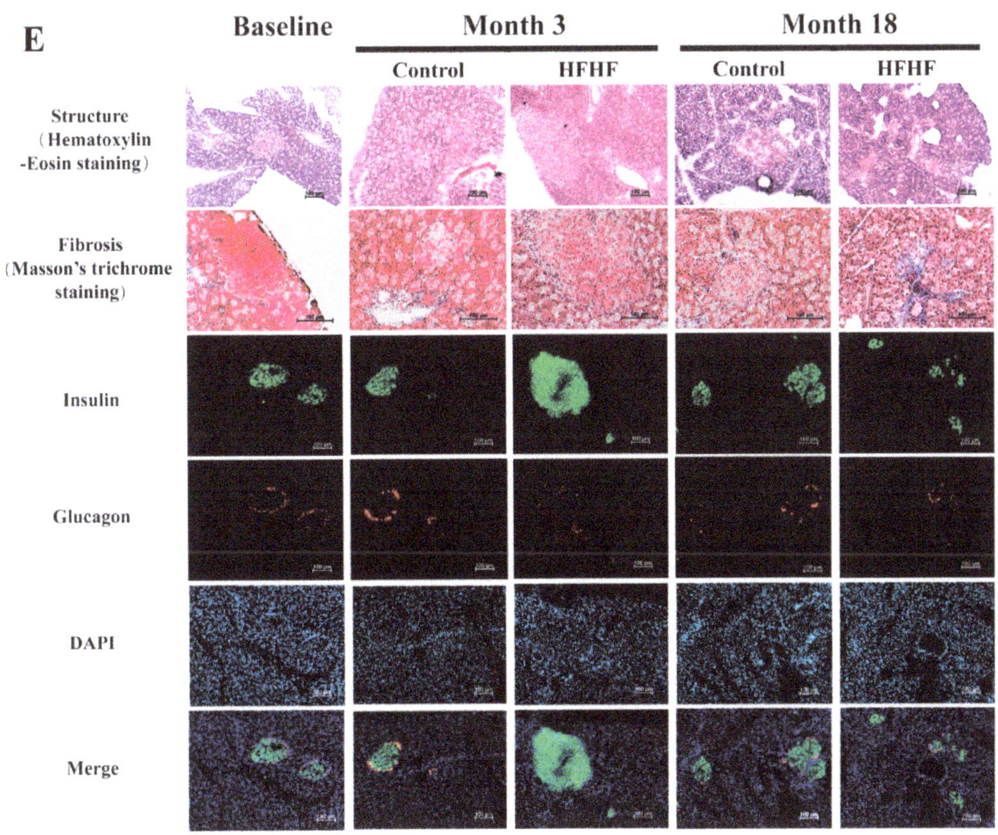

Figure 3. Structural and function changes of the pancreas in this study. (**A–D**) Fasting plasmatic insulin, C-peptide, proinsulin, and intact proinsulin levels during the study for the normal diet (control, blue) and high-fat, high-fructose diet (HFHF, red) groups at the beginning (baseline) and after 3 (month 3) and 18 (month 3) months of feeding. Results are shown as the mean ± SD. * Significant results versus the age-matched control group: * $p < 0.05$, ** $p < 0.01$, *** $p < 0.001$, **** $p < 0.0001$. (**E**) Changes in islets were visualized using hematoxylin/eosin staining, Masson's Trichrome staining, and insulin and glucagon immunofluorescence staining at the beginning (baseline) and after 3 (month 3) and 18 (month 18) months of normal (control) and high-fat, high-fructose (HFHF) diet feeding. Scale bar = 100 μm.

The islet structure was examined using HE staining. As shown in Figures 3E and S4, the size of islets in the pancreas of the HFHF group rats increased after 3 months. However, after 18 months of feeding, islet size decreased, and the morphology became irregular. This phenomenon was confirmed by immunofluorescence staining (Figure S4). At the same time, the pancreatic tissue of rats in this group was found to have been infiltrated by a large number of adipocytes (Figure S5).

3.3. HFHF Diet Increased the Plasmatic Degree of Oxidation

As shown in Figure 4A, 8-oxo-Gsn levels were significantly higher in the urine of the HFHF group rats than that of the age-matched control group (3 months, $p < 0.01$; 6 months, $p < 0.01$; 12 months, $p < 0.01$; 18 months, $p = 0.4$). A similar trend was observed in urinal 8-oxo-dGsn detection. After 12 months of HFHF treatment, the level of 8-oxo-dGsn in the HFHF group increased significantly compared to the age-matched control group (12 months, $p < 0.05$; 18 months, $p < 0.05$) (Figure 4B), while the amount of 8-oxo-dGsn in the normal diet group remained unchanged during the study. Thus, long-term consumption of an HFHF diet induces RNA and DNA oxidation, suggesting the presence of an over-oxidative environment.

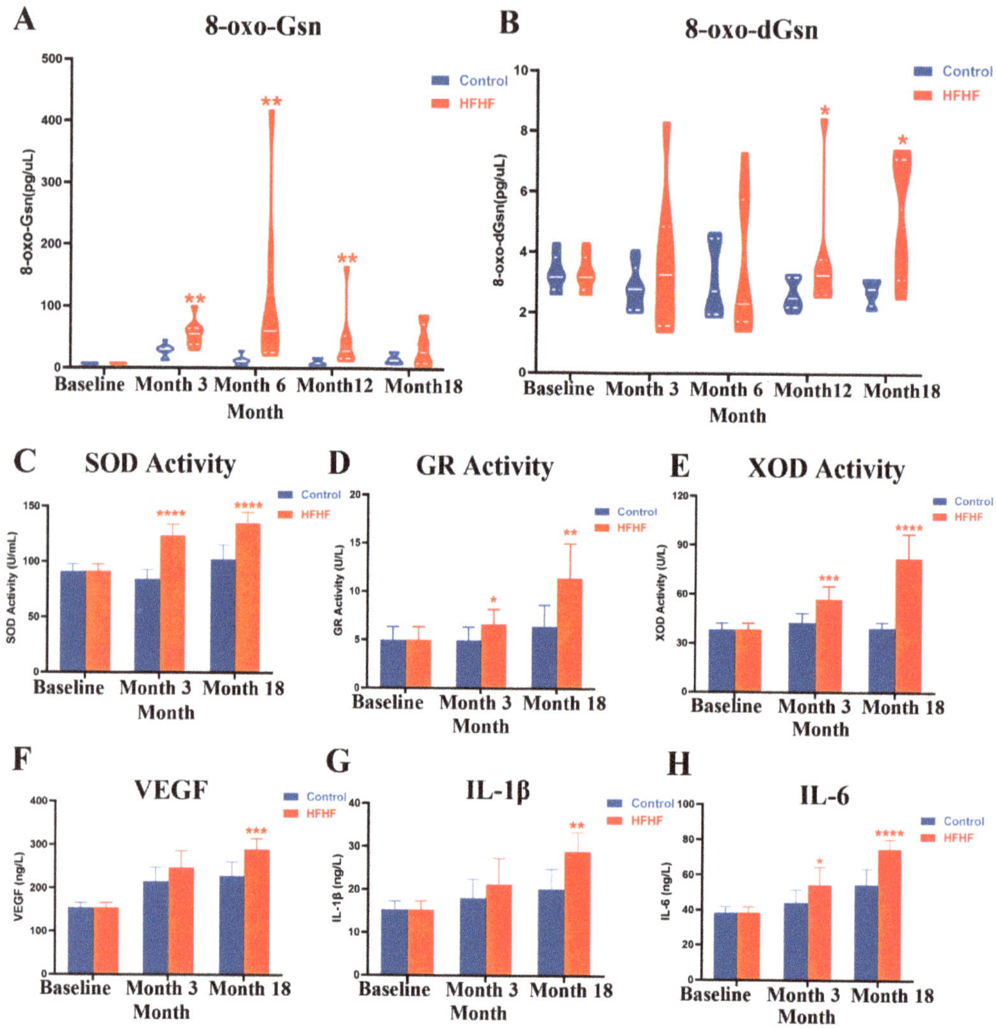

Figure 4. Changes in oxidation and inflammation levels during this study. Evolution of 8-oxo-Gsn (**A**) and 8-oxo-dGsn (**B**) during the study for the normal diet (control, blue) and high-fat, high-fructose diet (HFHF, red) groups. Results are shown as the median (interquartile range). (**C**,**D**) Effects of diet on SOD (**C**), GR (**D**), and XOD (**E**) activity at the beginning (baseline) and after 3 (month 3) and 18 (month 18) months of control and HFHF diet. (**F**–**H**). Effects of diet on VEGF (**F**), IL-1β (**G**), and IL-6 (**H**) during the study at the beginning (baseline) and after 3 (month 3) and 18 (month 18) months of treatments. Results are shown as the mean ± SD. * Significant results versus the age-matched control group: * $p < 0.05$, ** $p < 0.01$, *** $p < 0.001$, **** $p < 0.0001$.

After administering the HFHF diet for 3 months, the levels of plasma oxidative parameters were significantly higher than those in the age-matched control group (Figure 4C–E, Table S2). After another 15 months of the HFHF diet, the levels of antioxidant enzyme activity (GR and XOD) and T-AOC in the HFHF group were twice those of the control group, suggesting that plasma oxidation was further intensified (Figure 4D,E, Table S2). Moreover, the expression of GSH, which acts as a ROS scavenger, decreased gradually (Table S2). These results suggest that the degree of OS in rats increased with consumption of the HFHF diet.

3.4. HFHF Diet Caused Inflammation

After feeding an HFHF diet for 3 months, compared with the age-matched group, there was no significant change in the plasma inflammation indices, except for IL-6 ($p < 0.05$) (Figure 4F–H, Table S2). Importantly, after 18 months, all inflammatory factors, including IL-1β ($p < 0.01$), IL-6 ($p < 0.0001$), MCP-1 ($p < 0.0001$), TNF-α ($p < 0.05$), VEGF ($p < 0.001$), and ICAM1 ($p < 0.05$) showed significant differences, indicating that the long-term HFHF diet induced an elevation in inflammation.

3.5. Identification and Transcriptomic Pattern of Pancreatic Cells from Single-Cell Levels in HFHF Rats after 18 Months

Six rats divided into baseline (young), HFHF (HFHF-old), and age-matched control (old) groups after 18 months of treatment were sacrificed, and their pancreases were isolated for scRNA-seq. A total of 78,024 cells were processed and qualified. After filtering the abnormal cells, data from 76,383 cells were loaded into Seurat for subsequent analysis (Figure S6F).

Unsupervised clustering analysis separated the cells into 10 clusters (Figures 5A and S7A). Specific cell markers were used to recognize identities of cell clusters, including *Cd3e* and *Cd8a* (T cells), *Ccl24* and *Cd83* (macrophage cells), *Cpe* and *Pnliprp1* (secretory cells), *Postn* and *Ccdc80* (fibroblast cells), *Irf8* and *Ly86* (B cells), *Aqp1* and *Flt1* (endothelial cells), *Acta2* and *Myh11* (vascular smooth muscle cells), and *Sox9* and *EpCAM* (ductal epithelial cells) (Figure 5B). The number of cell types in the pancreas differed among the three groups (Figure 5A,C). GO analysis indicated that inflammation and immune-related pathways were significantly enriched in the HFHF-old group compared to the old group (Figure S7C–E), which was consistent with a significant increase in the T-cell count (Figure 5C). These observations indicate that long-term HFHF diet altered the pancreatic cellular composition.

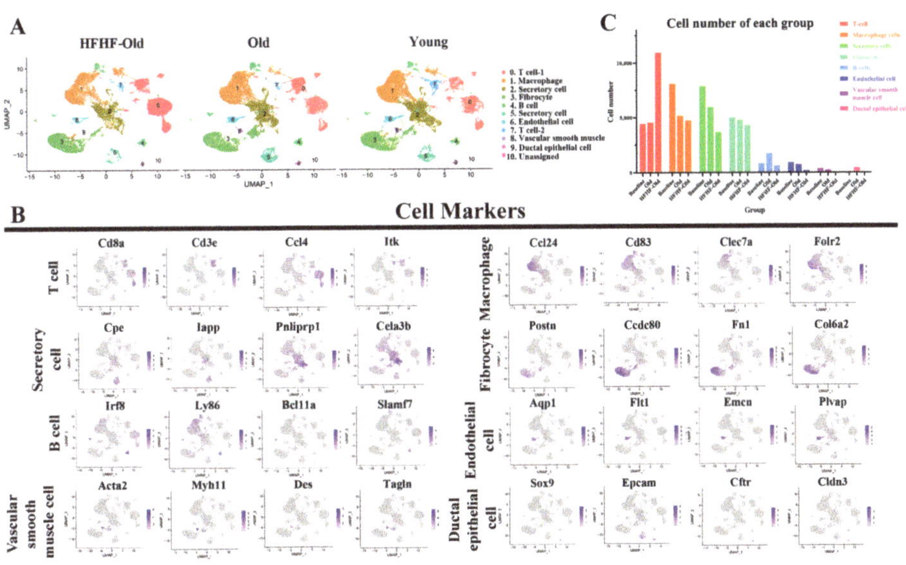

Figure 5. *Cont.*

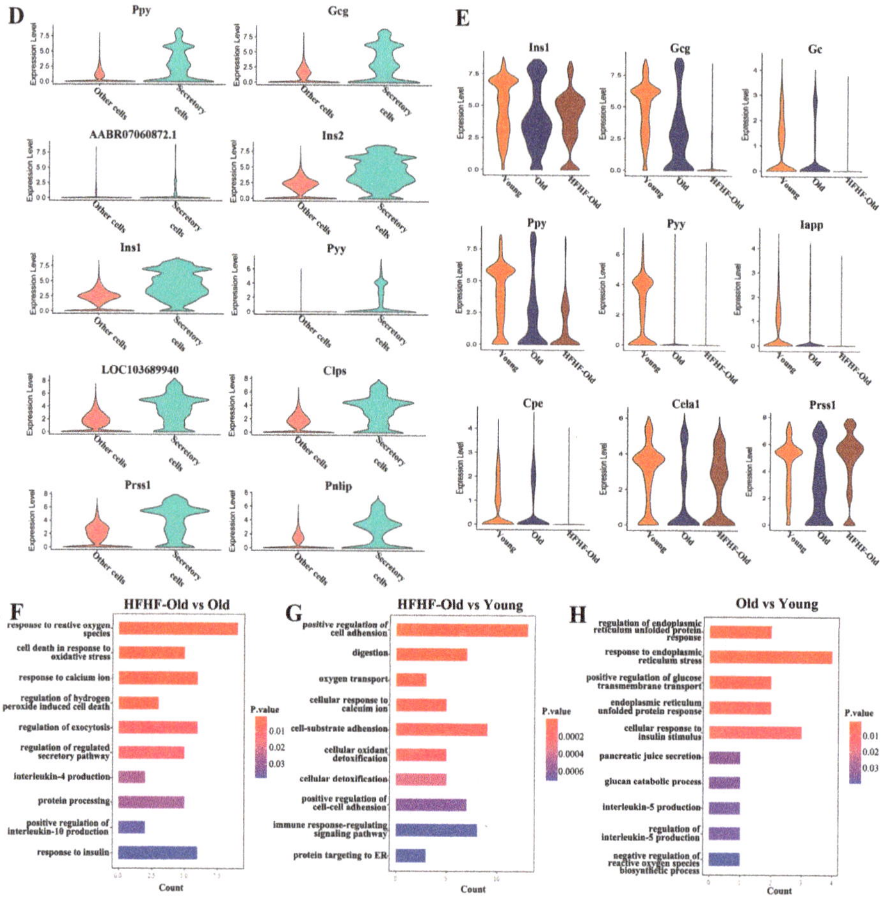

Figure 5. Distinct cell populations with transcriptional signatures in pancreatic secretory cells determined by a scRNA-Seq analysis of diabetic rats. (**A**) UMAP plot characterizing the differences in pancreatic cell distribution at the beginning (young) and after 18 months (HFHF-old, old) of diet feeding. Each point represents one cell. The distance between the two points is determined by the Pearson correlation of the 500 most highly expressed genes in each cell. Cell identity is recognized by the expression of marker genes. (**B**) UMAP plot of the marker gene expression in different cell types. (**C**) The number of different pancreatic cells at the beginning (young) and after 18 months (HFHF-old, old) of diet feeding. (**D**) Violin diagram of top 10 genes highly expressed in pancreatic secretory cells. (**E**) Violin diagram of secretory function-related gene expression in secretory cells among HFHF-old (HFHF, 18 months), old (control, 18 months), and young (baseline) groups. (**F–H**) Gene Ontology terms of upregulated enrichment obtained in secretory cells after a pairwise comparison of three groups ($p < 0.05$).

3.6. Long-Term HFHF Diet Affected the Expression of Characteristic Genes and Enriched Oxidative and Inflammatory Pathways in Secretory Cells

We then explored the changes in the expression of characteristic genes in secretory cells caused by the long-term HFHF diet and aging. As the cells in clusters 2 and 5 were classified as secretory cells (Figure 5A,B), we combined the two clusters and compared them with other types of cells. The top 10 differentially expressed genes in secretory cells are shown in Figure 5D, most of which were related to the synthesis of glucose-regulating hormones and pancreatic proteases. GO analysis indicated that peptide hormone secretion-

and digestion-related pathways were significantly enriched in the secretory cells (Figure S8). We found different expression patterns among the genes related to hormone synthesis in the three experimental groups. As shown in Figure 5E, the expression of genes related to glucose-regulating hormone synthesis decreased in the HFHF-old group, including *Ins1*, *Gcg*, *Ppy*, and *Pyy*. The total number of secretory cells also confirmed this among the groups, which showed that secretory cells number decreased in the old group, and the long-term HFHF diet exacerbated this decline (Figure 5C). In addition, GO analysis indicated that OS-related pathways were enriched in the secretory cell clusters of the HFHF-old group, while endoplasmic reticulum unfolded protein response-related pathways were mainly enriched in the old group (Figure 5F–H). Furthermore, inflammatory pathways were detected in the HFHF-old and old groups (Figure 5F–H). All of the above results verified that endoplasmic reticulum stress induced by unfolded proteins increases with age. A long-term HFHF diet downregulated the expression of glucose-regulating hormones and secretory cell numbers, while it potentially upregulated oxidative and inflammation-related pathways in pancreatic secretory cells.

3.7. The Long-Term HFHF Diet Affected the Number of Islet Cells and the Expression of Characteristic Genes

Four major endocrine cell types in islets and acinar cells were identified using specific cell markers: *Gcg* and *Gc* (alpha cells), *Ins1* and *Pcsk2* (beta cells), *Ppy* and *Pyy* (polypeptide (PP) cells), *Sst* (delta cell), and *Prss2* and *Cela1* (acinar cells) (Figures 6A,B and S9A,B). The number of islet cells, especially alpha, PP, and delta cells, was significantly decreased in the HFHF-old group, whereas that in the old group barely changed (Figure 6C). The number of beta cells decreased slightly in the long-term HFHF group compared to that in the old group. Since only eight delta cells were obtained in the long-term HFHF group, which was too few to analyze the differential expressed genes between groups, these islet cells were not included in subsequent transcriptional and functional studies.

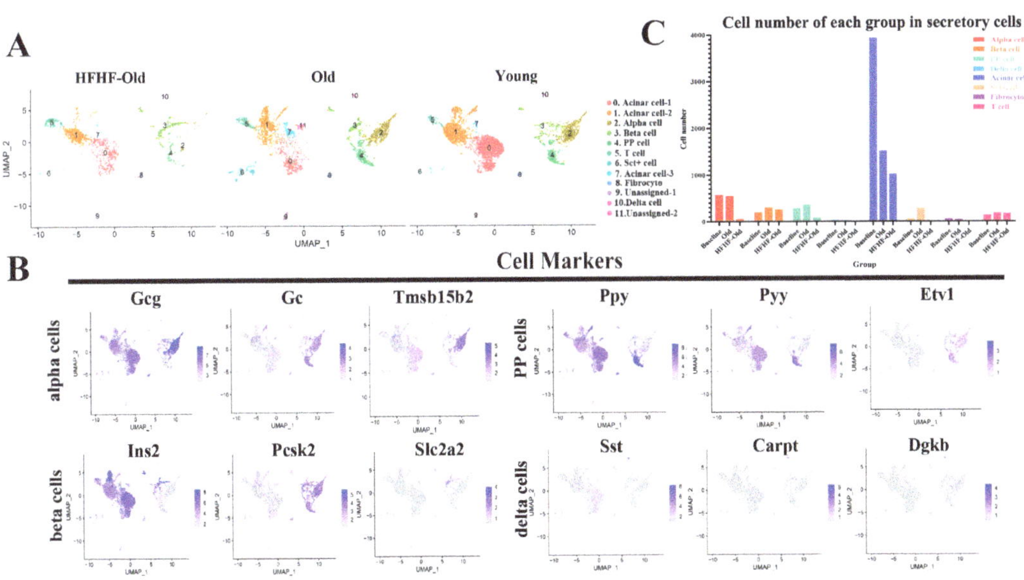

Figure 6. *Cont.*

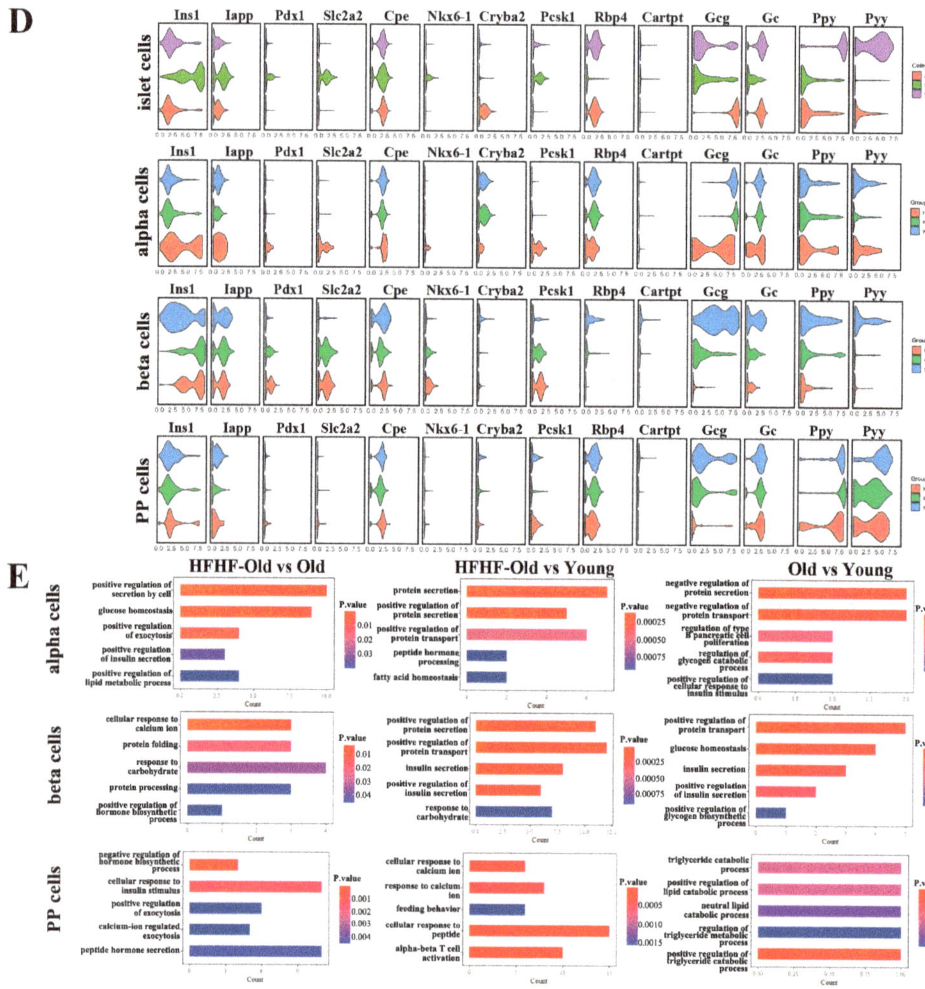

Figure 6. Distinct cell populations with transcriptional signatures in secreting islet cells. (**A**) UMAP plot characterizing the differences in secretory cell subpopulation distribution at the beginning (young) and after 18 months (HFHF-old, old) of diet feeding. Each point represents one cell. The distance between the two points is determined by the Pearson correlation of the 500 most highly expressed genes in each cell. Cell identity is recognized by the expression of marker genes. (**B**) UMAP plot of the marker gene expression in four types of islet cells. (**C**) The number of different pancreatic secretory cells at the beginning (young) and after 18 months (HFHF-old, old) of diet feeding. (**D**) Violin diagram characterizing the marker genes expression level in islet alpha, beta, and PP cells at the beginning (young) and after 18 months (HFHF-old, old) of diet feeding. (**E**) GO terms of upregulated secretory function-related enrichment obtained in islet alpha, beta, and PP cells after pairwise comparison of three groups ($p < 0.05$).

To further analyze the effects of long-term HFHF and the unavoidable aging on functional genes in islets, we compared the expression of several of the most common marker genes and transcriptional regulators of these three cells among the three experimental groups. Several known beta cell markers were also identified, including *Iapp* encoding islet amyloid polypeptide and *Slc2a2* (also known as *Glut2*) which is responsible for glucose uptake [55] (Figure 6D). During the aging process, the expression of major hormone-biosynthetic genes in islet cells, such as *Gcg* in alpha cells, *Ins1* in beta cells, and *Ppy* and *Pyy* in PP cells, was upregulated, but the long-term HFHF diet caused downregulation of these genes (Figure 6D). *Pdx1* (encoding pancreas/duodenum homeobox protein 1 and binding to the regulatory elements to upregulate insulin gene transcription [56]) and *Slc2a2* expression in beta cells was upregulated with aging, and the long-term HFHF diet further upregulated these genes. The expression of *Pdx1* and *Slc2a2* in alpha cells was also upregulated in the HFHF group. The long-term HFHF diet further exacerbated the downregulation of *Rbp4* (encoding retinol-binding protein 4, which is an adipokine involved in the development of obesity and insulin resistance [57]) expression in beta cells compared to the age-matched group. The expression of *Pcsk1*(encoding proprotein convertase subtilisin/kexin type 1, which has been identified in patients with early-onset obesity [58,59]) increased in alpha cells with a long-term HFHF diet. In addition, the expression of *Ins1* in alpha cells was increased in the HFHF group (Figure 6D). We further performed GO analysis to confirm functional changes in different islet cells, and the results showed that the long-term HFHF diet altered hormone secretion function as a positive regulation of protein secretion, while cellular response to calcium ion and protein processing pathways were significantly enriched (Figure 6E). In alpha cells, positive regulation of secretion, exocytosis, and peptide hormone processing were enriched in the HFHF-old group, suggesting that a long-term HFHF diet might increase function-related hormone processing and secretion in alpha cells. In beta cells, the positive regulation of protein secretion, insulin secretion, and positive regulation biosynthetic processes were also enriched in the HFHF-old and age-matched old groups. In PP cells, the protein hormone secretion and exocytosis pathways were enriched only in the HFHF-old group. Long-term HFHF treatment might downregulate the synthesis of glucose-regulating hormones but might increase hormone secretion in subpopulations of islets.

3.8. Long-Term HFHF Diet Increased the Oxidative Stress Level in Islet Cells

As the blood oxidative parameters were significantly increased in the HFHF group, we further investigated the oxidation-related genes and pathways that changed within islet subpopulations. The oxidoreductase gene expression differed among islet cell types (Figure 7A). The expression of *Atox1* (antioxidant 1 copper chaperone) and *Ndufs4* (ubiquinone oxidoreductase subunit S4) in alpha, beta, and PP cells gradually decreased with age, and the long-term HFHF diet further exacerbated the downregulation of these genes. The expression level of *Pyroxd1* (pyridine nucleotide-disulfide oxidoreductase domain 1) decreased in the old group, while it recovered in the HFHF-old group. Among alpha cells, *Ero1b* (endoplasmic reticulum oxidoreductase 1 beta) expression also decreased with age but was upregulated by the long-term HFHF diet. In contrast, in beta and PP cells, the expression of *Ero1b* was increased in the old group and further upregulated in the HFHF-old group. In alpha cells, the *Sod1* (superoxide dismutase 1) expression decreased with age, and long-term administration of an HFHF diet further exacerbated this decline. In beta cells, the expression of *Nos1* (nitric oxide synthase 1) was decreased in the old group, but the long-term HFHF diet induced partial recovery. In PP cells, the expression of *Gfod2* (glucose-fructose oxidoreductase domain containing 2) and *Foxred2* (FAD-dependent oxidoreductase domain containing 2) was higher in the HFHF-old group than in the old group. As the violin plots showed that a long-term HFHF diet increased the expression of various oxidoreductases in different subpopulations of islets, GO analyses were also proposed to determine changes in oxidative- and inflammation-related pathways in different islet cells. Oxidative-related biological processes were enriched in islet cells including beta and PP cells of HFHF-old group, such as response to reactive oxygen species biological processes and cellular response to oxidative stress compared with the old group (Figure 7B). However, negative regulation of reactive oxygen species was enriched in alpha and beta cells from the old group, suggesting that ROS could also be produced with normal aging. Combined with the slow change in plasma antioxidant parameters in the old group, this suggested that the level of ROS induced by aging was still within the normal range and did not cause OS. In addition, inflammation-related biological processes, including the response to IL-4 and IL-1, were only enriched in the HFHF-old group, suggesting that long-term HFHF diet caused inflammation in the alpha, beta, and PP cells of islets. On the basis of the above, a long-term HFHF diet changed the expression pattern of oxidoreductase genes and increased oxidation and inflammation in islet subpopulations.

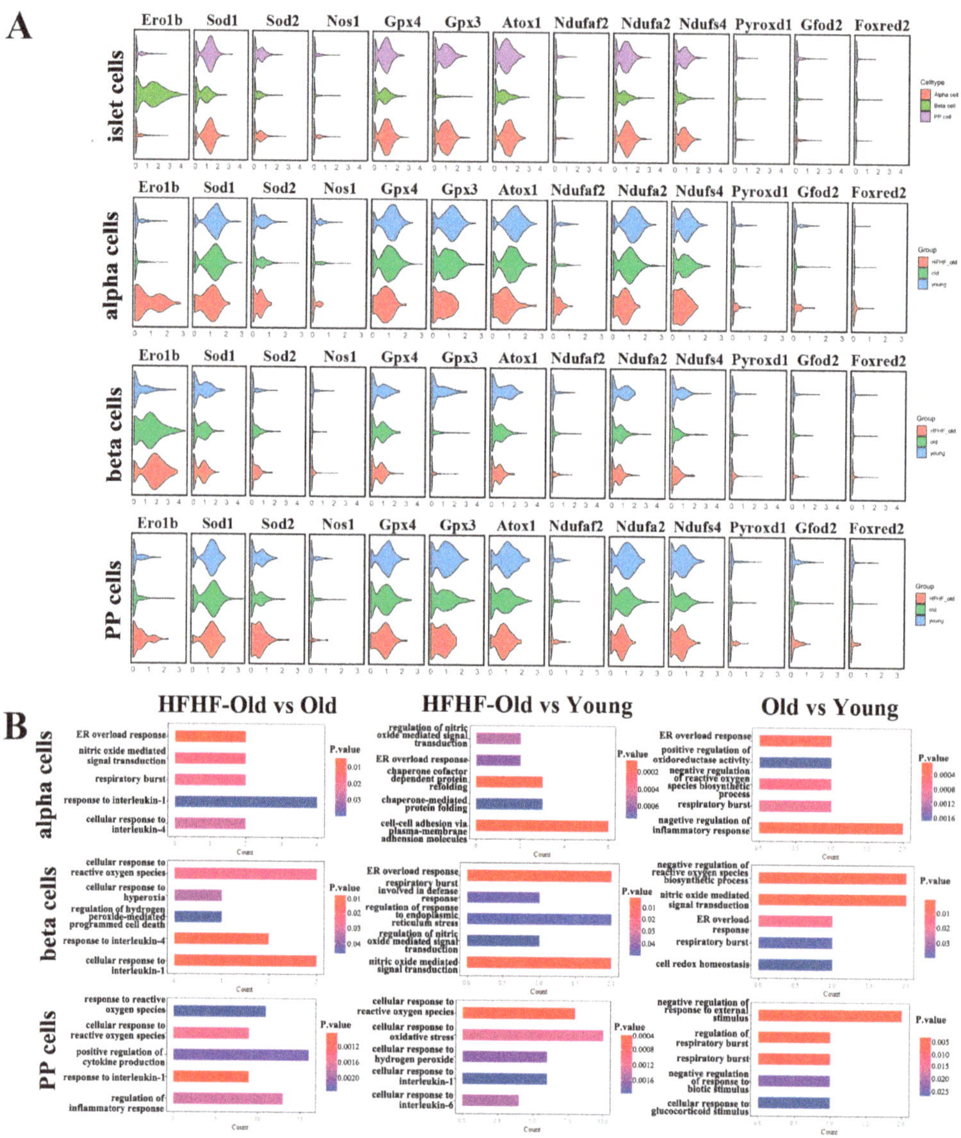

Figure 7. Distinct expression of oxidative-related oxidoreductase genes and Gene Ontology (GO) analysis findings among three groups. (**A**) Violin diagram characterizing the expression of oxidoreductase genes in islet alpha, beta, and PP cells at the beginning (young) and after 18 months (HFHF-old, old) of diet feeding. (**B**) GO terms of upregulated OS and inflammation-related enrichment obtained in islet alpha, beta, and PP cells after a pairwise comparison of three groups ($p < 0.05$).

4. Discussion

Various HFHF dietary patterns, which can induce diabetes in rats, have been reported [33–40,60]. However, the dose of fructose and fat administered to rodents in many studies was higher (50–60% of the diet) than that consumed by humans (10–15%) [61]. Therefore, we chose a model that more closely resembles the human diet (21.44% fat, 17.45% protein, 50.26% carbohydrate, and 20% fructose in water) [33]. Previous animal studies have shown that an HFHF diet induces persistent metabolic disorders and OS in pathogenesis, as well as hepatic and vascular complications [33–35]. We also found abnormal glucose and lipid metabolism in rats after 3 months of the HFHF diet, with significant abnormalities in glucose tolerance noted after 18 months. In our study, obvious vacuolar degeneration in the liver and an increase in oxidative parameters (such as GR, SOD, and 8-oxo-Gsn in urine) were also observed, consistent with previous human epidemiology [26–29] and animal [30,33–35] studies.

The function of islets is to secrete glucose-regulating hormones to maintain glucose homeostasis; however, this function is impaired in T2DM patients [48]. Therefore, we further explored changes in the structure and function of islets in a T2DM rat model induced by a long-term HFHF diet. After 3 months of an HFHF diet, the size of the islets increased. This morphological change was also verified in ovariectomized rats with an HFHF diet [43]. As only the level of insulin increased significantly and diabetes did not develop at this time, the change in islets might suggest that compensatory hyperplasia occurred. This result is consistent with previous research showing that beta-cell mass expansion is modulated indirectly by insulin via the nuclear hormone receptor peroxisome proliferator-activated receptor γ in response to obesity [62–66]. Notably, after 18 months of feeding, the islet volume decreased unexpectedly, and the morphology became irregular, which was consistent with the significant increases in fasting glucose levels, glucose tolerance, and levels of immature precursors of insulin. These results showed that an increase in islet mass occurred in the early compensatory phase in T2DM, while islet mass decreased in the late phase.

As the pancreatic structure and its secretory function were remarkably changed, scRNA-seq analysis was conducted to further explore changes in the cell composition and transcriptome patterns of pancreatic cells, secretory cells, and subpopulations of islet cells. A scRNA-seq analysis could help researchers to explore cell heterogeneity, identify cell states, and show changes in cell-type-specific genes during aging or disease development [67]. Previous studies have analyzed human islets via scRNA-seq, enabling a better understanding of islet pathology in aging and T2DM. Wang et al. found that alpha and beta cells from donors with T2DM had expression profiles with features observed in children, which indicated a partial dedifferentiation process [68]. Enge et al. found that an elevation of transcriptional noise and potential fate drift occurred in islet cells from elders [69]. Xin et al. identified 245 genes with disturbed expression in type 2 diabetes, 92% of which were not involved in islet cell function or growth in the previous study [70]. Rat, a popular animal model for studying T2DM, has a loosely dispersed mesenteric pancreas, which makes it difficult for researchers to collect and prepare single-cell suspensions for scRNA-seq studies. Thus, it would be worthwhile to explore methods and identify the characteristics and transcriptome patterns of rat islets and their subpopulations. In this study, cell-specific markers were used to identify secretory, immune, fibrocyte, endothelial, vascular smooth muscle, and ductal epithelial cells in the pancreas. We found that the secretory population decreased with age, and the long-term HFHF diet exacerbated this decline. Examination of the secretory cell subtypes showed that the number of islet cells, especially alpha, delta and PP cells, was significantly decreased in the HFHF-old group compared to the young group, while that of the old group was nearly unchanged. Among the three experimental groups, the expression of genes related to glucose-regulating hormone synthesis was upregulated during the aging process; however, the long-term HFHF diet caused downregulation. Nevertheless, GO analysis showed a positive regulation of hormone secretion induced by a long-term HFHF diet, while the positive regulation of protein secretion, exocytosis, peptide hormone processing, and positive regulation biosynthetic processes were enriched in islet cells of the HFHF group. These results suggest that a long-term HFHF diet may reduce the synthesis of glucose-regulating hormones while promoting the release of these hormones. Furthermore, the expression of *Ins1* in alpha cells increased with age and was further exacerbated by a long-term HFHF diet. This was confirmed by the increase in plasma insulin, proinsulin, and intact proinsulin levels, suggesting islet dysfunction.

Several studies have shown that the degree of oxidation increases in diabetic patients [24,26,27]. We also found elevated levels of plasma GR, SOD, XOD, GR, T-AOC, urine 8-oxo-Gsn, and 8-oxo-dGsn in rats fed an HFHF diet; however, the changes in the pancreas or islets were still not clear. Therefore, we used scRNA-seq to explore the transcriptome patterns of islet cells and their subgroups of oxidation-related genes and biological processes. In the present study, scRNA-seq showed that the long-term HFHF diet increased the degree of oxidation of secretory cells. GO analysis indicated that OS-related pathways were enriched in secretory cell clusters with the long-term HFHF diet, whereas the endoplasmic reticulum stress response induced by unfolded proteins was more significant during aging. In islet cells, different oxidoreductase gene expression in the HFHF-old group was upregulated in three subtypes of islet cells (alpha, beta, and PP cells), including *Pyroxd1*, *Ero1b*, *Nos1*, *Gfod2*, and *Foxred2*. Furthermore, compared with those of the old group, oxidative-related biological processes, such as the response to reactive oxygen species, were also enriched in the islet cells of the HFHF group. Moreover, the transcriptional level of oxidoreductase also changed in islet cells with aging; for example, *Sod1* decreased with age in alpha and beta cells, and a long-term HFHF diet further exacerbated this decline. All the above results verified that the long-term HFHF diet accelerated OS in the pancreas and islet subpopulations.

In addition, inflammatory pathways have also been regarded as the potential pathogenic mediators of overweight and DM [71,72]. Several reports support the notion that islet inflammatory processes are involved in the pathophysiological changes in T2DM [73]. We also found that inflammation levels were significantly increased in the long-term HFHF-treated rats. However, the cause of islet inflammation remains unclear [72]. We found that inflammation-related items, including the response to IL-4 and IL-1, were significantly enriched in alpha, beta, and PP cells of islets in the HFHF-old group according to GO analysis from the ScRNA-seq data. In addition, our plasma findings also confirmed that a long-term HFHF diet induced inflammation. After feeding the HFHF diet for 3 months, the levels of plasma oxidative parameters were significantly increased, while those of plasma inflammatory factors, including IL-1β, IL-6, MCP-1, VEGF, and ICAM1, were only changed after 18 months of feeding. The results showed that inflammation occurred after oxidative stress, suggesting that oxidative stress might be the cause of the high inflammation in T2DM patients.

5. Conclusions

In summary, our findings show that a long-term HFHF diet induces metabolic disorders, oxidation, and inflammation, as well as changes in islet size and irregular secretory functions in the pancreas. ScRNA-seq analysis of pancreatic cells revealed that the long-term HFHF diet resulted in decreased islet cell counts and the enrichment of hormone secretion biological processes, as well as of the oxidative and inflammatory pathways in islet alpha, beta, and PP cells. During the aging process, the expression of major hormone-synthesis genes is upregulated, and the transcriptional level of most oxidoreductases is decreased in islet cells. This study decoded the long-term HFHF diet and aging-associated structural and transcriptomic changes that underlie pancreatic islet functional decay at a single-cell resolution and indicated that HFHF diet-induced inflammation occurred after the accumulation of oxidative stress, suggesting an avenue to prevent improper diet-related diabetes.

Supplementary Materials: The following supporting information can be downloaded at https://www.mdpi.com/article/10.3390/nu14112181/s1. Figure S1: Evolution of weight in this study; Figure S2: Evolution of hepatic complications in study; Figure S3: Level of pancreas-related endocrine proteins during the intraperitoneal glucose tolerance test (IpGTT); Figure S4: Morphometric analysis of pancreatic islet size and density in this study; Figure S5: Adipocyte infiltration in the pancreas in this study. Figure S6: Basic information from scRNA-Seq Seurat Analysis of diabetic rats. Figure S7: Distinct cell populations with transcriptional signatures in pancreatic cells determined by scRNA-Seq analysis of diabetic rats. Figure S8: Gene Ontology (GO) terms of secretory cells; Figure S9: Distinct cell populations with transcriptional signatures in secretory cells determined by a scRNA-Seq analysis of diabetic rats. Table S1: Ingredients of the diet (%). Table S2: Evolution of physiological, metabolic, oxidative, and inflammatory parameters during the study.

Author Contributions: Conceptualization, J.-P.C.; methodology, Y.Z., J.-J.W., G.-Q.F. and Z.L.; software, Y.Z. and L.-T.Z.; validation, Y.Z., Q.-Y.W. and J.L.; formal analysis, Y.Z.; investigation, Y.Z. and J.-P.C.; resources, Y.Z. and J.-P.C.; data curation, Y.Z., Q.-Y.W., J.L. and J.-P.C.; writing—original draft preparation, Y.Z.; writing—review and editing, Y.Z. and J.-P.C.; visualization, Y.Z.; supervision, J.-P.C.; project administration, J.-P.C.; funding acquisition, J.-P.C. All authors have read and agreed to the published version of the manuscript.

Funding: This research was supported by the Ministry of Science and Technology (the National Key R&D Program of China 2018YFC2000300), CAMS Innovation Fund for Medical Sciences (No. 2021-1-I2M-050) and the National Natural Science Foundation of China (No. 82170856).

Institutional Review Board Statement: The animal study protocol was approved by the Animal Ethics and Welfare Committee of Beijing Stomatological Hospital (Approval No. KQYY-202103-001).

Informed Consent Statement: Not applicable.

Data Availability Statement: The data discussed in this publication were deposited in NCBI's Gene Expression Omnibus and are accessible through GEO Series accession number GSE201900 (https://www.ncbi.nlm.nih.gov/geo/query/acc.cgi?acc=GSE201900 (accessed on 19 May 2022)).

Conflicts of Interest: The authors declare no conflict of interest.

References

1. Global Report on Diabetes. Available online: https://www.who.int/publications/i/item/who-nmh-nvi-16.3 (accessed on 21 October 2016).
2. American Diabetes Association. Diagnosis and classification of diabetes mellitus. *Diabetes Care* **2013**, *36* (Suppl. 1), S67–S74. [CrossRef] [PubMed]
3. Saeedi, P.; Salpea, P.; Karuranga, S.; Petersohn, I.; Malanda, B.; Gregg, E.W.; Unwin, N.; Wild, S.H.; Williams, R. Mortality attributable to diabetes in 20–79 years old adults, 2019 estimates: Results from the International Diabetes Federation Diabetes Atlas. *Diabetes Res. Clin. Pract.* **2020**, *162*, 108086. [CrossRef] [PubMed]
4. Ley, S.H.; Hamdy, O.; Mohan, V.; Hu, F.B. Prevention and management of type 2 diabetes: Dietary components and nutritional strategies. *Lancet* **2014**, *383*, 1999–2007. [CrossRef]
5. Welsh, J.A.; Sharma, A.; Abramson, J.L.; Vaccarino, V.; Gillespie, C.; Vos, M.B. Caloric sweetener consumption and dyslipidemia among US adults. *JAMA* **2010**, *303*, 1490–1497. [CrossRef] [PubMed]
6. Softic, S.; Stanhope, K.L.; Boucher, J.; Divanovic, S.; Lanaspa, M.A.; Johnson, R.J.; Kahn, C.R. Fructose and hepatic insulin resistance. *Crit. Rev. Clin. Lab. Sci.* **2020**, *57*, 308–322. [CrossRef] [PubMed]
7. Tappy, L.; Le, K.A. Metabolic effects of fructose and the worldwide increase in obesity. *Physiol. Rev.* **2010**, *90*, 23–46. [CrossRef]
8. Ludwig, D.S. Dietary glycemic index and obesity. *J. Nutr.* **2000**, *130*, 280S–283S. [CrossRef]
9. Dekker, M.J.; Su, Q.; Baker, C.; Rutledge, A.C.; Adeli, K. Fructose: A highly lipogenic nutrient implicated in insulin resistance, hepatic steatosis, and the metabolic syndrome. *Am. J. Physiol. Endocrinol. Metab.* **2010**, *299*, E685–E694. [CrossRef]
10. Sun, X.; Han, F.; Lu, Q.; Li, X.; Ren, D.; Zhang, J.; Han, Y.; Xiang, Y.K.; Li, J. Empagliflozin Ameliorates Obesity-Related Cardiac Dysfunction by Regulating Sestrin2-Mediated AMPK-mTOR Signaling and Redox Homeostasis in High-Fat Diet-Induced Obese Mice. *Diabetes* **2020**, *69*, 1292–1305. [CrossRef]
11. DiNicolantonio, J.J.; Lucan, S.C.; O'Keefe, J.H. The Evidence for Saturated Fat and for Sugar Related to Coronary Heart Disease. *Prog. Cardiovasc. Dis.* **2016**, *58*, 464–472. [CrossRef]
12. Arisawa, K.; Uemura, H.; Yamaguchi, M.; Nakamoto, M.; Hiyoshi, M.; Sawachika, F.; Katsuura-Kamano, S. Associations of dietary patterns with metabolic syndrome and insulin resistance: A cross-sectional study in a Japanese population. *J. Med. Investig.* **2014**, *61*, 333–344. [CrossRef] [PubMed]

13. van Dam, R.M.; Grievink, L.; Ocke, M.C.; Feskens, E.J. Patterns of food consumption and risk factors for cardiovascular disease in the general Dutch population. *Am. J. Clin. Nutr.* **2003**, *77*, 1156–1163. [CrossRef] [PubMed]
14. Darani Zad, N.; Mohd Yusof, R.; Esmaili, H.; Jamaluddin, R.; Mohseni, F. Association of dietary pattern with biochemical blood profiles and bodyweight among adults with Type 2 diabetes mellitus in Tehran, Iran. *J. Diabetes Metab. Disord.* **2015**, *14*, 28. [CrossRef] [PubMed]
15. Esmaillzadeh, A.; Kimiagar, M.; Mehrabi, Y.; Azadbakht, L.; Hu, F.B.; Willett, W.C. Dietary patterns, insulin resistance, and prevalence of the metabolic syndrome in women. *Am. J. Clin. Nutr.* **2007**, *85*, 910–918. [CrossRef]
16. van Dam, R.M.; Rimm, E.B.; Willett, W.C.; Stampfer, M.J.; Hu, F.B. Dietary patterns and risk for type 2 diabetes mellitus in U.S. men. *Ann. Intern. Med.* **2002**, *136*, 201–209. [CrossRef]
17. Lutsey, P.L.; Steffen, L.M.; Stevens, J. Dietary intake and the development of the metabolic syndrome: The Atherosclerosis Risk in Communities study. *Circulation* **2008**, *117*, 754–761. [CrossRef]
18. Schulze, M.B.; Hoffmann, K.; Manson, J.E.; Willett, W.C.; Meigs, J.B.; Weikert, C.; Heidemann, C.; Colditz, G.A.; Hu, F.B. Dietary pattern, inflammation, and incidence of type 2 diabetes in women. *Am. J. Clin. Nutr.* **2005**, *82*, 675–684, quiz 714–675. [CrossRef]
19. Wirfalt, E.; Hedblad, B.; Gullberg, B.; Mattisson, I.; Andren, C.; Rosander, U.; Janzon, L.; Berglund, G. Food patterns and components of the metabolic syndrome in men and women: A cross-sectional study within the Malmo Diet and Cancer cohort. *Am. J. Epidemiol.* **2001**, *154*, 1150–1159. [CrossRef]
20. Medina-Remon, A.; Kirwan, R.; Lamuela-Raventos, R.M.; Estruch, R. Dietary patterns and the risk of obesity, type 2 diabetes mellitus, cardiovascular diseases, asthma, and neurodegenerative diseases. *Crit. Rev. Food Sci. Nutr.* **2018**, *58*, 262–296. [CrossRef]
21. Imamura, F.; O'Connor, L.; Ye, Z.; Mursu, J.; Hayashino, Y.; Bhupathiraju, S.N.; Forouhi, N.G. Consumption of sugar sweetened beverages, artificially sweetened beverages, and fruit juice and incidence of type 2 diabetes: Systematic review, meta-analysis, and estimation of population attributable fraction. *BMJ* **2015**, *351*, h3576. [CrossRef]
22. Malik, V.S.; Popkin, B.M.; Bray, G.A.; Despres, J.P.; Willett, W.C.; Hu, F.B. Sugar-sweetened beverages and risk of metabolic syndrome and type 2 diabetes: A meta-analysis. *Diabetes Care* **2010**, *33*, 2477–2483. [CrossRef] [PubMed]
23. InterAct, C.; Romaguera, D.; Norat, T.; Wark, P.A.; Vergnaud, A.C.; Schulze, M.B.; van Woudenbergh, G.J.; Drogan, D.; Amiano, P.; Molina-Montes, E.; et al. Consumption of sweet beverages and type 2 diabetes incidence in European adults: Results from EPIC-InterAct. *Diabetologia* **2013**, *56*, 1520–1530. [CrossRef]
24. Brownlee, M. Biochemistry and molecular cell biology of diabetic complications. *Nature* **2001**, *414*, 813–820. [CrossRef] [PubMed]
25. Betteridge, D.J. What is oxidative stress? *Metabolism* **2000**, *49*, 3–8. [CrossRef]
26. Gawlik, K.; Naskalski, J.W.; Fedak, D.; Pawlica-Gosiewska, D.; Grudzien, U.; Dumnicka, P.; Malecki, M.T.; Solnica, B. Markers of Antioxidant Defense in Patients with Type 2 Diabetes. *Oxidative Med. Cell. Longev.* **2016**, *2016*, 2352361. [CrossRef] [PubMed]
27. Bandeira Sde, M.; Guedes Gda, S.; da Fonseca, L.J.; Pires, A.S.; Gelain, D.P.; Moreira, J.C.; Rabelo, L.A.; Vasconcelos, S.M.; Goulart, M.O. Characterization of blood oxidative stress in type 2 diabetes mellitus patients: Increase in lipid peroxidation and SOD activity. *Oxidative Med. Cell. Longev.* **2012**, *2012*, 819310. [CrossRef] [PubMed]
28. Broedbaek, K.; Siersma, V.; Henriksen, T.; Weimann, A.; Petersen, M.; Andersen, J.T.; Jimenez-Solem, E.; Stovgaard, E.S.; Hansen, L.J.; Henriksen, J.E.; et al. Urinary markers of nucleic acid oxidation and long-term mortality of newly diagnosed type 2 diabetic patients. *Diabetes Care* **2011**, *34*, 2594–2596. [CrossRef]
29. Broedbaek, K.; Siersma, V.; Henriksen, T.; Weimann, A.; Petersen, M.; Andersen, J.T.; Jimenez-Solem, E.; Hansen, L.J.; Henriksen, J.E.; Bonnema, S.J.; et al. Association between urinary markers of nucleic acid oxidation and mortality in type 2 diabetes: A population-based cohort study. *Diabetes Care* **2013**, *36*, 669–676. [CrossRef]
30. Wang, W.X.; Luo, S.B.; Xia, M.M.; Mao, Y.H.; Zhou, X.Y.; Jiang, P.; Jiang, H.Y.; Dai, D.P.; Li, C.B.; Hu, G.X.; et al. Analysis of the oxidative damage of DNA, RNA, and their metabolites induced by hyperglycemia and related nephropathy in Sprague Dawley rats. *Free Radic. Res.* **2015**, *49*, 1199–1209. [CrossRef]
31. Al-Awar, A.; Kupai, K.; Veszelka, M.; Szucs, G.; Attieh, Z.; Murlasits, Z.; Torok, S.; Posa, A.; Varga, C. Experimental Diabetes Mellitus in Different Animal Models. *J. Diabetes Res.* **2016**, *2016*, 9051426. [CrossRef]
32. Lenzen, S. The mechanisms of alloxan- and streptozotocin-induced diabetes. *Diabetologia* **2008**, *51*, 216–226. [CrossRef] [PubMed]
33. Lozano, I.; Van der Werf, R.; Bietiger, W.; Seyfritz, E.; Peronet, C.; Pinget, M.; Jeandidier, N.; Maillard, E.; Marchioni, E.; Sigrist, S.; et al. High-fructose and high-fat diet-induced disorders in rats: Impact on diabetes risk, hepatic and vascular complications. *Nutr. Metab.* **2016**, *13*, 15. [CrossRef] [PubMed]
34. Dal, S.; Van der Werf, R.; Walter, C.; Bietiger, W.; Seyfritz, E.; Mura, C.; Peronet, C.; Legrandois, J.; Werner, D.; Ennahar, S.; et al. Treatment of NASH with Antioxidant Therapy: Beneficial Effect of Red Cabbage on Type 2 Diabetic Rats. *Oxidative Med. Cell. Longev.* **2018**, *2018*, 7019573. [CrossRef] [PubMed]
35. Van der Werf, R.; Walter, C.; Bietiger, W.; Seyfritz, E.; Mura, C.; Peronet, C.; Legrandois, J.; Werner, D.; Ennahar, S.; Digel, F.; et al. Beneficial effects of cherry consumption as a dietary intervention for metabolic, hepatic and vascular complications in type 2 diabetic rats. *Cardiovasc. Diabetol.* **2018**, *17*, 104. [CrossRef]
36. Liao, H.; Chou, L.M.; Chien, Y.W.; Wu, C.H.; Chang, J.S.; Lin, C.I.; Lin, S.H. Grape powder consumption affects the expression of neurodegeneration-related brain proteins in rats chronically fed a high-fructose-high-fat diet. *J. Nutr. Biochem.* **2017**, *43*, 132–140. [CrossRef]
37. Chou, L.M.; Lin, C.I.; Chen, Y.H.; Liao, H.; Lin, S.H. A diet containing grape powder ameliorates the cognitive decline in aged rats with a long-term high-fructose-high-fat dietary pattern. *J. Nutr. Biochem.* **2016**, *34*, 52–60. [CrossRef]

38. Afifi, N.A.; Ramadan, A.; Erian, E.Y.; Saleh, D.O.; Sedik, A.A.; Badawi, M.; El Hotaby, W. Trigonelline attenuates hepatic complications and molecular alterations in high-fat high-fructose diet-induced insulin resistance in rats. *Can. J. Physiol. Pharmacol.* **2017**, *95*, 427–436. [CrossRef]
39. Casagrande, B.P.; Gomes, M.F.P.; Moura, E.O.C.; Santos, A.C.C.; Kubota, M.C.; Ribeiro, D.A.; Pisani, L.P.; Medeiros, A.; Estadella, D. Age-dependent hepatic alterations induced by a high-fat high-fructose diet. *Inflamm. Res.* **2019**, *68*, 359–368. [CrossRef]
40. Vidal, E.; Lalarme, E.; Maire, M.A.; Febvret, V.; Gregoire, S.; Gambert, S.; Acar, N.; Bretillon, L. Early impairments in the retina of rats fed with high fructose/high fat diet are associated with glucose metabolism deregulation but not dyslipidaemia. *Sci. Rep.* **2019**, *9*, 5997. [CrossRef]
41. Liu, M.; Weiss, M.A.; Arunagiri, A.; Yong, J.; Rege, N.; Sun, J.; Haataja, L.; Kaufman, R.J.; Arvan, P. Biosynthesis, structure, and folding of the insulin precursor protein. *Diabetes Obes. Metab.* **2018**, *20* (Suppl. 2), 28–50. [CrossRef]
42. Rains, J.L.; Jain, S.K. Oxidative stress, insulin signaling, and diabetes. *Free Radic. Biol. Med.* **2011**, *50*, 567–575. [CrossRef] [PubMed]
43. Chansela, P.; Potip, B.; Weerachayaphorn, J.; Kangwanrangsan, N.; Chukijrungroat, N.; Saengsirisuwan, V. Morphological alteration of the pancreatic islet in ovariectomized rats fed a high-fat high-fructose diet. *Histochem. Cell Biol.* **2022**, *157*, 427–442. [CrossRef] [PubMed]
44. O'Dowd, J.F. The isolation and purification of rodent pancreatic islets of Langerhans. *Methods Mol. Biol.* **2009**, *560*, 37–42. [CrossRef] [PubMed]
45. Li, D.S.; Yuan, Y.H.; Tu, H.J.; Liang, Q.L.; Dai, L.J. A protocol for islet isolation from mouse pancreas. *Nat. Protoc.* **2009**, *4*, 1649–1652. [CrossRef]
46. Chang, N.; Tian, L.; Ji, X.; Zhou, X.; Hou, L.; Zhao, X.; Yang, Y.; Yang, L.; Li, L. Single-Cell Transcriptomes Reveal Characteristic Features of Mouse Hepatocytes with Liver Cholestatic Injury. *Cells* **2019**, *8*, 1069. [CrossRef]
47. Zhao, T.; Fu, Y.; Zhu, J.; Liu, Y.; Zhang, Q.; Yi, Z.; Chen, S.; Jiao, Z.; Xu, X.; Xu, J.; et al. Single-Cell RNA-Seq Reveals Dynamic Early Embryonic-like Programs during Chemical Reprogramming. *Cell Stem Cell* **2018**, *23*, 31–45e7. [CrossRef]
48. Leibiger, I.B.; Leibiger, B.; Berggren, P.O. Insulin signaling in the pancreatic beta-cell. *Annu. Rev. Nutr.* **2008**, *28*, 233–251. [CrossRef]
49. van der Kooi, J.B.; van Wanroy, P.J.; De Jonge, M.C.; Kornelis, J.A. Time separation between cough pulses in bladder, rectum and urethra in women. *J. Urol.* **1984**, *132*, 1275–1278. [CrossRef]
50. Sims, E.K.; Carr, A.L.J.; Oram, R.A.; DiMeglio, L.A.; Evans-Molina, C. 100 years of insulin: Celebrating the past, present and future of diabetes therapy. *Nat. Med.* **2021**, *27*, 1154–1164. [CrossRef]
51. Madsen, O.D.; Frank, B.H.; Steiner, D.F. Human proinsulin-specific antigenic determinants identified by monoclonal antibodies. *Diabetes* **1984**, *33*, 1012–1016. [CrossRef]
52. Gray, I.P.; Siddle, K.; Frank, B.H.; Hales, C.N. Characterization and use in immunoradiometric assay of monoclonal antibodies directed against human proinsulin. *Diabetes* **1987**, *36*, 684–688. [CrossRef] [PubMed]
53. Imai, S.; Takahashi, T.; Naito, S.; Yamauchi, A.; Okada, C.; Notsu, Y.; Sakikawa, I.; Hatanaka, M.; Iwasaki, T.; Morita, A.; et al. Development of a novel immunoassay specific for mouse intact proinsulin. *Anal. Biochem.* **2015**, *484*, 91–98. [CrossRef] [PubMed]
54. Ozawa, S.; Katsuta, H.; Suzuki, K.; Takahashi, K.; Tanaka, T.; Sumitani, Y.; Nishida, S.; Yoshimoto, K.; Ishida, H. Estimated proinsulin processing activity of prohormone convertase (PC) 1/3 rather than PC2 is decreased in pancreatic beta-cells of type 2 diabetic patients. *Endocr. J.* **2014**, *61*, 607–614. [CrossRef] [PubMed]
55. Martens, G.A.; Jiang, L.; Hellemans, K.H.; Stange, G.; Heimberg, H.; Nielsen, F.C.; Sand, O.; Van Helden, J.; Van Lommel, L.; Schuit, F.; et al. Clusters of conserved beta cell marker genes for assessment of beta cell phenotype. *PLoS ONE* **2011**, *6*, e24134. [CrossRef]
56. Zhu, Y.; Liu, Q.; Zhou, Z.; Ikeda, Y. PDX1, Neurogenin-3, and MAFA: Critical transcription regulators for beta cell development and regeneration. *Stem Cell Res. Ther.* **2017**, *8*, 240. [CrossRef]
57. Kotnik, P.; Fischer-Posovszky, P.; Wabitsch, M. RBP4: A controversial adipokine. *Eur. J. Endocrinol.* **2011**, *165*, 703–711. [CrossRef]
58. Yeo, G.S.H.; Chao, D.H.M.; Siegert, A.M.; Koerperich, Z.M.; Ericson, M.D.; Simonds, S.E.; Larson, C.M.; Luquet, S.; Clarke, I.; Sharma, S.; et al. The melanocortin pathway and energy homeostasis: From discovery to obesity therapy. *Mol. Metab.* **2021**, *48*, 101206. [CrossRef]
59. Jackson, R.S.; Creemers, J.W.; Ohagi, S.; Raffin-Sanson, M.L.; Sanders, L.; Montague, C.T.; Hutton, J.C.; O'Rahilly, S. Obesity and impaired prohormone processing associated with mutations in the human prohormone convertase 1 gene. *Nat. Genet.* **1997**, *16*, 303–306. [CrossRef]
60. Shum, M.; Pinard, S.; Guimond, M.O.; Labbe, S.M.; Roberge, C.; Baillargeon, J.P.; Langlois, M.F.; Alterman, M.; Wallinder, C.; Hallberg, A.; et al. Angiotensin II type 2 receptor promotes adipocyte differentiation and restores adipocyte size in high-fat/high-fructose diet-induced insulin resistance in rats. *Am. J. Physiol. Endocrinol. Metab.* **2013**, *304*, E197–E210. [CrossRef]
61. Odermatt, A. The Western-style diet: A major risk factor for impaired kidney function and chronic kidney disease. *Am. J. Physiol. Ren. Physiol.* **2011**, *301*, F919–F931. [CrossRef]
62. Hull, R.L.; Kodama, K.; Utzschneider, K.M.; Carr, D.B.; Prigeon, R.L.; Kahn, S.E. Dietary-fat-induced obesity in mice results in beta cell hyperplasia but not increased insulin release: Evidence for specificity of impaired beta cell adaptation. *Diabetologia* **2005**, *48*, 1350–1358. [CrossRef] [PubMed]

63. Linnemann, A.K.; Baan, M.; Davis, D.B. Pancreatic beta-cell proliferation in obesity. *Adv. Nutr.* **2014**, *5*, 278–288. [CrossRef] [PubMed]
64. Medina-Gomez, G.; Gray, S.L.; Yetukuri, L.; Shimomura, K.; Virtue, S.; Campbell, M.; Curtis, R.K.; Jimenez-Linan, M.; Blount, M.; Yeo, G.S.; et al. PPAR gamma 2 prevents lipotoxicity by controlling adipose tissue expandability and peripheral lipid metabolism. *PLoS Genet.* **2007**, *3*, e64. [CrossRef] [PubMed]
65. Rosen, E.D.; Kulkarni, R.N.; Sarraf, P.; Ozcan, U.; Okada, T.; Hsu, C.H.; Eisenman, D.; Magnuson, M.A.; Gonzalez, F.J.; Kahn, C.R.; et al. Targeted elimination of peroxisome proliferator-activated receptor gamma in beta cells leads to abnormalities in islet mass without compromising glucose homeostasis. *Mol. Cell Biol.* **2003**, *23*, 7222–7229. [CrossRef] [PubMed]
66. Vidal-Puig, A.; Jimenez-Linan, M.; Lowell, B.B.; Hamann, A.; Hu, E.; Spiegelman, B.; Flier, J.S.; Moller, D.E. Regulation of PPAR gamma gene expression by nutrition and obesity in rodents. *J. Clin. Investig.* **1996**, *97*, 2553–2561. [CrossRef] [PubMed]
67. He, X.; Memczak, S.; Qu, J.; Belmonte, J.C.I.; Liu, G.H. Single-cell omics in ageing: A young and growing field. *Nat. Metab.* **2020**, *2*, 293–302. [CrossRef]
68. Wang, Y.J.; Schug, J.; Won, K.J.; Liu, C.; Naji, A.; Avrahami, D.; Golson, M.L.; Kaestner, K.H. Single-Cell Transcriptomics of the Human Endocrine Pancreas. *Diabetes* **2016**, *65*, 3028–3038. [CrossRef]
69. Enge, M.; Arda, H.E.; Mignardi, M.; Beausang, J.; Bottino, R.; Kim, S.K.; Quake, S.R. Single-Cell Analysis of Human Pancreas Reveals Transcriptional Signatures of Aging and Somatic Mutation Patterns. *Cell* **2017**, *171*, 321–330.e14. [CrossRef]
70. Xin, Y.; Kim, J.; Okamoto, H.; Ni, M.; Wei, Y.; Adler, C.; Murphy, A.J.; Yancopoulos, G.D.; Lin, C.; Gromada, J. RNA Sequencing of Single Human Islet Cells Reveals Type 2 Diabetes Genes. *Cell Metab.* **2016**, *24*, 608–615. [CrossRef]
71. Shoelson, S.E.; Lee, J.; Goldfine, A.B. Inflammation and insulin resistance. *J. Clin. Investig.* **2006**, *116*, 1793–1801. [CrossRef]
72. Lontchi-Yimagou, E.; Sobngwi, E.; Matsha, T.E.; Kengne, A.P. Diabetes mellitus and inflammation. *Curr. Diabetes Rep.* **2013**, *13*, 435–444. [CrossRef] [PubMed]
73. Donath, M.Y.; Schumann, D.M.; Faulenbach, M.; Ellingsgaard, H.; Perren, A.; Ehses, J.A. Islet inflammation in type 2 diabetes: From metabolic stress to therapy. *Diabetes Care* **2008**, *31* (Suppl. 2), S161–S164. [CrossRef] [PubMed]

nutrients

MDPI

Article

A High-Fat Diet Induces Muscle Mitochondrial Dysfunction and Impairs Swimming Capacity in Zebrafish: A New Model of Sarcopenic Obesity

Yun-Yi Zou, Zhang-Lin Chen, Chen-Chen Sun, Dong Yang, Zuo-Qiong Zhou, Qin Xiao, Xi-Yang Peng * and Chang-Fa Tang *

Key Laboratory of Physical Fitness and Exercise Rehabilitation of Hunan Province, College of Physical Education, Hunan Normal University, Changsha 410012, China; 202020151346@hunnu.edu.cn (Y.-Y.Z.); zhanglinchen@hunnu.edu.cn (Z.-L.C.); sunchenchen1022@hunnu.edu.cn (C.-C.S.); yangdong@hunnu.edu.cn (D.Y.); zhouzuoqiong@hunnu.edu.cn (Z.-Q.Z.); qinzixiao715@163.com (Q.X.)

* Correspondence: xiyangpeng@hunnu.edu.cn (X.-Y.P.); changfatang@hunnu.edu.cn (C.-F.T.)

Abstract: Obesity is a highly prevalent disease that can induce metabolic syndrome and is associated with a greater risk of muscular atrophy. Mitochondria play central roles in regulating the physiological metabolism of skeletal muscle; however, whether a decreased mitochondrial function is associated with impaired muscle function is unclear. In this study, we evaluated the effects of a high-fat diet on muscle mitochondrial function in a zebrafish model of sarcopenic obesity (SOB). In SOB zebrafish, a significant decrease in exercise capacity and skeletal muscle fiber cross-sectional area was detected, accompanied by high expression of the atrophy-related markers Atrogin-1 and muscle RING-finger protein-1. Zebrafish with SOB exhibited inhibition of mitochondrial biogenesis and fatty acid oxidation as well as disruption of mitochondrial fusion and fission in atrophic muscle. Thus, our findings showed that muscle atrophy was associated with SOB-induced mitochondrial dysfunction. Overall, these results showed that the SOB zebrafish model established in this study may provide new insights into the development of therapeutic strategies to manage mitochondria-related muscular atrophy.

Keywords: high-fat diet; muscle; sarcopenic obesity; mitochondria; zebrafish

Citation: Zou, Y.-Y.; Chen, Z.-L.; Sun, C.-C.; Yang, D.; Zhou, Z.-Q.; Xiao, Q.; Peng, X.-Y.; Tang, C.-F. A High-Fat Diet Induces Muscle Mitochondrial Dysfunction and Impairs Swimming Capacity in Zebrafish: A New Model of Sarcopenic Obesity. *Nutrients* **2022**, *14*, 1975. https://doi.org/10.3390/nu14091975

Academic Editor: Susanne Klaus

Received: 31 March 2022
Accepted: 6 May 2022
Published: 9 May 2022

1. Introduction

Obesity is an epidemic disease in economically developed and developing countries and is associated with several chronic diseases, including diabetes mellitus and non-alcoholic fatty liver disease [1]. Unhealthy diets, such as a high-fat diet (HFD), are major causes of obesity [2,3]. Furthermore, HFDs can also contribute to metabolic disorders, pro-inflammatory cytokine expression, and insulin resistance in skeletal muscle [4], the latter of which is one of the primary causes of type 2 diabetes mellitus [5,6]. Long-term, high-fat dietary intake can induce myofibrillar protein degradation and myonuclear apoptosis and ultimately lead to sarcopenia [7]. Sarcopenic obesity (SOB) combined with the deterioration of skeletal muscle function can lead to decreased exercise capacity, accompanied by increased disability and frailty [8]. Thus, further studies of the mechanisms underlying the pathogenesis of SOB are necessary to facilitate the prevention and treatment of this muscle disorder.

Mitochondria are crucial organelles responsible for the ATP supply and metabolic status of skeletal muscle [9]. The reduction in muscle mass induced by SOB may contribute to dysregulation of the mitochondrial network and impairment of mitochondrial function [10,11]. Additionally, the dysregulation of mitochondrial function can accelerate lipid deposition and oxidative stress in muscle, thereby exacerbating the decrease in muscle mass [12]. Thus, the investigation of the molecular mechanisms regulating mitochondrial function and how obesity leads to mitochondrial dysfunction in sarcopenia may facilitate the development of strategies to manage SOB.

Many researchers have developed reliable zebrafish models to understand muscle degeneration and its pathophysiology [13,14]. We hypothesized that zebrafish could be an appropriate SOB model and that the molecular mechanism underlying the condition was related to mitochondrial function. In the current study, we evaluated the effects of HFD-induced SOB on muscle atrophy and exercise capacity in a zebrafish model. We also explored the causal relationship between muscle mitochondrial function and SOB in zebrafish. Our findings may provide insights into therapeutic targets and treatment strategies for SOB.

2. Materials and Methods

2.1. Animal Ethics Statement

All experiments were conducted under the Chinese guidelines for animal welfare and experimental protocols. Approval was obtained from the Animal Experiment Administration Committee of Hunan Normal University (Changsha, China; approval number: 2018/046).

2.2. Animals, Culture Conditions, and Dietary Experimental Design

Adult male zebrafish (AB strain, 4 months old) were raised under light at 28 °C for 14 h under standard husbandry conditions. Prior to the start of the experiment, zebrafish were fed a commercial diet (TP1FM21051; Trophic Animal Feed High-Tech Co., Ltd., Nantong, China) containing 6% fat for 1 week of adaptation. Subsequently, 90 healthy zebrafish were selected and randomly allocated to two diet groups (three tanks per dietary treatment, 15 zebrafish per tank): a normal diet (6% fat; ND) or an HFD (24% fat). During the 16-week feeding trial, experimental zebrafish were fed three times per day. At the end of the trial, the final body weight and body length of the fish in each tank were determined. Fifteen zebrafish from each dietary regimen group (5 fish/tank) were randomly selected for quantification of swimming capacity. Subsequently, zebrafish were randomly anesthetized with 20 mg/L tricaine methanesulfonate after fasting for 24 h, after which blood samples were collected from the caudal vein using 10 μL heparin-treated capillaries for blood glucose measurement. Liver and muscle samples were immediately collected for further experiments.

2.3. Biochemical Analyses

Blood glucose levels were measured using a blood glucometer purchased from Yuwell (Yuwell 580, Beijing, China). For measurement of muscle triglyceride, total cholesterol, catalase, and total superoxide dismutase levels, 150 mg of muscle tissue (from three zebrafish) from each group was homogenized in 0.9% normal saline (1:9, v/v) and centrifuged at 2000 rpm for 15 min. The supernatants were collected and aliquoted for further analysis. Biochemical parameters were measured using specific commercial kits (Jiancheng Biotech Co., Nanjing, China).

2.4. Histological Analysis

Liver samples from three zebrafish per group (one zebrafish per tank) were fixed in 4% paraformaldehyde solution for 24 h, embedded in paraffin, and sliced into 4-μm-thick sections for hematoxylin-eosin (H&E) staining and Masson staining.

Three fresh liver samples were fixed in 4% paraformaldehyde for 4 h and embedded in OCT. Frozen sections (8 μm thickness) were generated using a cryostat and the samples were fixed with 4% paraformaldehyde for an additional 30 min. The slides were washed in distilled water and stained with Oil Red O for 15 min. Oil Red O staining was performed to determine the lipid contents in liver tissue.

The muscle tissues (for H&E staining) of zebrafish (n = 3) were fixed in 4% paraformaldehyde, dehydrated through an ethanol series, dewaxed with xylene, and embedded in paraffin wax. The samples were transversely cut into 4 μm thick slices, dewaxed, dried, and stained with H&E under standard conditions. The images were captured using a microscope (Leica, Heidelberg, Germany) and analyzed using ImageJ (NIH, Bethesda, MD, USA).

2.5. Transmission Electron Microscopy

For transmission electron microscopy, tissues were fixed in 2.5% glutaraldehyde for 6–12 h. The fixative solution was discarded, and cells were transferred to phosphate-buffered saline. Next, cells were fixed in 1% osmic acid for 1–2 h, and dehydration was carried out by incubation in 30% ethanol for 10 min, 50% ethanol for 10 min, 70% uranyl acetate in ethanol (stained before embedding) for 3 h or overnight, 80% ethanol for 10 min, 95% ethanol for 15 min, 100% ethanol twice for 50 min each, and propylene oxide for 30 min. Next, samples were incubated in propylene oxide: epoxy resin (1:1) for 1–2 h and then in pure epoxy resin for 2–3 h. After embedding in pure epoxy resin, samples were baked in an oven at 40 °C for 12 h and then at 60 °C for 48 h. Samples were then cut into ultra-thin sections and placed on copper grids. Staining was then performed with lead and uranium stain, and images were acquired using a Japan Electronics JEM1400 transmission electron microscope and recorded with a Morada G3 digital camera.

2.6. Swimming Capacity and Oxygen Consumption Measurement

Analysis of the swimming performance and oxygen consumption of zebrafish was performed using a miniature swimming tunnel respirator (Loligo Systems, Viborg, Denmark). The following formula was used to calculate U_{crit} values for the swimming tests: U_{crit} = Uf + US × (Tf/TS), where Uf (cm/s) is the highest velocity, US (2.7 L) is the velocity increment, Tf (min) is the time elapsed at fatigue velocity, and TS (14 min) is the prescribed interval time. U_{crit} is expressed in terms of body lengths per second (BL/s). Maximal oxygen consumption (MO_2) was calculated using AutoResp 1 software (Loligo Systems, Viborg, Denmark). More detailed information can be found in our previous study [15].

2.7. Total RNA Extraction and Reverse Transcription-Quantitative Polymerase Chain Reaction (RT-qPCR)

The total RNA from zebrafish muscle tissue (one sample with two fish muscle tissues, n = 6) was extracted by homogenization in TRIzol solution (Thermo Fisher Scientific, Waltham, MA, USA) according to the manufacturer's protocol. RNA was reverse-transcribed into cDNA using a reverse transcription system kit (Takara, Tokyo, Japan). qPCR was performed using SYBR Green Master Mix (Thermo Fisher Scientific). Relative mRNA expression was determined using a Bio-Rad real-time PCR system (CFX96; Bio-Rad Laboratories, Hercules, CA, USA). Sangon Biotech synthesized primers for the detected genes and the reference gene, *gapdh*. Relative mRNA expression was determined using the $2^{-\Delta\Delta CT}$ method.

2.8. Western Blot

Zebrafish skeletal muscle tissue (n = 6) was lysed in cold RIPA buffer (Solarbio, Beijing, China) containing a mixture of protease and phosphatase inhibitors (Solarbio). Protein quantification was performed using a bicinchoninic acid protein assay kit (cat. no. E112-01/02; Vazyme, Nanjing, China), and the protein (30 mg) in each sample was separated by sodium dodecyl sulfate-polyacrylamide gel electrophoresis on 10% or 15% gels and then transferred to 0.45 or 0.22 mm polyvinylidene difluoride (PVDF) membranes. For analysis of phosphorylated proteins, membranes were blocked with 5% bovine serum albumin, whereas for analysis of other proteins, membranes were blocked with 5% fat-free milk. Membranes were incubated with primary antibodies overnight at 4 °C and washed with TBST. Subsequently, membranes were incubated with appropriate secondary horseradish peroxidase-conjugated antibodies. The proteins were detected using a gel imaging system (Tanon, Shanghai, China). The antibodies used were as follows: rabbit anti-β-actin antibody

(1:2000; Proteintech, Wuhan, China), rabbit anti-Atrogin-1 antibody (1:1000; Proteintech), rabbit anti-muscle RING-finger protein-1 (MuRF1) antibody (1:1000; Proteintech), rabbit anti-sirtuin 1 (SIRT1; 1:1500; Proteintech), rabbit anti-peroxisome proliferator-activated receptor γ coactivator 1-α (PGC1α) antibody (1:1000; Bioss, Beijing, China), rabbit anti-optic atrophy protein 1 (OPA1) antibody (1:1500; Proteintech), rabbit anti-mitofusin 2 (MFN2) antibody (1:1500; Proteintech), rabbit anti-dynamin-related protein 1 (DRP1) antibody (1:1500; Proteintech), and rabbit anti-phospho-AMP-activated protein kinase (AMPK; Thr172) antibody (1:2000; Cell Signaling Technology, Danvers, MA, USA). Protein expression was normalized to that of β-actin.

2.9. Statistical Analysis

Statistical analysis was performed using SPSS (version 22.0; IBM, Chicago, IL, USA) or the GraphPad Prism software (version 9.0; San Diego, CA, USA). Data were expressed as means $\pm$ standard deviations of three independent experiments. Unpaired *t*-tests were used to compare the mean values of the two groups. All experiments were repeated three times. Statistical significance was set at $p < 0.05$.

3. Results

3.1. Long-Term HFD Feeding Induced Obesity and Liver Injury in Zebrafish

As shown in Figure 1, zebrafish fed an HFD exhibited higher body weight compared with those in the ND group (Figure 1A), whereas no significant differences in body length were observed (Figure 1B). Fasting blood glucose and muscle triglyceride contents significantly increased in the HFD group (Figure 1C,D). After 16 weeks, HFD-fed zebrafish developed the characteristics of fatty liver with evident increases in lipid droplets, hepatic steatosis severity, and collagen contents (Figure 1E). These data indicate the successful establishment of HFD-induced obesity in zebrafish.

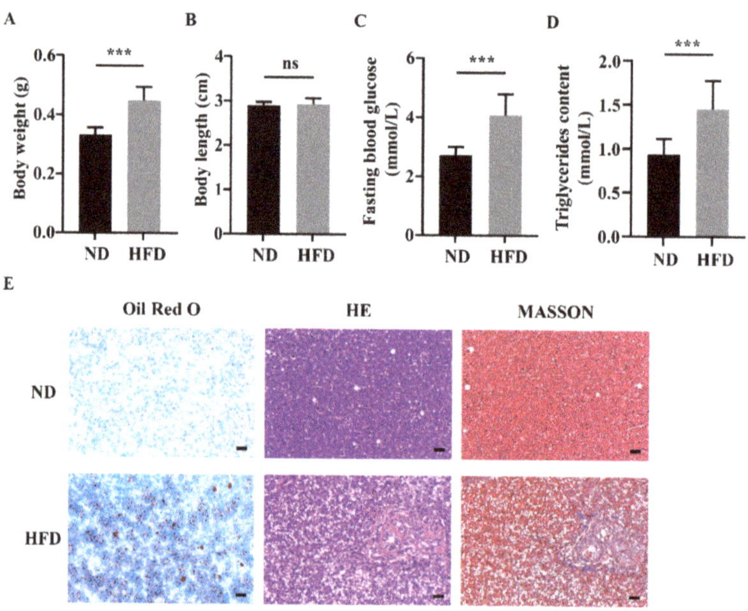

Figure 1. Long-term high-fat diet (HFD) feeding induced obesity and liver injury in zebrafish. (**A**) Body weight of zebrafish. (**B**) Body length of zebrafish. (**C**) Fasting blood glucose of zebrafish. (**D**) Muscle triglyceride contents. (**E**) Oil Red O staining, H&E staining, and Masson staining of zebrafish livers. ***, $p < 0.001$. Data represent means, and error bars represent standard errors of the means. Scale bar, 20 μm. ND, normal diet; HFD, high-fat diet.

3.2. Long-Term HFD Feeding Induced Skeletal Muscle Atrophy

After a 16-week feeding trial, we used transmission electron microscopy to examine the fiber morphology of muscles. Compared with the ND group, in which zebrafish muscle exhibited a normal structure, the muscles of zebrafish in the HFD group exhibited more lipid droplets. H&E staining indicated that the muscle fiber sizes of transverse sections in the HFD group were smaller than those in the ND group (Figure 2A,B). Furthermore, the expression of muscle atrophy markers (Atrogin-1 and MuRF1) was significantly increased in zebrafish in the HFD group (Figure 2C,D). These results demonstrated that long-term HFD feeding caused muscular atrophy in zebrafish.

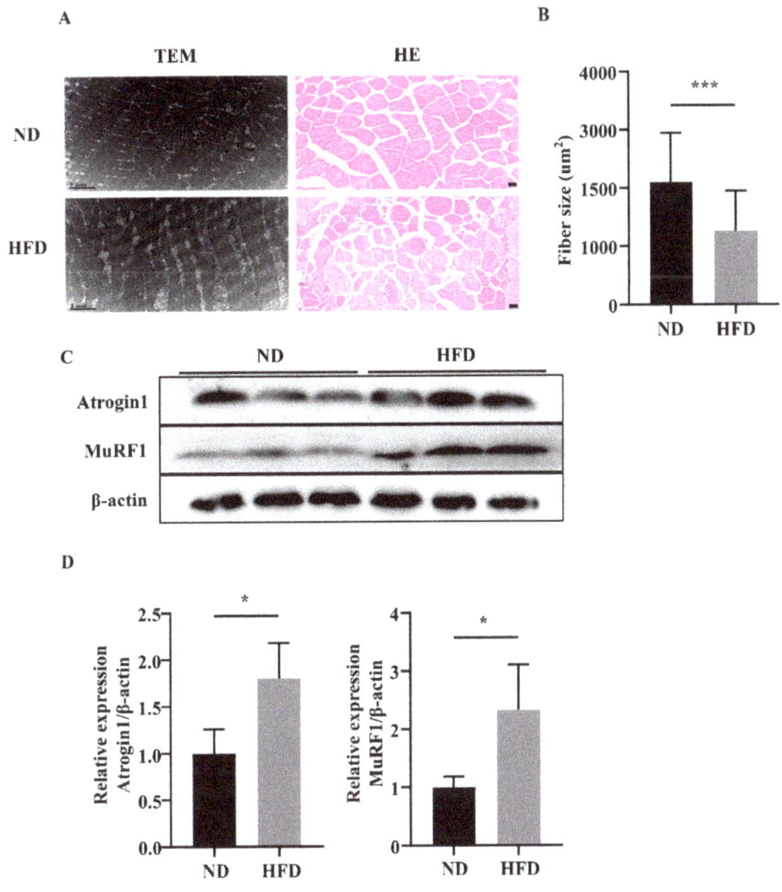

Figure 2. Comparison of skeletal muscle fiber size between zebrafish groups. (**A**) Representative photomicrographs of muscle sections stained H&E or imaged using transmission electron microscopy. (**B**) Average fiber size (based on H&E staining). (**C**,**D**) Atrogin-1 and MuRF1 protein expression. *, $p < 0.05$, ***, $p < 0.001$. Data represent means, and error bars represent standard errors of the means. Scale bars in transmission electron microscopy images, 2 μm. Scale bars in H&E-stained images, 20 μm. ND, normal diet; HFD, high-fat diet.

3.3. Long-Term HFD Feeding Impaired the Swimming Capacity of Zebrafish

We used exercise experimental protocols to measure the swimming capacity of the two groups of zebrafish. As shown in Figure 3A,B, as the water sped up beyond 6.2 BL/s, zebrafish in the ND group exhibited higher swimming capacity and MO_2 levels than those in the HFD group at each testing stage. Moreover, zebrafish in the HFD group exhibited decreased exhaustive swimming times and MO_{2max} values compared with those in the ND group (Figure 3C,D). Furthermore, zebrafish in the HFD group exhibited slower U_{crit} and $U_{crit\text{-}r}$ when compared with those in the ND group (Figure 3E,F). These results indicated that the swimming capacity of HFD-fed zebrafish was impaired.

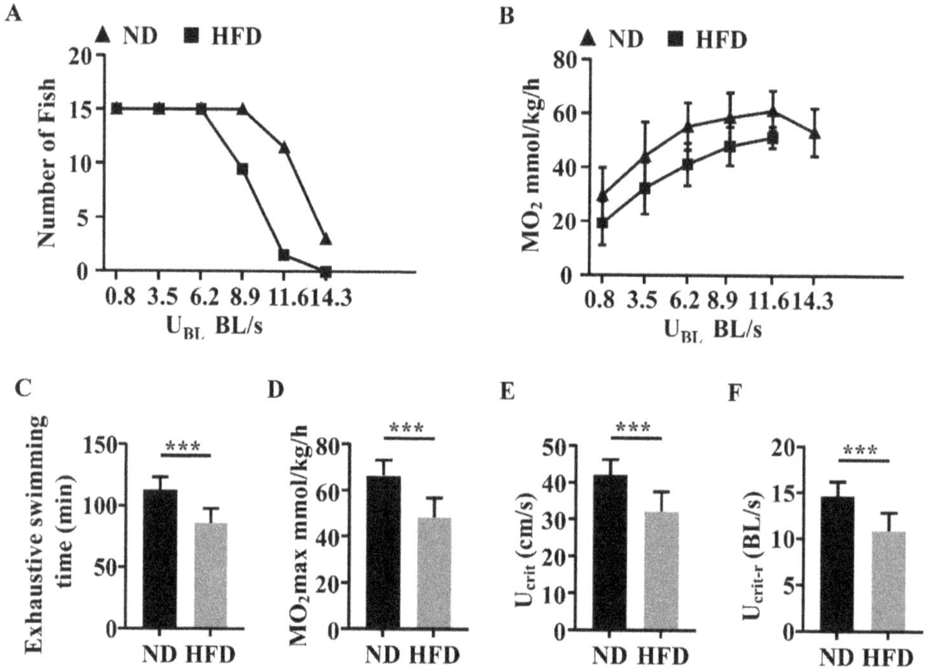

Figure 3. Comparison of swimming capacity tests between the two groups of zebrafish. (**A**) The number of zebrafish in the two groups at each speed test stage. (**B**) MO_2 levels in the two groups at each testing stage. (**C**) Exhaustive swimming times of zebrafish. (**D**) MO_{2max} of zebrafish. (**E**) U_{crit} of zebrafish. (**F**) $U_{crit\text{-}r}$ of zebrafish. ***, $p < 0.001$. Data represent means, and error bars represent standard errors of the means. Scale bar, 20 μm. ND, normal diet; HFD, high-fat diet; U_{crit}, critical swimming speed; MO_2, oxygen consumption.

3.4. Long-Term HFD Feeding Suppressed Skeletal Muscle Mitochondrial Biogenesis and Fatty Acid Oxidation-Related Gene Expression in Zebrafish

We studied the biogenesis and oxidation functions of muscle mitochondria to elucidate the potentially damaging effects of HFD feeding. AMPK, SIRT1, and PGC1α maintain normal mitochondrial biogenesis and fatty acid oxidation in skeletal muscle. As expected, we observed lower levels of phospho-AMPK, SIRT1, and PGC1α in the muscles of HFD-fed zebrafish compared with those in zebrafish in the ND group (Figure 4A,B). Moreover, we detected significant decreases in the expression of genes related to mitochondrial biogenesis (*pgc1α*, *nrf1*, and *tfam*), fatty oxidation (*pparab* and *cpt1a*), and electron transport chain (ETC) complex subunits (*sdha*, *uqcrc2b*, *cox4il*, and *atp5f*; Figure 4C). These data showed that HFD feeding may partially show a suppression of mitochondrial function.

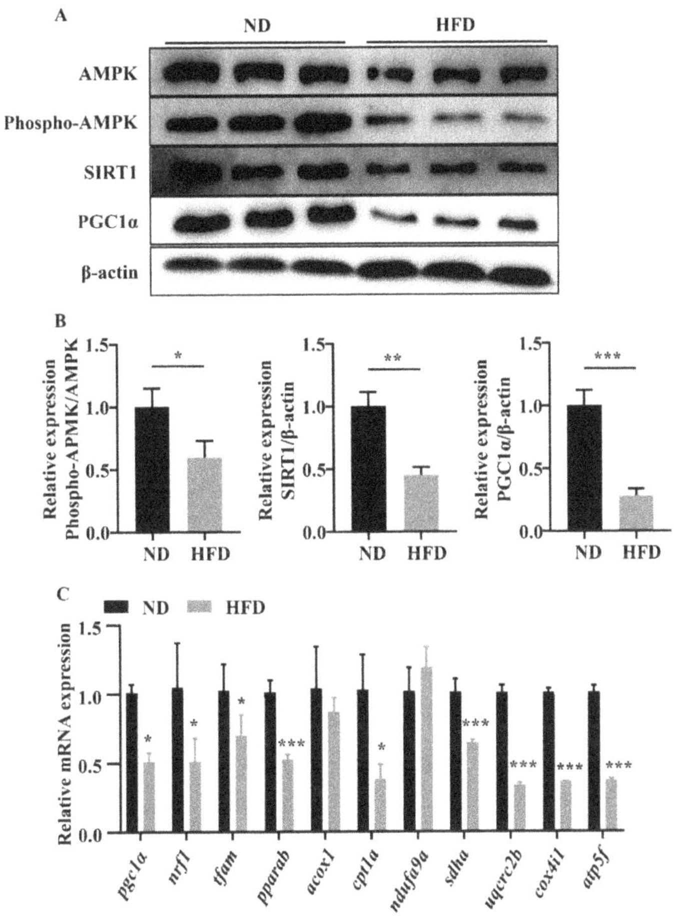

Figure 4. Long-term HFD feeding induced abnormal mitochondrial dysfunction. (**A**,**B**) Phospho-AMPK, SIRT1, and PGC1α protein levels. (**C**) mRNA expression of genes related to mitochondrial biogenesis, fatty oxidation, and ETC complexes subunits. *, $p < 0.05$, **, $p < 0.01$, ***, $p < 0.001$. Data represent means, and error bars represent standard errors of the means. ND, normal diet; HFD, high-fat diet; ETC, electron transport chain.

3.5. Long-Term HFD Feeding Induced Abnormal Mitochondrial Fusion and Fission in Zebrafish Skeletal Muscle

Abnormal mitochondrial fusion and fission are responsible for the impairment of mitochondrial function [16]. Compared with the ND group, skeletal muscles of zebrafish in the HFD group exhibited higher expression levels of mitochondrial fission-related proteins (e.g., DRP1) and lower levels of mitochondrial fusion-related proteins (e.g., OPA1 and MFN2; Figure 5A,B). Moreover, as shown in Figure 5C, most mitochondria in the HFD group were degenerated, enlarged, and swollen, and the mitochondrial cristae were broken or absent. These data suggested that long-term HFD feeding could promote mitochondrial fission and suppress mitochondrial fusion.

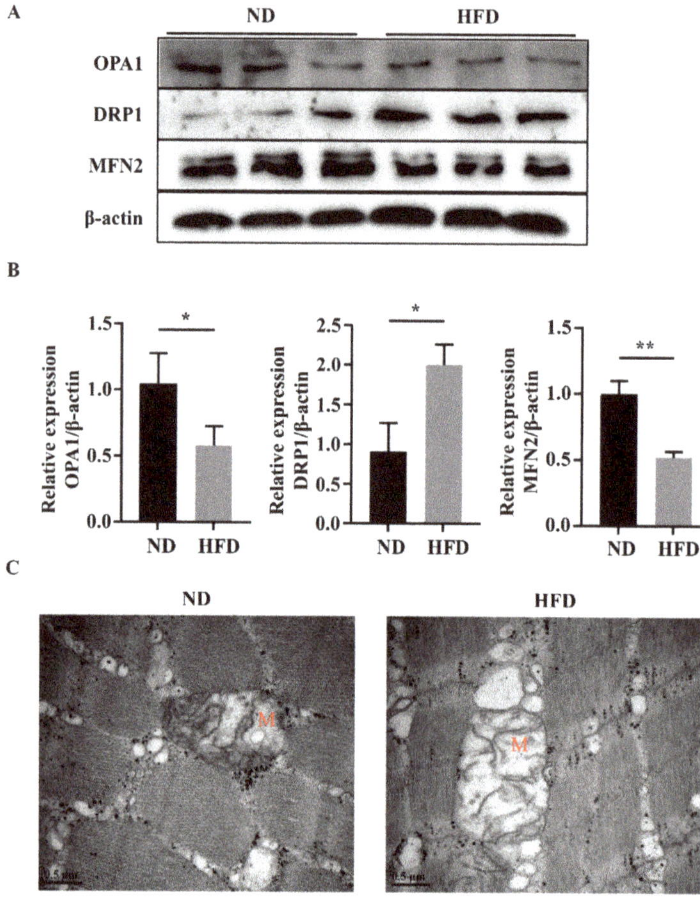

Figure 5. Long-term HFD feeding induced abnormal mitochondrial fusion and fission. (**A**,**B**) Protein expression of OPA1, MFN2, and DRP1. (**C**) Representative photomicrographs of muscle sections imaged using transmission electron microscopy. M, mitochondria. *, $p < 0.05$, **, $p < 0.01$. Data represent means, and error bars represent standard errors of the means. Scale bar in transmission electron microscopy images, 0.5 μm. ND, normal diet; HFD, high-fat diet.

4. Discussion

Obesity is an epidemic condition [17]. Severe obesity is accompanied by excessive accumulation of fat in visceral adipose tissue and normally lean tissues (e.g., the liver, heart, and skeletal muscle) [18]. Lipid overload in the skeletal muscle lowers muscle mass and induces sarcopenia [19,20]. Zebrafish is a promising model system for studying human diseases, such as obesity and sarcopenia, because of the functional conservation in substance and energy metabolism and the similar characteristics in muscular physiology. In this study, we fed adult zebrafish an HFD for 16 weeks to induce SOB and showed that long-term HFD feeding contributed to muscular atrophy and decreased swimming capacity in obese zebrafish. Moreover, we demonstrated that this outcome was related to mitochondrial dysfunction in the muscles of SOB-model zebrafish.

Obesity is primarily caused by the consumption of an unhealthy diet with a prolonged imbalance of energy intake and energy expenditure Here, we observed a clear increase in body weight and muscle triglyceride content in zebrafish in the HFD group. Additionally, zebrafish fed an HFD exhibited typical characteristics of fatty liver. These results indicated that zebrafish developed obesity after 16 weeks of HFD feeding. Glucose disposal activated by insulin is a vital metabolic function of skeletal muscle, and normal glucose metabolism is important for health [21]. Moreover, excessive fat accumulation in skeletal muscles can impair insulin signaling and glucose intake [22]. Most obese phenotypes are accompanied by the inhibition of insulin receptor substrate and AKT and increased insulin resistance [23]. The current findings further confirmed that long-term HFD feeding contributed to obesity in zebrafish.

SOB has become a subject of interest and research over the years [24]. SOB represents a combination between sarcopenia and obesity, and there are several common pathophysiological mechanisms between sarcopenia and obesity, resulting in increased disease severity when both conditions are present [19]. Skeletal muscle fiber atrophy is the major pathological trait observed in sarcopenia. In our study, we found that skeletal muscle fiber size was decreased in obese zebrafish and that the expression levels of muscle atrophy marker proteins (Atrogin-1 and MuRF1) were significantly elevated in obese skeletal muscles. These findings indicate that increased protein degradation and muscle atrophy were induced by HFD feeding in zebrafish. Moreover, we detected an evident decrease in swimming capacity in SOB zebrafish, with clear reductions in exhaustive swimming time, MO_{2max}, U_{crit}, and $U_{crit\text{-}r}$.

Mitochondria play essential roles in regulating fatty acid metabolism and ATP production, which are important for muscle contractibility and plasticity [25]. Impaired mitochondrial function in skeletal muscle is thought to promote muscular atrophy in subjects with obesity [10]. Indeed, a few weeks of HFD feeding can cause intramyocellular lipid accumulation and reduce mitochondrial function [26]. The AMPK/SIRT1 pathway maintains cellular energy stores in skeletal muscles [27]. In addition, as the main regulator of mitochondrial biogenesis, PGC1α can be stimulated by the AMPK/SIRT1 pathway [28]. The AMPK/SIRT1/PGC1α pathway is inhibited in the skeletal muscles of humans and mice with obesity [29,30]. These findings are also supported by our current results. In this study, we observed decreased levels of phospho-AMPK, SIRT1, and PGC1α proteins as well as downregulation of mRNA levels of various transcription factors (*nrf1* and *tfam*) in the muscles of SOB-model zebrafish. Moreover, mitochondrial biogenesis is responsible for oxidative phosphorylation and fatty acid β-oxidation [31]. This reduction in mitochondrial biogenesis may result in decreased ATP synthesis and fatty acid β-oxidation. In our study, we also observed the downregulation of genes involved in the mitochondrial complex and β-oxidation. These results indicated that skeletal muscle atrophy was associated with mitochondrial dysfunction.

Mitochondria are unique and highly dynamic organelles that exhibit continuous fission and fusion [32]. Fusion and fission processes regulate the length, size, and morphology of mitochondria, and these mitochondrial dynamics are associated with mitochondrial function. Under abnormal physiological conditions, mitochondria cannot maintain an appropriate balance between fission and fusion. Fusion ameliorates stress by mixing the contents of partially damaged mitochondria as a form of complementation, whereas fission is thought to induce oxidative stress and cell apoptosis [33]. Thus, mitochondrial homeostasis is critical for maintaining normal cellular events. MFN2 is responsible for outer membrane fusion, whereas OPA1 is required for inner membrane fusion. In the current study, we observed a strong inhibition of MFN and OPA1 in SOB model zebrafish. Moreover, we observed increased expression of the mitochondria fission marker DRP1, accompanied by abnormal mitochondrial structures in the HFD group. These findings indicated that a decreased mitochondrial quality occurred in the skeletal muscles and mitochondria of SOB model zebrafish.

In summary, in HFD-fed SOB model zebrafish, muscle atrophy was related to mitochondrial dysfunction. Considering the complex pathogenesis of SOB, such as lipotoxicity, mitochondrial dysfunction, inflammation, and insulin resistance, the current study only unveiled one aspect of it. Further comprehensive studies are necessary to investigate the detailed mechanisms linking the pathogenesis of SOB and mitochondria dysfunction and to assess the efficacy of pharmacological and exercise-based interventions targeting mitochondria to prevent or treat SOB.

Author Contributions: Y.-Y.Z. designed and performed the study; Z.-L.C., C.-C.S., Q.X. and D.Y. assisted with experimental operation guidance; Y.-Y.Z. wrote the manuscript; X.-Y.P., Z.-Q.Z. and C.-F.T. revised the manuscript. All authors have read and agreed to the published version of the manuscript.

Funding: This research was supported by the National Natural Science Foundation of China (grant nos. 81801392 and 32100919), Research Foundation of Hunan Province Education Department (grant no. 21B0066), Hunan Provincial Natural Science Foundation of China (grant no. 2019JJ50098), and Postgraduate Scientific Research Innovation Project of Hunan Province (grant nos. CX20200532 and CX20210483).

Institutional Review Board Statement: All aspects of this research were conducted in accordance with the Chinese guidelines for animal welfare and experimental protocols. Approval was obtained from the Animal Experiment Administration Committee of Hunan Normal University (Hunan, China; approval No.: 2018/046).

Informed Consent Statement: Not applicable.

Data Availability Statement: The data are available from the corresponding author.

Conflicts of Interest: The authors declare no conflict of interest.

References

1. Bluher, M. Obesity: Global epidemiology and pathogenesis. *Nat. Rev. Endocrinol.* **2019**, *15*, 288–298. [CrossRef] [PubMed]
2. Heydemann, A. An Overview of Murine High Fat Diet as a Model for Type 2 Diabetes Mellitus. *J. Diabetes Res.* **2016**, *2016*, 2902351. [CrossRef] [PubMed]
3. Recena Aydos, L.; Aparecida do Amaral, L.; Serafim de Souza, R.; Jacobowski, A.C.; Freitas Dos Santos, E.; Rodrigues Macedo, M.L. Nonalcoholic Fatty Liver Disease Induced by High-Fat Diet in C57bl/6 Models. *Nutrients* **2019**, *11*, 3067. [CrossRef] [PubMed]
4. Rasool, S.; Geetha, T.; Broderick, T.L.; Babu, J.R. High Fat with High Sucrose Diet Leads to Obesity and Induces Myodegeneration. *Front. Physiol.* **2018**, *9*, 1054. [CrossRef]
5. Hoeks, J.; Schrauwen, P. Muscle mitochondria and insulin resistance: A human perspective. *Trends Endocrinol. Metab.* **2012**, *23*, 444–450. [CrossRef]
6. DeFronzo, R.A.; Tripathy, D. Skeletal muscle insulin resistance is the primary defect in type 2 diabetes. *Diabetes Care* **2009**, *32* (Suppl. S2), S157–S163. [CrossRef]
7. Tournadre, A.; Vial, G.; Capel, F.; Soubrier, M.; Boirie, Y. Sarcopenia. *Jt. Bone Spine* **2019**, *86*, 309–314. [CrossRef]
8. Kalinkovich, A.; Livshits, G. Sarcopenic obesity or obese sarcopenia: A cross talk between age-associated adipose tissue and skeletal muscle inflammation as a main mechanism of the pathogenesis. *Ageing Res. Rev.* **2017**, *35*, 200–221. [CrossRef]
9. Hood, D.A.; Memme, J.M.; Oliveira, A.N.; Triolo, M. Maintenance of Skeletal Muscle Mitochondria in Health, Exercise, and Aging. *Annu. Rev. Physiol.* **2019**, *81*, 19–41. [CrossRef]
10. Pileggi, C.A.; Parmar, G.; Harper, M.E. The lifecycle of skeletal muscle mitochondria in obesity. *Obes. Rev.* **2021**, *22*, e13164. [CrossRef]
11. Huang, Y.; Zhu, X.; Chen, K.; Lang, H.; Zhang, Y.; Hou, P.; Ran, L.; Zhou, M.; Zheng, J.; Yi, L.; et al. Resveratrol prevents sarcopenic obesity by reversing mitochondrial dysfunction and oxidative stress via the PKA/LKB1/AMPK pathway. *Aging* **2019**, *11*, 2217–2240. [CrossRef] [PubMed]
12. Genders, A.J.; Holloway, G.P.; Bishop, D.J. Are Alterations in Skeletal Muscle Mitochondria a Cause or Consequence of Insulin Resistance? *Int. J. Mol. Sci.* **2020**, *21*, 6948. [CrossRef] [PubMed]
13. Daya, A.; Donaka, R.; Karasik, D. Zebrafish models of sarcopenia. *Dis. Model Mech.* **2020**, *13*, dmm042689. [CrossRef] [PubMed]
14. Christian, C.J.; Benian, G.M. Animal models of sarcopenia. *Aging Cell* **2020**, *19*, e13223. [CrossRef]
15. Chen, Z.; Zhou, Z.; Peng, X.; Sun, C.; Yang, D.; Li, C.; Zhu, R.; Zhang, P.; Zheng, L.; Tang, C. Cardioprotective responses to aerobic exercise-induced physiological hypertrophy in zebrafish heart. *J. Physiol. Sci.* **2021**, *71*, 33. [CrossRef]
16. Youle, R.J.; van der Bliek, A.M. Mitochondrial fission, fusion, and stress. *Science* **2012**, *337*, 1062–1065. [CrossRef]
17. Conway, B.; Rene, A. Obesity as a disease: No lightweight matter. *Obes. Rev.* **2004**, *5*, 145–151. [CrossRef]
18. Piche, M.E.; Tchernof, A.; Despres, J.P. Obesity Phenotypes, Diabetes, and Cardiovascular Diseases. *Circ. Res.* **2020**, *126*, 1477–1500. [CrossRef]
19. Choi, K.M. Sarcopenia and sarcopenic obesity. *Korean J. Intern. Med.* **2016**, *31*, 1054–1060. [CrossRef]
20. Wannamethee, S.G.; Atkins, J.L. Muscle loss and obesity: The health implications of sarcopenia and sarcopenic obesity. *Proc. Nutr. Soc.* **2015**, *74*, 405–412. [CrossRef]

21. Richter, E.A.; Hargreaves, M. Exercise, GLUT4, and skeletal muscle glucose uptake. *Physiol. Rev.* **2013**, *93*, 993–1017. [CrossRef] [PubMed]
22. Devarshi, P.P.; McNabney, S.M.; Henagan, T.M. Skeletal Muscle Nucleo-Mitochondrial Crosstalk in Obesity and Type 2 Diabetes. *Int. J. Mol. Sci.* **2017**, *18*, 831. [CrossRef] [PubMed]
23. Koh, J.H.; Johnson, M.L.; Dasari, S.; LeBrasseur, N.K.; Vuckovic, I.; Henderson, G.C.; Cooper, S.A.; Manjunatha, S.; Ruegsegger, G.N.; Shulman, G.I.; et al. TFAM Enhances Fat Oxidation and Attenuates High-Fat Diet-Induced Insulin Resistance in Skeletal Muscle. *Diabetes* **2019**, *68*, 1552–1564. [CrossRef]
24. Ribeiro, S.M.; Kehayias, J.J. Sarcopenia and the analysis of body composition. *Adv. Nutr.* **2014**, *5*, 260–267. [CrossRef] [PubMed]
25. Abrigo, J.; Simon, F.; Cabrera, D.; Vilos, C.; Cabello-Verrugio, C. Mitochondrial Dysfunction in Skeletal Muscle Pathologies. *Curr. Protein Pept. Sci.* **2019**, *20*, 536–546. [CrossRef] [PubMed]
26. Chen, P.B.; Yang, J.S.; Park, Y. Adaptations of Skeletal Muscle Mitochondria to Obesity, Exercise, and Polyunsaturated Fatty Acids. *Lipids* **2018**, *53*, 271–278. [CrossRef]
27. Morales-Alamo, D.; Calbet, J.A.L. AMPK signaling in skeletal muscle during exercise: Role of reactive oxygen and nitrogen species. *Free Radic. Biol. Med.* **2016**, *98*, 68–77. [CrossRef]
28. Tang, B.L. Sirt1 and the Mitochondria. *Mol. Cells* **2016**, *39*, 87–95. [CrossRef]
29. Mortensen, B.; Poulsen, P.; Wegner, L.; Stender-Petersen, K.L.; Ribel-Madsen, R.; Friedrichsen, M.; Birk, J.B.; Vaag, A.; Wojtaszewski, J.F. Genetic and metabolic effects on skeletal muscle AMPK in young and older twins. *Am. J. Physiol. Endocrinol. Metab.* **2009**, *297*, E956–E964. [CrossRef]
30. Abu Bakar, M.H.; Shariff, K.A.; Tan, J.S.; Lee, L.K. Celastrol attenuates inflammatory responses in adipose tissues and improves skeletal muscle mitochondrial functions in high fat diet-induced obese rats via upregulation of AMPK/SIRT1 signaling pathways. *Eur. J. Pharmacol.* **2020**, *883*, 173371. [CrossRef]
31. Dorn, G.W., II; Vega, R.B.; Kelly, D.P. Mitochondrial biogenesis and dynamics in the developing and diseased heart. *Genes Dev.* **2015**, *29*, 1981–1991. [CrossRef] [PubMed]
32. Lee, H.; Yoon, Y. Mitochondrial fission and fusion. *Biochem. Soc. Trans.* **2016**, *44*, 1725–1735. [CrossRef] [PubMed]
33. Chan, D.C. Mitochondrial Dynamics and Its Involvement in Disease. *Annu. Rev. Pathol.* **2020**, *15*, 235–259. [CrossRef] [PubMed]

nutrients

MDPI

Article

Bifidobacterium animalis subsp. *lactis* A6 Enhances Fatty Acid β-Oxidation of Adipose Tissue to Ameliorate the Development of Obesity in Mice

Yanxiong Huo [1], Guoping Zhao [2], Jinwang Li [2], Ran Wang [3], Fazheng Ren [1,3], Yixuan Li [1,3] and Xiaoyu Wang [1,*]

1 Key Laboratory of Precision Nutrition and Food Quality, College of Food Science & Nutritional Engineering, China Agricultural University, Beijing 100083, China; huoyanxiong195615@163.com (Y.H.); renfazheng@cau.edu.cn (F.R.); liyixuan@cau.edu.cn (Y.L.)
2 School of Food and Health, Beijing Technology and Business University, Beijing 100048, China; zhaoguoping525@126.com (G.Z.); sdlijinwang@sina.com (J.L.)
3 Key Laboratory of Functional Dairy, Co-Constructed by Ministry of Education and Beijing Municipality, Department of Nutrition and Health, China Agricultural University, Beijing 100083, China; wangran@cau.edu.cn
* Correspondence: xy.wang@cau.edu.cn; Tel.: +86-10-6273-6344

Abstract: Fatty acid β-oxidation (FAO) is confirmed to be impaired in obesity, especially in adipose tissues. We previously proved that *Bifidobacterium animalis* subsp. *lactis* A6 (BAA6) had protective effects against diet-induced obesity. However, whether BAA6 enhances FAO to ameliorate the development of obesity has not been explored. After being fed with high-fat diet (HFD) for 9 weeks, male C57BL/6J mice were fed HFD or BAA6 for 8 weeks. In vitro study was carried out using 3T3-L1 adipocytes to determine the effect of BAA6 culture supernatant (BAA6-CM). Here, we showed that administration of BAA6 to mice fed with HFD decreased body weight gain (by 5.03 g) and significantly up-regulated FAO in epididymal adipose tissues. In parallel, FAO in 3T3-L1 cells was increased after BAA6-CM treatment. Acetate was identified as a constituent of BAA6-CM that showed a similar effect to BAA6-CM. Furthermore, acetate treatment activated the GPR43-PPARα signaling, thereby promoting FAO in 3T3-L1 cells. The levels of acetate were also elevated in serum and feces (by 1.92- and 2.27-fold) of HFD-fed mice following BAA6 administration. The expression levels of GPR43 and PPARα were increased by 55.45% and 69.84% after BAA6 supplement in the epididymal fat of mice. Together, these data reveal that BAA6 promotes FAO of adipose tissues through the GPR43-PPARα signaling, mainly by increasing acetate levels, leading to alleviating the development of obesity.

Keywords: *Bifidobacterium animalis* subsp. *lactis* A6; obesity; fatty acid β-oxidation; acetate; GPR43; PPARα

Citation: Huo, Y.; Zhao, G.; Li, J.; Wang, R.; Ren, F.; Li, Y.; Wang, X. *Bifidobacterium animalis* subsp. *lactis* A6 Enhances Fatty Acid β-Oxidation of Adipose Tissue to Ameliorate the Development of Obesity in Mice. *Nutrients* **2022**, *14*, 598. https://doi.org/10.3390/nu14030598

Academic Editor: Riccardo Caccialanza

Received: 29 December 2021
Accepted: 25 January 2022
Published: 29 January 2022

1. Introduction

Obesity is a common health problem for individuals today, which could increase the risk of metabolic diseases [1]. It is characterized by elevating fat mass through imbalance of anabolism and catabolism, especially fatty acid β-oxidation (FAO) processes in adipocytes [2]. FAO is shown to be low in obesity, and lean persons have high FAO [3]. Thus, the reduction in FAO has been supposed to be a key driver of the lipid accumulation in obese individuals [4].

FAO is recognized as an essential factor in obesity [5,6]. It has been shown that obese patients have decreased FAO protein levels in adipose tissues [7]. A large body of evidence has shown that up-regulating the expression of FAO-related proteins could improve the obese phenotype [8]. Others have demonstrated that overexpression of FAO-related proteins, such as carnitine palmitoyltransferase-1 (CPT1), in adipocytes could decrease triglyceride (TG) accumulation in obesity [9]. Currently, approaches of drug and phytochemical modulation are used to promote FAO in obesity. Several drugs, such as rimonabant, have been used towards weight management by enhancing FAO [10]. However, the high cost and potential side effects for this kind of drug causes dissatisfaction to patients [11]. Although, phytochemicals such as phenolic compounds, polyunsaturated fatty acids, and plant sterols could also increase FAO [11]. Their applications are limited by weak stability [12]. Recently, probiotics are thought to possess various beneficial effects on human health with great safety [13,14]. More and more research has proved the beneficial effects of probiotics on enhancement of FAO in obesity, and this strategy has been regarded as a natural therapeutic agent to control metabolic syndrome [15]. Indeed, many reports reveal that *Akkermansia muciniphila* (AKK) could increase FAO to ameliorate obesity [16], which has been used as a uniquely promising strain for its high therapeutic effects [17]. Several findings have shown that administration of probiotic bacteria, such as *Lactobacillus* genus, to a high-fat diet (HFD)-fed mice promotes FAO in fat tissues, thereby protecting against obesity [18,19]. Further, probiotics such as *Lactobacillus* genus have been shown to be able to reduce fat percentage in obese people [20]. Notably, probiotics treatment to enhance FAO is considered as a means to alleviate obesity. However, the mechanism by which probiotics increase FAO has not been fully clarified.

G protein-coupled receptors (GPRs) play pivotal roles in cell signaling to maintain energy homeostasis [21,22]. They are known to regulate various adipocyte functions, including FAO and thermogenesis [23]. GPR43 is a key member for GPRs which is involved in adipocyte lipid metabolism [24,25]. Many studies have reported that GPR43 might up-regulate FAO-related genes expression [26]. Besides, peroxisomal proliferator-activated receptor α (PPARα) has a master function in the regulation of FAO [27]. Some findings have pointed out that PPARα activation might enhance gene expression of fatty acid-metabolizing proteins to promote FAO in adipose tissues [28]. Conversely, inhibition of PPARα expression could reduce FAO in adipocytes [29]. Hence, GPR43-PPARα signaling pathway is crucial to regulate FAO.

Bifidobacterium animalis subsp. *lactis* A6 (BAA6) is a probiotic which can stay active under gastric juice environment [30]. Additionally, our previous research suggested that administration of BAA6 to obese mice could reduce body weight gain [31]. However, the mechanism of BAA6 to ameliorate the development of obesity has not been clarified. Thus, the objective of this research was to test the hypothesis that BAA6 could promote FAO of adipose tissues to ameliorate the development of obesity through GPR43-PPARα signaling.

2. Materials and Methods

2.1. Preparation of Bacterial Cultures, BAA6-CM and Dead BAA6

BAA6 (CGMCC No. 9273) was obtained from the feces of a centenarian (Bama, Guangxi, China). BAA6 was incubated in none oxygen conditions, as described in previous study [31].

AKK (JCM 30893) was provided by Japan Collection of Microorganisms RIKEN BioResource Research Center (Ibaraki, Japan) and incubated in none oxygen condition, as presented previously [31].

BAA6 culture supernatant (BAA6-CM) was obtained as with the previous method [32]. Dead BAA6 was prepared as previously reported [33]. Briefly, BAA6 cells, grown to 10^9 colony-forming units (CFU)/mL, were centrifugated at 3000$\times$ g at 4 °C for 15 min. BAA6-CM was obtained from the cell-free supernatant. Then, the collected BAA6 cells were washed twice and reconstituted by physiological saline. Plate count was used to measure bacterial concentration. The dead BAA6 was prepared at 80 °C for 30 min.

2.2. Animals and Experimental Design

Male C57BL/6J mice (4-week-old, Body weight 18.63 $\pm$ 1.02 g) were provided by Beijing HFK Bioscience Co. Ltd. (Beijing, China). They were raised under a stable temperature (23 $\pm$ 1 °C) and 12 h light/dark cycle for 1 week. Mice were randomly assigned into two groups: normal diet (ND group, n = 6) and HFD (n = 22) for 9 weeks. Composition of diets was shown in Supplementary Table S1. Then, mice fed HFD (Body weight 30.62 $\pm$ 1.21 g) were randomly divided into three groups ($n = 6$): (1) HFD group; (2) HFD + BAA6 group, gavage of BAA6 at 10^9 CFU/kg per day for another 8 weeks; (3) HFD + AKK group (positive group), gavage of AKK at 10^9 CFU/kg per day for another 8 weeks. Mice were orally administrated with 0.9% saline solution in ND and HFD groups. Body weight and food intake were measured every week. When the experiment ended, mice were fasted for 12 h and then anesthetized by CO_2. Blood and fat tissues were collected. The fat tissues and serum were preserved at −80 °C for further detection. The experimental processes were approved by the Animal Experimentation Committee of China Agricultural University (AW18080202-1).

2.3. Cell Differentiation

The 3T3-L1 preadipocytes were obtained from American Type Culture Collection (Manassas, VA, USA). Cells were grown in six-well plates (2×10^5 cells/well) with Dulbecco's modified Eagle's medium (DMEM), containing 10% fetal bovine serum (FBS, Gibco Life Technologies, Rockville, MD, USA) at 37 °C and incubated to confluence. Subsequently, cells were differentiated, employing DMEM with FBS (10%), insulin (2 μg/mL), 3-isobutyl-1-methylxanthine (0.5 mmol/L), and dexamethasone (0.25 μmol/L) for 3 days. They continued incubation with insulin for another 5 days to fully differentiate into adipocytes [34].

2.4. Cell Viability Assay

The 3T3-L1 cell viability was detected by cell counting kit-8 assay (Beyotime Biotechnology, Beijing, China) according to the manufacturer's instructions. In brief, 3T3-L1 preadipocytes were grown in 96-well plates (1×10^4 cells/well). Subsequently, cells were differentiated into adipocytes. They were treated with BAA6-CM (0, 0.5, 1, 2, 4, and 8%) for 1–5 days and dead BAA6 (0, 10^4, 10^5, 10^6, 10^7, and 10^8 cells/mL) for 1–4 days, respectively. After treatment, CCK8 reagent was added and incubated for 1 h at 37 °C. Microplate reader (Bio-Rad, Hercules, CA, USA) was used to assess cell viability (at 450 nm) [35].

2.5. Biochemical Assay of Serum

Total cholesterol (TC), TG, and high-density lipoprotein cholesterol (HDL-C) were analyzed by No. 3 Hospital of Beijing University [36].

2.6. Metabolic Assessment

Body fat was assessed by Body Composition Analysis (MiniQMR23-060H-I, Shanghai Niumag Corporation, Shanghai, China). Oxygen consumption and respiratory exchange ratio (RER) were measured through mouse Comprehensive Laboratory Animal Monitoring System metabolic cages [34].

2.7. TG Quantification Assay in 3T3-L1 Cells

TG concentration was detected through commercial TG assay kit (Nanjing Jiancheng Bioengineering Institute, Nanjing, China) as reported previously [37].

2.8. Hematoxylin and Eosin and Oil Red O Staining

Hematoxylin and eosin assay, in epididymal adipose tissues, was performed as with the previous method [34]. Oil Red O (ORO) assay, in 3T3-L1 cells, was conducted as previously described [38]. Images were obtained through a fluorescent inverted microscope (DMi8, Leica, Weztlar, Germany) and analyzed by a blind observer using Image Pro Plus A6 (version 6.0.0.260, Media Cybernetics Corporation, Las Vegas, NV, USA).

2.9. β Oxidation Study in 3T3-L1 Cells

β oxidation assessment was estimated through a Seahorse Bioscience XF96 Analyzer (Seahorse Bioscience Inc., MA, USA), as described previously [39]. Briefly, oxygen consumption rate (OCR) was analyzed in FAO Assay Medium, followed by treatment with or without etomoxir (45 mmol/L). Then, 25 mmol/L XF bovine serum albumin (BSA) or 167 mmol/L XF Palmitate-BSA (#102720-100, Agilent Technologies, Wilmington, DE, USA) was added in response to oligomycin (1.2 mmol/L), fluoro-carbonyl cyanide phenylhydrazone (1 mmol/L), and rotenone/antimycin A (1 mmol/L) (#101706-100, Agilent Technologies, Wilmington, DE, USA) based on recommending processes. β oxidation was calculated according to Timper et al.'s protocol [39].

2.10. *Analysis for Acetate in BAA6-CM, Serum and Feces*

Acetate was detected by GC-MS/MS. Wuhan MetWare Biotechnology was in charge of extracting the sample and quantifying metabolites, as previously described [40]. Briefly, samples of BAA6-CM, serum, and feces were thawed and vortexed for 1 min prior to analysis. Samples (50 μL) were added to 100 μL of 36% phosphoric acid solution. The mixture was vortexed for 3 min and 150 μL methyl tert-butyl ether (containing 2 μg/mL 2-methylvaleric acid as an internal standard) solution was added. After that, centrifugal force at 12,000× *g* was used to centrifuge the mixture at 4 °C for 10 min. The supernatant was used for further analysis. Agilent 7890B instrument, coupled to a 7000D Triple Quadrupole gas chromatography mass spectrometry system with a DB-FFAP column (30 m × 0.25 mm × 0.25 μm, J&W Scientific, Folsom, CA, USA), was utilized. The carrier gas was helium, and the flow rate was 1.2 mL/min. The oven temperature was held at 90 °C for 1 min, risen to 100 °C at a rate of 25 °C/min, risen to 150 °C at a rate of 20 °C/min, held on 0.6 min, risen to 200 °C at a rate of 25 °C/min, and kept for 0.5 min after running for 3 min. The quantitation of acetate in BAA6-CM, serum, and feces was counted using constructing calibration curves.

2.11. *Analysis of Intracellular Calcium Concentration in 3T3-L1 Cells*

The content of intracellular calcium (Ca^{2+}) was measured as previously reported [41]. In brief, 3T3-L1 cells were cultured in 24-well plates (1×10^5 cells/well). Then, cells were treated with DMEM, including 5 μmol/L Fura-2/AM (Beyotime Biotechnology, Beijing, China) for 45 min at 37 °C. Then, the 3T3-L1 cells were incubated with phosphate-buffered saline containing BSA (0.2%) for 5 min. The ratio of fluorescence intensities was assessed by changing excitation wavelengths of between 340 and 380 nm, with emission at 510 nm. The concentration was calculated as in the previous study [41].

2.12. *Western Blot Analysis*

Western blots were carried out as previously described [31]. Image J software (version 2.0, Microsoft Corporation, Redmond, WA, USA) was used to assess the band signal intensities [42]. Primary antibodies were listed as follows: fatty acid synthase (FAS, 3189s), phosphorylated acetyl CoA carboxylase (p-ACC, 3661s), ACC (3676s), hormone-sensitive lipase (HSL, 4107s), adipose triglyceride lipase (ATGL, 2138s), AMP-activated protein kinase (AMPK, 2532s) and phosphorylated AMPK (p-AMPK, 2535s) antibodies (Cell Signaling Technology, Boston, MA, USA). Carnitine palmitoyltransferase-2 (CPT2, 26555-1-AP) and PPARα (66826-1-lg) antibodies were provided from Proteintech (Rosemont, IL, USA). CPT1 (YN3388), long-chain acyl-CoA dehydrogenase (ACADL, YT6498), medium-chain acyl-CoA dehydrogenase (ACADM, YT5024), calcium/calmodulin-dependent protein kinase β (CAMKKβ, YT0624), phosph-CAMKKβ (p-CAMKKβ, YP1285), and GPR41 (YT2020) antibodies were from Immunoway (Plano, TX, USA). GPR43 (GTX00823) antibody was obtained from Gene Tex (San Antonio, CA, USA). β-Actin (bs-10966R) antibody was provided by Bioss (Beijing, China).

2.13. *PPARα and GPR43 Knockdown*

PPARα or GPR43-specific small interfering RNA (siRNA) was used to transfect cells by Lipofectamine 2000 (Thermo Fisher Scientific, Waltham, MA, USA) (Table S2, Supplementary Materials), according to the manufacturer's instruction.

2.14. Statistical Analyses

All values were shown as mean ± standard deviation (SD). SPSS 21.0 (SPSS Inc., Chicago, IL, USA) was used to analyze data. The Shapiro–Wilk test was used to assess Gaussian distribution. Equal variance was analyzed by Levene's test. Statistical comparison was measured by Tukey's post-test after ANOVA analysis. Differences were considered significant when $p < 0.05$.

3. Results

3.1. BAA6 Decelerated Body Weight Gain and Lipid Accumulation in Obese Mice

To investigate the effect of BAA6 in alleviating the development of obesity in mice, the levels of body weight, food intake, fat mass, and serum lipid profiles of mice were measured. There was no remarkable difference in food intake of all mice (Table S3, Supplementary Materials). In HFD group, the body weight gain, final body weight, and relative fat weight were higher than ND group, which were reduced after gavage of BAA6 and AKK (positive control) (Figure 1A–C). Furthermore, the number of adipocytes increased, and the cell area reduced after gavage with BAA6 and AKK, compared to HFD group (Figure 1D). Meanwhile, HFD group displayed higher concentration of serum TG and TC, as well as lower concentrations of HDL-C than ND group (Figure 1E–G). Importantly, BAA6 and AKK treatment markedly increased HDL-C levels. Together, the above data indicated that BAA6 treatment could decrease fat mass and body weight, and BAA6 showed similar effects to AKK.

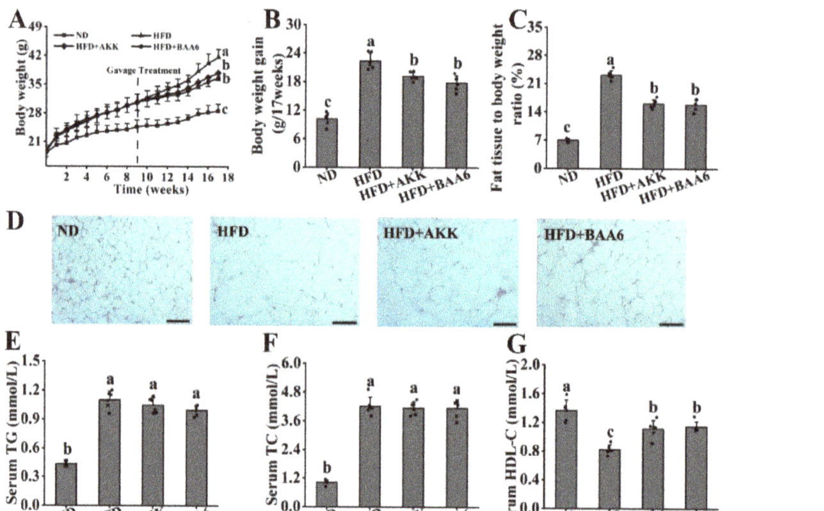

Figure 1. Effect of BAA6 administration on metabolic characteristics in high-fat diet (HFD)-fed mice. (**A**) Body weight, (**B**) body weight gain, (**C**) relative adipose tissue, (**D**) representative hematoxylin and eosin staining in epididymal adipose tissues (scale bar, 200 μm), serum levels of (**E**) triglyceride (TG), (**F**) total cholesterol (TC), and (**G**) high-density lipoprotein cholesterol (HDL-C) were analyzed. Values are shown as means ± SD ($n = 6$). Different lowercase letters mean remarkable differences between groups at $p < 0.05$. ND, fed normal diet; HFD, fed high-fat diet; HFD + BAA6, fed high-fat diet and administrated with *Bifidobacterium animalis* subsp. *lactis* A6 (BAA6); HFD + AKK, fed high-fat diet and administrated with *Akkermansia muciniphila* (AKK).

3.2. BAA6 Enhanced FAO in Adipose Tissues of Obese Mice

A great deal of research proved that low FAO contributed to obesity. The efficacy of BAA6, in alleviating the development of obesity, has been confirmed. Thus, the possible role of FAO was further explored. To investigate whether BAA6 alleviating the development of obesity was associated with enhancement of FAO in fat tissues, we firstly assessed the energy expenditure and FAO of obese mice in the resting state. Compared to ND group, energy expenditure (indicated by decreased oxygen consumption, VO_2) reduced in HFD group in both light and dark phases, while it was increased by BAA6 and AKK treatments (Figure 2A). Moreover, there was a remarkable decrease in FAO (indicated by increased RER) in HFD group during light and dark phases compared with ND group, which was recovered by BAA6 and AKK treatments during dark cycle (Figure 2B). Subsequently, proteins associated with lipid metabolism in the epididymal adipose tissues were measured. Western blotting of epididymal adipose tissues revealed the similar levels of lipid synthesis-related proteins, such as FAS and p-ACC, in BAA6 and AKK-treated mice relative to HFD mice, while lipid catabolism-related protein (HSL and ATGL) levels were significantly higher, after BAA6 and AKK supplementations, compared with HFD group (Figure S1, Supplementary Materials). Further, FAO-related proteins were assessed in the epididymal adipose tissues. The FAO-related protein levels (CPT1, CPT2, ACADL, and ACADM) in adipose tissues of HFD mice were lower than ND mice (Figure 2C), whereas these protein levels were increased, after BAA6 and AKK treatments, compared to HFD group. Our results indicated that BAA6 could enhance FAO in adipose tissues to alleviate the development of obesity.

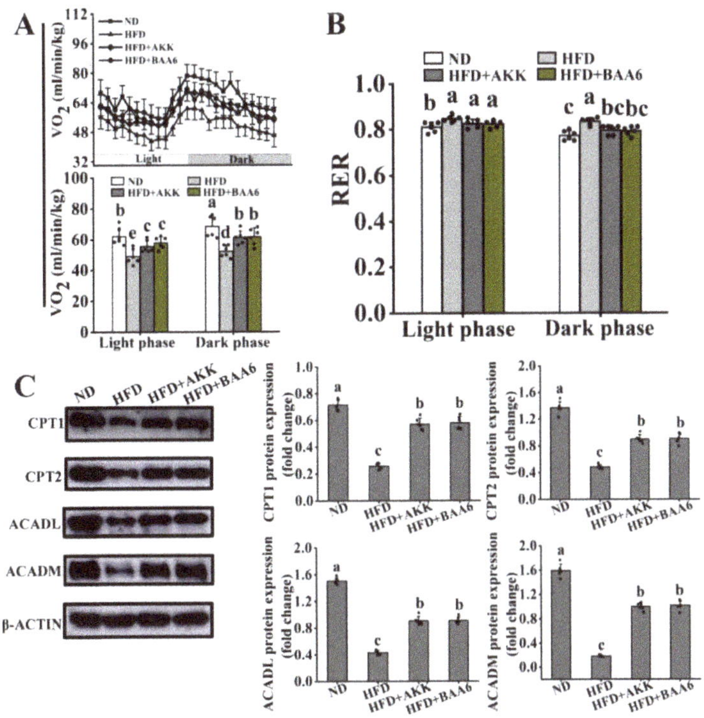

Figure 2. Effect of BAA6 on fatty acid β-oxidation (FAO) for epididymal adipose tissues in HFD-fed mice. (**A**) O_2 consumption and (**B**) respiratory exchange ratio (RER) were detected during a 12 h light/dark cycle in mice. (**C**) Expression levels of CPT1, CPT2, ACADL, and ACADM in epididymal adipose tissues were measured. Values are shown as means ± SD ($n = 6$). Different lowercase letters mean remarkable differences between groups at $p < 0.05$. CPT1, carnitine palmitoyltransferase-1; CPT2, carnitine palmitoyltransferase-2; ACADL, long-chain acyl-CoA dehydrogenase; ACADM, medium-chain acyl-CoA dehydrogenase.

3.3. BAA6-CM Promoted FAO in 3T3-L1 Adipocytes

To gain additional insights into the underlying mechanisms of BAA6 on FAO in adipose tissues, 3T3-L1 cells were chosen for further study. BAA6-CM and dead BAA6 were used to treat 3T3-L1 cells because it's unlikely that live BAA6 could reach adipose tissues in vivo. 3T3-L1 cells were exposed to BAA6-CM at 0–4% for 1–4 days, and dead BAA6 at 0–10^7 cells/mL for 1–3 days, to ensure cell viability (Figure S2A,B, Supplementary Materials). As shown in Figure 3A, no significant differences were observed in the TG content during various concentrations of BAA6-CM treatment for 1–2 days. However, TG content was markedly decreased with the increase concentration of BAA6-CM for 3–4 days, and there was no significant difference following BAA6-CM treatment between 3 and 4 days at 4% BAA6-CM. In addition, dead BAA6 treatment showed no remarkable effect on TG content (Figure S3, Supplementary Materials). Thus, BAA6-CM (0–4%) treatment for 3 days was used for further analyses. As visualized by ORO staining, the supplement of BAA6-CM diminished lipid accumulation of 3T3-L1 adipocytes in a concentration-dependent manner (Figure 3B). These data indicated that BAA6-CM had a positive effect on lipid metabolism in vitro.

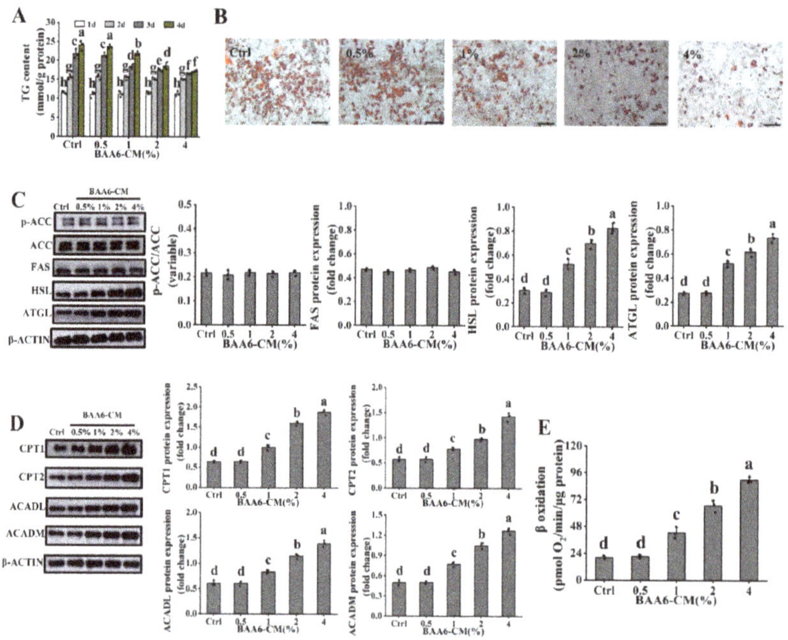

Figure 3. Effect of BAA6 culture supernatant (BAA6-CM) on TG content and FAO in 3T3-L1 cells. (**A**) TG concentration, in the presence of BAA6-CM (0–4%), for 1–4 days. (**B**) Photograph of Oil Red O (ORO) staining (scale bar, 200 µm) and western blot analysis, showing the levels of (**C**) p-ACC, FAS, HSL, and ATGL, and (**D**) levels of protein expression of CPT1, CPT2, ACADL, and ACADM after treatment with 0–4% BAA6-CM for 3 days. (**E**) β oxidation, measured by Seahorse XF platform, following various concentrations of 0–4% BAA6-CM for 3 days. Values are shown as means ± SD. Different lowercase letters mean remarkable differences between groups at $p < 0.05$.

To test the effects of FAO on TG reduction after BAA6-CM treatment, lipid metabolism-related proteins were firstly assessed. As displayed in Figure 3C, the lipid synthesis-related protein levels (FAS and p-ACC) had no remarkable change in BAA6-CM-treated cells relative to the control group. However, lipid catabolism-related protein (HSL and ATGL) contents were significantly higher in the BAA6-CM group in a concentration-dependent manner, indicating that BAA6-CM could promote lipid catabolism. Next, the key proteins in the processes of FAO were measured. BAA6-CM treatment significantly increased levels of FAO-related proteins (CPT1, CPT2, ACADL, and ACADM) by a dose-dependent way (Figure 3D). Further, FAO was evaluated using the Seahorse XF platform. As shown in Figure 3E, BAA6-CM treatment could significantly up-regulate FAO, compared to control group. These results implicated that BAA6-CM could promote FAO to decrease TG accumulation in vitro.

3.4. Acetate Is a Key BAA6-CM Metabolite That Increases FAO in 3T3-L1 Cells

We then wondered which kind of bioactive component in BAA6-CM was responsible for enhancing FAO in 3T3-L1 cells. Interestingly, accumulating evidence suggested that acetate could been generated by genus *Bifidobacterium* [43], and it was proved to enhance FAO in vivo and in vitro [44]. We found that acetate levels in BAA6-CM was higher compared with control media (Figure 4A). Subsequently, the effect of acetate on TG content was tested in vitro. As displayed in Figure 4B, 11.98 mmol/L acetate treatment (acetate concentration in 4% BAA6-CM group) had lower TG content in cells than control group, which was about 62% of the function, as compared to BAA6-CM treatment. Acetate treatment (23.96 mmol/L; two times of acetate concentration in 4% BAA6-CM group) existed a higher effect on TG content than that of BAA6-CM treatment. As visualized by ORO staining, the supplement of BAA6-CM and acetate could remarkably diminish lipid accumulation in vitro, compared to the control group (Figure 4C).

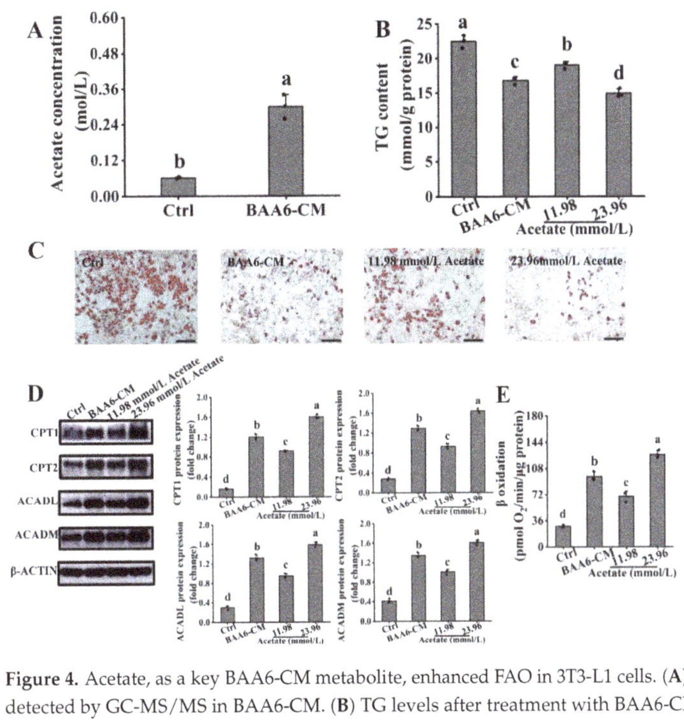

Figure 4. Acetate, as a key BAA6-CM metabolite, enhanced FAO in 3T3-L1 cells. (**A**) Acetate contents detected by GC-MS/MS in BAA6-CM. (**B**) TG levels after treatment with BAA6-CM (4%) or acetate (11.98 and 23.96 mmol/L; one and two times of acetate concentration in 4% BAA6-CM treatment) for 3 days. (**C**) Photograph of ORO staining (scale bar, 200 μm), and (**D**) levels of protein expression of CPT1, CPT2, ACADL, and ACADM after treatment with 4% BAA6-CM or acetate (11.98 and 23.96 mmol/L) for 3 days in 3T3-L1 cells. (**E**) β oxidation, measured by Seahorse XF platform, after treatment with 4% BAA6-CM or acetate (11.98 and 23.96 mmol/L) for 3 days. Values are shown as means ± SD. Different lowercase letters mean remarkable differences between groups at $p < 0.05$.

To assess the effect of acetate on FAO, the levels of key proteins involved in FAO were firstly measured. Figure 4D showed that acetate and BAA6-CM treatment remarkably elevated the expression of CPT1, CPT2, ACADL, and ACADM proteins relative to control group. Then, FAO was assessed using the Seahorse XF platform after acetate treatment. As shown in Figure 4E, FAO was remarkably enhanced by acetate and BAA6-CM treatments. Meanwhile, acetate (23.96 mmol/L) treatment obviously promoted FAO compared to the BAA6-CM group. Thus, acetate (23.96 mmol/L) treatment was used for further analyses. These data indicated that acetate was essential for BAA6-CM to enhance FAO in vitro.

3.5. PPARα Signaling Mediated the Regulation of Acetate on FAO in 3T3-L1 Cells

PPARα signaling has a master role in the regulation of FAO. Firstly, the expression of PPARα was investigated. The protein expression level of PPARα was significantly enhanced by acetate and BAA6-CM treatments, which was lower in the control group (Figure 5A). This data suggested that acetate could up-regulate PPARα expression. Then, to better clarify the role of PPARα in acetate and BAA6-CM-regulating FAO, a PPARα-specific siRNA was introduced into 3T3-L1 cells. The levels of PPARα expression were markedly declined after silencing treatment (Figure 5B). Moreover, acetate and BAA6-CM-induced increase in FAO key protein (CPT1, CPT2, ACADL, and ACADM) levels were diminished by PPARα knockdown (Figure 5C). In these experiments, the siRNA control group was not significantly different from the acetate and BAA6-only control, displaying no additional siRNA effects. What's more, knockdown of PPARα also disturbed the BAA6-CM and acetate-induced up-regulation of FAO (Figure 5D). These results indicated that acetate enhanced FAO through modulating PPARα signaling.

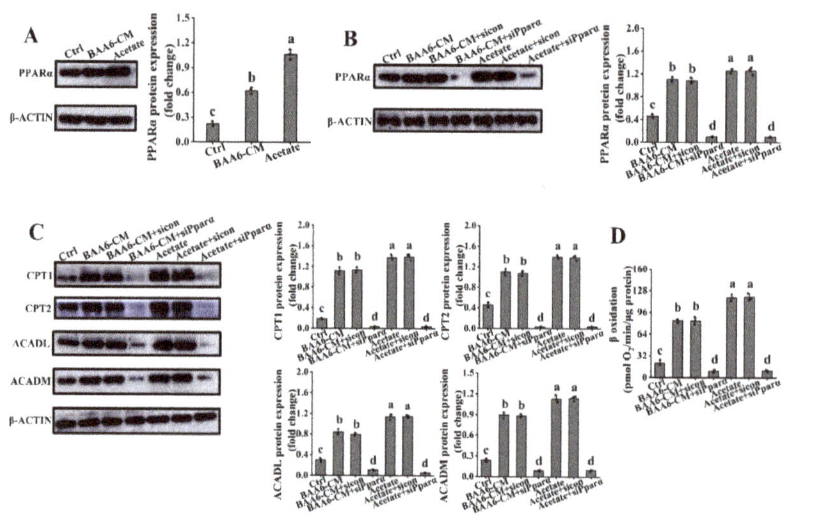

Figure 5. Effect of acetate or BAA6-CM on peroxisomal proliferator-activated receptor α (PPARα) signaling in 3T3-L1 cells. (**A**) Protein expression levels of PPARα after treatment with 4% BAA6-CM or 23.96 mmol/L acetate for 3 days. Levels of protein expression of (**B**) PPARα, and (**C**) CPT1, CPT2, ACADL, and ACADM after cell treated with control or PPARα-special small interfering RNA (siRNA), followed by addition of 4% BAA6-CM or 23.96 mmol/L acetate for 3 days. (**D**) β oxidation, measured by Seahorse XF platform after cell treatment, with control or PPARα-special siRNA, followed by addition of 4% BAA6-CM or 23.96 mmol/L acetate for 3 days. Values are shown as means ± SD. Different lowercase letters mean remarkable differences between groups at $p < 0.05$.

Several pieces of evidence verified that PPARα was the downstream target of AMPK [45]. Therefore, to explore whether activation of PPARα signaling was related to AMPK, we firstly determined the effects of acetate and BAA6-CM on AMPK activity in 3T3-L1 cells.

Acetate and BAA6-CM markedly improved the p-AMPK protein expression, compared to control group (Figure 6A). Then, when cells were pretreated with compound C (Com, an AMPK inhibitor), the levels of p-AMPK protein expression were markedly reduced (Figure 6B). Similarly, the up-regulation of acetate and BAA6-CM on PPARα, CPT1, CPT2, ACADL, and ACADM protein levels were markedly abrogated after the addition of Com (Figure 6B,C). Further, the increase in FAO for acetate and BAA6-CM was significantly destroyed following pretreatment with Com (Figure 6D). These findings revealed that AMPK played an important role in modulating PPARα signaling for acetate-mediated FAO in vitro.

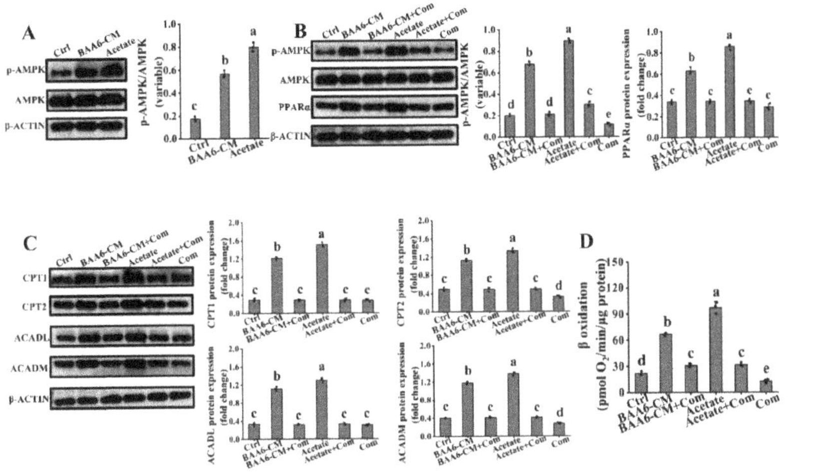

Figure 6. Effect of acetate or BAA6-CM on AMPK activity in 3T3-L1 cells. (**A**) Western blot analysis showing the levels of p-AMPK after treatment with 4% BAA6-CM or 23.96 mmol/L acetate for 3 days. Protein expression levels of (**B**) p-AMPK and PPARα and (**C**) CPT1, CPT2, ACADL, and ACADM in cells pretreated with 20 μmol/L compound C (Com), followed by treatment with 4% BAA6-CM or 23.96 mmol/L acetate for 3 days. (**D**) β oxidation, measured by Seahorse XF platform, after cells pre-treatment with 20 μmol/L Com, followed by treatment with 4% BAA6-CM or 23.96 mmol/L acetate for 3 days. Values are shown as means ± SD. Different lowercase letters mean remarkable differences between groups at $p < 0.05$. AMPK, AMP-activated protein kinase; p-AMPK, phosphorylated AMPK.

3.6. Acetate Activated GPR43 in 3T3-L1 Cells

GPRs had a pivotal role in cell signaling, which could regulate energy homeostasis [21]. GPR43 and GPR41 are important members of GPRs, which are receptors of acetate [46]. Therefore, we investigated which receptor was activated following acetate and BAA6-CM treatments. GPR43 protein expression was markedly increased in adipocytes after a supplement of acetate and BAA6-CM, compared with control group (Figure 7A). However, there was no significant difference on GPR41 expression after acetate and BAA6-CM treatments, compared to control group (Figure S4, Supplementary Materials). These results showed that acetate and BAA6-CM could activate GPR43. In addition, GPR43 might regulate AMPK activity by Ca^{2+}/CAMKKβ [47]. Thus, the concentration of intracellular Ca^{2+} and p-CAMKKβ protein expression level were further measured. In the presence of acetate and BAA6-CM, the levels of Ca^{2+} and p-CAMKKβ protein were remarkably increased, compared with control group (Figure 7B,C).

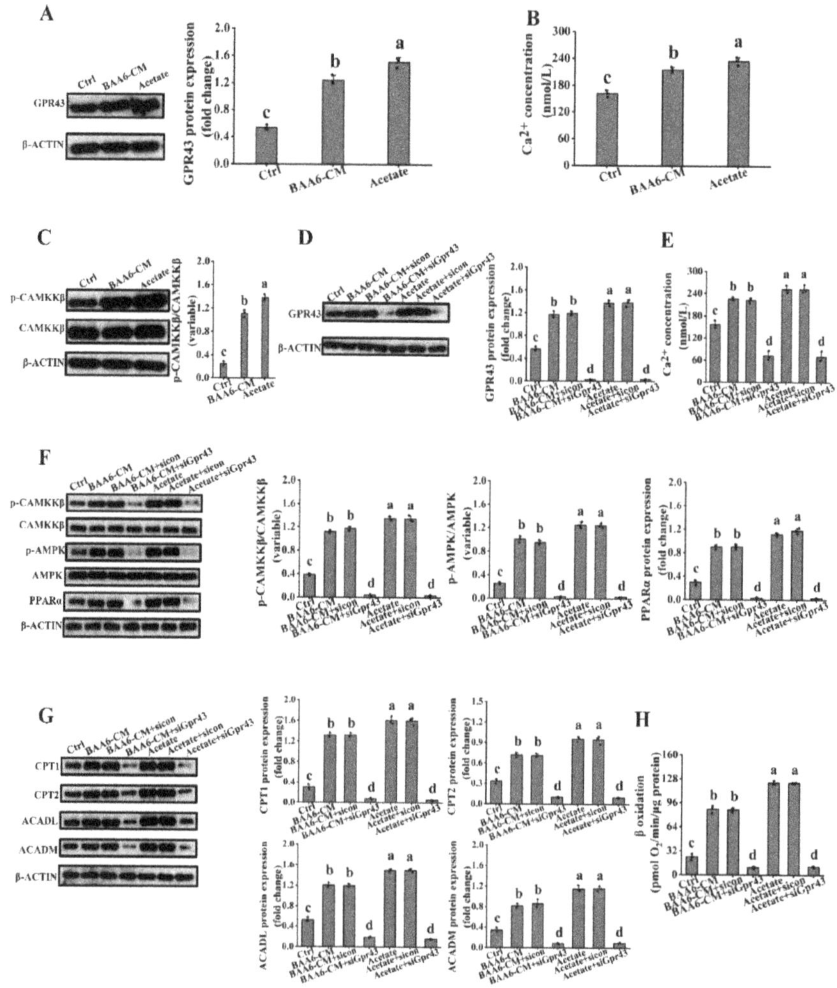

Figure 7. Effect of acetate or BAA6-CM on GPR43 signaling in 3T3-L1 cells. Western blot analysis showing the levels of (**A**) GPR43, (**B**) Ca^{2+} contents, and (**C**) p-CAMKKβ in 3T3-L1 cells after treatment with 4% BAA6-CM or 23.96 mmol/L acetate for 3 days. Protein expression levels of (**D**) GPR43 and (**E**) Ca^{2+} concentration, after the cell is treated with control or GPR43-special siRNA, followed by addition of 4% BAA6-CM or 23.96 mmol/L acetate for 3 days. (**F**) Protein expression levels of p-CAMKKβ, p-AMPK, PPARα, (**G**) protein levels of CPT1, CPT2, ACADL, ACADM, and (**H**) β oxidation after the cell is treated with control or GPR43-special siRNA, followed by addition of 4% BAA6-CM or 23.96 mmol/L acetate for 3 days. Values are shown as means ± SD. Different lowercase letters mean remarkable differences between groups at $p < 0.05$. GPR43, G protein-coupled receptor 43; p-CAMKKβ, calcium/calmodulin-dependent protein kinase β phosphorylation.

To verify whether acetate actually activated GPR43 to elicit downstream signals, the expression of GPR43 was silenced using GPR43 siRNA. Results showed that GPR43 siRNA was effective in decreasing the expression of GPR43 (Figure 7D). Consistently, acetate and BAA6-CM-induced up-regulation levels of intracellular Ca^{2+}, p-CAMKKβ, p-AMPK, and PPARα proteins were blunted by GPR43 knockdown (Figure 7E,F). Moreover, the acetate and BAA6-CM-stimulated expression of proteins (CPT1, CPT2, ACADL, and ACADM) and enhancement of FAO were also destroyed by the knockdown of GPR43 (Figure 7G,H).

Hence, these data indicated that activation of GPR43 signaling, by acetate, was crucial to regulate FAO in vitro.

3.7. BAA6 Activated GPR43-PPARα Signaling In Vivo

To determine whether GPR43-PPARα signaling was activated by BAA6 in adipose tissues of obese mice, we firstly explored the acetate contents in serum and feces. HFD mice had lower acetate concentration in serum and feces compared to ND mice. Meanwhile, levels of acetate in the HFD group were also lower than HFD + BAA6 group (Figure 8A,B). These results suggested that the raise of acetate level was due to BAA6. Then, we investigated the expression of GPR43, p-CAMKKβ, p-AMPK, and PPARα proteins. Results showed that HFD mice had lower levels of GPR43, p-CAMKKβ, p-AMPK, and PPARα proteins relative to ND mice. Compared with HFD group, BAA6 treatment of mice could remarkably up-regulate the expression of GPR43, p-CAMKKβ, p-AMPK, and PPARα proteins (Figure 8C). These findings suggested that GPR43-PPARα pathways in adipose tissues were also activated by BAA6 treatment in vivo.

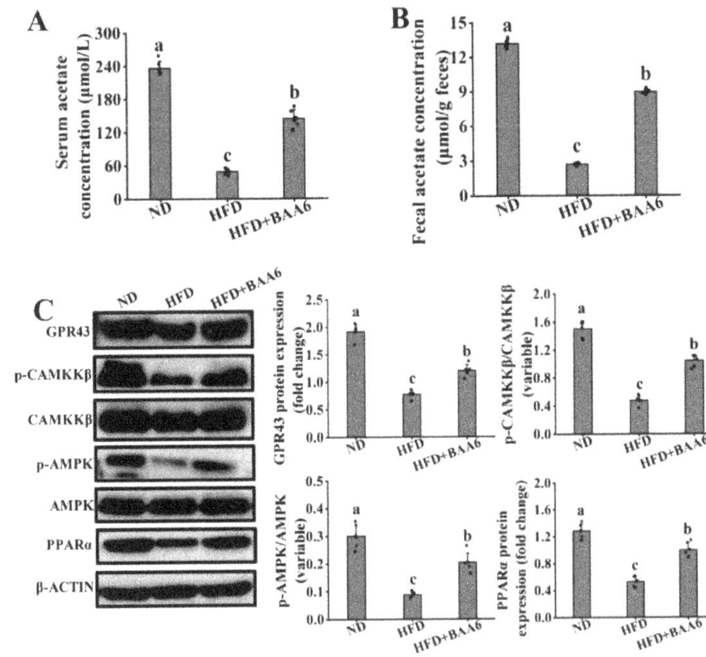

Figure 8. Effect of BAA6 administration on GPR43-PPARα signaling of epididymal adipose tissues in HFD-fed mice. Acetate concentration in serum (**A**) and feces (**B**). (**C**) Western blot results of GPR43, p-CAMKKβ, p-AMPK, and PPARα, following daily treatment with 10^9 CFU/kg of BAA6. Values are shown as means ± SD (n = 6). Different lowercase letters mean remarkable differences between groups at $p < 0.05$.

4. Discussion

Given abundant research indicating the positive influences for probiotics in the management of obesity, this approach is attracting more and more attention [15,48]. Our results showed that BAA6 treatment had a similar effect in alleviating the development of obesity to AKK, which was verified to ameliorate obesity in many studies [49]. Thus, in our study, we investigated the mechanism of BAA6 to alleviate the development of obesity.

It was reported that O_2 expenditure tended to decrease in HFD-fed mice, which exhibited a high RER [50]. Elevated RER was a characteristic of weight gain in obesity [51].

Meanwhile, RER was an indicator for FAO in obese mice, and low RER reflected the enhancement of FAO [52]. Thus, FAO could be a vital means to manage obesity [6]. In this study, these results suggested that BAA6 could remarkably increase O_2 expenditure and reduce RER during the dark cycle, which meant that BAA6 treatment was beneficial to enhance FAO in obese mice. Some studies have demonstrated that FAO in obesity is impaired, particularly in adipose tissues [53,54]. Adipose tissue is regarded as the largest organ of the body [55]. In lean individuals, adipose tissue accounted for 10% of body weight, but the proportion could increase up to 40% in obese subjects [55]. Meanwhile, FAO is one of the major biochemical factors to maintain function of fat tissues [56]. Unfortunately, the levels of CPT1, CPT2, ACADL, and ACADM proteins were decreased in the adipose tissues of obese mice, which were the major modulators for FAO [57]. A series of studies found that administration of probiotic bacteria, such as *Lactobacillus* genus, to HFD-fed mice promoted FAO in adipose tissues [18,19]. In addition, *Lactobacillus acidophilus* NS1 could enhance the FAO-related proteins expression and then, reduce body fat accumulation [58]. Our study also showed that the levels of CPT1, CPT2, ACADL, and ACADM proteins, in epididymal adipose tissues, were elevated (by 121.06%, 87.88%, 109.71%, and 452.62%, respectively) after BAA6 treatment. Above findings revealed that FAO, in the epididymal adipose tissues for obese mice, could be remarkably enhanced by BAA6.

To gain additional insights into the underlying mechanisms of BAA6 on FAO in adipose tissues, 3T3-L1 cells, as common models in vitro, were chosen for further study [59]. It might be unlikely for live probiotics to directly arrive to the adipose tissue in vivo [60]. In that case, the components of dead cells or metabolites might be used for adipocytes [60]. We observed that TG content in 3T3-L1 cells was significantly decreased after BAA6-CM supplementation, which was not remarkably different after dead BAA6 treatment. These phenomena indicated that BAA6-CM had positive roles on lipid metabolism in vitro. Some researchers found that culture supernatant of *Bifidobacterium* and *Lactobacillus* genera induced the expression of FAO-related proteins in cells [61]. Thus, probiotic metabolites were important factors for regulating FAO. Our study showed that the key proteins expression in FAO was significantly increased under BAA6-CM treatment in vitro. Furthermore, results from Seahorse XF platform also indicated that FAO was increased (by 1.11-, 2.30-, and 3.44- fold) after different concentration of BAA6-CM treatment. According to these findings, we assumed that BAA6-CM treatment could promote FAO in adipocytes.

However, it was necessary to confirm which bioactive components produced by BAA6 could enhance FAO. Many studies found that probiotics could generate short-chain fatty acids (SCFAs), which had positive effects on the host [62]. Acetate is the most abundant SCFA in humans, accounting for 50% of total SCFAs [63]. Accumulating studies suggested that acetate could be produced by genus *Bifidobacterium* [43]. Similarly, we found that the level of acetate in BAA6-CM was enhanced by 3.92-fold. Meanwhile, acetate concentration in serum and feces of HFD-fed mice was significantly elevated after administration with BAA6. These results indicated that the increased levels of acetate were due to BAA6. It was well known that probiotic metabolites could play roles in regulating the expression of metabolism-related proteins [64]. Dietary SCFAs supplementation might enhance the proteins expression that related to FAO in adipose tissues [65]. Furthermore, the expression levels of these proteins were also elevated, after acetate treatment, in vivo and in vitro [44]. As expected, our data further confirmed that acetate was the necessary metabolite of BAA6 to enhance FAO.

Evidence has pointed out that the PPARα pathway has a primary action in the regulation of FAO [27]. Virtually, expression of most of the proteins during FAO processes was under the control of PPARα [66]. Rakhshandehroo et al. found that propionate upregulated the PPARα expression in intestinal epithelial cells [66]. Our research obtained that acetate and BAA6-CM treatments could increase the protein expression of PPARα (by 16.51- and 3.84- fold) in vitro. Meanwhile, levels of PPARα protein were also enhanced, after BAA6 treatment, in HFD-fed mice. Multiple reports showed that deletion of PPARα would impair FAO in adipocytes of obese mice, resulting in fat accumulation [67]. In our study, acetate and BAA6-CM-induced enhancement of FAO were inhibited by PPARα siRNA. This result indicated that BAA6-generated acetate could regulate PPARα signaling to promote FAO. Besides, AMPK could mediate activation of PPARα signaling, which was the underlying mechanism for enhancement of FAO [45]. We observed that the levels of p-AMPK protein were significantly increased after acetate and BAA6-CM treatments. Furthermore, when 3T3-L1 cells were pretreated with Com (an inhibitor of AMPK), acetate and BAA6-CM-stimulated up-regulation of PPARα and FAO were diminished. Meanwhile, the levels of p-AMPK protein were also markedly increased after BAA6 supplementation in HFD-fed mice. These results proved that BAA6-generated acetate might activate AMPK-PPARα pathway signaling, which was associated with the up-regulation of FAO.

GPRs played an important role in cell signaling, especially in energy homeostasis, and regulated various adipocyte functions, such as FAO [21,23]. GPR43, as a key member of GPRs, might maintain the normal state of FAO through regulating related-genes expression [24,26]. Kimura et al. [68] reported that GPR43 knockout mice gained more weight compared with wild-type. Based on these, GPR43 has attracted attention as a potential therapeutic target for metabolic syndrome [69]. Low expression of GRP43 was found in the adipose tissues of obese mice, and dietary SCFAs supplementation significantly recovered its expression to decrease body weight [46]. As an important member of SCFAs, acetate could activate GPR43 signaling in adipocytes [70]. Consistently, in our study, acetate and BAA6-CM treatments could enhance the protein expression of GPR43 (by 181.55% and 131.78%) in vitro. Meanwhile, the levels of GPR43 protein were also remarkably raised in HFD-fed mice after administration with BAA6. In most cases, GPR43 is the main receptor of acetate [71]. Activation of GPR43 could regulate AMPK activity via Ca^{2+}/CAMKKβ pathway [47]. Several reports found that the levels of intracellular Ca^{2+} and p-CAMKKβ were increased after treatment with the GPR43 agonist [72]. In line with these studies, our data showed that acetate and BAA6-CM treatments could elevate levels of Ca^{2+} (by 45.85% and 33.76%) and p-CAMKKβ (by 17.61- and 4.60- fold) in vitro. Besides, inhibition of GPR43 signaling abrogated expression of p-AMPK protein [73]. To further confirm the role of GPR43 on downstream signals, the expression of GPR43 in 3T3-L1 cells was silenced using GPR43 siRNA. Results showed that knockdown of GPR43 significantly prevented acetate and BAA6-CM-elicited up-regulation of intracellular Ca^{2+}, p-CAMKKβ, p-AMPK, and PPARα. Moreover, acetate and BAA6-CM-induced enhancement of FAO were also destroyed by GPR43 knockdown. These findings indicated that BAA6-generated acetate could activate GPR43 signaling to promote FAO.

In conclusion, our study revealed that BAA6 supplementation could increase FAO in adipose tissues and alleviate the development of obesity. Acetate, an important metabolite of BAA6, was a pivotal factor to promote FAO through the GPR43-PPARα signaling. This study indicates that BAA6 may be a potential therapy to ameliorate diet-induced obesity. However, there are some limitations in our study. First, different age and sex of the mice need to be considered. In the future, clinical trials could be conducted to confirm the beneficial effect of BAA6 in obesity and obesity-related disorders.

Supplementary Materials: The following supporting information can be downloaded at: https://www.mdpi.com/article/10.3390/nu14030598/s1. Figure S1: Effect of BAA6 administration on lipid metabolism in HFD-fed mice, Figure S2: Effects of BAA6 culture supernatant (BAA6-CM) and dead BAA6 on cell viability in 3T3-L1 cells, Figure S3: Effect of dead BAA6 on TG concentration in 3T3-L1 cells, Figure S4: Effect of acetate or BAA6-CM on GPR41 expression levels of GPR41 in 3T3-L1 cells. Supplementary Table S1: Composition of diets, Supplementary Table S2: Small interfering RNA sequences applied in this experiment, Supplementary Table S3: Effect of oral administration of BAA6 on food intake in HFD-fed mice.

Author Contributions: Conceptualization, Y.H., Y.L., F.R. and X.W.; methodology, G.Z., J.L. and R.W.; data analyses, Y.H.; writing—original draft preparation, Y.H.; writing—review and editing, Y.H., G.Z., Y.L., F.R. and X.W. All authors have read and agreed to the published version of the manuscript.

Funding: This research was supported by the 111 project from the Education Ministry of China (No. B18053).

Institutional Review Board Statement: All animal studies were approved by the Animal Experimentation Ethics Committee of the China Agricultural University (Beijing, China) on 20 August 2020. The ethic approval code for animal studies were AW18080202-1.

Informed Consent Statement: Not applicable.

Data Availability Statement: Data presented in this study are available on request from the corresponding author.

Conflicts of Interest: The authors declare no conflict of interest.

References

1. Alberti, K.; Zimmet, P.; Shaw, J. The metabolic syndrome—A new worldwide definition. *Lancet* **2005**, *366*, 1059–1062. [CrossRef]
2. Lee, J.H.; Moon, M.H.; Jeong, J.K.; Park, Y.G.; Lee, Y.J.; Seol, J.W.; Park, S.Y. Sulforaphane induced adipolysis via hormone sensitive lipase activation, regulated by AMPK signaling pathway. *Biochem. Biophys. Res. Commun.* **2012**, *426*, 492–497. [CrossRef] [PubMed]
3. Fritzen, A.M.; Lundsgaard, A.M.; Kiens, B. Tuning fatty acid oxidation in skeletal muscle with dietary fat and exercise. *Nat. Rev. Endocrinol.* **2020**, *16*, 683–696. [CrossRef] [PubMed]
4. Samuel, V.T.; Petersen, K.F.; Shulman, G.I. Lipid-induced insulin resistance: Unravelling the mechanism. *Lancet* **2010**, *375*, 2267–2277. [CrossRef]
5. Botchlett, R.; Woo, S.L.; Liu, M.Y.; Pei, Y.; Guo, X.; Li, H.G.; Wu, C.D. Nutritional approaches for managing obesity-associated metabolic diseases. *J. Endocrinol.* **2017**, *233*, R145–R171. [CrossRef]
6. Park, K.S. Raspberry ketone increases both lipolysis and fatty acid oxidation in 3T3-L1 adipocytes. *Planta Med.* **2010**, *76*, 1654–1658. [CrossRef]
7. Ramirez, S.; Martins, L.; Jacas, J.; Carrasco, P.; Pozo, M.; Clotet, J.; Serra, D.; Hegardt, F.G.; Dieguez, C.; Lopez, M.; et al. Hypothalamic ceramide levels regulated by CPT1C mediate the orexigenic effect of ghrelin. *Diabetes* **2013**, *62*, 2329–2337. [CrossRef]
8. Fucho, R.; Casals, N.; Serra, D.; Herrero, L. Ceramides and mitochondrial fatty acid oxidation in obesity. *FASEB J.* **2017**, *31*, 1263–1272. [CrossRef]
9. Malandrino, M.I.; Fucho, R.; Weber, M.; Calderon-Dominguez, M.; Mir, J.F.; Valcarcel, L.; Escote, X.; Gomez-Serrano, M.; Peral, B.; Salvado, L.; et al. Enhanced fatty acid oxidation in adipocytes and macrophages reduces lipid-induced triglyceride accumulation and inflammation. *Am. J. Physiol.-Endocrinol. Metab.* **2015**, *308*, E756–E769. [CrossRef]
10. Backhouse, K.; Sarac, I.; Shojaee-Moradie, F.; Stolinski, M.; Robertson, M.D.; Frost, G.S.; Bell, J.D.; Thomas, E.L.; Wright, J.; Russell-Jones, D.; et al. Fatty acid flux and oxidation are increased by rimonabant in obese women. *Metabolism* **2012**, *61*, 1220–1223. [CrossRef]
11. Rupasinghe, H.P.V.; Sekhon-Loodu, S.; Mantso, T.; Panayiotidis, M.I. Phytochemicals in regulating fatty acid beta-oxidation: Potential underlying mechanisms and their involvement in obesity and weight loss. *Pharmacol. Ther.* **2016**, *165*, 153–163. [CrossRef]
12. De Vos, P.; Faas, M.M.; Spasojevic, M.; Sikkema, J. Encapsulation for preservation of functionality and targeted delivery of bioactive food components. *Int. Dairy J.* **2010**, *20*, 292–302. [CrossRef]
13. Kechagia, M.; Basoulis, D.; Konstantopoulou, S.; Dimitriadi, D.; Gyftopoulou, K.; Skarmoutsou, N.; Fakiri, E.M. Health benefits of probiotics: A review. *ISRN Nutr.* **2013**, *2013*, 481651. [CrossRef]
14. Van Baarlen, P.; Wells, J.M.; Kleerebezem, M. Regulation of intestinal homeostasis and immunity with probiotic *Lactobacilli*. *Trends Immunol.* **2013**, *34*, 208–215. [CrossRef]
15. Reid, G.; Younes, J.A.; Van der Mei, H.C.; Gloor, G.B.; Knight, R.; Busscher, H.J. Microbiota restoration: Natural and supplemented recovery of human microbial communities. *Nat. Rev. Microbiol.* **2011**, *9*, 27–38. [CrossRef]
16. Cani, P.D.; De Vos, W.M. Next-generation beneficial microbes: The case of *Akkermansia muciniphila*. *Front. Microbiol.* **2017**, *8*, 1765. [CrossRef]
17. Yan, J.; Sheng, L.L.; Li, H.K. *Akkermansia muciniphila*: Is it the Holy Grail for ameliorating metabolic diseases? *Gut Microbes* **2021**, *13*, 1984104. [CrossRef]
18. Kobyliak, N.; Conte, C.; Cammarota, G.; Haley, A.P.; Styriak, I.; Gaspar, L.; Fusek, J.; Rodrigo, L.; Kruzliak, P. Probiotics in prevention and treatment of obesity: A critical view. *Nut. Metab.* **2016**, *13*, 14. [CrossRef]
19. Sidossis, L.S.; Porter, C.; Saraf, M.K.; Borsheim, E.; Radhakrishnan, R.S.; Chao, T.; Ali, A.; Chondronikola, M.; Mlcak, R.; Finnerty, C.C.; et al. Browning of subcutaneous white adipose tissue in humans after severe adrenergic stress. *Cell Metab.* **2015**, *22*, 219–227. [CrossRef]
20. Green, M.; Arora, K.; Prakash, S. Microbial medicine: Prebiotic and probiotic functional foods to target obesity and metabolic syndrome. *Int. J. Mol. Sci.* **2020**, *21*, 2890. [CrossRef]
21. Moran, B.M.; Flatt, P.R.; McKillop, A.M. G protein-coupled receptors: Signalling and regulation by lipid agonists for improved glucose homoeostasis. *Acta Diabetol.* **2016**, *53*, 177–188. [CrossRef]
22. Wang, J.; Xiao, R.P. G protein-coupled receptors in energy homeostasis. *Sci. China Life Sci.* **2014**, *57*, 672–680. [CrossRef]

23. Barella, L.F.; Jain, S.; Kimura, T.; Pydi, S.P. Metabolic roles of G protein-coupled receptor signaling in obesity and type 2 diabetes. *FEBS J.* **2021**, *288*, 2622–2644. [CrossRef]
24. Lu, J.; Fang, B.C.; Zheng, Y.Y.; Yu, X.; Huang, G.R.; Wang, Z.N.; Deng, X.M.; Guan, S. 1,3-dichloro-2-propanol induced lipid accumulation in HepG2 cells through cAMP/protein kinase A and AMP-activated protein kinase pathways via Gi/o-coupled receptors. *Environ. Toxicol. Pharmacol.* **2017**, *55*, 118–126. [CrossRef]
25. Horiuchi, H.; Kamikado, K.; Aoki, R.; Suganuma, N.; Nishijima, T.; Nakatani, A.; Kimura, I. *Bifidobacterium animalis* subsp. *lactis* GCL2505 modulates host energy metabolism via the short-chain fatty acid receptor GPR43. *Sci. Rep.* **2020**, *10*, 4158. [CrossRef]
26. Kimura, I.; Inoue, D.; Hirano, K.; Tsujimoto, G. The SCFA receptor GPR43 and energy metabolism. *Front. Endocrinol.* **2014**, *5*, 85. [CrossRef]
27. Park, H.S.; Jang, J.E.; Ko, M.S.; Woo, S.H.; Kim, B.J.; Kim, H.S.; Park, H.S.; Park, I.S.; Koh, E.H.; Lee, K.U. Statins increase mitochondrial and peroxisomal fatty acid oxidation in the liver and prevent non-alcoholic steatohepatitis in mice. *Diabetes Metab. J.* **2016**, *40*, 376–385. [CrossRef]
28. Finck, B.N.; Bernal-Mizrachi, C.; Han, D.H.; Coleman, T.; Sambandam, N.; LaRiviere, L.L.; Holloszy, J.O.; Semenkovich, C.F.; Kelly, D.P. A potential link between muscle peroxisome proliferator-activated receptor-alpha signaling and obesity-related diabetes. *Cell Metab.* **2005**, *1*, 133–144. [CrossRef]
29. Wei, D.; Liao, L.; Wang, H.J.; Zhang, W.; Wang, T.T.; Xu, Z.P. Canagliflozin ameliorates obesity by improving mitochondrial function and fatty acid oxidation via PPAR alpha in vivo and in vitro. *Life Sci.* **2020**, *247*, 117414. [CrossRef]
30. Sun, E.N.; Zhao, L.; Ren, F.Z.; Liu, S.L.; Zhang, M.; Guo, H.Y. Complete genome sequence of *Bifidobacterium animalis* subsp *lactis* A6, a probiotic strain with high acid resistance ability. *J. Biotechnol.* **2015**, *200*, 8–9. [CrossRef]
31. Huo, Y.X.; Lu, X.H.; Wang, X.Y.; Wang, X.F.; Chen, L.L.; Guo, H.Y.; Zhang, M.; Li, Y.X. *Bifidobacterium animalis* subsp. *lactis A6* alleviates obesity associated with promoting mitochondrial biogenesis and function of adipose tissue in mice. *Molecules* **2020**, *25*, 1490. [CrossRef]
32. Park, S.S.; Lee, Y.J.; Kang, H.; Yang, G.; Hong, E.J.; Lim, J.Y.; Oh, S.; Kim, E. *Lactobacillus amylovorus* KU4 ameliorates diet-induced obesity in mice by promoting adipose browning through PPAR gamma signaling. *Sci. Rep.* **2019**, *9*, 20152. [CrossRef] [PubMed]
33. An, M.; Park, Y.H.; Lim, Y.H. Antiobesity and antidiabetic effects of the dairy bacterium *Propionibacterium freudenreichii* MJ2 in high-fat diet-induced obese mice by modulating lipid metabolism. *Sci. Rep.* **2021**, *11*, 2481. [CrossRef] [PubMed]
34. Xie, S.Y.; Li, Y.; Teng, W.D.; Du, M.; Li, Y.X.; Sun, B.G. Liensinine inhibits beige adipocytes recovering to white adipocytes through blocking mitophagy flux in vitro and in vivo. *Nutrients* **2019**, *11*, 1640. [CrossRef]
35. Teng, W.D.; Li, Y.; Du, M.; Lei, X.E.; Xie, S.Y.; Ren, F.Z. Sulforaphane prevents hepatic insulin resistance by blocking serine palmitoyltransferase 3-mediated ceramide biosynthesis. *Nutrients* **2019**, *11*, 1185. [CrossRef]
36. Fang, B.; Li, J.W.; Zhang, M.; Ren, F.Z.; Pang, G.F. Chronic chlorpyrifos exposure elicits diet-specific effects on metabolism and the gut microbiome in rats. *Food Chem. Toxicol.* **2018**, *111*, 144–152. [CrossRef]
37. Lin, S.L.; Wang, Z.Y.; Lin, Y.L.; Ge, S.H.; Hamzah, S.S.; Hu, J.M. Bound phenolics from fresh lotus seeds exert anti-obesity effects in 3T3-L1 adipocytes and high-fat diet-fed mice by activation of AMPK. *J. Funct. Foods* **2019**, *58*, 74–84. [CrossRef]
38. Luo, H.K.; Guo, Y.C.; Liu, Y.T.; Wang, Y.; Zheng, R.X.; Ban, Y.; Peng, L.; Yuan, Q.; Liu, W.Q. Growth differentiation factor 11 inhibits adipogenic differentiation by activating TGF-beta/smad signalling pathway. *Cell Proliferat.* **2019**, *52*, e12631. [CrossRef]
39. Timper, K.; Del Rio-Martin, A.; Cremer, A.L.; Bremser, S.; Alber, J.; Giavalisco, P.; Varela, L.; Heilinger, C.; Nolte, H.; Trifunovic, A.; et al. GLP-1 receptor signaling in astrocytes regulates fatty acid oxidation, mitochondrial integrity, and function. *Cell Metab.* **2020**, *31*, 1189–1205.e13. [CrossRef]
40. Chen, L.L.; Zhou, X.Y.; Wang, Y.W.; Wang, D.K.; Ke, Y.S.; Zeng, X.L. Propionate and butyrate produced by gut microbiota after probiotic supplementation attenuate lung metastasis of melanoma cells in mice. *Mol. Nutr. Food Res.* **2021**, *65*, 2100096. [CrossRef]
41. He, D.Y.; Wang, N.; Sai, X.; Li, X.Y.; Xu, Y.P. Camellia euphlebia protects against corticosterone-induced apoptosis in differentiated PC12 cells by regulating the mitochondrial apoptotic pathway and PKA/CREB/BDNF signaling pathway. *Food Chem. Toxicol.* **2019**, *126*, 211–222. [CrossRef] [PubMed]
42. Hong, J.; Hwang, E.; Kim, H.; Jeong, Y.; Lee, I. Artemisia capillaris inhibits lipid accumulation in 3T3-L1 adipocytes and obesity in C57BL/6J mice fed a high fat diet. *J. Med. Food* **2009**, *12*, 736–745. [CrossRef] [PubMed]
43. Fukuda, S.; Toh, H.; Hase, K.; Oshima, K.; Nakanishi, Y.; Yoshimura, K.; Tobe, T.; Clarke, J.M.; Topping, D.L.; Suzuki, T.; et al. *Bifidobacteria* can protect from enteropathogenic infection through production of acetate. *Nature* **2011**, *469*, 543–547. [CrossRef]
44. Den Besten, G.; Bleeker, A.; Gerding, A.; Van Eunen, K.; Havinga, R.; Van Dijk, T.H.; Oosterveer, M.H.; Jonker, J.W.; Groen, A.K.; Reijngoud, D.J.; et al. Short-chain fatty acids protect against high-fat diet-induced obesity via a PPAR-dependent switch from lipogenesis to fat oxidation. *Diabetes* **2015**, *64*, 2398–2408. [CrossRef] [PubMed]
45. Diniz, T.A.; De Lima, E.A.; Teixeira, A.A.; Biondo, L.A.; Rocha, L.A.F.; Valadao, I.C.; Silveira, L.S.; Cabral-Santos, C.; De Souza, C.O.; Rosa, J.C. Aerobic training improves NAFLD markers and insulin resistance through AMPK-PPAR-alpha signaling in obese mice. *Life Sci.* **2021**, *266*, 118868. [CrossRef]
46. Lu, Y.Y.; Fan, C.N.; Li, P.; Lu, Y.F.; Chang, X.L.; Qi, K.M. Short chain fatty acids prevent high-fat-diet-induced obesity in mice by regulating G protein-coupled receptors and gut microbiota. *Sci. Rep.* **2016**, *6*, 37589. [CrossRef]
47. Yoshida, H.; Ishii, M.; Akagawa, M. Propionate suppresses hepatic gluconeogenesis via GPR43/AMPK signaling pathway. *Arch. Biochem. Biophys.* **2019**, *672*, 108057. [CrossRef]

48. Kong, C.; Gao, R.Y.; Yan, X.B.; Huang, L.S.; Qin, H.L. Probiotics improve gut microbiota dysbiosis in obese mice fed a high fat or high-sucrose diet. *Nutrition* **2019**, *60*, 175–184. [CrossRef]
49. Molinaro, F.; Paschetta, E.; Cassader, M.; Gambino, R.; Musso, G. Probiotics, prebiotics, energy balance, and obesity mechanistic insights and therapeutic implications. *Gastroenterol. Clin.* **2012**, *41*, 843–854. [CrossRef]
50. Attane, C.; Foussal, C.; Le Gonidec, S.; Benani, A.; Daviaud, D.; Wanecq, E.; Guzman-Ruiz, R.; Dray, C.; Bezaire, V.; Rancoule, C.; et al. Apelin treatment increases complete fatty acid oxidation, mitochondrial oxidative capacity, and biogenesis in muscle of insulin-resistant mice. *Diabetes* **2012**, *61*, 310–320. [CrossRef]
51. Marra, M.; Scalfi, L.; Covino, A.; Esposito-Del Puente, A.; Contaldo, F. Fasting respiratory quotient as a predictor of weight changes in non-obese women. *Int. J. Obes.* **1998**, *22*, 601–603. [CrossRef] [PubMed]
52. Cavaliere, G.; Trinchese, G.; Bergamo, P.; De Filippo, C.; Raso, G.M.; Gifuni, G.; Putti, R.; Moni, B.H.; Canani, R.B.; Meli, R.; et al. Polyunsaturated fatty acids attenuate diet induced obesity and insulin resistance, modulating mitochondrial respiratory uncoupling in rat skeletal muscle. *PLoS ONE* **2016**, *11*, e0149033. [CrossRef] [PubMed]
53. Kusminski, C.M.; Scherer, P.E. Mitochondrial dysfunction in white adipose tissue. *Trends Endocrinol. Metab.* **2012**, *23*, 435–443. [CrossRef] [PubMed]
54. Auguet, T.; Guiu-Jurado, E.; Berlanga, A.; Terra, X.; Martinez, S.; Porras, J.A.; Ceausu, A.; Sabench, F.; Hernandez, M.; Aguilar, C.; et al. Downregulation of lipogenesis and fatty acid oxidation in the subcutaneous adipose tissue of morbidly obese women. *Obesity* **2014**, *22*, 2032–2038. [CrossRef]
55. Hausman, D.B.; DiGirolamo, M.; Bartness, T.J.; Hausman, G.J.; Martin, R.J. The biology of white adipocyte proliferation. *Obes. Rev.* **2001**, *2*, 239–254. [CrossRef]
56. Townsend, K.L.; An, D.; Lynes, M.D.; Huang, T.L.; Zhang, H.B.; Goodyear, L.J.; Tseng, Y.H. Increased mitochondrial activity in BMP7-treated brown adipocytes, due to increased CPT1- and CD36-mediated fatty acid uptake. *Antioxid. Redox Sig.* **2013**, *19*, 243–257. [CrossRef]
57. Mansuy-Aubert, V.; Zhou, Q.L.; Xie, X.Y.; Gong, Z.W.; Huang, J.Y.; Khan, A.R.; Aubert, G.; Candelaria, K.; Thomas, S.; Shin, D.J.; et al. Imbalance between neutrophil elastase and its inhibitor alpha(1)-antitrypsin in obesity alters insulin sensitivity, Inflammation, and energy expenditure. *Cell Metab.* **2013**, *17*, 534–548. [CrossRef]
58. Park, S.S.; Lee, Y.J.; Song, S.; Kim, B.; Kang, H.; Oh, S.; Kim, E. *Lactobacillus acidophilus* NS1 attenuates diet-induced obesity and fatty liver. *J. Endocrinol.* **2018**, *237*, 87–100. [CrossRef]
59. Ruiz-Ojeda, F.J.; Ruperez, A.I.; Gomez-Llorente, C.; Gil, A.; Aguilera, C.M. Cell models and their application for studying adipogenic differentiation in relation to obesity: A review. *Int. J. Mol. Sci.* **2016**, *17*, 1040. [CrossRef]
60. Lee, E.; Jung, S.R.; Lee, S.Y.; Lee, N.K.; Paik, H.D.; Lim, S.I. *Lactobacillus plantarum strain* Ln4 attenuates diet-induced obesity, insulin resistance, and changes in hepatic mRNA levels associated with glucose and lipid metabolism. *Nutrients* **2018**, *10*, 643. [CrossRef]
61. Hossain, M.; Park, D.S.; Rahman, M.S.; Ki, S.J.; Lee, Y.R.; Imran, K.M.; Yoon, D.; Heo, J.; Lee, T.J.; Kim, Y.S. *Bifidobacterium longum* DS0956 and *Lactobacillus rhamnosus* DS0508 culture-supernatant ameliorate obesity by inducing thermogenesis in obese-mice. *Benef. Microbes* **2020**, *11*, 361–373. [CrossRef] [PubMed]
62. Van der Hee, B.; Wells, J.M. Microbial regulation of host physiology by short-chain fatty acids. *Trends Microbiol.* **2021**, *29*, 700–712. [CrossRef] [PubMed]
63. Zietek, M.; Celewicz, Z.; Szczuko, M. Short-chain fatty acids, maternal microbiota and metabolism in pregnancy. *Nutrients* **2021**, *13*, 1244. [CrossRef] [PubMed]
64. Husted, A.S.; Trauelsen, M.; Rudenko, O.; Hjorth, S.A.; Schwartz, T.W. GPCR-mediated signaling of metabolites. *Cell Metab.* **2017**, *25*, 777–796. [CrossRef] [PubMed]
65. Krishnan, J.; Danzer, C.; Simka, T.; Ukropec, J.; Walter, K.M.; Kumpf, S.; Mirtschink, P.; Ukropcova, B.; Gasperikova, D.; Pedrazzini, T.; et al. Dietary obesity-associated Hif1 alpha activation in adipocytes restricts fatty acid oxidation and energy expenditure via suppression of the Sirt2-NAD(+) system. *Genes Dev.* **2012**, *26*, 259–270. [CrossRef] [PubMed]
66. Rakhshandehroo, M.; Knoch, B.; Muller, M.; Kersten, S. Peroxisome proliferator-activated receptor alpha target genes. *Ppar Res.* **2010**, *2010*, 612089. [CrossRef] [PubMed]
67. Gao, Q.; Jia, Y.Z.; Yang, G.S.; Zhang, X.H.; Boddu, P.C.; Petersen, B.; Narsingam, S.; Zhu, Y.J.; Thimmapaya, B.; Kanwar, Y.S.; et al. PPAR alpha-deficient ob/ob obese mice become more obese and manifest severe hepatic steatosis due to decreased fatty acid oxidation. *Am. J. Pathol.* **2015**, *185*, 1396–1408. [CrossRef]
68. Kimura, I.; Ozawa, K.; Inoue, D.; Imamura, T.; Kimura, K.; Maeda, T.; Terasawa, K.; Kashihara, D.; Hirano, K.; Tani, T.; et al. The gut microbiota suppresses insulin-mediated fat accumulation via the short-chain fatty acid receptor GPR43. *Nat. Commun.* **2013**, *4*, 1829. [CrossRef]
69. Tiwari, A. GPR43: An emerging target for the potential treatment of type 2 diabetes, obesity and insulin resistance. *Curr. Opin. Investig. Drugs* **2010**, *11*, 385–393.
70. Aoki, R.; Kamikado, K.; Suda, W.; Takii, H.; Mikami, Y.; Suganuma, N.; Hattori, M.; Koga, Y. A proliferative probiotic *Bifidobacterium* strain in the gut ameliorates progression of metabolic disorders via microbiota modulation and acetate elevation. *Sci. Rep.* **2017**, *7*, 43522. [CrossRef]

71. Macia, L.; Tan, J.; Vieira, A.T.; Leach, K.; Stanley, D.; Luong, S.; Maruya, M.; McKenzie, C.I.; Hijikata, A.; Wong, C.; et al. Metabolite-sensing receptors GPR43 and GPR109A facilitate dietary fibre-induced gut homeostasis through regulation of the inflammasome. *Nat. Commun.* **2015**, *6*, 6734. [CrossRef] [PubMed]
72. McNelis, J.C.; Lee, Y.S.; Mayoral, R.; Van der Kant, R.; Johnson, A.M.F.; Wollam, J.; Olefsky, J.M. GPR43 potentiates β-Cell function in obesity. *Diabetes* **2015**, *64*, 3203. [CrossRef] [PubMed]
73. Kopp, C.; Singh, S.P.; Regenhard, P.; Muller, U.; Sauerwein, H.; Mielenz, M. Trans-cinnamic acid increases adiponectin and the phosphorylation of AMP-activated protein kinase through G-protein-coupled receptor signaling in 3T3-L1 adipocytes. *Int. J. Mol. Sci.* **2014**, *15*, 2906–2915. [CrossRef] [PubMed]

nutrients

MDPI

Article

Nova1 or Bim Deficiency in Pancreatic β-Cells Does Not Alter Multiple Low-Dose Streptozotocin-Induced Diabetes and Diet-Induced Obesity in Mice

Manoja K. Brahma [1], Peng Xiao [2], Madalina Popa [1], Javier Negueruela [1], Valerie Vandenbempt [1], Stéphane Demine [1], Alessandra K. Cardozo [2] and Esteban N. Gurzov [1,*]

[1] Signal Transduction and Metabolism Laboratory, Laboratoire de Gastroentérologie Expérimental et Endotools, Université Libre de Bruxelles, 1070 Brussels, Belgium

[2] Inflammatory and Cell Death Signaling in Diabetes, Laboratoire de Gastroentérologie Expérimental et Endotools, Université Libre de Bruxelles, 1070 Brussels, Belgium

* Correspondence: esteban.gurzov@ulb.be; Tel.: +32-2-555-8908

Abstract: The loss of functional pancreatic β-cell mass is an important hallmark of both type 1 and type 2 diabetes. The RNA-binding protein NOVA1 is expressed in human and rodent pancreatic β-cells. Previous in vitro studies indicated that NOVA1 is necessary for glucose-stimulated insulin secretion and its deficiency-enhanced cytokine-induced apoptosis. Moreover, Bim, a proapoptotic protein, is differentially spliced and potentiates apoptosis in NOVA1-deficient β-cells in culture. We generated two novel mouse models by Cre-Lox technology lacking Nova1 (βNova1$^{-/-}$) or Bim (βBim$^{-/-}$) in β-cells. To test the impact of Nova1 or Bim deletion on β-cell function, mice were subjected to multiple low-dose streptozotocin (MLD-STZ)-induced diabetes or high-fat diet-induced insulin resistance. β-cell-specific Nova1 or Bim deficiency failed to affect diabetes development in response to MLD-STZ-induced β-cell dysfunction and death evidenced by unaltered blood glucose levels and pancreatic insulin content. In addition, body composition, glucose and insulin tolerance test, and pancreatic insulin content were indistinguishable between control and βNova1$^{-/-}$ or βBim$^{-/-}$ mice on a high fat diet. Thus, Nova1 or Bim deletion in β-cells does not impact on glucose homeostasis or diabetes development in mice. Together, these data argue against an in vivo role for the Nova1-Bim axis in β-cells.

Keywords: pancreatic β-cells; Nova1; Bim; obesity; diabetes

Citation: Brahma, M.K.; Xiao, P.; Popa, M.; Negueruela, J.; Vandenbempt, V.; Demine, S.; Cardozo, A.K.; Gurzov, E.N. Nova1 or Bim Deficiency in Pancreatic β-Cells Does Not Alter Multiple Low-Dose Streptozotocin-Induced Diabetes and Diet-Induced Obesity in Mice. *Nutrients* **2022**, *14*, 3866. https://doi.org/10.3390/nu14183866

Academic Editor: Martina Barchitta

Received: 16 December 2021
Accepted: 13 January 2022
Published: 18 September 2022

1. Introduction

Humans and vertebrates respond to circulating glucose levels by secreting insulin from β-cells in the pancreas. The loss of functional β-cell mass is an important hallmark of type 1 diabetes (T1D) and can contribute to the development of type 2 diabetes (T2D). This decline in β-cell mass is associated with apoptosis [1,2]. Genome Wide Association Studies (GWAS) allowed the understanding of transcriptional regulation and identified several candidate genes linked to β-cell dysfunction in T1D and T2D [3]. However, the gene networks contributing to dysregulated insulin secretion and the pathogenesis of diabetes is not fully understood. During transcription, the cellular machinery processes pre-mRNAs through a complex series of events before a mature mRNA is formed. Alternative mRNA splicing is one of the important post-transcriptional mechanisms in which a single gene produces multiple mRNA transcripts that generate different protein isoforms [4]. Abnormal alternative splicing has been associated with many diseases in the adipose tissue, liver, and brain [5–7]. In the pancreas, alternative splicing in β-cells has recently gained momentum with multiple lines of evidence suggesting its association with β-cell dysfunction in both T1D and T2D [8–12].

Alternative splicing of mRNA is achieved through multiple RNA-binding proteins (RBPs) that bind to pre-mRNAs to facilitate intron removal by defining the exon slice recognition site. Impaired RBP function has been linked to diabetes development [13]. RBPs are enriched and play an important role in β-cell function and survival [8,14]. Interestingly, RNA-Sequencing (RNA-Seq) of human tissues has revealed that pancreatic islets express several neuron-specific splicing factors [15], indicating that (i) β-cells and neurons share similar splicing signature and factors and (ii) there is a crucial role of neuron-enriched splicing factors present in β-cell function [16]. The Neuro-Oncological Ventral Antigens (NOVA) are a family of two RBPs (NOVA1 and NOVA2) that are associated with approximately 700 alternative splicing events in neurons, both in the cytoplasm and nucleus [17]. By using a combination of RNA-Seq, siRNA technology, and functional studies, Villate O et al. showed that Nova1 is necessary for glucose-stimulated insulin secretion in rodent INS-1E β-cells. Furthermore, the silencing of NOVA1 in rodent β-cells and dispersed human islets induces both basal and cytokine-induced apoptosis [18]. Moreover, the authors demonstrated that Bcl-2 Interacting Mediator of cell death (Bim), a proapoptotic protein [19], was differentially spliced and potentiated apoptosis in Nova1-deficient β-cells. In addition, Bim has been shown to modulate several apoptotic events in β-cells under stress [2,19–22]. While results from this in vitro finding strongly suggest a crucial activity of NOVA1/BIM in β-cell function and survival, the role of NOVA1 in β-cell functioning in vivo remains elusive. Interestingly, a recent finding by Vernia S et al. showed that adipocyte-specific Nova1-deficient mice exhibit augmented thermogenesis and improved glycemia, suggesting an important physiological metabolic role of Nova1 in vivo [23]. Collectively, based on these findings, we hypothesized that β-cell-specific deletion of Nova1-Bim will have a major impact on β-cell function and survival in vivo. To address this, we used Cre-recombinase technology to delete Nova1 or Bim specifically in β-cells in mice and characterized the impact of deletion of the genes on glucose homeostasis in vivo.

2. Materials and Methods

2.1. *Animal Models and In Vivo Procedures*

Mice were studied in accordance with the ethics protocol approved by the Commision d'Ethicque du Bien-Être Animal (CEBEA), Faculté de Médecine, Université libre de Bruxelles (dossier No. 732). β-cell-specific suppression of Nova1 or Bim were generated using Cre-LoxP technology on C57BL/6N background. Homozygous Nova1floxed mice ($Nova1^{fl/fl}$ Genome way, Lyon, France) were generated by insertion of floxed sequence in addition to the endogenous exons 1–3 of Nova1 in chromosome 12 and kindly provided by the ULB Center for Diabetes Research (Brussels, Belgium). $Nova1^{fl/fl}$ mice were crossed with mice harboring one allele of Cre-recombinase under the mouse Ins1 promoter (Ins1Cre; Jackson Laboratory, Bar Harbor, ME, USA) to generate mice heterozygous for the floxed Nova1 gene and expressed the Ins1Cre transgene ($Nova1^{+/fl/Cre}$). $Nova1^{+/fl/Cre}$ mice were crossed with $Nova1^{fl/fl}$ mice to generate β-cell-specific Nova1 knockout mice ($\beta Nova1^{-/-}$). β-cell-specific Bim knockout mice ($\beta Bim^{-/-}$) were generated by using a similar approach. Thus, $Bim^{fl/fl}$ mice [24,25] (kindly provided by Philippe Bouillet, The Walter and Eliza Hall Institute, Melbourne, Australia) were crossed with Ins1Cre to generate $\beta Bim^{-/-}$ mice. Animals were housed at 22 °C on a 12:12-h light-dark cycle with ad libitum access to food and water. We have used littermates for our study.

2.2. *Diabetes Induction and High Fat Diet Treatment*

To induce diabetes in mice, we utilized a mouse multiple low-dose streptozotocin (MLD-STZ) mouse model. Male mice were injected with streptozotocin (45 mg/kg body weight) prepared in a 0.1 M citrate buffer (pH 4.5) for 5 consecutive days. Body weight and blood glucose (blood drawn from tail vein) were measured every week. Body composition was measured before STZ injection and every 2 weeks. To further test the impact of the loss of Nova1 or Bim, we used a second mouse model of β-cell stress, where we fed mice with a high fat diet (HFD, 60% kcal fat, D16042106i Research Diets, New Brunswick, NJ, USA) for 12 weeks. Tissues were collected under non-fasted conditions. Body fat, lean mass, and total water mass were determined by using EchoMRI™3-in-1 (Houston, TX, USA). We have used the following cohorts for the experiments: 11–12-week-old male Nova1$^{fl/fl}$ and βNova1$^{-/-}$ for MLD-STZ; 10–11-week-old male Nova1$^{fl/fl}$ and βNova1$^{-/-}$ for high fat diet; 9–10-week-old male Bim$^{fl/fl}$ and βBim$^{-/-}$ for MLD-STZ; 10–11-week-old male Bim$^{fl/fl}$ and βBim$^{-/-}$ for high fat diet.

2.3. *Isolation of Mouse Islet and Fluorescence-Activated Cell Sorting (FACS) of β-Cells*

Mouse islets were isolated by digesting the pancreas with collagenase (Sigma-Aldrich, St Louis, MI, USA; cat. #C9263) prepared in serum-free M199 media (Gibco-Thermo Fisher, Waltham, MA, USA; cat #22350) at a concentration of 0.8 mg/mL. Isolated islets were hand-picked under a stereomicroscope two or three times, until a population of pure islets was obtained.

For FACS-sorting of β-cells, islets were first washed with cell dissociation medium (124 mM NaCl, 5.4 mM KCl, 0.8 mM $MgSO_4$, 1 mM Na_2PO_4, and 10 mM HEPES). Following the washing, pancreatic islets were disaggregated by gentle continuous pipetting in trypsin (1 mg/mL; Sigma-Aldrich; cat. #T9935) and DNase I (Sigma-Aldrich; cat. #10104159001; 1 mg/mL) in a water bath at 37 °C. Dispersed islet cells were centrifuged, and the cell pellet was resuspended with FACS buffer, filtered through 35 μm nylon mesh cell strainer-caped tubes (Fisher Scientific, Hampton, NH, USA; cat. #10100151), and β-cells were purified by light scatter activity and autofluorescence by FACS (BD FACSAria™ III, BD Biosciences, Franklin Lakes, NJ, USA) as previously described [26,27]. After sorting, β-cell fraction was collected in a tube containing RPMI 1640 (Gibco-Thermo Fisher, cat. #61870044). Cells were centrifuged and the pellet was lysed using RIPA buffer and processed for immunoblotting.

2.4. *Glucose Stimulated Insulin Secretion*

Isolated islets were allowed to recover for 1–2 h at 37 °C (5% CO_2) in M199 medium supplemented with streptomycin, penicillin, and 10% FBS. Following incubation, groups of ten size-matched islets were incubated with 1 mL of Krebs buffer containing 1.6 mM glucose at 37 °C in 5% CO_2 incubator for 30 min. For stimulation, islets were switched to a solution of the KRB buffer containing either 2.8 mM (control) or 16.7 mM (stimulated) glucose for 60 min at 37 °C. The supernatant fraction was assayed for secreted insulin by ELISA (Crystal Chem Cat 90080, Crystal Chem Inc., IL, USA) and the islets were collected for measurements of protein for normalization.

2.5. *Western Blotting*

Total protein lysates were isolated in a RIPA buffer (150 mM NaCl, 1 mM EDTA, 50 mM Tris/HCl, 0.5% SDS, 1% NP-40), plus a Halt™ protease and phosphatase inhibitor cocktail (Thermo Fisher, cat #78442). Protein was quantified by a BCA protein assay kit (Thermo Fisher; cat. #PI23227). 50–100 μg protein lysate was separated on polyacrylamide gels and transferred to a 0.22 μM nitrocellulose membrane (Bio-Rad, Hercules, CA, USA). Membranes were immunoblotted with primary antibodies for NOVA1 (MilliporeSigma, Burlington, MA, USA; cat. #07-637), NOVA2 (Proteintech, Rosemont, IL, USA; cat. #55002-1-AP), BIM (Cell Signaling Technology Inc., Danvers, MA, USA; cat. #2933, clone C34C5), or GAPDH (Trevigen, Wiesbaden, Germany; cat. #2275-PC-100) in a milk blocking buffer. Proteins were detected using goat anti-rabbit IgG (Dako Agilent, Santa Clara, CA, USA; cat. #P0448) secondary antibody labelled with HRP followed by signal visualization western blot imaging system (Amersham ImageQuant 800 western blot imaging system, Cytiva Life Science, Marlborough, MA, USA).

2.6. *Histological Studies and Immunohistochemical Staining*

The pancreata were harvested from the sacrificed mice after dissection and were washed with saline. The specimens were fixed in 10% buffered formalin (pH 7.4) and embedded into paraffin blocks. The blocks were cut into (7 μm) paraffin sections by a rotator microtome. The sections were stained with Hematoxylin and Eosin (H&E) and with Masson trichrome stains as indicated [28,29].

7 μm thick paraffin sections were mounted on positively charged slides and subjected to immunohistochemical procedures. Paraffin sections were dewaxed, and antigen unmasking was performed using a heated citrate buffer (10 mM). Sections were permeabilized using triton (0.1%) followed by blocking with 2% milk (15 min) and 10% normal goat serum (30 min room temperature) was used to block non-specific binding sites. Specimens were incubated with primary antibodies overnight (Insulin, Dako Agilent; cat. #A0564; Glucagon, Sigma-Aldrich; cat. #G2654; CD68, Cell Signaling; cat. #97778) at 4 °C followed by 1 h incubation with the secondary antibody conjugated to the fluorochrome for insulin (Goat anti-Guinea Pig IgG Thermo Fisher, cat. #A21435 or #A11073), glucagon (Donkey anti-Mouse IgG, Sigma-Aldrich, cat. #A21202 or #A32794), and CD68 (Donkey anti-Rabbit IgG, Sigma-Aldrich, cat. #A31571). Specimen slices incubated only with secondary antibody were used as negative controls. Nuclei were counterstained with DAPI (Vector Laboratories, Burlingarme, CA, USA; cat. #H-1200) before mounting. Images were analyzed on a fluorescent microscope (Axio Observer D1, Carl Zeiss, Oberkochen, Germany). The quantification of the percentage of insulin and glucagon positive cells per islet was performed within the islets, which were analyzed using Cell profiler automated counting software (Broad Institute, Cambridge, MA, USA).

2.7. Pancreatic Insulin Content

Pancreas pieces close to the intestine and liver were collected followed by homogenization with ethanol/water/acid (75/23.5/1.5, % *v/v*). Homogenized samples were incubated on a tube rotator for 24 h at 4 °C and centrifuged at 1500 g for 30 min. Supernatant was collected and used for insulin measurement using a commercial Insulin ELISA kit (Crystal Chem Inc., Chicago, IL, USA; cat. #90080). Insulin content was expressed as μg insulin/g pancreas weight (μg/g).

2.8. Intraperitoneal Glucose and Insulin Tolerance Tests (IPGTT and ITT)

An intraperitoneal glucose tolerance test (IPGTT) was carried out after a 6 h fast in age-matched male mice, and 2 g/Kg glucose/body weight was administered by intraperitoneal injection. For MLD-STZ studies, 1 g/Kg body weight of glucose was administered. For the insulin tolerance test (ITT), regular human insulin (Sigma-Aldrich; cat. #I9278; 0.65 U/kg in saline) was administered intraperitoneally to 4 h fasted mice. Blood was collected from the tail vein before (basal, 0 min) and at 15, 30, 60, 90, and 120 min after glucose or insulin administration using a commercial glucometer (Accu-Check Perfoma, Roche, Basel, Switzerland).

2.9. Statistical Analysis

Analyses were performed with GraphPad Prism 8.4 software (Prism, San Diego, CA, USA). All the results are expressed as mean ± SEM. Student's t-tests or, where appropriate, two-way analysis of variance (ANOVA) was performed. Statistical analysis was performed using the criterion for significance of $p < 0.05$ for all comparisons.

3. Results

3.1. Generation of Mice with β-Cell-Specific Loss of Nova1 (βNova1$^{-/-}$) or Bim (βBim$^{-/-}$)

To directly investigate the effects of β-cell-specific Nova1 knockout in mice, we used the Cre-LoxP recombination system. Briefly, Nova1$^{fl/fl}$ mice were crossed with transgenic mice expressing the Cre recombinase under the control of the mouse insulin promoter (Ins1Cre). Nova1 deletion was confirmed by PCR analysis of genomic DNA isolated from ear notches (data not shown) and immunoblot analysis of Nova1 protein expression in isolated islets. As shown in Figure 1A (left panel and right panel), immunoblot analysis confirmed a robust deletion (85%) of Nova1 in isolated islets. Since Nova1 is a neuron-specific alternative splicing factor, we confirmed that Nova1 deletion was specific to β-cells and its expression not altered in the brain (Figure 1A). In addition, Nova2 expression in pancreatic islets was not affected by Nova1 deletion in β-cells (Figure S1).

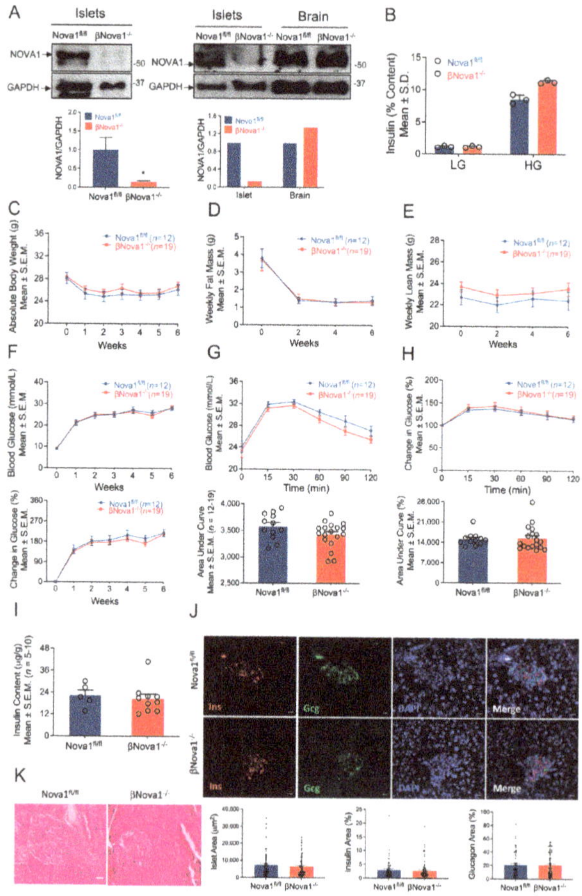

Figure 1. Impact of β-cell-specific Nova1 deletion on glucose homeostasis. Nova1 was deleted in β-cells of mice using Cre-LoxP technology. Nova1 deletion was confirmed in isolated islets from Nova1$^{fl/fl}$ or βNova1$^{-/-}$ mice by immunoblotting using anti-NOVA1 antibody (**A**). Ex vivo glucose stimulated insulin secretion was performed in isolation from Nova1$^{fl/fl}$ or βNova1$^{-/-}$ mice in the presence of low or high glucose (**B**). To test the impact of diabetes on glucose homeostasis Nova1$^{fl/fl}$ or βNova1$^{-/-}$ mice were injected with 5-days of MLD-STZ and followed for 6 weeks. Body weight (**C**), body composition (**D**,**E**), and blood glucose (**F**; **top and bottom panel**) were monitored following STZ injection. GTT was performed in 6 h fasted mice after 6 weeks of STZ administration. Blood glucose was measured at indicated time points and expressed as mmol/L (**G**) or % change in blood glucose normalized to time 0 (**H**). Figure **G** (**bottom panel**) and **H** (**bottom panel**) represent glucose areas under the curve. Pancreatic insulin content was measured in acid–ethanol extracted fraction using a commercial ELISA kit (**I**). Pancreatic sections were stained for insulin, glucagon, and DNA (**J**). Scale bar, 20 μm. Representative of 3 mice per genotype (**J**). **Bottom panel** (**J**) showing islet area, insulin, and glucagon positive cells (calculated as % of total islet area). Histology of mouse pancreas stained by Hematoxylin and Eosin (**K**). Destruction and distortion of endocrine cells in STZ-diabetic mice with the presence of dead and inflammatory cells. Scale bar, 25 μm. Data are presented as Mean ± SEM. * $p < 0.05$ vs. Nova1$^{fl/fl}$ mice.

3.2. *In Vivo Glucose Homeostasis Was Unchanged in βNova1$^{-/-}$ Mice in Response to MLD-STZ-Induced Diabetes*

To test the impact of Nova1 deletion on β-cells, we first performed in vitro glucose stimulated insulin secretion in the isolated islets from βNova1$^{-/-}$ mice. While basal insulin secretion was unchanged between 2 genotypes, a trend towards increase (+32%) in glucose-stimulated insulin secretion was observed in βNova1$^{-/-}$ mice compared to Nova1$^{fl/fl}$ mice when stimulated with high glucose (Figure 1B). Next, to test our hypothesis that Nova1 deletion exacerbates the impaired glucose tolerance in diabetes, we subjected βNova1$^{-/-}$ and littermate control mice to a model of inflammatory-mediated diabetes by injecting MLD-STZ and followed the glycemia, body composition, and body weight. Nova1 deletion in β-cells did not alter body weight (Figure 1C) or composition (Figures 1D,E and S2A–H). Similarly, blood glucose measured every week was similar in both genotypes (Figure 1F; top and bottom panels) and the mice had no differences in elevated blood glucose (Nova1$^{fl/fl}$: 26 ± 1.1 mmol/L, βNova1$^{-/-}$: 24.6 ± 1.1 mmol/L). At the end of 6 weeks, IPGTT performed in βNova1$^{-/-}$ and littermate controls revealed no change between genotypes in glucose tolerance indicated by blood glucose levels (Figure 1G; top and bottom panels) and % change in blood glucose (Figure 1H; top and bottom panels) after glucose injection. No change in insulin content in the pancreas was observed in these mice (Figure 1I). Finally, we performed immunofluorescent staining with the pancreas sections of the mice to analyze both insulin-secreting β-cells and glucagon-secreting α-cells. Most islets were disrupted in both genotypes and no difference in the islet area or insulin area, or in the glucagon area or macrophage numbers, were observed (Figures 1J and S3A,B). Similarly, histology of pancreas stained by H&E showed no difference between two genotypes (Figure 1K). There are several caveats and concerns about β-cell specific Cre mouse lines, and results should be carefully interpreted with these models [30–32]. Importantly, the Ins1Cre mouse model used in the present study does not affect body weights and glucose homeostasis when treated with MLD-STZ compared to C57BL/6N control mice (Supplementary Figure S4A–F) [32]. In addition, we confirmed that the Cre transgene alone does not affect body composition (Supplementary Figure S5A–H). Collectively, these results show that Nova1 deletion in β-cells does not impair glucose homeostasis.

3.3. *In Vivo Glucose Homeostasis Was Unchanged in βNova1$^{-/-}$ Mice un Response to High-Fat Diet*

To further understand the impact of Nova1 deletion in a model of moderate β-cell stress, we induced metabolic stress (obesity and insulin resistance) by feeding the mice a high-fat diet for 12 weeks [25,33]. The high-fat diet (HFD) increased body weight (Figure 2A,B) and fat mass in both βNova1$^{-/-}$ and control mice (Figure 2C,D) after 12 weeks. Before starting the HFD, the glucose tolerance test was not different between βNova1$^{-/-}$ and Nova1$^{fl/fl}$ control mice (Figure 2E,F). After 12 weeks of the HFD, fasting blood glucose was elevated in both Nova1$^{fl/fl}$ (Before HFD: 9.6 ± 0.3, after HFD: 11.9 ± 0.8) and βNova1$^{-/-}$ mice (Before HFD: 10.2 ± 0.3, after HFD: 12.1 ± 0.3). Interestingly, βNova1$^{-/-}$ mice demonstrated a moderate improvement in glucose tolerance at 60 and 120 min after glucose injection. However, the glucose area under curve demonstrated the only trend towards reduction in (p = 0.056) in βNova1$^{-/-}$ mice (Figure 2G,H). To assess insulin action, an intraperitoneal insulin tolerance test was performed 1 week after the IPGTT. No change in ITT was observed between the 2 groups (Figure 2I,J). Moreover, pancreatic insulin content was also indistinguishable between Nova1$^{fl/fl}$ and βNova1$^{-/-}$ mice (Figure 2K). To further determine if Nova1 deletion altered insulin expression at the histological level, we investigated the histology of the pancreas via H&E and immunofluorescence analysis on pancreatic sections from high-fat fed Nova1$^{fl/fl}$ and βNova1$^{-/-}$ mice. Interestingly, as depicted in Figure 2L (bottom panel), pancreatic sections from βNova1$^{-/-}$ mice showed a significant reduction in the percentage insulin area (p = 0.0163) despite the unchanged islet area between Nova1$^{fl/fl}$ and βNova1$^{-/-}$ mice. However, the H&E staining did not show any notable change in the islet morphology between the two genotypes (Figure 2M). This suggests that although Nova1 deletion in β-cells reduces insulin positive cells, it is not sufficient to bring change in the overall glucose homeostasis at the whole-body level.

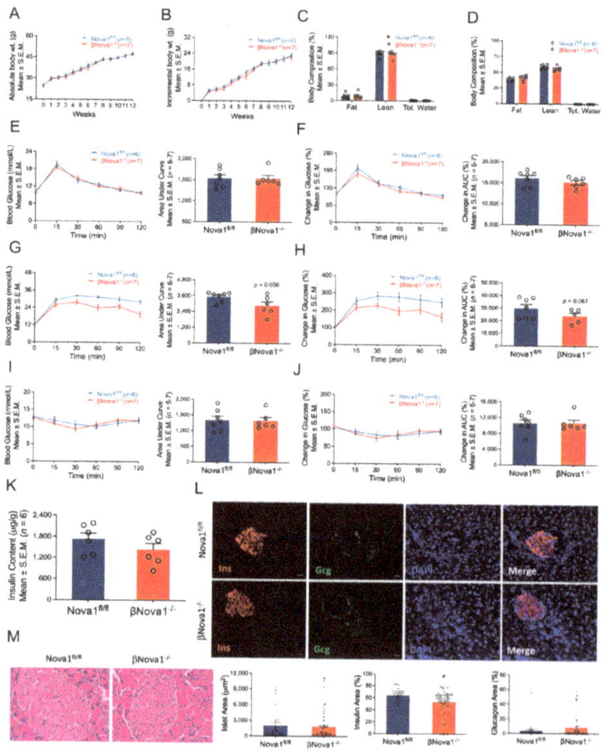

Figure 2. Effect of high-fat diet on glucose homeostasis of βNova1$^{-/-}$ mice. βNova1$^{-/-}$ mice were fed a high-fat diet for 12 weeks. Body weight was measured weekly (**A**,**B**). Incremental body weight (**B**) was calculated as cumulative gain in body weight (g) every week. Body composition (fat mass, lean mass, and total water) was measured before and after 12 weeks of high-fat diet feeding (**C**,**D**). A glucose tolerance test performed before high-fat diet feeding did not show any change in βNova1$^{-/-}$ mice compared to Nova1$^{fl/fl}$ mice (**E**,**F**). After 12 weeks following the high-fat diet, βNova1$^{-/-}$ mice exhibited a trend towards improved glucose tolerance compared to Nova1$^{fl/fl}$ mice (**G**,**H**). However, an insulin tolerance test measured one week after the IPGTT did not show any difference between the two genotypes (**I**,**J**). Pancreatic insulin content was measured in an acid–ethanol extracted fraction using a commercial ELISA kit (**K**). Immunofluorescence staining of pancreatic sections (insulin, glucagon, and DNA) from both high-fat-fed and control Nova1$^{fl/fl}$ and βNova1$^{-/-}$ mice (**L**). Insulin (red staining) is contained in the core zone of the islet, while glucagon (green staining) is found in the mantle zone. Scale bar, 20 μm. Representative of 3 mice per genotype (**L**). **Bottom panel** (**L**) showing islet area, insulin, and glucagon positive cells (calculated as % of total islet area). Histology of mouse pancreas stained by Hematoxylin and Eosin (**M**). Scale bar, 25 μm. Data are presented as Mean ± SEM. * $p < 0.05$ vs. Nova1$^{fl/fl}$ mice.

3.4. Bim Deletion in β-Cells Did Not Alter In Vivo Glucose Homeostasis in Response to MLD-STZ-Induced T1D or High-Fat Diet-Induced Obesity

The Bcl-2 homology domain 3 (BH3)-only protein Bim is a critical mediator of cytokine-induced β-cell death and it has been shown that Bim is involved in β-cell apoptosis induced by β-cell Nova1 deficiency [18]. However, the β-cell-specific role of Bim in vivo remains elusive. Therefore, we generated a mouse model lacking Bim specifically in β-cells using a

Cre-Lox approach similar to that used for β$Nova1^{-/-}$ mice. Bim deletion was confirmed in the β-cells of β$Bim^{-/-}$ mice (Figure 3A). Bim deficiency in the β-cells did not alter diabetes development induced by MLD-STZ, body weight or composition (Figure 3B–D and Figure S6A–H), and weekly blood glucose (Figure 3E; top and bottom panels). At the end of 6 weeks, IPGTT performed in β$Bim^{-/-}$ and littermate control mice did not show any change in glucose tolerance indicated by blood glucose levels (Figure 3F; top and bottom panels) and % change in blood glucose (Figure 3G; top and bottom panels) after glucose bolus injection during IPGTT. The insulin content measured in the pancreas of these mice was indistinguishable between the groups (Figure 3H). Immunofluorescence analysis of pancreatic sections showed that the percentage insulin area was slightly reduced ($p = 0.03$) in β$Bim^{-/-}$ mice (Figure 3I; top and bottom panels). However, H&E staining did not show any apparent changes between the two genotypes (Figure 3J). Collectively, these results show that Bim deletion in β-cells does not impair glucose homeostasis. Next, we subjected these mice to a 12-week HFD regimen to test the in vivo impact of Bim deletion in response to obesity-induced metabolic stress. After 12 weeks of the HFD, body weight (Figure 4A,B) and body composition (Figure 4C,D) were similar in β$Bim^{-/-}$ and littermate control mice. Before starting the HFD, the glucose tolerance test was unchanged between $Bim^{fl/fl}$ and β$Bim^{-/-}$ mice (Figure 4E,F). The HFD feeding did not affect the blood glucose levels measured during IPGTT (Figure 4G,H) or ITT (Figure 4I,J), between $Bim^{fl/fl}$ and β$Bim^{-/-}$ obese mice. Similarly, pancreatic insulin content (Figure 4K), immunofluorescence staining (Figure 4L), and histological analysis of H&E (Figure 4M) did not show any major change between high-fat diet fed $Bim^{fl/fl}$ and β$Bim^{-/-}$ littermate mice. Taken together, these results show that Bim deletion in β-cells does not improve glucose homeostasis in obese or MLD-STZ-mediated diabetic mice.

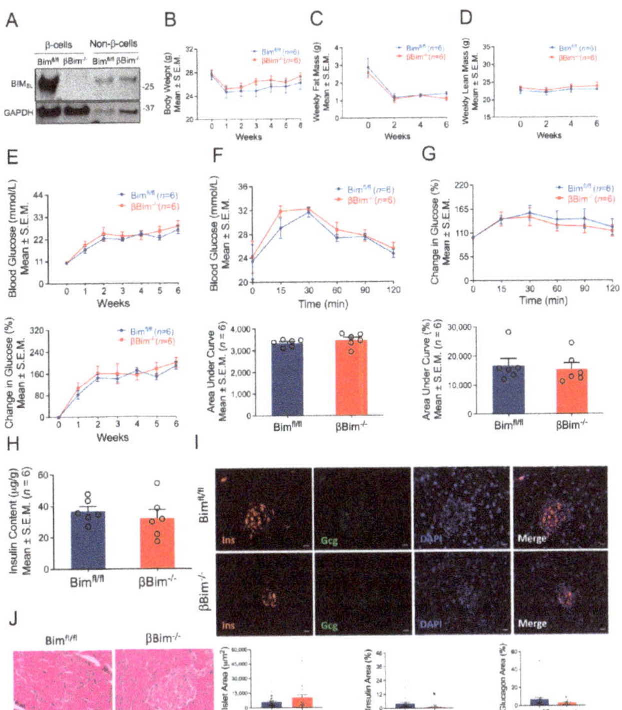

Figure 3. Impact of β-cell-specific Bim deletion on glucose homeostasis. Bim was deleted in β-cells of mice using Cre-LoxP technology. Bim deletion was confirmed in FACS-purified β-cells isolated from islets of $Bim^{fl/fl}$ or β$Bim^{-/-}$ mice by immunoblotting using anti-Bim antibody (**A**). To test the

impact of diabetes on glucose homeostasis, $Bim^{fl/fl}$ or $\beta Bim^{-/-}$ mice were injected with 5-days of MLD-STZ and followed for 6 weeks. Body weight (**B**), body composition (**C**,**D**), and blood glucose (**E**; top and bottom panel) were monitored following STZ injection. GTT was performed in 6 h fasted mice after 6 weeks of STZ administration. Blood glucose was measured at indicated time points and expressed as mmol/L (**F**) or % change in blood glucose normalized to time 0 (**G**). Figure (**F**) (**bottom panel**) and **G** (**bottom panel**) represent glucose areas under the curve. Pancreatic insulin content was measured in an acid–ethanol extracted fraction using a commercial ELISA kit (**H**). Destruction and distortion of endocrine cells was observed in STZ-diabetic mice with the presence of dead and inflammatory cells. Pancreatic sections were stained for insulin and glucagon and DNA. Scale bar, 20 μm. Representative of 3 mice per genotype (**I**). **Bottom panel** (**I**) showing islet area, insulin, and glucagon positive cells (calculated as % of total islet area). Histology of mouse pancreas stained by Hematoxylin and Eosin (**J**). Scale bar, 25 μm. Data are presented as Mean ± SEM. * $p < 0.05$ vs. $Bim^{fl/fl}$ mice.

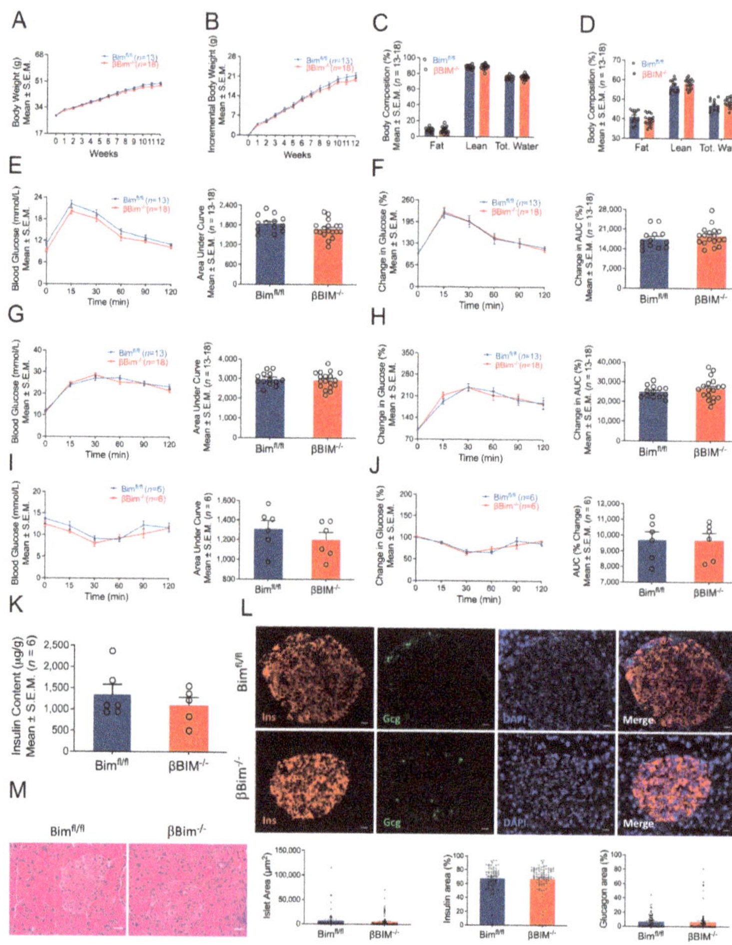

Figure 4. Effect of high-fat diet on glucose homeostasis of $\beta Bim^{-/-}$ mice. $\beta Bim^{-/-}$ mice were subjected to a high-fat diet regimen for 12 weeks. Body weight was measured weekly (**A**,**B**). Body

(fat mass, lean mass, and total water) was measured before and after 12 weeks of high-fat diet feeding (**C,D**). Glucose tolerance tests measured before (**E,F**) and 12 weeks after high-fat diet feeding (**G,H**) did not show change in βBim$^{-/-}$ mice compared to Bim$^{fl/fl}$ mice. Similarly, an insulin tolerance test measured one week after the IPGTT did not show any difference between the two genotypes (**I,J**). Pancreatic insulin content was measured in an acid–ethanol extracted fraction using a commercial ELISA kit (**K**). Pancreatic sections were stained for insulin, glucagon, and DNA (**L**). **Bottom panel** (**L**) showing islet area, insulin, and glucagon positive cells (calculated as % of total islet area). Representative of 3 mice per genotype. Scale bar, 20 μm. Histology of mouse pancreas stained by Hematoxylin and Eosin (**M**). Scale bar, 25 μm. Data are presented as Mean ± SEM.

4. Discussion

The importance of alternative splicing in diabetes and pancreatic β-cell function is gaining interest. Previous findings have suggested an important role for NOVA1 in pancreatic β-cell function and survival including the following: (i) studies showing expression of NOVA1 in human pancreatic islets [15] and purified rat β-cells [34]; (ii) NOVA1 knockdown in vitro induced apoptosis (basal and cytokine-induced) in human islets impairs glucose-stimulated insulin secretion in INS-1E cells and regulates transcripts involved in exocytosis, apoptosis, insulin signaling, splicing, and transcription in FACS-purified rat β-cells [18]. Despite this convincing in vitro evidence, the in vivo function and physiological relevance of NOVA1 in β-cells remain unclear. Here, we determined the in vivo metabolic effect of β-cell-specific Nova1 deletion in mice to directly test the hypothesis that the loss of Nova1 in β-cells will adversely affect glucose homeostasis. Surprisingly, we did not detect any changes in glucose-stimulated insulin secretion in mice with a β-cell specific loss of Nova1. Conversely, βNova1$^{-/-}$ mice islets demonstrated a trend towards increased—albeit not significant—insulin secretion under high glucose stimulation. Nova1 is known to be indispensable for neuronal survival as Nova1-null mice die postnatally due to neuronal apoptosis [35] and motor neuron dysfunction [36]. However, βNova1$^{-/-}$ mice did not show any adverse phenotypes, suggesting that Nova1 may not be necessary for β-cell survival or function. Since we did not observe any spontaneous adverse phenotypes in βNova1$^{-/-}$ mice, we reasoned that Nova1 may be necessary under pathological conditions. Therefore, we challenged Nova1 deficient β-cells with inflammatory stress (MLD-STZ) or moderate metabolic stress (insulin resistance) conditions. To address this question, we subjected these mice to two independent models of metabolic stress with impaired β-cell function. First, we used MLD-STZ-induced diabetes, which is a widely used model for studying β-cell inflammation and death. We injected STZ to Nova1$^{fl/fl}$ and βNova1$^{-/-}$ mice and followed them for 6 weeks. We did not see any changes in blood glucose, glucose tolerance (only a trend towards a reduction at later time points after glucose bolus), or pancreatic insulin content between the two genotypes. Second, it is established that a long-term high-fat diet enhances β-cell proliferation and mass [37]. Therefore, to test the impact of Nova1 loss in β-cells, we fed βNova1$^{-/-}$ mice a high-fat diet. No difference in body composition and pancreatic levels was observed between the two genotypes. Surprisingly, βNova1$^{-/-}$ mice exhibited a moderate improvement in HFD-induced impaired glucose tolerance compared to Nova1$^{fl/fl}$ mice. Overall, from two metabolic stress mouse models, our data clearly argue against a β-cell-specific physiological/pathological role of Nova1.

It has also been shown that Nova1 knockdown in INS-1E cells induces basal and cytokine-induced apoptosis and this cell death is mediated by the BH3-only protein Bim [18]. Both in vitro and in in vivo studies [38,39] have previously shown that Bim is downstream Nova1 and plays an important role in β-cell apoptosis in diabetes. Moreover, Bim activates apoptosis in several models of β-cell stress and high mRNA levels of Bim is detected in islets of T2D patients [21]. These findings collectively suggest that BIM plays a physiological metabolic role. However, the impact of β-cell-specific role of Bim in glucose homeostasis in vivo has not been previously studied. We generated a mouse model with β-cell-specific deletion of Bim and found indistinguishable parameters both under physiological condition and in response to inflammatory-mediated diabetes or HFD-induced obesity. Hence, our finding shows for the first time that the loss of Bim in β-cells is not sufficient to improve β-cell function and/or survival in vivo.

Thus, the question remains: why did the loss of Nova1-Bim in β-cells have no effect on glucose homeostasis in vivo? Several plausible reasons emerge that may explain the discrepancy between previous strong in vitro data supporting a beneficial role of Nova1 in β-cells and our in vivo findings in this study. First, though NOVA1 expression has been detected in human pancreatic islets [15] and purified rat β-cells [34], their expression pattern in the diseased context is largely unknown. Detailed characterization of pancreatic NOVA1 expression in diabetic patients or rodent models of diabetes would help to determine if altered NOVA1 expression is adaptive or maladaptive during diabetes. Second, it is possible that β-cells-specific Nova1 knockout induces a compensatory increase of alternative RBPs other than Nova2. This is not in line with the indispensable role of Nova1 in neurons. Global knockout mice of Nova1 die postnatally from a motor deficit associated with apoptotic death of spinal and brainstem neurons [35]. It is conceivable that deletion of both Nova1 and Nova2 expression is necessary to induce β-cell dysfunction in vivo. In line with this, a recent study using adipocyte-specific deletion of Nova1 and Nova2 has shown an additive effect, promoting adipose tissue thermogenesis and improving glycemia [23]. Finally, to our knowledge, the study by Villate O et al. is the only study that has used multiple in vitro models to show the importance of Nova1 in β-cell function and survival [18]. Therefore, it is important to consider that in vitro studies may provide molecular insights into cellular processes, but they do not often completely recapitulate the complexity of in vivo physio and pathophysiology, which makes it difficult to translate the relevance of the findings.

A limitation of our studies is that we focused our in vivo work on adult male mice, but the role of Nova1 and Bim in β-cell function should also be clarified in the development of obesity and diabetes at different ages and in females [40]. In addition, we have not performed mechanistic/gene sequencing studies to identify altered gene expression or post-transcriptional gene modifications by knocking out Nova1 or Bim in β-cells. In the present work, our main goal was to determine whether deficiency of Nova1 or Bim in β-cells has an in vivo metabolic impact and whether previous in vitro findings can be translated to an in vivo set up. In summary, our data provide evidence against a role for Nova1 and Bim in β-cell function.

5. Conclusions

The loss of functional pancreatic β-cell mass is an important hallmark of both T1D and T2D. Our data from two independent mouse models provide evidence against a critical role for Nova1 and Bim in β-cells and do not support their modulation to improve β-cell survival. It is important to note, however, that our data do not completely rule out a potential contribution of NOVA1/BIM to β-cell dysfunction in human T1D or T2D.

Supplementary Materials: The following supporting information can be downloaded at: https://www.mdpi.com/article/10.3390/nu14183866/s1, Supplementary Figure S1: Nova2 expression is not affected in β$Nova1^{-/-}$ islets, Supplementary Figure S2: Change in body composition of β$Nova1^{-/-}$ mice after MLD-STZ administration, Supplementary Figure S3: Expression of insulin, glucagon and the macrophage marker CD68 in pancreas of MLD-STZ-treated β$NOVA1^{-/-}$, β$Bim^{-/-}$ and control mice, Supplementary Figure S4: Transgenic expression of the Cre gene did not affect impaired glucose homeostasis induced by MLD-STZ, Supplementary Figure S5: Change in body composition of Ins1Cre mice after MLD-STZ administration, Supplementary Figure S6: Change in body composition of β$Bim^{-/-}$ mice after MLD-STZ administration.

Author Contributions: M.K.B., P.X., M.P., J.N., V.V. and S.D. collected and analyzed data. M.K.B. and E.N.G. wrote the manuscript. A.K.C. contributed to data analysis, experimental design, and reviewed and edited the manuscript. E.N.G. researched data, contributed to designed experiments, and reviewed and edited the manuscript. E.N.G. is the guarantor of this work and, as such, had full access to all the data in the study and takes responsibility for the integrity of the data and the accuracy of the data analysis. All authors have read and agreed to the published version of the manuscript.

Funding: This work was supported by a Fonds National de la Recherche Scientifique (FNRS)-MIS grant (33650793), an FNRS-CDR grant (35275350), a European Research Council (ERC) Consolidator grant METAPTPs (GA817940), and a JDRF Career Development Award (CDA-2019-758-A-N). V.V. is supported by a FNRS PhD Aspirant scholarship. E.N.G. is a Research Associate of the FNRS, Belgium. Work in the AKC group was supported by the Excellence of Science grant (FNRS, Belgium, convention 30826052).

Institutional Review Board Statement: The animal study protocol was approved by the Commision d'Ethicque du Bien-Être Animal (CEBEA), Faculté de Médecine, Université libre de Bruxelles (dossier N° 732).

Informed Consent Statement: Not applicable.

Data Availability Statement: The data presented in this study are available within this article and supplementary figures. The data presented in this study are available on request from the corresponding author.

Acknowledgments: We thank André Dias, Shruti Mohan, Ao Li, Grégory Vegh, Pascale Lybaert and personnel from the ULB Center for Diabetes Research (Université libre de Bruxelles) for experimental assistance and technical support, Decio Eizirik for the Nova1$^{fl/fl}$ mice, and Eduardo Gilglioni for help with the graphical abstract.

Conflicts of Interest: The authors declare no conflict of interest.

References

1. Gurzov, E.N.; Ke, P.C.; Ahlgren, U.; Garcia Ribeiro, R.S.; Gotthardt, M. Novel Strategies to Protect and Visualize Pancreatic beta Cells in Diabetes. *Trends Endocrinol. Metab.* **2020**, *31*, 905–917. [CrossRef] [PubMed]
2. Gurzov, E.N.; Eizirik, D.L. Bcl-2 proteins in diabetes: Mitochondrial pathways of beta-cell death and dysfunction. *Trends Cell Biol.* **2011**, *21*, 424–431. [CrossRef] [PubMed]
3. Storling, J.; Pociot, F. Type 1 Diabetes Candidate Genes Linked to Pancreatic Islet Cell Inflammation and Beta-Cell Apoptosis. *Genes* **2017**, *8*, 72. [CrossRef] [PubMed]
4. Licatalosi, D.D.; Darnell, R.B. RNA processing and its regulation: Global insights into biological networks. *Nat. Rev. Genet.* **2010**, *11*, 75–87. [CrossRef]
5. Wong, C.M.; Xu, L.; Yau, M.Y. Alternative mRNA Splicing in the Pathogenesis of Obesity. *Int. J. Mol. Sci.* **2018**, *19*, 632. [CrossRef]
6. Webster, N.J.G. Alternative RNA Splicing in the Pathogenesis of Liver Disease. *Front. Endocrinol.* **2017**, *8*, 133. [CrossRef]
7. Lipscombe, D.; Lopez Soto, E.J. Alternative splicing of neuronal genes: New mechanisms and new therapies. *Curr. Opin. Neurobiol.* **2019**, *57*, 26–31. [CrossRef]
8. Alvelos, M.I.; Bruggemann, M.; Sutandy, F.R.; Juan-Mateu, J.; Colli, M.L.; Busch, A.; Lopes, M.; Castela, A.; Aartsma-Rus, A.; Konig, J.; et al. The RNA-binding profile of the splicing factor SRSF6 in immortalized human pancreatic beta-cells. *Life Sci. Alliance* **2021**, *4*. [CrossRef]
9. Marcheva, B.; Perelis, M.; Weidemann, B.J.; Taguchi, A.; Lin, H.; Omura, C.; Kobayashi, Y.; Newman, M.V.; Wyatt, E.J.; McNally, E.M.; et al. A role for alternative splicing in circadian control of exocytosis and glucose homeostasis. *Genes Dev.* **2020**, *34*, 1089–1105. [CrossRef]
10. Yau, B.; Blood, Z.; An, Y.; Su, Z.; Kebede, M.A. Type 2 diabetes-associated single nucleotide polymorphism in Sorcs1 gene results in alternative processing of the Sorcs1 protein in INS1 beta-cells. *Sci. Rep.* **2019**, *9*, 19466. [CrossRef]
11. Cnop, M.; Abdulkarim, B.; Bottu, G.; Cunha, D.A.; Igoillo-Esteve, M.; Masini, M.; Turatsinze, J.V.; Griebel, T.; Villate, O.; Santin, I.; et al. RNA sequencing identifies dysregulation of the human pancreatic islet transcriptome by the saturated fatty acid palmitate. *Diabetes* **2014**, *63*, 1978–1993. [CrossRef] [PubMed]
12. Nogueira, T.C.; Paula, F.M.; Villate, O.; Colli, M.L.; Moura, R.F.; Cunha, D.A.; Marselli, L.; Marchetti, P.; Cnop, M.; Julier, C.; et al. GLIS3, a susceptibility gene for type 1 and type 2 diabetes, modulates pancreatic beta cell apoptosis via regulation of a splice variant of the BH3-only protein Bim. *PLoS Genet.* **2013**, *9*, e1003532. [CrossRef] [PubMed]
13. Nutter, C.A.; Kuyumcu-Martinez, M.N. Emerging roles of RNA-binding proteins in diabetes and their therapeutic potential in diabetic complications. *Wiley Interdiscip. Rev. RNA* **2018**, *9*, e1459. [CrossRef] [PubMed]

14. Juan-Mateu, J.; Alvelos, M.I.; Turatsinze, J.V.; Villate, O.; Lizarraga-Mollinedo, E.; Grieco, F.A.; Marroqui, L.; Bugliani, M.; Marchetti, P.; Eizirik, D.L. SRp55 Regulates a Splicing Network That Controls Human Pancreatic beta-Cell Function and Survival. *Diabetes* **2018**, *67*, 423–436. [CrossRef]
15. Eizirik, D.L.; Sammeth, M.; Bouckenooghe, T.; Bottu, G.; Sisino, G.; Igoillo-Esteve, M.; Ortis, F.; Santin, I.; Colli, M.L.; Barthson, J.; et al. The human pancreatic islet transcriptome: Expression of candidate genes for type 1 diabetes and the impact of pro-inflammatory cytokines. *PLoS Genet.* **2012**, *8*, e1002552. [CrossRef]
16. Juan-Mateu, J.; Rech, T.H.; Villate, O.; Lizarraga-Mollinedo, E.; Wendt, A.; Turatsinze, J.V.; Brondani, L.A.; Nardelli, T.R.; Nogueira, T.C.; Esguerra, J.L.S.; et al. Neuron-enriched RNA-binding Proteins Regulate Pancreatic Beta Cell Function and Survival. *J. Biol. Chem.* **2017**, *292*, 3466–3480. [CrossRef]
17. Meldolesi, J. Alternative Splicing by NOVA Factors: From Gene Expression to Cell Physiology and Pathology. *Int. J. Mol. Sci.* **2020**, *21*, 3941. [CrossRef]
18. Villate, O.; Turatsinze, J.V.; Mascali, L.G.; Grieco, F.A.; Nogueira, T.C.; Cunha, D.A.; Nardelli, T.R.; Sammeth, M.; Salunkhe, V.A.; Esguerra, J.L.; et al. Nova1 is a master regulator of alternative splicing in pancreatic beta cells. *Nucleic Acids Res.* **2014**, *42*, 11818–11830. [CrossRef]
19. Barthson, J.; Germano, C.M.; Moore, F.; Maida, A.; Drucker, D.J.; Marchetti, P.; Gysemans, C.; Mathieu, C.; Nunez, G.; Jurisicova, A.; et al. Cytokines tumor necrosis factor-alpha and interferon-gamma induce pancreatic beta-cell apoptosis through STAT1-mediated Bim protein activation. *J. Biol. Chem.* **2011**, *286*, 39632–39643. [CrossRef]
20. Santin, I.; Moore, F.; Colli, M.L.; Gurzov, E.N.; Marselli, L.; Marchetti, P.; Eizirik, D.L. PTPN2, a candidate gene for type 1 diabetes, modulates pancreatic beta-cell apoptosis via regulation of the BH3-only protein Bim. *Diabetes* **2011**, *60*, 3279–3288. [CrossRef]
21. Wali, J.A.; Rondas, D.; McKenzie, M.D.; Zhao, Y.; Elkerbout, L.; Fynch, S.; Gurzov, E.N.; Akira, S.; Mathieu, C.; Kay, T.W.; et al. The proapoptotic BH3-only proteins Bim and Puma are downstream of endoplasmic reticulum and mitochondrial oxidative stress in pancreatic islets in response to glucotoxicity. *Cell Death Dis.* **2014**, *5*, e1124. [CrossRef] [PubMed]
22. Miani, M.; Elvira, B.; Gurzov, E.N. Sweet Killing in Obesity and Diabetes: The Metabolic Role of the BH3-only Protein BIM. *J. Mol. Biol.* **2018**, *430*, 3041–3050. [CrossRef] [PubMed]
23. Vernia, S.; Edwards, Y.J.; Han, M.S.; Cavanagh-Kyros, J.; Barrett, T.; Kim, J.K.; Davis, R.J. An alternative splicing program promotes adipose tissue thermogenesis. *eLife* **2016**, *5*, e17672. [CrossRef] [PubMed]
24. Raynor, J.; Sholl, A.; Plas, D.R.; Bouillet, P.; Chougnet, C.A.; Hildeman, D.A. IL-15 Fosters Age-Driven Regulatory T Cell Accrual in the Face of Declining IL-2 Levels. *Front. Immunol.* **2013**, *4*, 161. [CrossRef]
25. Litwak, S.A.; Pang, L.; Galic, S.; Igoillo-Esteve, M.; Stanley, W.J.; Turatsinze, J.V.; Loh, K.; Thomas, H.E.; Sharma, A.; Trepo, E.; et al. JNK Activation of BIM Promotes Hepatic Oxidative Stress, Steatosis, and Insulin Resistance in Obesity. *Diabetes* **2017**, *66*, 2973–2986. [CrossRef]
26. Pipeleers, D.G.; In't Veld, P.A.; Van de Winkel, M.; Maes, E.; Schuit, F.C.; Gepts, W. A new in vitro model for the study of pancreatic A and B cells. *Endocrinology* **1985**, *117*, 806–816. [CrossRef]
27. Schaschkow, A.; Pang, L.; Vandenbempt, V.; Elvira, B.; Litwak, S.A.; Vekeriotaite, B.; Maillard, E.; Vermeersch, M.; Paula, F.M.M.; Pinget, M.; et al. STAT3 Regulates Mitochondrial Gene Expression in Pancreatic beta-Cells and Its Deficiency Induces Glucose Intolerance in Obesity. *Diabetes* **2021**, *70*, 2026–2041. [CrossRef]
28. Litwak, S.A.; Loh, K.; Stanley, W.J.; Pappas, E.G.; Wali, J.A.; Selck, C.; Strasser, A.; Thomas, H.E.; Gurzov, E.N. p53-upregulated-modulator-of-apoptosis (PUMA) deficiency affects food intake but does not impact on body weight or glucose homeostasis in diet-induced obesity. *Sci. Rep.* **2016**, *6*, 23802. [CrossRef]
29. Litwak, S.A.; Wali, J.A.; Pappas, E.G.; Saadi, H.; Stanley, W.J.; Varanasi, L.C.; Kay, T.W.; Thomas, H.E.; Gurzov, E.N. Lipotoxic Stress Induces Pancreatic beta-Cell Apoptosis through Modulation of Bcl-2 Proteins by the Ubiquitin-Proteasome System. *J. Diabetes Res.* **2015**, *2015*, 280615. [CrossRef]
30. Magnuson, M.A.; Osipovich, A.B. Pancreas-specific Cre driver lines and considerations for their prudent use. *Cell Metab.* **2013**, *18*, 9–20. [CrossRef]
31. Estall, J.L.; Screaton, R.A. Of Mice and Men, Redux: Modern Challenges in beta Cell Gene Targeting. *Endocrinology* **2020**, *161*, 1–3. [CrossRef] [PubMed]
32. Oropeza, D.; Jouvet, N.; Budry, L.; Campbell, J.E.; Bouyakdan, K.; Lacombe, J.; Perron, G.; Bergeron, V.; Neuman, J.C.; Brar, H.K.; et al. Phenotypic Characterization of MIP-CreERT1Lphi Mice With Transgene-Driven Islet Expression of Human Growth Hormone. *Diabetes* **2015**, *64*, 3798–3807. [CrossRef] [PubMed]
33. Gurzov, E.N.; Tran, M.; Fernandez-Rojo, M.A.; Merry, T.L.; Zhang, X.; Xu, Y.; Fukushima, A.; Waters, M.J.; Watt, M.J.; Andrikopoulos, S.; et al. Hepatic oxidative stress promotes insulin-STAT-5 signaling and obesity by inactivating protein tyrosine phosphatase N2. *Cell Metab.* **2014**, *20*, 85–102. [CrossRef]
34. Kutlu, B.; Burdick, D.; Baxter, D.; Rasschaert, J.; Flamez, D.; Eizirik, D.L.; Welsh, N.; Goodman, N.; Hood, L. Detailed transcriptome atlas of the pancreatic beta cell. *BMC Med. Genom.* **2009**, *2*, 3. [CrossRef] [PubMed]
35. Jensen, K.B.; Dredge, B.K.; Stefani, G.; Zhong, R.; Buckanovich, R.J.; Okano, H.J.; Yang, Y.Y.; Darnell, R.B. Nova-1 regulates neuron-specific alternative splicing and is essential for neuronal viability. *Neuron* **2000**, *25*, 359–371. [CrossRef]
36. Ruggiu, M.; Herbst, R.; Kim, N.; Jevsek, M.; Fak, J.J.; Mann, M.A.; Fischbach, G.; Burden, S.J.; Darnell, R.B. Rescuing Z+ agrin splicing in Nova null mice restores synapse formation and unmasks a physiologic defect in motor neuron firing. *Proc. Natl. Acad. Sci. USA* **2009**, *106*, 3513–3518. [CrossRef]

37. Mosser, R.E.; Maulis, M.F.; Moullé, V.S.; Dunn, J.C.; Carboneau, B.A.; Arasi, K.; Pappan, K.; Poitout, V.; Gannon, M. High-fat diet-induced β-cell proliferation occurs prior to insulin resistance in C57Bl/6J male mice. *Am. J. Physiol. Endocrinol. Metab.* **2015**, *308*, E573–E582. [CrossRef] [PubMed]
38. Ren, D.; Sun, J.; Wang, C.; Ye, H.; Mao, L.; Cheng, E.H.; Bell, G.I.; Polonsky, K.S. Role of BH3-only molecules Bim and Puma in β-cell death in Pdx1 deficiency. *Diabetes* **2014**, *63*, 2744–2750. [CrossRef]
39. Ren, D.; Sun, J.; Mao, L.; Ye, H.; Polonsky, K.S. BH3-only molecule Bim mediates β-cell death in IRS2 deficiency. *Diabetes* **2014**, *63*, 3378–3387. [CrossRef]
40. Shansky, R.M. Are hormones a "female problem" for animal research? *Science* **2019**, *364*, 825–826. [CrossRef]

MDPI
St. Alban-Anlage 66
4052 Basel
Switzerland
www.mdpi.com

Nutrients Editorial Office
E-mail: nutrients@mdpi.com
www.mdpi.com/journal/nutrients

www.ingramcontent.com/pod-product-compliance
Lightning Source LLC
Chambersburg PA
CBHW040816250926
51850CB00007BA/1180

* 9 7 8 3 7 2 5 8 1 2 6 5 3 *